Birkhäuser

Applied and Numerical Harmonic Analysis

More information about this series at http://www.springer.com/series/4968

Matthew Hirn • Shidong Li • Kasso A. Okoudjou
Sandra Saliani • Özgür Yilmaz
Editors

Excursions in Harmonic Analysis, Volume 6

In Honor of John Benedetto's 80th Birthday

Birkhäuser

Editors
Matthew Hirn
Department of Computational Mathematics,
Science & Engineering
Michigan State University
East Lansing, MI, USA

Shidong Li
Department of Mathematics
San Francisco State University
San Francisco, CA, USA

Kasso A. Okoudjou
Department of Mathematics
Tufts University
Medford, MA, USA

Sandra Saliani
Department of Mathematics
Computer Science and Economics
University of Basilicata
Potenza, PZ, Italy

Özgür Yilmaz
Department of Mathematics
University of British Columbia
Vancouver, BC, Canada

ISSN 2296-5009 ISSN 2296-5017 (electronic)
Applied and Numerical Harmonic Analysis
ISBN 978-3-030-69636-8 ISBN 978-3-030-69637-5 (eBook)
https://doi.org/10.1007/978-3-030-69637-5

Mathematics Subject Classification: 43-XX

This book is published under the imprint Birkhäuser, www.birkhauser-science.com by the registered company Springer Nature Switzerland AG.
The registered company address is: Gewerbestrasse 11, 6330 Cham, Switzerland

Dedicated to Cathy

ANHA Series Preface

The *Applied and Numerical Harmonic Analysis (ANHA)* book series aims to provide the engineering, mathematical, and scientific communities with significant developments in harmonic analysis, ranging from abstract harmonic analysis to basic applications. The title of the series reflects the importance of applications and numerical implementation, but richness and relevance of applications and implementation depend fundamentally on the structure and depth of theoretical underpinnings. Thus, from our point of view, the interleaving of theory and applications and their creative symbiotic evolution is axiomatic.

Harmonic analysis is a wellspring of ideas and applicability that has flourished, developed, and deepened over time within many disciplines and by means of creative cross-fertilization with diverse areas. The intricate and fundamental relationship between harmonic analysis and fields such as signal processing, partial differential equations (PDEs), and image processing is reflected in our state-of-the-art *ANHA* series.

Our vision of modern harmonic analysis includes mathematical areas such as wavelet theory, Banach algebras, classical Fourier analysis, time-frequency analysis, and fractal geometry, as well as the diverse topics that impinge on them.

For example, wavelet theory can be considered an appropriate tool to deal with some basic problems in digital signal processing, speech and image processing, geophysics, pattern recognition, biomedical engineering, and turbulence. These areas implement the latest technology from sampling methods on surfaces to fast algorithms and computer vision methods. The underlying mathematics of wavelet theory depends not only on classical Fourier analysis, but also on ideas from abstract harmonic analysis, including von Neumann algebras and the affine group. This leads to a study of the Heisenberg group and its relationship to Gabor systems, and of the metaplectic group for a meaningful interaction of signal decomposition methods. The unifying influence of wavelet theory in the aforementioned topics illustrates the justification for providing a means for centralizing and disseminating information from the broader, but still focused, area of harmonic analysis. This will be a key role of *ANHA*. We intend to publish with the scope and interaction that such a host of issues demands.

Along with our commitment to publish mathematically significant works at the frontiers of harmonic analysis, we have a comparably strong commitment to publish major advances in the following applicable topics in which harmonic analysis plays a substantial role:

Antenna theory	*Prediction theory*
Biomedical signal processing	*Radar applications*
Digital signal processing	*Sampling theory*
Fast algorithms	*Spectral estimation*
Gabor theory and applications	*Speech processing*
Image processing	*Time-frequency and*
Numerical partial differential equations	*Time-scale analysis*
	Wavelet theory

The above point of view for the *ANHA* book series is inspired by the history of Fourier analysis itself, whose tentacles reach into so many fields.

In the last two centuries Fourier analysis has had a major impact on the development of mathematics, on the understanding of many engineering and scientific phenomena, and on the solution of some of the most important problems in mathematics and the sciences. Historically, Fourier series were developed in the analysis of some of the classical PDEs of mathematical physics; these series were used to solve such equations. In order to understand Fourier series and the kinds of solutions they could represent, some of the most basic notions of analysis were defined, e.g., the concept of "function." Since the coefficients of Fourier series are integrals, it is no surprise that Riemann integrals were conceived to deal with uniqueness properties of trigonometric series. Cantor's set theory was also developed because of such uniqueness questions.

A basic problem in Fourier analysis is to show how complicated phenomena, such as sound waves, can be described in terms of elementary harmonics. There are two aspects of this problem: first, to find, or even define properly, the harmonics or spectrum of a given phenomenon, e.g., the spectroscopy problem in optics; second, to determine which phenomena can be constructed from given classes of harmonics, as done, for example, by the mechanical synthesizers in tidal analysis.

Fourier analysis is also the natural setting for many other problems in engineering, mathematics, and the sciences. For example, Wiener's Tauberian theorem in Fourier analysis not only characterizes the behavior of the prime numbers, but also provides the proper notion of spectrum for phenomena such as white light; this latter process leads to the Fourier analysis associated with correlation functions in filtering and prediction problems, and these problems, in turn, deal naturally with Hardy spaces in the theory of complex variables.

Nowadays, some of the theory of PDEs has given way to the study of Fourier integral operators. Problems in antenna theory are studied in terms of unimodular trigonometric polynomials. Applications of Fourier analysis abound in signal processing, whether with the fast Fourier transform (FFT), or filter design, or the adaptive modeling inherent in time-frequency-scale methods such as wavelet theory. The coherent states of mathematical physics are translated and modulated Fourier transforms, and these are used, in conjunction with the uncertainty principle, for dealing with signal reconstruction in communications theory. We are back to the raison d'être of the *ANHA* series!

University of Maryland
College Park

John J. Benedetto
Series Editor

Foreword

John Benedetto was the first doctoral thesis student I supervised. It is sort of inadvisable to start out with a graduate student like that: it may saddle you, going forward, with unrealistically high expectations.

But, we were a good pair: John bright-eyed and brash in his twenties, just getting into the thesis student's role, and me bright-eyed and brash in my thirties, just getting into the thesis adviser's role. Actually, John was intent not only on writing a thesis on analysis and Banach spaces with me but also on learning all about Thomist philosophy from the eminent French philosopher Étienne Gilson on the other side of the Toronto campus. I did not know about this double ambition of his at the time—John reminisced about it to me only some years later, having found meanwhile that it was a bit much to juggle the two specialties at once. If I had known I was dealing with a mathematician cum medieval philosopher, actually, it would have made John seem even more akin, because I at the same stage of my education had set out not only to become a mathematician in the image of George Mackey and Garrett Birkhoff but at the same time to become a composer in the image of Irving Fine. And, I would not have been let down when John gave up his philosophical moonlighting, for I had had to give up my concentration on music in the same way.

So, John was declared Doctor of Philosophy and launched on his professional career with my blessing and that of the University of Toronto. I applauded his service in New York, Maryland, Pisa, and elsewhere. It may seem that our research emphases diverged a bit, but it does not feel to me that we got out of touch. In particular, we both welcomed the rise of wavelet theory with enthusiasm and without needing to consult each other. But, I remained mostly a spectator, while John threw himself into the amazing development of applied Fourier analysis. He became one of the leaders in forming the field and in making it known to a wider public, and he leads a large phalanx of creative Fourier analysts in the next generation. I have been duly appreciative of the achievements of the Norbert Wiener Center, though I have viewed them mostly from afar.

One has no right to take pride in the work of one's students and grandstudents, but I confess to feeling that pride in this case, however, unjustified. May they carry on whatever in my own life deserves to be carried on.

Toronto, ON, Canada
May 2020

Chandler Davis

Preface

"John J. Benedetto has had a profound influence not only on the direction of harmonic analysis and its applications, but also on the entire community of people involved in the field." This statement can be found in the preface of the volume celebrating John's 60th birthday and holds true even more so today. During the 20 years that follows, the world has witnessed that the breadth and depth of John's influence continue to expand. Besides his enormously impactful scientific research contributions, John's influence also lies in, for instance, advising 61 Ph.D. students (so far) and nurturing many other junior scholars; founding the Journal of Fourier Analysis and Applications (JFAA), and the book series of Applied and Numerical Harmonic Analysis (ANHA); establishing the renowned Norbert Winner Center, and fostering a wide range of highly relevant health and scientific research. All in all, John's most profound influence lies in his building of a worldwide community of scholars in harmonic analysis and its applications. Advancing beautiful mathematical ideas and applications is an underlying theme of John's illustrious career and is continuing in the latest forum of the annual February Fourier Talks (FFT). A full account of John's influence on the field of harmonic analysis would require volumes.

In honor of John's 80th birthday, this book is another assemblage of community's appreciation to John's deep impact on the field of harmonic analysis and applications and to the scientific community. Needless to say, the original articles collected in this volume are all highly relevant and written by prominent, well-respected scholars in the field. This volume covers an invited chapter and the following five parts:

1. John Benedetto's Mathematical Work,
2. Harmonic Analysis,
3. Wavelets and Frames,
4. Sampling and Signal Processing,
5. Compressed Sensing and Optimizations.

As such, this book shall be once again an excellent reference and resource for graduate students and professionals in the field. Contributors of the volume include A. Abtahi, A. Aldroubi, C. Cabrelli, P.G. Casazza, J. Cahill, D.-C. Chang,

E. Cordero, W. Czaja, S.B. Damelin, S. Data, M. Dorfler, N. Dyn, M. de Gosson, Y. Han, C. Hegde, C. Heil, J.A. Hogan, Y. Hu, R. Johnson, D. Joyner, F. Keinert, J.D. Lakey, C. Leonard, W. Li, Y. Li, R.D. Martin, F. Marvasti, I. Medri, K.D. Merrill, D.G. Mixon, U. Molter, F. Nicola, A. Olevskii, M. Pekala, I.Z. Pesenson, A. Petrosyan, D.L. Ragozin, T. Strohmer, J. Stueck, T.T. Tran, A. Ulanovskii, E.S. Weber, M. Werman, T. Wertz, X. Wu, S. Zheng, and X. Zhuang.

To close, we would like to thank Radu V. Balan, Wojciech Czaja, Luke Evans, Alfredo Nava-Tudela, and Kasso Okoudjou, for organizing the conference celebrating John's birthday, and Jean-Pierre Gabardo, Christopher Heil, Emily King, Götz Pfander, and David Walnut for putting together an outstanding scientific program. We also acknowledge the financial support of the Institute for Mathematics and its Applications and the Department of Mathematics at the University of Maryland.

East Lansing, MI, USA — Matthew Hirn

San Francisco, CA, USA — Shidong Li

Medford, MA, USA — Kasso A. Okoudjou

Potenza, PZ, Italy — Sandra Saliani

Vancouver, BC, Canada — Özgür Yilmaz

November 2020

Contents

Part I Introduction

John Benedetto's Mathematical Work 3
David Joyner

Part II Harmonic Analysis

Absolute Continuity and the Banach–Zaretsky Theorem 27
Christopher Heil

Spectral Synthesis and $H^1(\mathbb{R})$ 53
Raymond Johnson

Universal Upper Bound on the Blowup Rate of Nonlinear Schrödinger Equation with Rotation 59
Yi Hu, Christopher Leonard, and Shijun Zheng

Almost Eigenvalues and Eigenvectors of Almost Mathieu Operators 77
Thomas Strohmer and Timothy Wertz

Spatio–Spectral Limiting on Redundant Cubes: A Case Study 97
Jeffrey A. Hogan and Joseph D. Lakey

Part III Wavelets and Frames

A Notion of Optimal Packings of Subspaces with Mixed-Rank and Solutions 119
Peter G. Casazza, Joshua Stueck, and Tin T. Tran

Construction of Frames Using Calderón–Zygmund Operator Theory 145
Der-Chen Chang, Yongsheng Han, and Xinfeng Wu

Equiangular Frames and Their Duals 163
Somantika Datta

Wavelet Sets for Crystallographic Groups 185
Kathy D. Merrill

Discrete Translates in Function Spaces 199
Alexander Olevskii and Alexander Ulanovskii

Part IV Sampling and Signal Processing

Local-to-Global Frames and Applications to the Dynamical Sampling Problem 211
Akram Aldroubi, Carlos Cabrelli, Ursula Molter, and Armenak Petrosyan

Signal Analysis Using Born–Jordan-Type Distributions 221
Elena Cordero, Maurice de Gosson, Monika Dörfler, and Fabio Nicola

Sampling by Averages and Average Splines on Dirichlet Spaces and on Combinatorial Graphs 243
Isaac Z. Pesenson

Dynamical Sampling: A View from Control Theory 269
Rocío Díaz Martín, Ivan Medri, and Ursula Molter

Linear Multiscale Transforms Based on Even-Reversible Subdivision Operators 297
Nira Dyn and Xiaosheng Zhuang

Part V Compressed Sensing and Optimization

Sparsity-Based MIMO Radars 323
Azra Abtahi and Farokh Marvasti

Robust Width: A Characterization of Uniformly Stable and Robust Compressed Sensing 343
Jameson Cahill and Dustin G. Mixon

On Min-Max Affine Approximants of Convex or Concave Real-Valued Functions from $\mathbb{R}^k$, Chebyshev Equioscillation and Graphics 373
Steven B. Damelin, David L. Ragozin and Michael Werman

A Kaczmarz Algorithm for Solving Tree Based Distributed Systems of Equations 385
Chinmay Hegde, Fritz Keinert, and Eric S. Weber

Maximal Function Pooling with Applications 413
Wojciech Czaja, Weilin Li, Yiran Li, and Mike Pekala

Index 431

Contributors

Azra Abtahi Advanced Communications Research Institute (ACRI), Electrical Engineering Department, Sharif University of Technology, Tehran, Iran

Akram Aldroubi Department of Mathematics, Vanderbilt University, Nashville, TN, USA

Carlos Cabrelli Departamento de Matemática, FCEyN, UBA and IMAS CONICET, Buenos Aires, Argentina

Jameson Cahill University of North Carolina Wilmington, Wilmington, NC, USA

Peter G. Casazza Department of Mathematics, University of Missouri, Columbia, MO, USA

Der-Chen Chang Department of Mathematics and Statistics, Georgetown University, Washington, DC, USA
Graduate Institute of Business Administration, College of Management, Fu Jen Catholic University, New Taipei City, Taiwan

Elena Cordero Università di Torino, Dipartimento di Matematica, Torino, Italy

Wojciech Czaja Norbert Wiener Center, Department of Mathematics, University of Maryland College Park, College Park, MD, USA

Steven B. Damelin Department of Mathematics, University of Michigan, Ann Arbor, MI, USA

Somantika Datta Department of Mathematics, University of Idaho, Moscow, ID, USA

Maurice de Gosson University of Vienna, Faculty of Mathematics, Wien, Austria

Rocío Díaz Martín IAM – CONICET, CABA, Argentina
Universidad Nacional de Córdoba, Córdoba, Argentina

Monika Dörfler University of Vienna, Faculty of Mathematics, Wien, Austria

Nira Dyn School of Mathematical Sciences, Tel-Aviv University, Tel-Aviv, Israel

Yongsheng Han Department of Mathematics and Statistics, Auburn University, Auburn, AL, USA

Chinmay Hegde Electrical and Computer Engineering, Iowa State University, Ames, IA, USA
Electrical and Computer Engineering, New York University, New York, NY, USA

Christopher Heil School of Mathematics, Georgia Tech, Atlanta, GA, USA

Jeffrey A. Hogan School of Mathematical and Physical Sciences, University of Newcastle, Callaghan, NSW, Australia

Yi Hu Department of Mathematical Sciences, Georgia Southern University, Statesboro, GA, USA

Raymond Johnson University of Maryland, College Park, MD, USA

David Joyner Department of Mathematics, U.S. Naval Academy (Retired), Annapolis, MD, USA

Fritz Keinert Department of Mathematics, Iowa State University, Ames, IA, USA

Joseph D. Lakey New Mexico State University, Las Cruces, NM, USA

Christopher Leonard Department of Mathematics, North Carolina State University, Raleigh, NC, USA

Weilin Li Courant Institute of Mathematical Sciences, New York University, New York, NY, USA

Yiran Li Norbert Wiener Center, Department of Mathematics, University of Maryland College Park, College Park, MD, USA

Farokh Marvasti Advanced Communications Research Institute (ACRI), Electrical Engineering Department, Sharif University of Technology, Tehran, Iran

Ivan Medri Department of Mathematics, Vanderbilt University, Nashville, TN, USA

Kathy D. Merrill Department of Mathematics, Colorado College, Colorado Springs, CO, USA

Dustin G. Mixon The Ohio State University, Columbus, OH, USA

Ursula Molter Dto. de Matemática, FCEyN, Universidad de Buenos Aires, Buenos Aires, Argentina

Fabio Nicola Politecnico di Torino, Dipartimento di Scienze Matematiche, Torino, Italy

Alexander Olevskii Tel Aviv University, Tel Aviv, Israel

Mike Pekala Norbert Wiener Center, Department of Mathematics, University of Maryland College Park, College Park, MD, USA

Isaac Z. Pesenson Department of Mathematics, Temple University, Philadelphia, PA, USA

Armenak Petrosyan Computational and Applied Mathematics Group, Oak Ridge National Laboratory, Oak Ridge, TN, USA

David L. Ragozin Department of Mathematics, University of Washington, Seattle, WA, USA

Thomas Strohmer University of California, Davis, CA, USA

Joshua Stueck Department of Mathematics, University of Missouri, Columbia, MO, USA

Tin T. Tran Department of Mathematics, University of Missouri, Columbia, MO, USA

Alexander Ulanovskii Stavanger University, Stavanger, Norway

Eric S. Weber Department of Mathematics, Iowa State University, Ames, IA, USA

Michael Werman Department of Computer Science, The Hebrew University, Jerusalem, Israel

Timothy Wertz Yale-NUS College, Singapore, Singapore

Xinfeng Wu Department of Mathematics, China University of Mining and Technology, Beijing, P.R. China

Shijun Zheng Department of Mathematical Sciences, Georgia Southern University, Statesboro, GA, USA

Xiaosheng Zhuang Department of Mathematics, City University of Hong Kong, Kowloon, Hong Kong

Acronyms

AC	Absolutely continuous
ACO	Approximately controllable
AOB	Approximately observable
ART	Algebraic Reconstruction Technique: a method of image reconstruction in computerized tomography
BEC	Bose–Einstein condensation
BIBD	Balanced incomplete block design
BJ	Born–Jordan distribution
BJDn	Born–Jordan distribution of order n
BV	Bounded variation
CNN	Convolutional neural network
CPU	Central processing unit
CS	Compressive sensing
CSC	Convolutional sparse coding
DCP	Deep coding problem
DCPP	Deep coding problem with pooling
DOA/DOD	Directions of arrival/departure
DS	Dynamical sampling
ECO	Exactly controllable
EF	Equiangular frame
ENR	Ratio of total transmitted energy to the noise energy
EOB	Exactly observable
ETF	Equiangular tight frame
ETFF	Equiangular tight fusion frame
FFT	February Fourier Talks
FTC	Fundamental Theorem of Calculus
GFT	Graph Fourier transform
IBM	International Business Machines
JB	John Benedetto
Lip	Lipschitz continuous
LMS	License in Mediaeval Studies

MIMO Radar	Multiple-input multiple-output radar
MIT	Massachusetts Institute of Technology
ML	Maximum likelihood
MRA	Multiresolution analysis
MTER	Multiscale transform based on an even-reversible subdivision
MUBs	Mutually unbiased bases
NLS	Nonlinear Schrödinger equation
NP-hard	Nondeterministic polynomial-time hard
NWC	Norbert Wiener Center for Harmonic Analysis and Applications
NYU	New York University
PDE	Partial differential equation
RCA	Radio Corporation of America
RGS	Reynolds Gauss–Seidel method
ROC	Receiver operating characteristic
RNLS	Rotational nonlinear Schrödinger equation
RS	Random sampling
SNR	Signal-to-noise ratio
SOR	Successive over-relaxation method
SPIE	Society of Photographic Instrumentation Engineers
SSL	Spatio-spectral limiting
SSN	Spectral neutral neighbor
SVM	Support vector machine
TA	Teaching assistant
UMCP	University of Maryland at College Park

Part I
Introduction

The first part of this volume serves as its introduction and contains a single chapter in which D. Joyner summarizes the mathematical work of John, including an exhaustive list of his students and publications.

John Benedetto's Mathematical Work

David Joyner

Abstract John Joseph Benedetto (JB) has been at the University of Maryland, College Park, since 1965. In this chapter, I will submit data that attests to JB's (a) large number of PhD students, (b) large number of papers (As a linear regression computation shows, the number of PhD students (per year) he advises and the number of papers (published per year) are both increasing, on average. See below.), and (c) remarkable outreach into the business sector, inviting cooperation between industry and his group of UMCP mathematicians that became the Norbert Wiener Center.

1 Brief Biography

On June 17, 1933, Vienna DiTonno married John ("Zip") Benedetto in Wakefield, Mass., the working class town just north of Boston where they were born and raised. Zip and Vienna were children of the depression and never got past 8th grade in school. Their only child, JB was born there six years later, on July 16, 1939.

Zip ran a pool hall in downtown Wakefield. While JB was an excellent student, in high school he got no further than trigonometry and solid geometry, as they did not teach calculus at the time. After school, to his mom's dismay, JB would visit the pool hall almost daily to help his dad run his business (and to play a little pool!). Another person who frequented Zip's pool hall was Robert McCloskey, a Harvard professor[1] and a collegiate billiards champion as an undergraduate. Seeing JB's academic talent, McCloskey told Zip[2] to encourage JB to apply to Harvard. However, after JB graduated from Malden Catholic High School in 1956, he applied

[1]According to archives of "The Crimson," McCloskey was appointed Chair of the Government Department at Harvard in 1958.

[2]Sadly, Zip passed away at age 44 in May of 1956, when JB was 16.

D. Joyner (✉)
Department of Mathematics, U.S. Naval Academy (Retired), Annapolis, MD, USA

M. Hirn et al. (eds.), *Excursions in Harmonic Analysis, Volume 6*, Applied and Numerical Harmonic Analysis, https://doi.org/10.1007/978-3-030-69637-5_1

(and was accepted) to Boston College instead of Harvard. At Boston College, he had inspirational teachers for his first- and second-year mathematics courses, convincing JB to major in mathematics. As a nod perhaps to McCloskey, as a senior, JB applied to Harvard for graduate school, and nowhere else. Fortunately for mathematics, he was accepted and, after graduating from Boston College in 1960, began to take courses from Gleason (real analysis), Widder (Laplace transforms), Mackey, and Walsh (of Walsh functions fame), among others. His master's degree was awarded by Harvard in 1962.

In the fall of 1962, JB left Harvard for the University of Toronto, where he studied with Chandler Davis,[3] who JB did not know of at Harvard. The reason for this move to Canada is not as simple as it sounds. It has really nothing to do with the fact that both Walsh and Davis had advisors in the Birkhoff family (and both on the Harvard faculty). At Boston College and Harvard, JB was very interested in Thomistic philosophy (the philosophy of Thomas Aquinas), and he knew the Pontifical Institute of Medieval Studies at St. Michael's College was a subset of the University of Toronto. His plan was to get a PhD in mathematics in 1964 and an LMS from the Pontifical Institute[4] along the way. JB even knew what he wanted to work on for his PhD: the Laplace transform of distributions and topological vector spaces.[5] So, in the summer of 1962, JB is a man who knows what he wants. However, once JB arrived in Toronto that fall, life had other plans. First, JB was assigned as a TA to Chandler Davis. That is how they met and started working together. Second, he started taking classes at the Pontifical Institute, but after the first philosophy course dropped his plan to get an LMS. In fact, JB was Chandler Davis' first PhD student, and they got along very well.[6] JB's PhD degree was awarded by the University of Toronto a few years later in 1964, and a revised version of his thesis was published in [1966b] (Fig. 1).

In 1964, after graduating, JB took a tenure-track job at New York University.[7] While a graduate student, during the summers JB worked at RCA in Burlington, MA. However, starting the summer of 1964 and part time during the academic year, JB worked at IBM Cambridge, instead of RCA. On a whim, JB left NYU for a tenure-track position at UMCP the following year. Except for visiting positions at MIT, the Mittag-Leffler Institute, and Scuola Normale Superiore, JB has been at UMCP since 1965. Once at UMCP, JB continued to consult for industry but, of course, eventually this work came under the umbrella of the Norbert Wiener Center (more on that below).

[3]From the it's-a-small-world department, Chandler Davis' PhD advisor was Garrett Birkhoff, son of George David Birkhoff, who was Joseph Walsh's PhD advisor.

[4]The LMS, a License in Mediaeval Studies or "Licentiate," is a kind of post-graduate degree awarded by The Pontifical Institute of Mediaeval Studies. There is no analogous degree offered in the United States.

[5]Inspired by Widder's course on the Laplace transform and the reading course with Mackey on distributions and topological vector space that he took at Harvard.

[6]JB has written about his connection with Chandler Davis in [2014a].

[7]That year, the Courant Institute moved to its current location, in Weaver Hall.

Fig. 1 JB getting his PhD, with mom Vienna and grandpa Mr. DiTonno in 1964

As far as the arc of his career is concerned, JB's main mathematical inspirations are:

- Chandler Davis (PhD advisor) (Fig. 2),
- N. Wiener (whom JB never met),
- A. Beurling (whom JB never met),
- A. Gleason, one of his teachers at Harvard.

While JB has told me that many of the ideas he gets for papers are from thinking about mathematics while on a walk or traveling, I know there is another source: lots of hand computations. To illustrate this, I will tell a story connected with my PhD thesis (in 1983, shortly after the publication of my favorite paper of his, [1980a]). As a graduate student, he assigned me a problem connected with his Mathematische Annalen 1980 paper. I do not remember the problem, but I remember that after I solved it, I did not want to take credit for it if he already solved it but just did not, for whatever reason, add it to his paper. So one day, we had a meeting in his office about this psychological problem I was having. He said to resolve the matter, I could read the notes he made while writing the paper. Apparently, for each paper JB writes, he keeps his notes (or at least, did at the time) in a notebook. So, JB pulls out this massive notebook (the kind with the extra large rings) full of hand-written computations. That JB keep such a massive set of detailed notes for each paper was amazing to me at the time, and still is.

In his career, JB has been a Senior Fulbright-Hays Scholar, a SPIE Wavelet Pioneer, a Fellow of the American Mathematical Society, and a SIAM Fellow. His

Fig. 2 Chandler Davis and JB in 1964

paper [1989b] won MITRE's Best Paper Award, and he was named Distinguished Scholar-Teacher by the University of Maryland in 1999.

Currently, JB is the Director of the Norbert Wiener Center for Harmonic Analysis and Applications (NWC), which he founded in 2004. It serves as an interface between funding agencies and industry with problems that can be solved using harmonic analysis by mathematicians at the NWC.[8] In its 15 years of existence, the mathematicians at the NWC have brought in over 7 million dollars in grants and have worked with over 15 industrial partners.[9] Besides dollar grants, many of these industrial partners have also supported numerous student internships. Hundreds have spoken at or attended the annual NWC conference, the February Fourier Talks, or FFT. The NWC is also connected with the *Journal of Fourier Analysis and its Applications*[10] and the *Applied and Numerical Harmonic Analysis* book series.[11]

As of summer 2019, JB has directed 58 PhD students (with several more in the pipeline). As of this writing, JB is in the top 100 of all PhD advisors worldwide.[12] JB does not co-author published PhD theses of his PhD students. Nonetheless, he has over 200 publications, of this writing, and over 80 co-authors (many of which are his former PhD students, if they do research with JB going beyond their thesis). What is even more impressive is that none of JB's academic publications were co-authored until 1983. At the time of this writing, JB's most frequent co-author (by far) is his UMCP colleague, Wojciech Czaja.

[8]Currently, JB, Radu Balan, Wojciech Czaja, and Kasso Okoudjou.

[9]For example, NIH, AFOSR, Siemens, MITRE, DARPA, ONR, NSF, and many more.

[10]For which JB is the Founding Editor-in-Chief.

[11]For which JB is the Series Editor.

[12]According to the database "Mathematics Genealogy Project."

2 Coda

In summary, JB's piercing intellectual curiosity has led to over 200 hundred refereed publications and about 60 PhD students, so far. Which reminds me of the old joke, "Great mathematicians never die, they just tend to infinity."

3 PhD Theses

Here is a list of the 61 (and counting) PhD students that JB has advised.

1. 1971, George Benke, *Sidon sets and the growth of L^p norms*
2. 1977a, Wan-Chen Hsieh, *Topologies for spectral synthesis of the space of bounded functions*
3. 1977b, Fulvio Ricci, *Support preserving multiplication of pseudo-measures*
4. 1980, Ward Evans,[13] *Beurling's spectral analysis and continuous pseudo-measures*
5. 1983, W. David Joyner, *The harmonic analysis of Dirichlet series and the Riemann zeta function*
   NSF Post-Doc IAS 1984.
6. 1987, Jean-Pierre Gabardo, *Spectral gaps and uniqueness problems in Fourier analysis*
   Sloan Dissertation Fellowship 1986.
7. 1989, David Walnut, *Weyl–Heisenberg wavelet expansions: existence and stability in weighted spaces*
   Sloan Dissertation Fellowship 1988
8. 1990a, Christopher Heil, *Wiener amalgam spaces in generalized harmonic analysis and wavelet theory*
   NSF Post-Doc MIT 1990
9. 1990b, Rodney Kerby, *Correlation function and the Wiener–Wintner theorem in higher dimensions*
10. 1990c, George Yang, *Applications of Wiener–Tauberian theorem to a filtering problem and convolution equations*
11. 1991a, William Heller, *Frames of exponentials and applications*
12. 1991b, Joseph Lakey, *Weighted norm inequalities for the Fourier transform*
13. 1992, Erica Bernstein, *Generalized Riesz products and pyramidal schemes*
14. 1993a, Shidong Li, *The theory of frame multiresolution analysis and filter design*
15. 1993b, Sandra Saliani, *Nonlinear wavelet packets*
16. 1993c, Anthony Teolis, *Discrete signal representation*
17. 1994, Georg Zimmermann, *Projective multiresolution analysis and generalized sampling*

[13] Now named Celia Evans.

18. 1998a, Melissa Harrison, *Frames and irregular sampling from a computational perspective*
19. 1998b, Hui-Chuan Wu, *Multidimensional irregular sampling in terms of frames*
20. 1999a, Manuel Leon, *Minimally supported frequency wavelets*
21. 1999b, Götz Pfander, *Periodic wavelet transforms and periodicity detection*
22. 1999c, Oliver Treiber, *Affine data representations and filter banks*
23. 2000, Sherry Scott, *Spectral analysis of fractal noise in terms of Wiener's generalized harmonic analysis and wavelet theory*
24. 2001a, Matthew Fickus, *Finite normalized tight frames and spherical equidistribution*
25. 2001b, Ioannis Konstantinidis, *The characterization of multiscale generalized Riesz product measures*
26. 2002a, Anwar A. Saleh, *A finite dimensional model for the inverse frame operator*
27. 2002b, Jeffrey Sieracki, *Greedy adaptive discrimination: Signal component analysis by simultaneous matching pursuit with application to ECoG signature detection*
28. 2002c, Songkiat Sumetkijakan, *A fractal set constructed from a class of wavelet sets*
29. 2003a, Alexander M. Powell, *The uncertainty principle in harmonic analysis and Bourgain's theorem*
    Dissertation Fellowship 2000
30. 2003b, Shijun Zheng, *Besov spaces for the Schrödinger operator with barrier potential*
    Dissertation Fellowship 2000
31. 2004, Joseph Kolesar, Σ–Δ *modulation and correlation criteria for the construction of finite frames arising in communication theory*
32. 2005a, Andrew Kebo, *Quantum detection and finite frames*
33. 2005b, Juan Romero, *Generalized multiresolution analysis: construction and measure-theoretic characterization*
34. 2006a, Abdelkrim Bourouihiya, *Beurling weighted spaces, product-convolution operators, and the tensor product of frames*
35. 2006b, Aram Tangboondouangjit, *Sigma-Delta quantization: number-theoretic aspects of refining error estimates*
36. 2007a, Somantika Datta, *Wiener's generalized harmonic analysis and waveform design*
37. 2007b, Onur Oktay, *Frame quantization theory and equiangular tight frames*
38. 2008, David Widemann, *Dimensionality reduction for hyperspectral data* (Co-adviser, W. Czaja)
39. 2009a, Matthew Hirn, *Enumeration of harmonic frames and frame based dimension reduction* (Co-adviser, K. Okoudjou)
    Wylie Dissertation Fellowship 2009
40. 2009b, Emily King, *Wavelet and frame theory: frame bound gaps, generalized shearlets, Grassmannian fusion frames, and p-adic wavelets* (Co-adviser, W. Czaja)
    Wylie Dissertation Fellowship 2008

41. 2010, Christopher Flake, *The multiplicative Zak transform, dimension reduction, and wavelet analysis of LIDAR data* (Co-adviser, W. Czaja)
42. 2011a, Enrico Au-Yeung, *Balayage of Fourier transforms and the theory of frames*
43. 2011b, Avner Halevy, *Extensions of Laplacian eigenmaps for manifold learning* (Co-adviser W. Czaja)
44. 2011c, Nathaniel Strawn, *Geometric structures and optimization on finite frames* (Co-adviser, R. Balan)
45. 2012a, Kevin Duke, *A study of the relationship between spectrum and geometry through Fourier frames and Laplacian eigenmaps*
46. 2012b, Alfredo Nava-Tudela, *Image representation and compression via sparse solutions of systems of linear equations*
47. 2013, Rongrong Wang, *Global geometric conditions on dictionaries for the convergence of ℓ^1 minimization problems* (Co-adviser W. Czaja)
48. 2014a, Travis Andrews, *Frame multiplication theory for vector-valued harmonic analysis*
49. 2014b, Alex Cloninger, *Exploiting data-dependent structure for improving sensor acquisition and integration* (Co-adviser W. Czaja)
    Wylie Dissertation Fellowship 2013
    NSF Postdoctoral Fellowship to Yale
50. 2014c, Tim Doster, *Harmonic analysis inspired data fusion with applications in remote sensing* (Co-adviser W. Czaja)
51. 2014d, Wei-Hsuan Yu, *Spherical two-distance sets and related topics in harmonic analysis* (Co-adviser A. Barg)
52. 2015a, Gokhan Civan, *Identification of operators on elementary locally compact abelian groups*
53. 2015b, Paul Koprowski, *Graph theoretic uncertainty principles*
54. 2015c, James Murphy, *Anisotropic harmonic analysis and integration of remotely sensed data* (Co-advisor W. Czaja)
55. 2016, Matthew Begué, *Expedition in data and harmonic analysis on graphs* (Co-advisor K. Okoudjou)
56. 2018a, Weilin Li, *Topics on harmonic analysis, sparse representations, and data analysis* (Co-advisor W. Czaja)
    Wylie Dissertation Fellowship 2017
57. 2018b, Mark Magsino, *Constant amplitude zero-autocorrelation sequences and single pixel camera imaging*
58. 2018c, Franck Njeunje, *Computational methods in machine learning: transport model, Haar wavelet, DNA classification, and MRI* (Co-advisor W. Czaja)
59. 2020a, Shujie Kang, *Generalized frame potential and problems related to SIC-POVMs* (Co-advisor K. Okoudjou)
60. 2020b, Chenzhi Zhao, *Non-harmonic Fourier analysis and applications*
61. 2020c, Kung-Ching Lin, *Nonlinear sampling theory and efficient signal recovery*

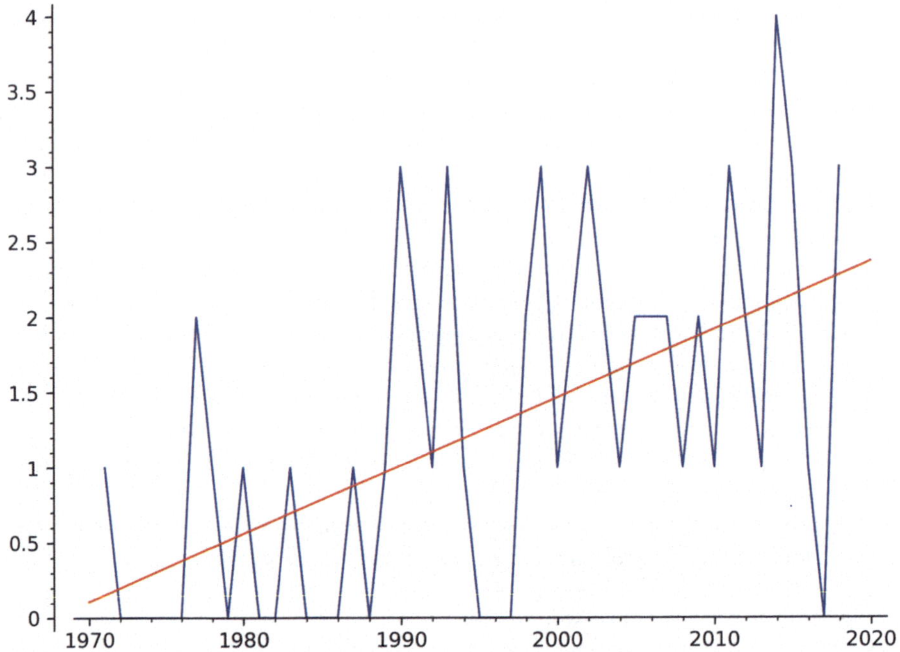

Fig. 3 Linear regression on the number of JB's PhD students graduating per year

This is an average of about 1.2 PhD students per year. The list of pairs (year, number of JB's PhD students graduating that year) between 1971 and 2017 has best linear fit[14] $y = ax + b$, where $a = 0.0451\ldots$, $b = -88.8937\ldots$. In rough terms, the number of PhD students JB graduates per year increases by about 4.5% per year. The graph is in Fig. 3.

4 Papers

The majority of mathematical papers by JB deal with the representation of an "arbitrary" function[15] (typically on $\mathbb{R}$ or $\mathbb{R}^n$ and subject to some conditions), in one way or another (by a Fourier series, wavelet expansion, integral transform, and so on). The functions JB considers can be pretty general, but the point is that he represents them for us in a nice way and then uses such a representation to derive

[14] Again, thanks to SageMath.

[15] Of course, the question "what's a function?" immediately arises. Here, we include both "generalized functions" (e.g., a distribution in the sense of Schwartz) and Radon measures as functions.

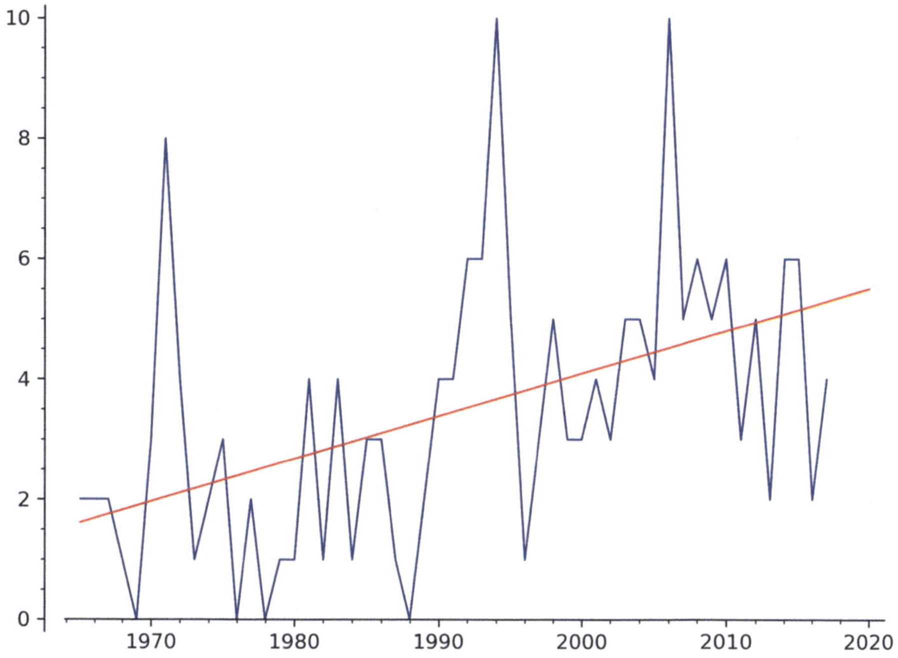

Fig. 4 Linear regression on the number of JB's papers published per year

something useful. In many of his papers, JB takes such a representation and either (a) analyzes it to obtain estimates of a related quantity, or (b) applies it to an engineering problem, or (c) uses it to investigate a question in another field such as graph theory or analytic number theory.

Firstly, the list below includes some repetition (which I have tried to indicate). For example, some "technical reports" were revised and then submitted to a journal for publication. Secondly, some technical reports were not even submitted (e.g., they might have a more expository flavor). Finally, we note that some papers have very similar, or even identical, titles but are essentially unrelated (unless indicated).

Numerically, there is an average of about 3.48 papers per year. The list of pairs (year, number of papers published that year) between 1965 and 2017 has best linear fit[16] $y = ax + b$, where $a = 0.0708\ldots$, $b = -137.6162\ldots$. In rough terms, the number of papers JB publishes per year increases by about 7% per year. The graph is in Fig. 4.

[16]Thanks to the SageMath command `find_fit`.

List of John J. Benedetto's Published Papers

1965a. *Representation theorem for the Fourier transform of integrable functions*, Bull. Soc. Roy. Sci. de Liege 9–10 (1965) 601–604.
1965b. *Onto Criterion for adjoint maps*, Bull. Soc. Roy. Sci. de Liege 9–10 (1965) 605–609.
1966a. **Generalized Functions**, Institute for Fluid Dynamics and Applied Math., BN–431 (1966) 1–380.
1966b. *The Laplace transform of generalized functions*, Canad. J. Math. 18 (1966), 357–374.
1967a. *Tauberian translation algebras*, Ann. di Mat. 74 (1967) 255–282.
1967b. *Analytic representation of generalized functions,* Math. Zeit. 97 (1967), 303–319.
1968a. **Pseudo-measures and Harmonic Synthesis**, University of Maryland, Department of Mathematics Lecture Notes 5 (1968) 1–316.
1970a. *A strong form of spectral resolution*, Ann. di Mat. 86 (1970), 313–324.
1970b. *Sets without true distributions*, Bull. Soc. Roy. Sci. de Liege 7–8 (1970) 434–437.
1970c. *Support preserving measure algebras and spectral synthesis,* Math. Zeit. 118 (1970), 271–280.
1971a. **Harmonic Analysis on Totally Disconnected Sets**, Lecture Notes in Mathematics, 202, Springer-Verlag, 1971.
1971b. *Dirichlet series, spectral synthesis, and algebraic number fields*, University of Maryland, Department of Mathematics, TR 71–41 (1971) 1–23. (Also referenced as *Dirichlet series, spectral synthesis, and algebraic number fields, Part I.* There isn't a part II, so I've used the shorter title.)
1971c. *Trigonometric sums associated with pseudo-measures*, Ann. Scuola Norm. Sup., Pisa 25 (1971) 229–248.
1971d. *Sui problemi di sintesi spettrale*, Rend. Sem. Mat., Milano 4l (1971) 55–61.
1971e. *Il Problema degli insiemi Helson-S*, Rend. Sem. Mat., Milano 41 (1971) 63–68.
1971f. *(LF) spaces and distributions on compact groups and spectral synthesis on* $\mathbb{R}/2\pi\mathbb{Z}$, Math. Ann.194 (1971) 52–67.
1972a. *Measure zero: Two case studies*, TR, 1972, 26 pages.
1972b. *Ensembles de Helson et synthese spectrale*, CRAS, Paris 274 (1972) 169–170.
1972c. *Construction de fonctionnelles multiplicatives discontinues sur des algebres metriques*, CRAS, Paris 274 (1972) 254–256.
1972d. *A support preserving Hahn-Banach property to determine Helson-S Sets*, Inventiones Math. 16 (1972) 214–228
1973a. *Idele characters in spectral synthesis on* $\mathbb{R}/2\pi\mathbb{Z}$, Ann. Inst. Fourier 23 (1973) 45–64.
1974a. *Pseudo-measure energy and spectral synthesis*, Can. J. Math. 26 (1974) 985–1001.

1974b. *Tauberian theorems, Wiener's spectrum, and spectral synthesis*, Rend. Sem. Mat., Milano 44 (1974) 63–73.
1975a. **Spectral Synthesis**, Pure and Applied Mathematics series, vol. 66, Academic Press, N.Y., 1975.
1975b. *Zeta functions for idelic pseudo-measures*, University of Maryland, Department of Mathematics, TR 74–55 (1975) 1–46.
(Appeared as [1979a].)
1975c. *The Wiener spectrum in spectral synthesis*, Studies in Applied Math. (MIT) 54 (1975) 91–115
1977a. *Analytic properties of idelic pseudo-measures*, University of Maryland, Department of Mathematics, TR 77–62 (1977) 1–33.
1977b. *Idelic pseudo-measures and Dirichlet series*, Symposia Mathematica, Academic Press, 1976 Conference on Harmonic Analysis, Rome 22 (1977) 205–222.
1979a. *Zeta functions for idelic pseudo-measures*, Ann. Scuola Norm. Sup., Pisa 6 (1979) 367–377.
1980a. *Fourier analysis of Riemann distributions and explicit formulas*, Math. Ann. 252 (1980) 141–164.
1981a. *The role of Wiener's Tauberian theorem in power spectrum computation*, University of Maryland, Department of Mathematics, TR 81–41 (1981) 1–44.
1981b. *Spectral deconvolution*, University of Maryland, Department of Mathematics, TR 81–63 (1981) 1–25.
1981c. *The theory of constructive signal analysis*, Studies in Applied Math. (MIT) 65 (1981) 37–80.
1981d. *Wiener's Tauberian theorem and the uncertainty principle*, Proc. of Modern Harmonic Analysis Conference 1982, Torino-Milano, (1983) 863–887.
1982a. *A closure problem for signals in semigroup invariant systems*, SIAM J. Math. Analysis 13 (1982) 180–207.
1983a. *Estimation problems and stochastic image analysis* (with S. Belbas), University of Maryland, Interdisciplinary Applied Mathematics Program, TR89–67 (1983) 1–15.
(Note: While TR89-67 typically suggests this was written in 1989, this report was written in 1983.)
1983b. *Harmonic analysis and spectral estimation*, J. Math. Analysis and Applications 91 (1983) 444–509.
1983c. *Weighted Hardy spaces and the Laplace transform* (with H. Heinig), Cortona Conference 1982, Lecture Notes in Mathematics, 992, Springer-Verlag, (1983) 240–277.
1983d. ' *Wiener's Tauberian theorem and the uncertainty principle*, Proc. of Modern Harmonic Analysis Conference 1982, Torino-Milano, (1983) 863–887.
1984a. *A local uncertainty principle*, SIAM, J. Math. Analysis 15 (1984) 988–995.
1985a. *An inequality associated with the uncertainty principle*, Rend. Circ. Mat. di Palermo 34 (1985) 407–421.

1985b. *Some mathematical methods for spectrum estimation*, in **Fourier Techniques and Applications**, J.F. Price, editor, Plenum Publishing (1985) 73–100.

1985c. *Fourier uniqueness criteria and spectrum estimation theorems*, in **Fourier Techniques and Applications**, J.F. Price, editor, Plenum Publishing (1985) 149–172.

1986a. *Inequalities for spectrum estimation*, Linear Algebra and Applications 84 (1986) 377–383.

1986b. *Weighted Hardy spaces and the Laplace transform II* (with H. Heinig and R. Johnson), Math. Nachrichten (Triebel commemorative volume), 132 (1987) 29–55.

1986c. *Fourier inequalities with A_p weights* (with H. Heinig and R. Johnson), General Inequalities 5, Oberwolfach, (1986), ISNM 80 (1987) 217–232.

1987a. *A quantitative maximum entropy theorem for the real line*, Integral Equations and Operator Theory 10 (1987) 761–779.

1989a. *Gabor representations and wavelets*, AMS Contemporary Mathematics, 91 (1989) 9–27.

1989b. *The Wiener-Plancherel formula in Euclidean space*, (with G. Benke and W. Evans), Advances in Applied Math., 10 (1989) 457–487.
(Note: This won the The Best Paper Award from the MITRE Corporation.)

1990a. *Heisenberg wavelets and the uncertainty principle*, Prometheus Inc., TR (1990) 1–3.

1990b. *Wavelet auditory models and irregular sampling*, Prometheus Inc., TR (1990) 1–6.

1990c. *Uncertainty principle inequalities and spectrum estimation*, NATO-ASI, in **Fourier Analysis and its Applications 1989**, J. Byrnes, editor, Kluwer Publishers, The Netherlands, Series C, 315 (1990) 143–182.

1990d. *Irregular sampling and the theory of frames, I* (with W. Heller), Mat. Note, 10, Suppl. no.1 (1990) 103–125.
(Note: There is no part II. This paper is mostly independent of [1992a] and [1992f].)

1991a. *Support dependent Fourier transform norm inequalities*, (with C. Karanikas), Rend. Sem. Mat., Roma, 11 (1991) 157–174.

1991b. *The spherical Wiener-Plancherel formula and spectral estimation*, SIAM Math. Analysis, 22 (1991) 1110–1130.

1991c *Fourier transform inequalities with measure weights, II*
(with H. Heinig), Second International Conference on Function Spaces 1989,
Poznan, Poland, Teubner Texte zur Mathematik series, 120 (1991) 140–151.

1991d. *A multidimensional Wiener-Wintner theorem and spectrum estimation*, Trans. AMS, 327 (1991) 833–852.

1992a. *Irregular sampling and the theory of frames*, in **The Role of Wavelets in Signal Processing Applications**, AFT Science and Research Center, Wright-Patterson Air Force Base, TR (1992) 21–44.

1992b. *Stationary frames and spectral estimation*, NATO-ASI, in **Probabilistic and Stochastic Methods in Analysis, with Applications 1991**, J. Byrnes, editor, Kluwer Publishers, The Netherlands, Series C, (1992) 117–161.

1992c. *Uncertainty principles for time-frequency operators* (with C. Heil and D. Walnut), Operator Theory: Advances and Applications, Birkhäuser, 58 (1992) 1–25.

1992d. *An auditory motivated time-scale signal representation* (with A. Teolis), IEEE-SP International Symposium on Time-Frequency and Time-Scale Analysis, (1992) 49–52.

1992e. *Fourier transform inequalities with measure weights* (with H. Heinig), Advances in Math., 96 (1992) 194–225.

1992f. *Irregular sampling and frames*, in **Wavelets - a Tutorial in Theory and Applications**, C. Chui, editor, Academic Press, Boston, (1992) 445–507.

1993a. *Multiresolution analysis frames with applications* (with S. Li), IEEE - ICASSP, III (1993) 304–307.

1993b. *On frames and filter banks* (with S. Li), Conference on Information Sciences and Systems, at The Johns Hopkins University, Technical Co-Sponsorship with IEEE, 1993, invited.

1993c. *A wavelet auditory model and data compression* (with A. Teolis), Applied and Computational Harmonic Analysis, 1(1993) 3–28.

1993d. *Local frames* (with A. Teolis), SPIE, Mathematical Imaging: Wavelet Applications in Signal and Image Processing, 2034 (1993) 310–321.

1993e. *Wavelets and sampling*, in **Wavelets and their Applications**, sponsored/published by IEEE - South Australian Section, University of Adelaide, The Flinders University, and the Research Centre for Sensor Signal and Information Processing, (1993) 1–37.

1993f. *From a wavelet auditory model to definitions of the Fourier transform*, NATO-ASI, in Wavelets and their Applications 1992, J. Byrnes, editor, Kluwer Publishers, The Netherlands, Series C, (1993).

1994a. *Noise reduction filters in mathematical models of biological systems*, The MITRE Corp., TR (1994) 1–6.

1994b. *Noise reduction using frames, irregular sampling, and wavelets*, ORD Signal Processing, Fort Meade, MD, 1994.

1994c. *Frame decompositions, sampling, and uncertainty principle inequalities*, in **Wavelets: Mathematics and Applications**, J. Benedetto and M. Frazier, editors, CRC Press, Boca Raton, FL, (1994) 247–304.

1994d. *Gabor frames for L^2 and related spaces* (with D. Walnut), in **Wavelets: Mathematics and Applications**, J. Benedetto and M. Frazier, editors, CRC Press, Boca Raton, FL (1994) 97–162.

1994e. *Noise suppression using a wavelet model* (with A. Teolis), IEEE - ICASSP (1994).

1994f. *Subband coding for sigmoidal nonlinear operations* (with S. Saliani), SPIE, Wavelet Applications, 2242 (1994) 19–27.

1994g. *Subband coding and noise reduction in multiresolution analysis frames* (with S. Li), SPIE, Wavelet Applications in Signal and Image Processing II, 2303 (1994) 154–165.

1994h. *The definition of the Fourier transform for weighted inequalities* (with J. Lakey), J. of Functional Analysis, 120 (1994) 403–439.

1994i. *Analysis and feature extraction of epileptiform EEG waveforms* (with D. Colella, G. Jacyna, et al.), Fifth International Cleveland Clinic - Bethel Epilepsy Symposium, 1994, Poster.

1994j. *Narrow band frame multiresolution analysis with perfect reconstruction* (with Shidong Li), IEEE - SP International Symposium on Time-Frequency and Time-Scale Analysis, (1994) 36–39.

1995a. *Pyramidal Riesz products associated with subband coding and self-similarity* (with E. Bernstein), SPIE, Wavelet Applications for Dual Use, 2491 (1995) 212–221, invited.

1995b. *Poisson's summation formula in the construction of wavelet bases*, with G. Zimmermann, Proceedings of ICIAM, Hamburg, 1995, invited.

1995c. *Wavelet-based analysis of EEG signals for detection and localization of epileptic seizures* (with G. Benke, M. Bozek-Kuzmicki, D. Colella, G. Jacyna), SPIE, Wavelet Applications for Dual Use, 2491 (1995) 760–769.

1995d. *Differentiation and the Balian-Low theorem* (with C. Heil and D. Walnut), J. of Fourier Analysis and Applications, 1 (1995) 355–402.

1995e. *Wavelet analysis of spectrogram seizure chirps* (with D. Colella), SPIE, Wavelet Applications in Signal and Image Processing III, 2569 (1995) 512–521, invited.

1995f. *Local frames and noise reduction* (with A. Teolis), Signal Processing, 45 (1995) 369–387.

1996a. *Frame signal processing applied to biolectric data*, in Wavelets in Biology and Medicine, A. Aldroubi and M. Unser, editors, CRC Press, Inc., Boca Raton, FL, 1996, Chapter 18, pages 467–486, invited.

1997a. *Generalized harmonic analysis and Gabor and wavelet systems*, AMS Proceedings of Symposia in Applied Mathematics, volume 52, Wiener Centenary Volume, 1997, pages 85–113, invited.

1997b. *Sampling operators and the Poisson summation formula*, (with G. Zimmermann), Journal of Fourier Analysis and Applications, 3 (1997) 505–523.

1997c *Wavelet detection of periodic behavior in EEG and ECoG data* (with G. Pfander), Proceedings of 15th IMACS World Congress on Scientific Computation, Modelling, and Applied Mathematics, Berlin, 1 (1997) 75–80, invited.

1998a. *Gabor systems and the Balian-Low Theorem* (with C. Heil and D. Walnut), in **Gabor Analysis and Algorithms, Theory and Applications**, H. G. Feichtinger and T. Strohmer, editors, Birkhäuser, Boston, 1998, pages 85–122, invited.

1998b. *Noise reduction in terms of the theory of frames*, in **Signal and Image Representation in Combined Spaces**, J. Zeevi and R. Coifman, editors, Academic Press, New York, 1998, pages 259–284, invited.

1998c. *Self-similar pyramidal structures and signal reconstruction* (with M. Leon and S. Saliani), SPIE, Wavelet Applications V, 3391 (1998) 304–314, invited.

1998d. *The theory of multiresolution frames and applications to filter banks* (with S. Li), Applied and Computational Harmonic Analysis, 5 (1998), 389–427.

1998e. *Frames, sampling, and seizure prediction*, in **Advances in Wavelets**, K.-S. Lau, editor, Springer-Verlag, New York, 1998, Chapter 1, pages 1–25, invited.

1998f. *Wavelet periodicity detection algorithms* (with G. Pfander), SPIE, Wavelet Applications in Signal and Image Processing VI, 3458 (1998), 48–55, invited.

1999a. *A multidimensional irregular sampling algorithm and applications* (with H.-C. Wu), IEEE - ICASSP, Phoenix, Special Session on Recent Advances in Sampling Theory and Applications, 4 (1999), 4 pages, invited.

1999b. *The construction of multiple dyadic minimally supported frequency wavelets on* $\mathbb{R}^d$ (with M. Leon), AMS Contemporary Math. Series, 247 (1999) 43–74.

1999c. *A Beurling covering theorem and multidimensional irregular sampling* (with H.-C. Wu), in **Sampling Theory and Applications**, Loen, Norway, sponsored/published by Norwegian University of Science and Technology, (1999) 142–148, invited.

2000a. *Sampling theory and wavelets*, NATO-ASI, in **Signal Processing for Multimedia 1998**, J. Byrnes, editor, Kluwer Publishers, The Netherlands, 2000, invited.

2000b. *Ten books on wavelets*, SIAM Review, 42 (2000) 127–138. (Although not a research paper, JB was asked to write an extensive review of recent books on wavelet theory. The result involved input from several of his graduate students.)

2000c. *Non-uniform sampling theory and spiral MRI reconstruction* (with H.-C. Wu), SPIE, Wavelet Applications in Signal and Image Processing VIII, 4119 (2000), invited.

2001a. *The classical sampling theorem, and non-uniform sampling and frames* (with P.S.J.G. Ferreira), Chapter 1 of **Modern Sampling Theory: Mathematics and Applications**, J.J. Benedetto and P.S.J.G. Ferreira, editors, Birkhäuser Boston, 2001.

2001b. *Frames, irregular sampling, and a wavelet auditory model* (with S. Scott), Chapter 14 in **Sampling Theory and Practice**, F. Marvasti, editor, Kluwer Academic/Plenum Publishers, New York, 2001, invited.

2001c. *Wavelet frames: multiresolution analysis and extension principles* (with O. Treiber), Chapter 1 of **Wavelet Transforms and Time-Frequency Signal Analysis**, L. Debnath, editor, Birkhäuser, Boston, 2001, 3–36, invited.

2001d. *The construction of single wavelets in d-dimensions* (with M. Leon), J. Geometric Analysis, 11 (2001) 1–15.

2002a. *MRI signal reconstruction by Fourier frames on interleaving spirals* (with A. Powell and H.-C. Wu), IEEE - ISBI 2002, 4 pages, invited.

2002b. *A fractal set constructed from a class of wavelet sets* (with S. Sumetkijakan), AMS Contemporary Math. Series, 313 (2002) 19–35.

2002c. *Periodic wavelet transforms and periodicity detection* (with G. Pfander), SIAM J. Applied Math., 62 (2002) 1329–1368.

2003a. *Finite normalized tight frames* (with M. Fickus), Advances in Computational Math., 18 (2003) 357–385.

2003b. *The Balian-Low theorem and regularity of Gabor systems* (with W. Czaja, P. Gadziński, and A. Powell), J. Geometric Analysis, 13 (2003) 239–254.

2003c. *Weighted Fourier inequalities: new proof and generalization* (with H. Heinig), J. Fourier Analysis and Applications, 9 (2003) 1–37.

2003d. *The Balian-Low theorem for symplectic forms* (with W. Czaja and A. Maltsev), Journal of Mathematical Physics, 44 (2003) 1735–1750.

2003e. *Local sampling for regular wavelet and Gabor expansions* (with N. Atreas and C. Karanikas), Sampling Theory in Signal and Image processing, 2 (2003) 1–24

2003f. *A Wiener-Wintner theorem for $1/f$ power spectra* (with R. Kerby and S. Scott), J. Math. Analysis and Applications, 279 (2003) 740–755.

2004a. Software package for CAZAC code generators and Doppler shift analysis (with J. Donatelli and J. Ryan), 2004, see http://www.math.umd.edu/~jjb/cazac.

2004b. *Prologue* for **Sampling, Wavelets, and Tomography**, J. J. Benedetto and A. Zayed, editors, Birkhäuser, Boston, MA, 2004. (Although not a research paper, this is longer than most prologues and contains new information on sampling techniques and Claude Shannon.)

2004c. *Constructive approximation in waveform design* (invited) in **Advances in Constructive Approximation Theory**, M. Neamtu and E. B. Saff, editors, Nashboro Press, (2004) 89–108.

2004d. *A wavelet theory for local fields and related groups* (with R. L. Benedetto), J. Geometric Analysis, 14(3) (2004) 423–456

2004e. *Sigma-Delta quantization and finite frames* (with A. Powell and Ö. Yilmaz), ICASSP, Montreal, 2004, invited.

2005a. *Multiscale Riesz products and their support properties* (with E. Bernstein and I. Konstantinidis), Acta Applicandae Math, 88(2) (2005) 201–227.

2005b. *Analog to digital conversion for finite frames* (with A. Powell and Ö. Yilmaz), SPIE, Wavelet Applications in Signal and Image Processing (2005), invited.

2005c. *Greedy adaptive discrimination: component analysis by simultaneous sparse approximation*, (with J. Sieracki), SPIE, Wavelet Applications in Signal and Image Processing (2005).

2005d. *A (p,q)-version of Bourgain's theorem* (with A. Powell), Trans. Amer. Math. Soc., 358 (2005) 2489–2505.

2006a. *Tight frames and geometric properties of wavelet sets* (with S. Sumetkijakan), Advances in Computational Math., 24 (2006) 35–56.

2006b. *Geometrical properties of Grassmannian frames for* $\mathbb{R}^2$ *and* $\mathbb{R}^3$ (with J. Kolesar), EURASIP J. Applied Signal Processing, Special Issue on Frames and Overcomplete Representations in Signal Processing (2006) 17 pages.

2006c. *Sigma-Delta quantization and finite frames* (with A. Powell and Ö. Yilmaz), IEEE Trans. Information Theory, 52(5) (2006) 1990–2005.

2006d. *Introduction* for **Fundamental Papers in Wavelet Theory**, edited by C. Heil and D. F. Walnut, Princeton University Press, 2006. (Although not a research paper, this is an extensive, 20 page introduction for an important volume.)

2006e. *An endpoint* $(1, \infty)$ *Balian-Low theorem* (with W. Czaja, A. Powell, and J. Sterbenz), Math. Research Letters, 13 (2006) 467–474.

2006f. *An optimal example for the Balian-Low uncertainty principle*, (with W. Czaja and A. Powell), SIAM Journal of Mathematical Analysis, 38 (2006) 333–345.

2006g. *Zero autocorrelation waveforms: a Doppler statistic and multifunction problems* (with J. Donatelli, I. Konstantinidis, and C. Shaw), ICASSP, Toulouse, 2006, invited.

2006h. *A Doppler statistic for zero autocorrelation waveforms* (with J. Donatelli, I. Konstantinidis, and C. Shaw), Conference on Information Sciences and Systems, at Princeton University, Technical Co-Sponsorship with IEEE, (2006), pages 1403–1407, invited.

2006i. *Frame expansions for Gabor multipliers* (with G. Pfander), Applied and Computational Harmonic Analysis, 20 (2006) 26–40.

2006j. *Second order Sigma-Delta quantization of finite frame expansions* (with A. Powell and Ö. Yilmaz), Applied and Computational Harmonic Analysis, 20 (2006) 128–148.

2007a. *Ambiguity and sidelobe behavior of CAZAC waveforms* (with A. Kebo, I. Konstantinidis, M. Dellomo, J. Sieracki), IEEE Radar Conference, Boston (2007).

2007b. *The construction of d-dimensional MRA frames* (with J. Romero), J. Applied Functional Analysis, 2 (2007) 403–426.

2007c. *Ambiguity function and frame theoretic properties of periodic zero autocorrelation functions* (with J. Donatelli), IEEE J. of Selected Topics in Signal Processing, 1 (2007) 6–20.

2007d. *Target tracking using particle filtering and CAZAC sequences* (with I. Kyriakides, I. Konstantinidis, D. Morrell, A. Papandreou-Suppappola), IEEE International Waveform Diversity and Design, Pisa (2007), invited.

2007e. *Concatenating codes for improved ambiguity behavior* (with A. Bourouihiya, I. Konstantinidis, and K. Okoudjou), Adaptive Waveform Technology for Futuristic Communications, Radar, and Navigation Systems, International Conference on Electromagnetics in Advanced Applications (ICEAA), Torino, 2007, invited.

2008a. *The role of frame force in quantum detection* (with A. Kebo), J. Fourier Analysis and Applications, 14 (2008) 443–474.

2008b. *PCM-Σ–Δ comparison and sparse representation quantization* (with O. Oktay), Conference of Information Science and Systems, Princeton, 2008, 6 pages, invited.

2008c. *Multiple target tracking using particle filtering and multicarrier phase-coded CAZAC sequences* (with I. Kyriakides, A. Papandreou-Suppappola, D. Morrell, and I. Konstantinidis), Sensor, Signal and Information Processing Workshop (SenSIP) 2008, Sedona AZ.

2008d. *Human electrocortigraphic signature determination by eGAD sparse approximation* (with N. Crone and J. Sieracki), Sensor, Signal and Information Processing Workshop (SenSIP) 2008, Sedona AZ.

2008e. *Complex Sigma-Delta quantization algorithms for finite frames* (with O. Oktay and A. Tangboondouangjit), AMS Contemporary Mathematics, 464 (2008) 27–49.

2008f. *Frames and a vector-valued ambiguity function* (with J. Donatelli), IEEE - Asilomar, 2008, invited.

2009a. *Phase coded waveforms and their design - the role of the ambiguity function* (with I. Konstantinidis and M. Rangaswamy), IEEE Signal Processing Magazine (invited), 26 (2009) 22–31.

2009b. *Hadamard matrices and infinite unimodular sequences with* 0*-autocorrelation* (with S. Datta), IEEE International Waveform Diversity and Design 2009, Orlando, invited, but withdrawn since the authors couldn't attend the conference.
(Note: See [2010c].)

2009c. *Frame based kernel methods for automatic classification in hyperspectral data* (with W. Czaja, C. Flake, and M. Hirn), Proceedings IEEE-IGARSS 2009, invited.

2009d. *Smooth functions associated with wavelet sets on $\mathbb{R}^d$, $d \geq 1$, and frame bound gaps* (with E. King), Acta Applicandae Math, 107 (2009) 121–142.

2009e. *Geometric properties of Shapiro-Rudin polynomials* (with J. Sugar-Moore), Involve - A Journal of Mathematics, 2(4)(2009) 449–468.

2010a. *Besov spaces for the Schrödinger operator with barrier potential* (with S. Zheng), Complex Analysis and Operator Theory, 4(4)(2010) 777–811.

2010b. *Pointwise comparison of PCM and Sigma-Delta quantization* (with O. Oktay), Constructive Approximation, 32(1)(2010) 131–158.

2010c. *Construction of infinite unimodular sequences with zero autocorrelation* (with S. Datta), Advances in Computational Mathematics, 32 (2010) 191–207.

2010d. *Wavelet packets for multi- and hyper-spectral imagery* (with W. Czaja, M. Ehler, C. Flake, and M. Hirn), Wavelet Applications in Industrial Processing VII, Proc. SPIE San Jose, 7535 (2010) 8–11.

2010e. *Frame potential classification algorithm for retinal data* (with W. Czaja and M. Ehler), 26th Southern Biomedical Engineering Conference, College Park, MD, 2010.

2010f. *Maximally separated frames for automatic classification in hyperspectral data* (with W. Czaja, M. Ehler, and N. Strawn), IGARSS 2010, Honolulu, preprint.

2011a. *The construction of wavelet sets* (with R. L. Benedetto) in **Wavelets and Multiscale Analysis**, J. Cohen and A. I. Zayed, editors, Springer-Birkhäuser, 2011, Chapter 2, pages 17–56.

2011b. *Discrete autocorrelation-based multiplicative MRAs and sampling on $\mathbb{R}$* (with. S. Datta), Sampling Theory in Signal and Image Processing 10 (2011) 111–133.

2011c. *Intrinsic wavelet and frame applications* (with T. Andrews), invited paper, SPIE 2011, Orlando.

2012a. *Image representation and compression via sparse solutions of systems of linear equations* (with A. Nava-Tudela), Technical Report, 2012. (Revised and published in [2014f].)

2012b. *Optimal ambiguity functions and Weil's exponential sums bound* (with R. L. Benedetto and J. Woodworth), J. Fourier Analysis and Applications, 18 (2012) 471–487.

2012c. *Constructions and a generalization of perfect autocorrelation sequences on $\mathbb{Z}$* (with S. Datta), invited chapter in volume dedicated to Gil Walter, X. Shen and A. Zayed, eds., Chapter 8, Springer (2012) 183–207.

2012d. *Integration of heterogeneous data for classification in hyperspectral satellite imagery* (with W. Czaja, J. Dobrosotskaya, T. Doster, K. Duke, and D. Gillis), SPIE 2012, Baltimore.

2012e. *Semi-supervised learning of heterogeneous data in remote sensing imagery* (with W. Czaja, J. Dobrosotskaya, T. Doster, K. Duke, and D. Gillis), invited paper SPIE 2012, Baltimore.

2013a. *Balayage and short time Fourier transform frames* (with E. Au-Yeung), SampTA 2013 at Bremen, invited.

2013b. *Wavelet packets for time-frequency analysis of multi-spectral images* (with W. Czaja and M. Ehler), International J. of Geomathematics, 4 (2013) 137–154.

2014a. *Chandler Davis as mentor*, Mathematical Intelligencer, 36 (2014) 20–21.

2014b. *Wavelet packets and nonlinear manifold learning for analysis of hyperspectral data* (with W. Czaja, T. Doster, and C. Schwartz), SPIE 2014, Baltimore.

2014c. *Operator-based integration of information in multimodal radiological search mission with applications to anomaly detection* (with A. Cloninger, W. Czaja, T. Doster, B. Manning, T. McCullough, K. Kochersbeger, and M. McLean), Proc. SPIE 9073, Chemical, Biological, Radiological, Nuclear, and Explosives (CBRNE) Sensing XV, Baltimore, 2014.

2014d. *Nonlinear dimensionality reduction via the ENH-LTSA method for hyperspectral image classification* (with W. Czaja, A. Halevy, W. Li, C. Liu, B. Shi, W. Sun, R. Wang), IEEE Journal of Selected Topics in Applied Earth Observations and Remote Sensing (IEEE JSTARS), 7 (2014) 375–388.

2014e. *UL-Isomap based dimensionality reduction for hyperspectral imagery classification* (with W. Czaja, A. Halevy, W. Li, C. Liu, B. Shi, W. Sun, H. Wu), Journal of Photogrammetry and Remote Sensing (ISPRS), 89 (2014) 25–36.

2014f. *Sampling in image representation and compression* (with A. Nava-Tudela), Sampling Theory in Honor of Paul Bützer on his 85th Birthday, A. Zayed, editor, invited Chapter, Springer-Birkhäuser, 2014, Chapter 7, pages 149–188.

2015a. *Fourier operators in applied harmonic analysis* (with M. Begué), in **Sampling Theory – a Renaissance**, G. Pfander, editor, invited Chapter, Springer-Birkhäuser, Chapter 5, 2015, 185–216.

2015b. *The HRT conjecture for functions with certain behavior at infinity* (with A. Bourouihiya), J. Geometric Analysis 25 (2015) 226–254.

2015c. *Dimension reduction and remote sensing using modern harmonic analysis* (with W. Czaja), in **Handbook of Geomathematics**, edited by W. Freeden, Z. Nashed, and T. Sonar), invited Chapter, Springer, New York, 2015, pages 2609–2632.

2015d. *Generalized Fourier frames in terms of balayage* (with E. Au-Yeung), J. Fourier Analysis and Applications, 21 (2015) 472–508.

2015e. *Graph theoretic uncertainty principles* (with P. Koprowski), SampTA 2015 Washington D.C., invited, 2015.

2015f. *Balayage and pseudo-differential equation frame inequalities* (with E. Au-Yeung), SampTA 2015 Washington D.C., 2015.

2016a. *Spatial-spectral operator theoretic methods for hyperspectral image classification* (with W. Czaja, J. Dobrosotskaya, T. Doster, and K. Duke), International Journal on Geomathematics (GEM) 7 (2016) 275–297.

2016b. *Preface for* ***Finite Frames***, K. Okoudjou, editor, AMS Symposia in Mathematics (based on AMS Workshop at 2015 JMM), 2016.

2017a. *Uncertainty principles and weighted norm inequalities* (with M. Dellatorre), invited chapter in **Functional Analysis, Harmonic Analysis, and Image Processing: a Collection of Papers in Honor of Björn Jawerth**, Michael Cwikel and Mario Milman, editors, AMS Contemporary Mathematics 693 (2017) 55–78.

2017b. *A frame reconstruction algorithm with applications to MRI* (with A. Nava-Tudela, A. Powell, and Y. Wang), invited chapter, Chapter 9 in **Frames and Other Bases in Abstract and Function Spaces: Novel Methods in Harmonic Analysis, Volume 1**, I. Pesenson, et al. editors, Springer-Birkhäuser, New York, 2017, pages 185–214.

2018a. *Frames of translates for number-theoretic groups* (with R. L. Benedetto), accepted in J. Geometrical Analysis, to appear.
Online publication 2018, arXiv: 1804.07783v2.

2019a. *Frame multiplication theory and a vector-valued DFT and ambiguity function* (with T. Andrews and J. Donatelli), J. Fourier Analysis and Appl. 25(4) (2019), 1975–1854.
Online publication 2017, arXiv: 1706.05579v1

2019b. *CAZAC sequences and Haagerup's characterization of cyclic N-roots* (with K. Cordwell and M. Magsino), invited chapter in **New Trends in Applied Harmonic Analysis, Volume 2: Harmonic Analysis, Geometric Measure Theory, and Applications,** C. Cabrelli, U. Molter, et al., editors, Springer-Birkhäuser, New York, 2019, Chapter 1, pages 1–43.

2019c. *Reactive sensing and multiplicative frame super-resolution* (with M. Dellomo), 39 pages. Submitted 2019 to IEEE-IT.
Online publication 2019, arXiv: 1903.05677

2020a. *Super-resolution by means of Beurling minimal extrapolation* (with W. Li), Applied and Computational Harmonic Analysis 48 (1) (2020) 218–241.
Online publication 2016, arXiv: 1601.05761v3

2020b. *A generalization of Gleason's frame function for quantum measurement* (with P. Koprowski and J. Nolan), 38 pages. Submitted 2020 to J. Math. Physics.
Online publication 2020, arXiv: 2001.06738v1

2020c. *Haar approximation from within for* $L^p(\mathbb{R}^d)$, $0 < p < 1$ (with F. Njeunje), 33 pages. Submitted 2020 to J. of Sampling Theory, Signal Processing, and Data Analysis.

Part II
Harmonic Analysis

The second part of this volume contains contributions that can be viewed as part of Harmonic Analysis as broadly construed. In the first chapter, Heil discusses the notion of absolute continuity and the Banach–Zaretsky Theorem. In the second chapter, Johnson presents some of his recent results on the still open spectral synthesis problem. The third chapter written by Hu, Leonard, and Zheng deals with the nonlinear Schrödinger equation. In the fourth chapter of this part, Strohmer and Wertz present some results on the almost eigenvalues of the Almost Mathieu operators. The last chapter of this part, is a contribution by Hogan and Lakey providing a case study of the notion of spatio-spectral limiting on redundant cubes.

Absolute Continuity and the Banach–Zaretsky Theorem

Christopher Heil

Abstract The Banach–Zaretsky Theorem is a fundamental but often overlooked result that characterizes the functions that are absolutely continuous. This chapter presents basic results on differentiability, absolute continuity, and the Fundamental Theorem of Calculus with an emphasis on the role of the Banach–Zaretsky Theorem.

1 Introduction

I started as a graduate student at the University of Maryland in 1982, and John Benedetto became my Ph.D. advisor around 1986. While working with John, I came across his book "Real Variable and Integration" [1]. Unfortunately, the text was out of print, but I checked the book out of the library and made a "grad student copy" (meaning I xeroxed the text). This became an important reference for me. Much later there was a limited reprinting of the book, and I am proud to own a copy autographed by John himself, dated May 2000. More recently, John and Wojciech Czaja collaborated to create a revised and expanded version of the text [2], published in 2009. An especially appealing aspect of both books is the inclusion of many historical references and discussions of the historical development of the results.

Quoting from the Zentralblatt review (Zbl. 0336.26001) of [1],

> At the heart of the differential and integral calculus is the so-called fundamental theorem of the calculus ... In the theory of Lebesgue integration this result takes on the form that a function is the indefinite integral of a Lebesgue integrable function if and only if it is absolutely continuous. – In the introduction to the book under review the author states that the main mathematical reason for having written the book is that none of the other texts

This work was partially supported by a grant from the Simons Foundation.

C. Heil (✉)
School of Mathematics, Georgia Tech, Atlanta, GA, USA
e-mail: heil@math.gatech.edu

M. Hirn et al. (eds.), *Excursions in Harmonic Analysis, Volume 6*, Applied and Numerical Harmonic Analysis, https://doi.org/10.1007/978-3-030-69637-5_2

> in this area of analysis and at this level stress the importance of the notion of absolute continuity as it pertains to such fundamental results .... It is the reviewer's opinion that the author has written a beautiful and extremely useful textbook on real variable and integration theory. ... All in all the author is to be congratulated to have written such a fine addition to the literature. W.A.J. Luxemburg.

Eventually, it became time for me to write my own book on real analysis ([6], which appeared recently). My primary goal was to write a classroom text for students taking their first course on Lebesgue measure and the Lebesgue integral. There are many books on real analysis available, but I wanted to write the book that I wished my instructor had used when I first took that course. One of my principles of writing for this book was that each proof should be both rigorous and *enlightening* (which often, though not always, entails finding a "simple proof"). The two chapters on differentiation and absolute continuity were especially difficult in this regard. In the standard texts, there are a variety of approaches to the proofs of the basic theorems in this area, but most of the standard proofs are very technical, far from "enlightening" for the beginning student. In this chapter (which may be called a *mathematical essay*), I will outline how I dealt with these issues. The highlight is that the *Banach–Zaretsky Theorem*, an elegant but often-overlooked result that I first learned about from John's book (of course), plays a key role in making the presentation "enlightening." Unfortunately, I was still left with one theorem whose proof I was unable to simplify to my satisfaction, and I will discuss this failure as well.

Of course, what is "simple," "enlightening," "elegant," or "obvious" is, almost by definition, in the eye of the beholder. Consequently, the reader may disagree with my opinions of what is important, elegant, or enlightening. I would be happy to engage in further discussion about how best to present absolute continuity, differentiation, and related results to the first-time student.

1.1 The Fundamental Theorem of Calculus

Our fundamental goal is to fully understand the Fundamental Theorem of Calculus (which we sometimes refer to using the acronym *FTC*). The version that we often learn in an undergraduate calculus course is that if a function f is differentiable at every point in a closed finite interval $[a, b]$ and if f' is continuous on $[a, b]$, then the Fundamental Theorem of Calculus holds, and it tells us that

$$\int_a^x f'(t)\, dt = f(x) - f(a), \qquad \text{for all } x \in [a, b]. \tag{1.1}$$

Since we have assumed that f' is continuous, the integral on the line above exists as a Riemann integral. Does the Fundamental Theorem of Calculus hold if we assume only that f' is Lebesgue integrable? Precisely:

Fig. 1 The Cantor–Lebesgue function (also known as the *Devil's staircase*)

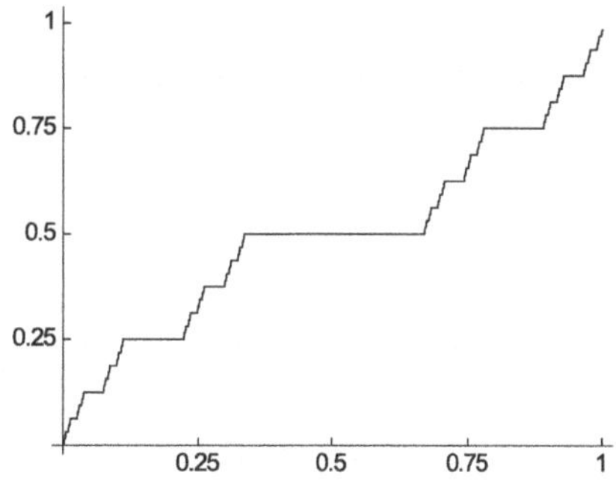

If $f'(x)$ exists for a.e. x and f' is integrable, must Eq. (1.1) *hold?*

The construction of the classical *Cantor–Lebesgue function* (pictured in Fig. 1) shows that the answer to this question is *no* in general. The Cantor–Lebesgue function is *singular* in the following sense.

Definition 1 (Singular Function) A function f on $[a, b]$ is *singular* if f is differentiable at almost every point in $[a, b]$ and $f' = 0$ almost everywhere (a.e.) on $[a, b]$. $\diamond$

For a singular function, we have $\int_a^b f' = 0$, which need not equal $f(b) - f(a)$. For which functions does the FTC hold? We will consider this in the coming pages.

Remark 1 For simplicity of presentation, we will focus on real-valued functions. However, nearly all of the results carry over without change to complex-valued functions. One result where there is a small difference is in the statement of the Banach–Zaretsky Theorem. We will explain the necessary change in hypotheses when we discuss that theorem. $\diamond$

1.2 Notation

We use the standard notations of real analysis. We mention some specific terminology that we will use.

- If E is a subset of the Euclidean space $\mathbb{R}^d$, then $|E|_e$ denotes the exterior Lebesgue measure of E. If E is a Lebesgue measurable set, then we write this measure as $|E|$.
- The Euclidean norm of a point $x \in \mathbb{R}^d$ is denoted by $\|x\|$.
- If E is a measurable subset of $\mathbb{R}^d$ and $1 \leq p < \infty$, then $L^p(E)$ denotes the Lebesgue space of all p-integrable functions on E, and $\|f\|_p = \left(\int_E |f(t)|^p\,dt\right)^{1/p}$ denotes the L^p-norm of an element of $L^p(E)$. If $E = [a, b]$, then we denote this Lebesgue space by $L^p[a, b]$.

- If E is a measurable subset of $\mathbb{R}^d$, then $L^\infty(E)$ denotes the Lebesgue space of all essentially bounded functions on E, and $\|f\|_\infty = \operatorname{ess\,sup}_{t\in E} |f(t)|$ denotes the L^∞-norm of an element of $L^\infty(E)$. If $E = [a, b]$, then we denote this Lebesgue space by $L^\infty[a, b]$.

2 Differentiation

2.1 Functions of Bounded Variation

We will briefly review the notion of bounded variation. As with most of the results and definitions that we will discuss, more extensive discussion and motivation can be found in [6]. The idea behind this definition is that if we fix finitely many points $a = x_0 < \cdots < x_n = b$ in the interval $[a, b]$, then the quantity $\sum_{j=1}^n |f(x_j) - f(x_{j-1})|$ is an approximation to the "total variation" in the height of f across the interval. A function has bounded variation if there is an upper bound to such approximations, and the total variation is the supremum of all of those approximations.

Definition 2 (Bounded Variation) Let $f : [a, b] \to \mathbb{R}$ be given. For each finite partition

$$\Gamma = \bigl\{a = x_0 < \cdots < x_n = b\bigr\}$$

of $[a, b]$, set

$$S_\Gamma[f] = \sum_{j=1}^n |f(x_j) - f(x_{j-1})|.$$

The *total variation* of f over $[a, b]$ (or simply the *variation* of f, for short) is

$$V[f] = \sup\bigl\{S_\Gamma[f] : \Gamma \text{ is a partition of } [a, b]\bigr\}.$$

We say that f has *bounded variation* on $[a, b]$ if $V[f] < \infty$. We collect the functions that have bounded variation on $[a, b]$ to form the space

$$\mathrm{BV}[a, b] = \bigl\{f : [a, b] \to \mathbb{R} : f \text{ has bounded variation}\bigr\}. \qquad \diamond.$$

Letting $\mathrm{Lip}[a, b]$ denote the space of functions that are Lipschitz continuous on $[a, b]$, and $C^1[a, b]$ the space of functions that are differentiable on $[a, b]$ and whose derivative is continuous on $[a, b]$, we have the proper inclusions

$$C^1[a, b] \subsetneq \mathrm{Lip}[a, b] \subsetneq \mathrm{BV}[a, b].$$

The following theorem gives a fundamental characterization of functions in BV$[a, b]$. The proof is nontrivial, but it is (in our opinion) "clear and enlightening" for the student who works through it, and so we will be content to refer to [6] for details. For us, a *monotone increasing* function on $[a, b]$ is a function f such that $f(x) \le f(y)$ whenever $a \le x \le y \le b$.

Theorem 3 (Jordan Decomposition) *If* $f: [a, b] \to \mathbb{R}$, *then* $f \in \mathrm{BV}[a, b]$ *if and only if there exist monotone increasing functions* g *and* h *such that* $f = g - h$. ◊

Thus, many questions (though not all) can be reduced to questions about monotone increasing functions.

2.2 *Our Failure: Differentiability of Monotone Increasing Functions*

How hard can it be to understand monotone increasing functions? At first glance, monotone functions seem to be "simple and easy." Indeed, here is one easy fact.

Lemma 4 (Discontinuities of Monotone Functions) *If* $f: [a, b] \to \mathbb{R}$ *is monotone increasing, then it has at most countably many discontinuities, and they are all jump discontinuities.*

Proof Since f is monotone increasing and takes real values at each point of $[a, b]$, it follows that $f(a) \le f(x) \le f(b)$ for every x, so f is bounded. Further, the one-sided limits

$$f(x-) = \lim_{y \to x^-} f(y) \qquad \text{and} \qquad f(x+) = \lim_{y \to x^+} f(y)$$

exist at every point $x \in (a, b)$. The appropriate one-sided limits also exist at the endpoints a and b. Consequently, each point of discontinuity of f must be a jump discontinuity. Since f is bounded, if we fix a positive integer k, there can be at most finitely many points $x \in [a, b]$ such that

$$f(x+) - f(x-) \ge \frac{1}{k}.$$

Every jump discontinuity must satisfy this inequality for some positive integer k, so there can be at most countably many discontinuities. □

With only countably many discontinuities, all of which are jump discontinuities, just how complicated could a monotone increasing function be? Should not it be differentiable at all but those countably many discontinuities? In fact, it is much more complicated than that. Things would be easier if the discontinuities were separated, but there is no reason that they have to be. Here is a monotone increasing function that is discontinuous at every rational point.

Example Let $\{r_k\}_{k\in\mathbb{N}}$ be an enumeration of the rational points in (0, 1). Then, the function

$$f(x) = \sum_{n=1}^{\infty} 2^{-n} \chi_{[r_n,1]}(x), \qquad x \in [0, 1]$$

is monotone increasing on [0, 1], right-continuous at every point in [0, 1], discontinuous at every rational point in (0, 1), and continuous at every irrational point in (0, 1). ◇

Thus, monotone increasing functions are more complicated than we may initially suspect. However, our next theorem states that every monotone increasing function on $[a, b]$ is differentiable *at almost every point.*

Theorem 5 (Differentiability of Monotone Functions) *If a function $f\colon [a,b] \to \mathbb{R}$ is monotone increasing, then $f'(x)$ exists for almost every $x \in [a,b]$.* ◇

Surprisingly (at least to me), Theorem 5 appears to be a deep and difficult result. I am not aware of any proof that I would call "simple," or even any proof that I would call "enlightening" for the beginning student. I have not been able to construct any such proof myself. There are a number of different standard approaches to the proof; the reader may find it interesting to compare the proofs given in [3, Thm. 7.5], [5, Thm. 3.23], [11, Thm. 3.3.14], or [13, Thm. 7.5], for example. In the end, for the proof presented in my text, I adopted a standard technique based on the *Vitali Covering Lemma*. In my opinion, Theorem 5 has the dubious distinction of being the theorem in [6] whose proof is the least enlightening. I will not present that proof here; for this chapter, we will simply accept Theorem 5 as given and move onward. In comparison to that theorem, the proof of the next corollary is much easier.

Corollary 6 *If $f\colon [a,b] \to \mathbb{R}$ is monotone increasing on $[a,b]$, then f' is measurable, $f' \ge 0$ a.e., $f' \in L^1[a,b]$, and*

$$0 \le \int_a^b f' \le f(b) - f(a).$$

Proof For simplicity of presentation, extend the domain of f to the entire real line by setting $f(x) = f(a)$ for $x < a$ and $f(x) = f(b)$ for $x > b$.

(a) Since f is differentiable at almost every point, we know that the functions

$$f_n(x) = \frac{f(x+\frac{1}{n}) - f(x)}{\frac{1}{n}} = n\bigl(f(x+\tfrac{1}{n}) - f(x)\bigr), \qquad x \in \mathbb{R}$$

converge pointwise a.e. to $f'(x)$ on $[a,b]$ as $n \to \infty$. Furthermore, each f_n is measurable and nonnegative (because f is monotone increasing), so f' is measurable and $f' \ge 0$ a.e.

(b) Since the functions f_n are nonnegative and converge pointwise a.e. to f', we can apply Fatou's Lemma to obtain

$$\int_a^b f' = \int_a^b \liminf_{n\to\infty} f_n \le \liminf_{n\to\infty} \int_a^b f_n.$$

On the other hand, recalling how we extended the domain of f to $\mathbb{R}$, for each individual n we compute that

$$\begin{aligned}
\int_a^b f_n &= n\int_{a+\frac{1}{n}}^{b+\frac{1}{n}} f - n\int_a^b f && \text{(by the definition of } f_n) \\
&= n\int_b^{b+\frac{1}{n}} f - n\int_a^{a+\frac{1}{n}} f \\
&= n\int_b^{b+\frac{1}{n}} f(b) - n\int_a^{a+\frac{1}{n}} f && \text{(since } f \text{ is constant on } [b,\infty)) \\
&\le n\int_b^{b+\frac{1}{n}} f(b) - n\int_a^{a+\frac{1}{n}} f(a) && \text{(since } f \text{ is monotone increasing)} \\
&= f(b) - f(a).
\end{aligned}$$

Therefore,

$$\int_a^b f' \le \liminf_{n\to\infty} \int_a^b f_n \le f(b) - f(a) < \infty,$$

so f' is integrable. □

Since every function with bounded variation is the difference of two monotone increasing functions, an immediate consequence of Corollary 6 is that if f belongs to $\mathrm{BV}[a,b]$ then f is differentiable at almost every point and $f' \in L^1[a,b]$.

2.3 The Simple Vitali Lemma

We mentioned the Vitali Covering Lemma earlier, perhaps implicitly deriding its use in our earlier statements. In fact, this is a fundamental result of great importance. Unfortunately, its proof is (in my opinion) less "enlightening" than many others. However, there is a more restricted version that is sufficient for many proofs, and this "simple" version does have an elegant and enlightening proof that we present here. The proof that we give is close to the one that I learned from Folland's text [5] (which is one of the great reference texts in analysis). Essentially, this *Simple Vitali Lemma* states that if we are given any collection of open balls in $\mathbb{R}^d$, then we can find *finitely many disjoint balls* from the collection that cover a fixed fraction of the measure of the union of the original balls. Up to an ε, this fraction is 3^{-d} (so in dimension $d = 1$, we can choose disjoint open intervals that cover about $1/3$ of the original collection). The proof is an example of a *greedy algorithm*—basically, we choose a ball B_1 that has the largest possible radius, then choose B_2 to be the largest possible ball disjoint from B_1, and so forth.

Theorem 7 (Simple Vitali Lemma) *Let $\mathcal{B}$ be any nonempty collection of open balls in $\mathbb{R}^d$. Let U be the union of all of the balls in $\mathcal{B}$, and fix $0 < c < |U|$. Then there exist finitely many* disjoint *balls $B_1, \dots, B_N \in \mathcal{B}$ such that*

$$\sum_{k=1}^{N} |B_k| > \frac{c}{3^d}.$$

Proof Since $c < |U|$, the number c is finite, and therefore, there exists a compact set $K \subseteq U$ whose measure satisfies $c < |K| < |U|$. Since $\mathcal{B}$ is an open cover of the compact set K, we can find finitely many balls $A_1, \dots, A_m \in \mathcal{B}$ such that

$$K \subseteq \bigcup_{j=1}^{m} A_j.$$

Let B_1 be an A_j ball that has maximal radius.

If there are no A_j balls that are disjoint from B_1, then we set $N = 1$ and stop. Otherwise, let B_2 be an A_j ball with largest radius that is disjoint from B_1 (if there is more than one such ball, just choose one of them). We then repeat this process, which must eventually stop, to select disjoint balls $B_1, \dots, B_N$ from $A_1, \dots, A_m$. These balls need not cover K, but we hope that they will cover an appropriate portion of K.

To prove this, let B_k^* denote the open ball that has the same center as B_k, but with radius three times larger. Suppose that $1 \le j \le m$, but A_j is not one of $B_1, \dots, B_N$.

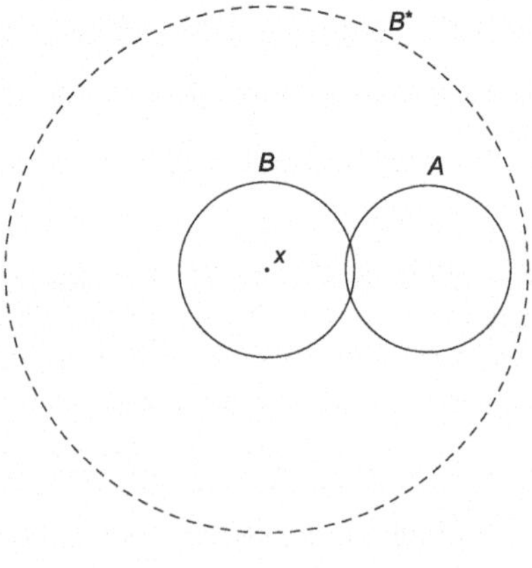

Fig. 2 Circle B has radius 1, circle A has radius 0.95, and circle B^* (which has the same center x as circle B) has radius 3

Then, A_j must intersect at least one of the balls $B_1, \dots, B_N$. Let k be the smallest index such that $A_j \cap B_k \neq \varnothing$. By construction,

$$\text{radius}(A_j) \leq \text{radius}(B_k).$$

It follows from this that $A_j \subseteq B_k^*$ (for a "proof of picture," see Fig. 2). Thus, every set A_j that is not one of $B_1, \dots, B_N$ is contained in some B_k^*. Hence,

$$K \subseteq \bigcup_{j=1}^{m} A_j \subseteq \bigcup_{k=1}^{N} B_k^*,$$

and therefore,

$$c < |K| \leq \sum_{k=1}^{N} |B_k^*| = 3^d \sum_{k=1}^{N} |B_k|. \qquad \square$$

The Vitali Covering Lemma is a more refined version of this result, and the proof requires a more refined, but still essentially "greedy," approach. One full proof can be found in [6, Thm. 5.3.3].

2.4 *The Lebesgue Differentiation Theorem*

If a function $f: \mathbb{R}^d \to \mathbb{R}$ is continuous at a point x, then f "does not vary much" over a small ball $B_h(x) = \{t \in \mathbb{R}^d : \|t\| < h\}$ of radius h centered at x. We can write this average in several forms. In one form, it is the convolution $f * k_h$ of f with the normalized characteristic function

$$k_h(t) = \frac{1}{|B_h(0)|} \chi_{B_h(0)}(t).$$

This function k_h takes the value $1/|B_h(0)|$ when t belongs to the ball $B_h(0)$, and is zero outside the ball. Letting $f_h(x)$ denote the average of f over the ball $Bh(0)$, and noting that the Lebesgue measure of a ball is $|B_h(x)| = C_d h^d$ where C_d is a constant that depends only on the dimension d, we can write the average of f explicitly in several forms:

$$\begin{aligned}
\widetilde{f}_h(x) &= \frac{1}{|B_h(0)|} \int_{B_h(0)} f(t)\,dt && \text{(definition of average)}\\
&= \frac{1}{C_d h^d} \int_{B_h(0)} f(x-t)\,dt && \text{(by change of variable)}\\
&= \int_{\mathbb{R}^d} f(x-t)\,k_h(t)\,dt && \text{(definition of } k_h\text{)}\\
&= (f * k_h)(x) && \text{(definition of convolution).}
\end{aligned}$$

An easy exercise shows that if f is continuous at x then $\widetilde{f}_h(x) \to f(x)$ as $h \to 0$. Much more surprising is the following deep and important result, which shows that these averages converge for almost every x even if we only assume that f is locally integrable (i.e., f is integrable on every compact subset of $\mathbb{R}^d$).

Theorem 8 (Lebesgue Differentiation Theorem) *If f is locally integrable on $\mathbb{R}^d$, then for almost every $x \in \mathbb{R}^d$ we have*

$$\lim_{h\to 0} \frac{1}{|B_h(x)|} \int_{B_h(x)} |f(x) - f(t)|\,dt \;=\; 0,$$

and

$$\lim_{h\to 0} \widetilde{f}_h(x) \;=\; \lim_{h\to 0} \frac{1}{|B_h(x)|} \int_{B_h(x)} f(t)\,dt \;=\; f(x).$$

In particular, $\widetilde{f}_h(x) \to f(x)$ at almost every point. $\diamond$

Here is a corollary that we will need.

Corollary 9 *If f is locally integrable on $\mathbb{R}$, then for almost every $x \in \mathbb{R}$ we have*

$$f(x) \;=\; \lim_{h\to 0} \frac{1}{h} \int_x^{x+h} f(t)\,dt. \qquad \diamond$$

The proof of Theorem 8 does take considerable work, but it is rewarding for the student to read and understand. One approach uses the Simple Vitali Lemma to prove the *Hardy–Littlewood Maximal Theorem*, which deals with the *supremum* of averages over balls instead of the limit of those averages. That proof is too long to include here. On the other hand, it is also true that if f is integrable on $\mathbb{R}^d$, then the averages of f over the balls $B_h(x)$ converge to f *in L^1-norm* as $h \to 0$. We will give the short (and enlightening) proof of this fact, which is actually a special case of more general facts about *approximate identities* for convolution (discussed in more detail in [6, Ch. 9]).

Theorem 10 *If $f \in L^1(\mathbb{R}^d)$, then $\widetilde{f}_h \to f$ in L^1-norm. That is,*

$$\lim_{h \to 0} \|f - \widetilde{f}_h\|_1 = \lim_{h \to 0} \int_{\mathbb{R}^d} |f(x) - \widetilde{f}_h(x)|\, dx = 0.$$

Proof The function k_h has been defined so that $\int k_h = 1$ for every h. Using Tonelli's Theorem to interchange the order of integration and noting that k_h is only nonzero on $B_h(0)$, we can therefore estimate the L^1-norm of the difference $f - \widetilde{f}_h$ as follows:

$$\begin{aligned}
\|f - \widetilde{f}_h\|_1 &= \int_{\mathbb{R}^d} |f(x) - \widetilde{f}_h(x)|\, dx \\
&= \int_{\mathbb{R}^d} \left| f(x) \int_{\mathbb{R}^d} k_h(t)\, dt - \int_{\mathbb{R}^d} f(x-t)\, k_h(t)\, dt \right| dx \\
&\le \int_{\mathbb{R}^d} \int_{\mathbb{R}^d} |f(x) - f(x-t)|\, k_h(t)\, dt\, dx \\
&= \frac{1}{C_d\, h^d} \int_{B_h(0)} \left(\int_{\mathbb{R}^d} |f(x) - f(x-t)|\, dx \right) dt \\
&= \frac{1}{C_d\, h^d} \int_{\|t\| < h} \|f - T_t f\|_1\, dt,
\end{aligned}$$

where $T_t f(x) = f(x-t)$ denotes the translation of f by t. However, the Lebesgue integral has a "strong continuity" property with respect to translation. Specifically,

$$\lim_{t \to 0} \|f - T_t f\|_1 = 0.$$

Therefore, if we fix an $\varepsilon > 0$, then there is some $\delta > 0$ such that $\|f - T_t f\|_1 < \varepsilon$ whenever $\|t\| < \delta$. Consequently, for all $0 < h < \delta$, we have

$$\|f - \widetilde{f}_h\|_1 \le \frac{1}{C_d\, h^d} \int_{\|t\| < h} \|f - T_t f\|_1\, dt \le \frac{1}{C_d\, h^d} \int_{\|t\| < h} \varepsilon\, dt = \varepsilon. \qquad \square$$

3 The Fundamental Theorem of Calculus

3.1 The Simple Version of the FTC

Here is a reformulation of the simple version of the FTC. This result follows directly from the fact that all continuous functions on $[a, b]$ are uniformly continuous (and its proof is a good exercise for the student). In the statement of this theorem, *differentiable everywhere on* $[a, b]$ means differentiable in the usual sense at each point of (a, b), differentiable from the right at a, and differentiable from the left at b.

Theorem 11 (Simple Version of the FTC) *If g is a continuous function g on $[a, b]$, then its indefinite integral*

$$G(x) = \int_a^x g(t)\, dt, \qquad x \in [a, b]$$

has the following properties:

(a) *G is differentiable everywhere on $[a, b]$,*

(b) *$G'(x) = g(x)$ for every $x \in [a, b]$,*

(c) *$G \in C^1[a, b]$, and consequently, G is Lipschitz and has bounded variation on $[a, b]$.* ◇

Thus, if g is continuous, then its indefinite integral G is differentiable at every point, and it is an *antiderivative* of g because $G' = g$. What happens if we assume only that the function g is *integrable*? Here is a partial answer. The proof is short but "instructive and enlightening."

Lemma 12 *If $g \in L^1[a, b]$, then its indefinite integral*

$$G(x) = \int_a^x g(t)\, dt, \qquad x \in [a, b]$$

has the following properties:

(a) *G is continuous on $[a, b]$,*

(b) *$G \in \mathrm{BV}[a, b]$, and*

(c) *the total variation of G is bounded by the L^1-norm of g, i.e.,*

$$V[G] \le \int_a^b |g(t)|\, dt = \|g\|_1.$$

Proof

(a) Fix any point $x \in (a, b)$. If $h > 0$ is small enough that $x + h$ belongs to $[a, b]$, then

$$G(x+h) - G(x) = \int_x^{x+h} g(t)\, dt = \int_a^b g(t)\, \chi_{[x,x+h]}(t)\, dt.$$

The integrand $g \cdot \chi_{[x,x+h]}$ is bounded by the integrable function $|g|$, and it converges pointwise a.e. to zero as $h \to 0^+$. The Dominated Convergence Theorem therefore implies that $G(x + h) - G(x) \to 0$ as $h \to 0^+$. Combining this with a similar argument for $h \to 0^-$, we see that G is continuous at x. Similar one-sided arguments show that G is continuous from the right at $x = a$ and continuous from the left at $x = b$, so G is continuous on the interval $[a, b]$.

(b), (c) If $\Gamma = \{a = x_0 < \cdots < x_n = b\}$ is a partition of $[a, b]$, then

$$S_\Gamma[G] = \sum_{j=1}^{n} |G(x_j) - G(x_{j-1})|$$

$$\leq \sum_{j=1}^{n} \int_{x_{j-1}}^{x_j} |g(t)|\,dt = \int_a^b |g(t)|\,dt = \|g\|_1.$$

Taking the supremum over all such partitions Γ, we see that G has bounded variation and $V[G] \leq \|g\|_1$. □

Unfortunately, Lemma 12 is not very satisfactory when compared to Theorem 11. We are still left with the following questions.

- If $g \in L^1[a, b]$, is the indefinite integral $G(x) = \int_a^x g(t)\,dt$ a differentiable function of x?
- If the indefinite integral G is differentiable, is it the antiderivative of g? That is, is it true that $G' = g$?

My own intuition suggests that the answers to both questions should be *yes*. And we will eventually see that this is true *in a qualified sense*, but not literally true in the same sense as in Theorem 6 . In particular, although the beginning student may not see an obvious counterexample, it is not true that the indefinite integral G must be differentiable *at every point* x (challenge—what is one such function?). Likewise, it is not true that $G'(x)$ must equal $g(x)$ *at every point* x. Instead, we will prove—after some work—that the following statements hold.

- G is an *absolutely continuous* function and, as a consequence, it is differentiable at *almost every* point in $[a, b]$, and
- $G'(x) = g(x)$ for *almost every* $x \in [a, b]$.

3.2 *Absolute Continuity*

To motivate the definition of absolute continuity, recall that a function $f : [a, b] \to \mathbb{R}$ is *uniformly continuous* on $[a, b]$ if for every $\varepsilon > 0$ there exists a $\delta > 0$ such that

$$|x - y| < \delta \implies |f(x) - f(y)| < \varepsilon.$$

Absolutely continuous functions satisfy a similar but more stringent requirement, given next. In the statement of this definition, a collection of intervals is *nonover-*

lapping if any two intervals in the collection intersect at most at their boundaries (hence, their interiors are disjoint).

Definition 13 (Absolutely Continuous Function) We say that a function $f : [a,b] \to \mathbb{R}$ is *absolutely continuous on* $[a,b]$ if for every $\varepsilon > 0$ there exists a $\delta > 0$ such that for any finite or countably infinite collection of nonoverlapping subintervals $\{[a_j, b_j]\}_j$ of $[a,b]$, we have

$$\sum_j (b_j - a_j) < \delta \implies \sum_j |f(b_j) - f(a_j)| < \varepsilon.$$

We denote the class of absolutely continuous functions on $[a,b]$ by

$$\mathrm{AC}[a,b] = \{f : [a,b] \to \mathbb{R} : f \text{ is absolutely continuous on } [a,b]\}. \qquad \diamondsuit$$

Every absolutely continuous function has bounded variation. Using the spaces we have introduced previously, we have the inclusions

$$C^1[a,b] \subsetneq \mathrm{Lip}[a,b] \subsetneq \mathrm{AC}[a,b] \subsetneq \mathrm{BV}[a,b] \subsetneq L^\infty[a,b] \subsetneq L^1[a,b].$$

The next lemma answers one of the questions that we posed immediately after the proof of Lemma 12.

Lemma 14 *If* $g \in L^1[a,b]$, *then its indefinite integral*

$$G(x) = \int_a^x g(t)\,dt, \qquad x \in [a,b],$$

has the following properties:

(a) *G is absolutely continuous on* $[a,b]$,

(b) *G is differentiable at almost every point of* $[a,b]$, *and*

(c) $G' \in L^1[a,b]$.

Proof Fix any $\varepsilon > 0$. Since g is integrable, there exists a constant $\delta > 0$ such that $\int_E |g| < \varepsilon$ for every measurable set $E \subseteq [a,b]$ whose measure satisfies $|E| < \delta$ (this is called the "absolute continuity" property of the Lebesgue integral; see [6, Exer. 4.5.5] for one derivation). Let $\{[a_j, b_j]\}_j$ be any countable collection of nonoverlapping subintervals of $[a,b]$ that satisfies $\sum (b_j - a_j) < \delta$, and set $E = \cup (a_j, b_j)$. Then $|E| < \delta$, so

$$\sum_j |G(b_j) - G(a_j)| = \sum_j \left| \int_{a_j}^{b_j} g \right| \le \sum_j \int_{a_j}^{b_j} |g| = \int_E |g| < \varepsilon.$$

Thus $G \in \mathrm{AC}[a, b]$. Since G is absolutely continuous, it has bounded variation. Applying the Jordan Decomposition Theorem (Theorem 1), G can be written as the difference of two monotone increasing functions, say $G = u - v$. Each of u and v is differentiable at almost every point, so G is differentiable a.e. Further, u' and v' are both integrable by Corollary 6, so $G' = u' - v'$ is integrable as well. □

3.3 The Growth Lemmas

In Sect. 3.4, we will prove the Banach–Zaretsky Theorem, which gives a reformulation of absolute continuity that is related to the issue of whether a function maps sets with measure zero to sets with measure zero. To do this, we need to understand how much a continuous function can "blow up" the measure of a set. That is, can we estimate the exterior Lebesgue measure $|f(E)|_e$ in terms of the measure $|E|_e$ of E and some properties of f? (Although we are assuming that f is continuous, it is not true that a continuous function must map measurable sets to measurable sets, which is why we are using exterior measure here.)

If we impose enough conditions on f, then it is easy to formulate such a "growth" result. For example, suppose that f is Lipschitz on $[a, b]$ with Lipschitz constant K, i.e.,

$$|f(x) - f(y)| \leq K\,|x - y|, \qquad \text{for all } x,\ y \in [a, b].$$

In this case, if we choose x and y from a subinterval $[c, d]$, then we will always have $|f(x) - f(y)| \leq K\,(d - c)$. Thus, the diameter of the image of $[c, d]$ under f is at most $K\,(d - c)$, and therefore $|f([c, d])|_e \leq K\,|[c, d]|$. That is, the measure of the image of $[c, d]$ under f is at most K times the measure of $[c, d]$. Since (in one dimension) the exterior measure of a set is fundamentally based on intervals, it is not surprising then that we can extend that inequality and show that $|f(E)|_e \leq K\,|E|_e$ for every subset E of $[a, b]$ (this is a nice exercise for the student). In particular, if f is differentiable everywhere on $[a, b]$ and f' is bounded on $[a, b]$, then f is Lipschitz and $K = \|f'\|_\infty$ is a Lipschitz constant, where $\|\cdot\|_\infty$ denotes the L^∞-norm. However, in order to prove the Banach–Zaretsky Theorem, we will need to show that if f' is bounded *on a single subset* E then the estimate $|f(E)|_e \leq K\,|E|_e$ holds *for that set* E (with $K = \sup_{x \in E} |f'(x)|$). We need to obtain this estimate without assuming that f' is bounded on all of $[a, b]$, or that f is Lipschitz on $[a, b]$. Our next (very enlightening!) result takes a more sophisticated approach to derive an elegant estimate.

Remark 2 Naturally, I first learned this "Growth Lemma" from John's book [1]. A proof was published by Varberg in the comparatively "recent" paper [12], and in [6], I state that this is the earliest published proof of which I am aware. (also compare the related paper [10] by Serrin and Varberg). Interestingly enough, Varberg himself comments that this theorem is "an elegant inequality which the author discovered lying buried as an innocent problem in Natanson's book [7]" (but it is not proved in that text). Recently, I took another look at John's book, and to my surprise, I saw that he cites the text by Saks [9] (whose first edition appeared in 1937) for a proof; see [9, Lem. VII.6.3]. Thus, the history of this result seems to trace back farther than I knew. Yet the modern literature seems to be largely unaware of this elegant lemma. One text that does present the result is by Bruckner, Bruckner, and Thomson [3] (though they give no references). In fact, they prove a more general theorem there. Our proof is inspired by the proof given in [3]. ◇

Lemma 15 (Growth Lemma I) *Let E be any subset of $[a,b]$. If $f\colon [a,b] \to \mathbb{R}$ is differentiable at* every *point of E and if*

$$M_E = \sup_{x \in E} |f'(x)| < \infty,$$

then

$$|f(E)|_e \le M_E \, |E|_e.$$

Proof Choose any $\varepsilon > 0$. If $x \in E$, then

$$\lim_{\substack{y \to x, \\ y \in [a,b]}} \frac{|f(x) - f(y)|}{|x - y|} = |f'(x)| \le M_E.$$

Therefore, there exists an integer $n_x \in \mathbb{N}$ such that

$$y \in [a,b], \ |x - y| < \frac{1}{n_x} \implies |f(x) - f(y)| \le (M_E + \varepsilon)\,|x - y|. \tag{3.1}$$

For each $n \in \mathbb{N}$, let

$$E_n = \bigl\{x \in E : n_x \le n\bigr\}.$$

The sets E_n are nested increasing ($E_1 \subseteq E_2 \subseteq \cdots$), and their union is E. We do not know whether E_n is a measurable set, but fortunately continuity from below does hold for *exterior* Lebesgue measure (this is Problem 2.4.8 in [6]; in contrast, continuity from above can fail if the sets are not Lebesgue measurable!). Therefore,

$$|E|_e = \lim_{n \to \infty} |E_n|_e. \tag{3.2}$$

The images $f(E_n)$ are also nested increasing and their union is $f(E)$, so we likewise have

$$|f(E)|_e = \lim_{n \to \infty} |f(E_n)|_e. \tag{3.3}$$

Fix any particular integer n. By the definition of exterior Lebesgue measure, there exists a collection of countably many closed intervals $\{I_n^k\}_k$ such that

$$E_n \subseteq \bigcup_k I_n^k \qquad \text{and} \qquad \sum_k |I_n^k| \le |E_n|_e + \varepsilon. \tag{3.4}$$

By replacing I_n^k with $I_n^k \cap [a,b]$, we may assume that $I_n^k \subseteq [a,b]$ for each n and k. Further, by subdividing if necessary, we may assume that each interval I_n^k has length less than $1/n$.

Suppose that x and y are any two points in $E_n \cap I_n^k$. Then, since $x \in E_n$, we have $n_x \le n$. Also, since x and y belong to I_n^k, whose length is less than $1/n$,

$$|x-y| < \frac{1}{n} \le \frac{1}{n_x}.$$

It therefore follows from Eq. (3.1) that

$$|f(x)-f(y)| \le (M_E+\varepsilon)\,|x-y| \le (M_E+\varepsilon)\,|I_n^k|.$$

Since this is true for all $x,\ y \in E_n \cap I_n^k$, we conclude that

$$\operatorname{diam}\bigl(f(E_n \cap I_n^k)\bigr) = \sup\bigl\{|f(x)-f(y)| : x,\ y \in E_n \cap I_n^k\bigr\} \le (M_E+\varepsilon)\,|I_n^k|.$$

This implies that $f(E_n \cap I_n^k)$ is contained in an interval with length at most $(M_E+\varepsilon)\,|I_n^k|$. Hence,

$$|f(E_n \cap I_n^k)|_e \le (M_E+\varepsilon)\,|I_n^k|. \tag{3.5}$$

Consequently,

$$\begin{aligned}
|f(E_n)|_e &= \Bigl|\textstyle\bigcup_k f(E_n \cap I_n^k)\Bigr|_e && \text{(by Eq. (3.4))}\\
&\le \sum_k |f(E_n \cap I_n^k)|_e && \text{(by subadditivity)}\\
&\le (M_E+\varepsilon) \sum_k |I_n^k| && \text{(by Eq. (3.5))}\\
&\le (M_E+\varepsilon)\bigl(|E_n|_e+\varepsilon\bigr) && \text{(by Eq. (3.4)).}
\end{aligned}$$

Therefore, by applying Eqs. (3.2) and (3.3), we see that

$$\begin{aligned}
|f(E)|_e &= \lim_{n\to\infty} |f(E_n)|_e\\
&\le (M_E+\varepsilon) \lim_{n\to\infty} \bigl(|E_n|_e+\varepsilon\bigr)\\
&= (M_E+\varepsilon)\bigl(|E|_e+\varepsilon\bigr).
\end{aligned}$$

Since ε is arbitrary, the result follows. □

Our second Growth Lemma also appears in Varberg's paper [12], and our proof is based on the one given there. This result relates the exterior measure of $f(E)$ to the integral of $|f'|$ on E. A fun observation is that a key step in the proof is nothing less than the fact that $k = (k-1)+1$. Sometimes the most elementary facts have large consequences.

Lemma 16 (Growth Lemma II) *Assume that* $f\colon [a,b] \to \mathbb{R}$ *is measurable. If* E *is a measurable subset of* $[a,b]$ *and* f *is differentiable at* every *point of* E*, then*

$$|f(E)|_e \le \int_E |f'|.$$

Proof An exercise for the student is to show that the derivative $f'\colon E \to \mathbb{R}$ is a measurable function on E. Hence, $\int_E |f'|$ exists, though it could be infinite. Fix any $\varepsilon > 0$, and for each $k \in \mathbb{N}$, define

$$E_k = \bigl\{x \in E : (k-1)\varepsilon \le |f'(x)| < k\varepsilon\bigr\}.$$

The sets E_k are measurable and disjoint, and since f is differentiable everywhere on E, we have $E = \cup E_k$. Since Lebesgue measure is countably additive, it follows that

$$|E| = \sum_{k=1}^{\infty} |E_k|.$$

Lemma 15 implies that $|f(E_k)|_e \le k\varepsilon\,|E_k|$, so we see that

$$\begin{aligned}
|f(E)|_e = \biggl|\bigcup_{k=1}^{\infty} f(E_k)\biggr|_e &\le \sum_{k=1}^{\infty} |f(E_k)|_e \\
&\le \sum_{k=1}^{\infty} k\varepsilon\,|E_k| \\
&= \sum_{k=1}^{\infty} (k-1)\varepsilon\,|E_k| + \sum_{k=1}^{\infty} \varepsilon\,|E_k| \\
&\le \sum_{k=1}^{\infty} \int_{E_k} |f'| + \varepsilon\,|E| \\
&= \int_E |f'| + \varepsilon\,|E|.
\end{aligned}$$

Since ε is arbitrary and $|E| < \infty$, the result follows. □

The two Growth Lemmas are striking in their own right and can be used to prove a number of interesting corollaries. We refer to [6] for more details.

3.4 The Banach–Zaretsky Theorem

Now we reach the highlight of this chapter, which is the often-overlooked *Banach–Zaretsky Theorem*. This result tells us what properties that a function $f: [a, b] \to \mathbb{R}$ needs to possess in addition to continuity in order to be absolutely continuous. Specifically, f must map sets with measure zero to sets with measure zero, and we must also know either that f has bounded variation, or that f is differentiable almost everywhere and f' is integrable. (The hypothesis that $|A| = 0$ implies $|f(A)| = 0$ is sometimes referred to as *Luzin's condition*.) The proof is a nice application of the Growth Lemmas.

Theorem 17 (Banach–Zaretsky Theorem) *If $f: [a, b] \to \mathbb{R}$, then the following three statements are equivalent.*

(a) $f \in \mathrm{AC}[a, b]$.

(b) *f is continuous, $f \in \mathrm{BV}[a, b]$, and*

$$A \subseteq [a, b], \ |A| = 0 \implies |f(A)| = 0.$$

(c) *f is continuous, f is differentiable a.e., $f' \in L^1[a, b]$, and*

$$A \subseteq [a, b], \ |A| = 0 \implies |f(A)| = 0.$$

Proof (a) $\Rightarrow$ (b). Every absolutely continuous function is continuous and has bounded variation, so our task is to show that f maps sets with measure zero to sets with measure zero.

Suppose that A is a subset of $[a, b]$ that has measure zero. Since the two-element set $\{a, b\}$ has measure zero and its image $\{f(a), f(b)\}$ also has measure zero, it suffices to assume that A is contained within the open interval (a, b). Fix $\varepsilon > 0$. By the definition of absolute continuity, there exists some $\delta > 0$ such that if $\{[a_j, b_j]\}_j$ is any countable collection of nonoverlapping subintervals of $[a, b]$ that satisfy $\sum (b_j - a_j) < \delta$, then $\sum |f(b_j) - f(a_j)| < \varepsilon$.

By basic properties of Lebesgue measure, there is an open set $U \supseteq A$ whose measure satisfies

$$|U| < |A| + \delta = \delta.$$

By replacing U with the open set $U \cap (a, b)$, we may assume that $U \subseteq (a, b)$. Since U is open, we can write it as a union of countably many disjoint open intervals contained in (a, b), say

$$U = \bigcup_j (a_j, b_j).$$

Fix any particular j. Since f is continuous on the closed interval $[a_j, b_j]$, there is a point in $[a_j, b_j]$ where f attains its minimum value on $[a_j, b_j]$, and another point where f attains its maximum. Let c_j and d_j be points in $[a_j, b_j]$ such that f has a max at one point and a min at the other. By interchanging their roles if necessary, we may assume that $c_j \le d_j$. Because f is continuous, the Intermediate Value Theorem implies that the image of $[a_j, b_j]$ under f is the set of all points between $f(c_j)$ and $f(d_j)$. Hence, the exterior Lebesgue measure of this image is

$$\big|f([a_j, b_j])\big|_e = |f(d_j) - f(c_j)|.$$

Now, $[c_j, d_j] \subseteq [a_j, b_j]$, so $\big\{[c_j, d_j]\big\}_j$ is a collection of nonoverlapping subintervals of $[a, b]$. Moreover,

$$\sum_j |d_j - c_j| \le \sum_j (b_j - a_j) = |U| < \delta.$$

Therefore, $\sum |f(d_j) - f(c_j)| < \varepsilon$, and hence,

$$|f(A)|_e \le |f(U)|_e \le \sum_j \big|f([a_j, b_j])\big|_e = \sum_j \big|f(d_j) - f(c_j)\big| < \varepsilon.$$

Since ε is arbitrary, we conclude that $|f(A)| = 0$.

(b) $\Rightarrow$ (c). This follows from Corollary 6.

(c) $\Rightarrow$ (a). Assume that statement (c) holds, and let D be the set of points where f is differentiable. By hypothesis, $Z = [a, b] \setminus D$ has measure zero, so $D = [a, b] \setminus Z$ is a measurable set.

Let $[c, d]$ be an arbitrary subinterval of $[a, b]$. Since f is continuous, the Intermediate Value Theorem implies that f must take every value between $f(c)$ and $f(d)$. Therefore $f([c, d])$, the image of $[c, d]$ under f, must contain an interval of length $|f(d) - f(c)|$. Define

$$B = [c, d] \cap D \qquad \text{and} \qquad A = [c, d] \setminus D.$$

The set A has measure zero, so $|f(A)| = 0$ by hypothesis. Since f is differentiable at every point of B, we therefore compute that

$$
\begin{aligned}
|f(d) - f(c)| &\le \big|f([c,d])\big|_e \\
&= \big|f(B) \cup f(A)\big|_e && \text{(since } [c,d] = B \cup A\text{)} \\
&\le |f(B)|_e + |f(A)|_e && \text{(by subadditivity)} \\
&\le \int_B |f'| + 0 && \text{(by Lemma 16 and hypotheses)} \\
&\le \int_c^d |f'| && \text{(since } B \subseteq [c,d]\text{).}
\end{aligned}
\tag{3.6}
$$

This calculation holds for every subinterval $[c, d]$ of $[a, b]$.

Now fix $\varepsilon > 0$. Because f' is integrable, there is some $\delta > 0$ such that for every measurable set $E \subseteq [a, b]$ we have

$$|E| < \delta \implies \int_E |f'| < \varepsilon.$$

Let $\{[a_j, b_j]\}_j$ be any countable collection of nonoverlapping subintervals of $[a, b]$ such that $\sum (b_j - a_j) < \delta$. Then, $E = \cup [a_j, b_j]$ is a measurable subset of $[a, b]$ and $|E| < \delta$, so $\int_E |f'| < \varepsilon$. Applying equation (3.6) to each subinterval $[a_j, b_j]$, it follows that

$$\sum_j \big|f(b_j) - f(a_j)\big| \le \sum_j \int_{a_j}^{b_j} |f'| = \int_E |f'| < \varepsilon.$$

Hence, f is absolutely continuous on $[a, b]$. □

Few books seem to mention the Banach–Zaretsky Theorem (also known as the Banach–Zarecki Theorem). Two that do are [1] and [3].

Remark 3 The statement of the Banach–Zaretsky Theorem for complex-valued functions is similar, except that both the real and imaginary parts of f must map sets of measure zero to sets of measure zero (see [6] for details). Specifically, if $f\colon [a, b] \to \mathbb{C}$ and we write $f = f_r + i f_i$, where f_r and f_i are real-valued, then statements (a)–(c) of Theorem 17 are equivalent if we replace the hypothesis "$|f(A)| = 0$" by "$|f_r(A)| = |f_i(A)| = 0$." ◊

3.5 Corollaries

We give several easy and immediate implications of the Banach–Zaretsky Theorem.

Corollary 18 *Absolutely continuous functions map sets of measure zero to sets of measure zero, and they map measurable sets to measurable sets.*

Proof Continuous functions map compact sets to compact sets. An F_σ-set is a countable union of compact sets, so continuous functions map F_σ-sets to F_σ-sets. If f is absolutely continuous, then it also maps sets of measure zero to sets of measure zero. But every measurable set can be written as the union of an F_σ-set and a set of measure zero, so f maps measurable sets to measurable sets. □

To motivate our second implication, we recall that if f is differentiable everywhere on $[a, b]$ and f' is bounded, then f is Lipschitz and therefore absolutely continuous (for one proof, see [6, Lem. 5.2.5]). What happens if f is differentiable everywhere on $[a, b]$ but we only know that f' is *integrable*? Although such a function need not be Lipschitz, the next corollary shows that f will be absolutely continuous.

Corollary 19 *If* $f\colon [a, b] \to \mathbb{R}$ *is differentiable everywhere on* $[a, b]$ *and* $f' \in L^1[a, b]$, *then* $f \in \mathrm{AC}[a, b]$.

Proof Let A be any subset of $[a, b]$ that has measure zero. Since f is differentiable everywhere, it is continuous and hence measurable. Because A is a measurable set, we can therefore apply Lemma 16 to obtain the estimate

$$|f(A)|_e \leq \int_A |f'| = 0.$$

Consequently, the Banach–Zaretsky Theorem implies that f is absolutely continuous. □

Since all countable sets have measure zero, we immediately obtain the following refinement of Corollary 19.

Corollary 20 *Suppose that* $f\colon [a, b] \to \mathbb{R}$ *is continuous and is differentiable at all but countably many points of* $[a, b]$. *If* $f' \in L^1[a, b]$, *then* $f \in \mathrm{AC}[a, b]$.

Proof If we let $Z = \{x \in [a, b] : f'(x) \text{ does not exist}\}$, then Z is countable by hypothesis. Suppose that $A \subseteq [0, 1]$ satisfies $|A| = 0$. Then, $|A \setminus Z| = 0$. Since f is measurable and differentiable at every point of $A \setminus Z$, Growth Lemma II implies that

$$|f(A \setminus Z)|_e \leq \int_{A \setminus Z} |f'| = 0.$$

On the other hand, the set $A \cap Z$ is countable, so $f(A \cap Z)$ is also countable, and therefore, $|f(A \cap Z)|_e = 0$. Applying countable subadditivity for exterior measure, we see that

$$|f(A)|_e \leq |f(A \setminus Z)|_e + |f(A \cap Z)|_e = 0.$$

The Banach–Zaretsky Theorem therefore implies that $f \in \mathrm{AC}[a, b]$. □

Considering the Cantor–Lebesgue function, we see that we cannot extend Corollary 20 to functions that are differentiable at almost every point. Indeed, the Cantor–Lebesgue function φ is continuous and differentiable a.e., but it is not absolutely continuous.

Remark 4 We easily obtained Corollary 20 from the Banach–Zaretsky Theorem. In contrast, in the notes to Chapter 3 of [5], Folland writes,

> It is a highly nontrivial theorem that if F is continuous on $[a, b]$, $F'(x)$ exists for every $x \in [a, b] \setminus A$ where A is *countable*, and $F' \in L^1$, then F is absolutely continuous and hence can be recovered from F' by integration. A proof can be found in Cohn [4, Sec. 6.3]; see also Rudin [8, Theorem 7.26] for the somewhat easier case when $A = \varnothing$.

To us, this highlights the fundamental but overlooked nature of the Banach–Zaretsky Theorem. ◇

Our final implication uses the Banach–Zaretsky Theorem (and the Growth Lemmas) to show that the only functions that are both absolutely continuous and singular are constant functions.

Corollary 21 (AC + Singular Implies Constant) *If* $f\colon [a, b] \to \mathbb{R}$ *is both absolutely continuous and singular, then* f *is constant.*

Proof Suppose that $f \in \mathrm{AC}[a, b]$ and $f' = 0$ a.e., and define

$$E = \{x =\in [a, b] : f'(x) = 0\} \qquad \text{and} \qquad Z = [a, b] \setminus E.$$

Since $|Z| = 0$, the Banach–Zaretsky Theorem implies that $|f(Z)| = 0$. Since E is measurable and f is differentiable on E, Growth Lemma II implies that

$$|f(E)|_e \le \int_E |f'| = 0.$$

Therefore, the range of f has measure zero because

$$|\mathrm{range}(f)|_e = \big|f([a, b])\big|_e = |f(E) \cup f(Z)|_e \le |f(E)|_e + |f(Z)|_e = 0.$$

However, f is continuous and $[a, b]$ is compact, so the Intermediate Value Theorem implies that the range of f is either a single point or a closed interval $[c, d]$. Since $\mathrm{range}(f)$ has measure zero, we conclude that it is a single point, and therefore f is constant. □

One standard proof of Corollary 21 uses the Vitali Covering Lemma (e.g., see the exposition in [13, Thm. 7.28]). By using the Banach–Zaretsky Theorem, we obtain a much simpler and more enlightening proof. However, the Vitali Covering Lemma does still play a role, since we use it to prove that monotone increasing functions are differentiable a.e. (Theorem 5).

3.6 The Fundamental Theorem of Calculus

Following Lemma 12, we asked two questions: First, is the indefinite integral G of an integrable function g differentiable? Second, if G is differentiable, does $G' = g$? The first question was answered affirmatively in Lemma 14, and the next lemma will show that $G' = g$ a.e.

Lemma 22 *If $g \in L^1[a, b]$, then its indefinite integral*

$$G(x) = \int_a^x g(t)\,dt, \qquad x \in [a, b],$$

is absolutely continuous and satisfies $G' = g$ a.e.

Proof Lemma 14 implies that G is absolutely continuous. Applying Corollary 9 (extend g by zero outside of $[a, b]$, so that it is locally integrable on $\mathbb{R}$), we also see that, for almost every $x \in [a, b]$,

$$\frac{G(x+h) - G(x)}{h} = \frac{1}{h}\int_x^{x+h} g(t)\,dt \to g(x) \quad \text{as } h \to 0.$$

Therefore, G is differentiable and $G'(x) = g(x)$ for almost every x. □

Now we give a simple proof of the Fundamental Theorem of Calculus.

Theorem 23 (Fundamental Theorem of Calculus) *If $f : [a, b] \to \mathbb{R}$, then the following three statements are equivalent.*

(a) $f \in \mathrm{AC}[a, b]$.

(b) *There exists a function $g \in L^1[a, b]$ such that*

$$f(x) - f(a) = \int_a^x g(t)\,dt, \qquad \text{for all } x \in [a, b].$$

(c) *f is differentiable almost everywhere on $[a, b]$, $f' \in L^1[a, b]$, and*

$$f(x) - f(a) = \int_a^x f'(t)\,dt, \qquad \text{for all } x \in [a, b].$$

Proof (a) ⇒ (c). Suppose that f is absolutely continuous on $[a, b]$. Then f has bounded variation, so we know that f' exists a.e. and is integrable. Lemma 22 implies that the indefinite integral

$$F(x) = \int_a^x f'(t)\,dt$$

is absolutely continuous and satisfies $F' = f'$ a.e. Hence, $(F - f)' = 0$ a.e., so the function $F - f$ is both absolutely continuous and singular. Applying Corollary 21, we conclude that $F - f$ is constant. Consequently, for all $x \in [a, b]$,

$$F(x) - f(x) = F(a) - f(a) = 0 - f(a) = -f(a).$$

(c) $\Rightarrow$ (b). This follows by taking $g = f'$.

(b) $\Rightarrow$ (a). This follows from Lemma 22. □

Combining Theorem 23 with the Banach–Zaretsky Theorem gives us a remarkable list of equivalent characterizations of absolute continuity of functions on $[a, b]$. Many other results follow from this, and we refer to [6] for a continuation of this story.

Acknowledgments This chapter draws extensively from Chapters 5 and 6 of [6]. That material is used with permission of Springer. Many classic and recent volumes influenced the writing, the choice of topics, the proofs, and the selection of problems in [6]. We would like to explicitly acknowledge those tests that had the most profound influence. These include Benedetto and Czaja [2], Bruckner, Bruckner, and Thomson [3], Folland [5], Rudin [8], Stein and Shakarchi [11], and Wheeden and Zygmund [13]. I greatly appreciate all of these texts and encourage the reader to consult them. Many additional texts and papers are listed in the references to [6].

References

1. Benedetto, J. J.: Real Variable and Integration. B. G. Teubner, Stuttgart (1976)
2. Benedetto, J. J., Czaja, W.: Integration and Modern Analysis. Birkhäuser, Boston (2009)
3. Bruckner, A. M., Bruckner, J. B., Thomson, B. S.: Real Analysis. Prentice Hall, Upper Saddle River, NJ (1997)
4. Cohn, D. L.: Measure theory. Birkhäuser, Boston (1980)
5. Folland, G. B.: Real Analysis, Second Edition. Wiley, New York (1999)
6. Heil, C.: Introduction to Real Analysis. Springer, Cham (2019)
7. Natanson, I. P.: Theory of Functions of a Real Variable. Translated from the Russian by Leo F. Boron with the collaboration of Edwin Hewitt. Frederick Ungar Publishing Co., New York (1955)
8. Rudin, W.: Real and Complex Analysis, Third Edition. McGraw-Hill, New York (1987).
9. Saks, S.: Theory of the Integral, Second revised edition. English translation by L. C. Young, Dover, New York (1964)
10. Serrin J., Varberg, D. E.: A general chain rule for derivatives and the change of variables formula for the Lebesgue integral. Amer. Math. Monthly **76**, 514–520 (1969)
11. Stein, E. M., Shakarchi, R.: Real Analysis. Princeton University Press, Princeton, NJ (2005)
12. Varberg, D. E.: On absolutely continuous functions. Amer. Math. Monthly **72**, 831–841 (1965)
13. Wheeden, R. L., Zygmund, A.: Measure and Integral. Marcel Dekker, New York-Basel (1977)

Spectral Synthesis and $H^1(\mathbb{R})$

Raymond Johnson

Abstract I will describe joint work with the late Bob Warner. We knew that $H^1(\mathbb{R})$ gave better results for singular integrals than $L^1(\mathbb{R})$; our question was would the same be true for spectral synthesis.

1 Introduction

I will describe joint work with the late Bob Warner. We knew that $H^1(\mathbb{R})$ gave better results for singular integrals than $L^1(\mathbb{R})$; our question was would the same be true for spectral synthesis. Now, spectral synthesis is a very old problem that was very hot in the sixties as part of a circle of questions about thin sets. In 1972 Korner [1, p. 266], gave counterexamples to many questions about thin sets and there was not much left for mathematicians to work on; they largely abandoned the area.

I will work on $\mathbb{R}$ and let E be a closed subset of $\mathbb{R}$ (terminology is taken from [10]). The largest ideal of functions whose Fourier transforms vanish on E is

$$I(E) = \{f \in L^1(\mathbb{R}) | \widehat{f}(\xi) = 0, \xi \in E\}$$

and the set

$$j(E) = \{f \in L^1(\mathbb{R}) | \widehat{f}(\xi) = 0, \text{ in a neighborhood of } E\}$$

is contained if $I(E)$ and its closure, $I_0(E)$ is the smallest closed ideal of functions whose Fourier transform has zero set E. E is a set of spectral synthesis if $I(E) = I_0(E)$. Calderón introduced a class of C-sets defined in Rudin [8] by E is a C-set if for any $f \in L^1(\mathbb{R})$, $f \in I(E)$, and any $\varepsilon > 0$, there is a $g \in L^1(\mathbb{R})$ with $\widehat{g}$ having

R. Johnson (✉)
University of Maryland, College Park, MD, USA
e-mail: rlj@math.umd.edu

M. Hirn et al. (eds.), *Excursions in Harmonic Analysis, Volume 6*, Applied and Numerical Harmonic Analysis, https://doi.org/10.1007/978-3-030-69637-5_3

compact support disjoint from E, such that $\|f - f * g\| < \varepsilon$. C-sets are S-sets but the converse is one of the older problems left. They are also called Calderón sets or Calderón-Ditkin sets, or Ditkin sets.

2 Main Results

There are still three big questions remaining in spectral synthesis:

1. Every C-set is an S-set from the definition; is the converse true? [1, p. 168]
   Some of you may know that Bob devoted most of his work to consideration around this C-set, S-set problem. I did not get drawn into it after I asked Bob what he really thought and he said that on Monday, Wednesday, and Friday, he thought it was true and he had a proof. On Tuesday, Thursday, and Saturday, he thought he had a counterexample. On Sunday, he tried not to think about it. We were primarily interested in the next two questions.
2. The boundary problem. If ∂E is an S-set, is E an S-set?
   The Schwartz counterexample [3, p. 313], [8, p. 165], [7, p. 49] showed that there was a closed S-set whose boundary was not an S-set; this asks if the converse might be true.
3. The union problem. If E_1, E_2 are S-sets, is $E_1 \cup E_2$ a S-set?
   A bit is known about this problem: if $E_1 \cap E_2 = \emptyset$ it is true and if $E_1 \cap E_2$ is a C-set, it is true (due to Warner [1, p. 172]).

We studied the dual of a form entering in the Beurling-Pollard theorem in L^∞ [2, 6, 10],

$$B(f, E) = \int_{E^c} \frac{|\widehat{f}(t)|}{\rho(t, E)}\, dt,$$

where $\rho(t, E) = \inf\{|t - s| : s \in E\}$, and

$$Q(E) \cap H^1 = \{f \in H^1(\mathbb{R}) | B(f, E) < \infty\}.$$

We were able to show by classical methods that $j(E) \subseteq Q(E) \subseteq I_0(E)$. If Q were closed in H^1, or complete, it would follow that E is a S-set. The elements of $Q(E)$ have nice properties, but completeness in H^1 is not among them. B is a norm on Q, but we cannot show that Q is complete in B or in H^1. We have a normed space intersected with a Banach space, which need not be Banach, but by Fatou, it actually is.

Theorem 2.1 *$Q(E) \cap H^1$ is a Banach space under its norm $B(f, E) + \|f\|_{H^1}$.*

Proof Suppose that $\{f_n\}$ is a Cauchy sequence in Q, and let $\varepsilon > 0$. There is a N such that if $m, n > N$, we have

$$B(f_n - f_m, E) + \|f_n - f_m\|_{H^1} < \varepsilon/2.$$

Since H^1 is a Banach space, there is a $g \in H^1$ such that $f_n \to g \in H^1$. Thus $\widehat{f_n} \to \widehat{g}$ uniformly in the continuous functions, and if we fix $n > N$, and let $m \to \infty$, it follows from Fatou that

$$B(f_n - g, E) = \int_{E^c} \frac{\lim_m |\widehat{f_n}(t) - \widehat{f_m}(t)|}{\rho(t, E)} dt \leq \liminf_m B(f_n - f_m, E).$$

Since $\liminf x_n + \liminf y_n \leq \liminf(x_n + y_n)$ (Royden, Second Edition, p. 36), it follows that

$$B(f_n - g, E) + \|f_n - g\|_{H^1} \leq \liminf_m B(f_n - f_m, E) + \liminf \|f_n - f_m\|_{H^1},$$

and by the result cited from Royden, this is

$$\leq \liminf(B(f_n - f_m, E) + \|f_n - f_m\|_{H^1}).$$

It follows that if $n > N$,

$$\begin{aligned} |||f_n - g|||_Q &= B(f_n - g, E) + \|f_n - g\|_{H^1} \\ &\leq \liminf_m (B(f_n - f_m, E) + \|f_n - f_m\|_{H^1}) \leq \varepsilon/2 \\ &< \varepsilon, \end{aligned}$$

which completes the proof. □

We wish to study convergence in H^1 and to know when $Q \cap H^1$ is complete in H^1. Fortunately, the open mapping theorem gives a simple criterion.

Theorem 2.2 *Suppose*

- *X is a Banach space and Y is a normed space*
- *$X \cap Y$ is a Banach space with its natural norm.*

The following are equivalent

1. *$X \cap Y$ is a Banach space under $\|\cdot\|_X$*
2. *$\exists D > 0$ such that $||x||_Y \leq D||x||_X$, for every $x \in X \cap Y$.*

When this inequality holds,

$$\|\cdot\|_{X\cap Y} \equiv \|\cdot\|_X.$$

Proof (2) $\Rightarrow$ (1). If the Y norm is dominated by the X norm on $X \cap Y$, and $X \cap Y$ is complete under its natural norm, it is also complete under the X norm. Assume the inequality and suppose x_n is a sequence in $X \cap Y$ which is Cauchy in the X norm. It is also Cauchy in $X \cap Y$ since

$$\|x\|_{X\cap Y} = \|x\|_X + \|x\|_Y \leq (D + 1)\|x\|_X,$$

and since $X \cap Y$ is complete, $x_n \to x \in X \cap Y$, while

$$\|x_n - x\|_X \leq \|x_n - x\|_{X \cap Y}$$

proves that any Cauchy sequence in $X \cap Y$ in the X norm, converges in the X norm. The norm equivalence follows because we have shown that when completeness holds,

$$\| \cdot \|_X \leq \| \cdot \|_{X \cap Y} \leq (D + 1)\| \cdot \|_X.$$

(1) $\Rightarrow$ (2) This is an easy application of the closed graph/open mapping theorem. If $X \cap Y$ is a Banach space under the X norm, $X \cap Y$ is a Banach space under two norms, and

$$\|x\|_X \leq \|x\|_{X \cap Y}, x \in X \cap Y.$$

This implies $i : X \cap Y \mapsto X \cap Y$ is continuous, where the second space has the X norm. Since it is also 1:1 and onto, by the Open Mapping theorem, the inverse map is also continuous and there is a constant $D > 0$ such that

$$\|x\|_X + \|x\|_Y \leq D\|x\|_X, x \in X \cap Y.$$

Hence, $\|x\|_Y \leq D\|x\|_X, x \in X \cap Y$. □

In our case, $X = H^1, Y = Q(E)$, $Q \cap H^1$ will be complete in H^1 if and only if $B(f, E) \leq D\|f\|_{H^1}, f \in H^1$. We call those sets Q-sets [5]. It is easy to see that Q sets satisfy the boundary and union property. If ∂E is a Q-set, then E is a Q-set because $B(f, E) \leq B(f, \partial E)$, and if E_1, E_2 are Q-sets, then $E_1 \cup E_2$ is a Q-set because $B(f, E_1 \cup E_2) \leq B(f, E_1) + B(f, E_2)$. They are dilation invariant, but not obviously translation invariant. By a result of Hardy, $\{0\}$ is a Q-set because [9, p. 128]

$$B(f, \{0\}) = \int_{\mathbb{R}} \frac{|\widehat{f}(t)|}{\rho(t, \{0\})} dt = \int_{\mathbb{R}} \frac{|\widehat{f}(t)|}{|t|} dt \leq C\|f\|_{H^1}.$$

It takes quite a bit more work, but any singleton $\{a\}$ is a Q-set, and therefore so is any finite union of points. Therefore intervals and finite unions of intervals are Q-sets (the boundary of an interval is two points), and the union of Q-sets is a Q-set. It is also easy to see from the definition that $\emptyset$ and $\mathbb{R}$ are also Q-set.

Half lines in $\mathbb{R}$, $[a, \infty)$, $(-\infty, b]$ are also Q-sets because their boundary is a singleton. However, something interesting happens. Bob kept looking for some residue of analyticity in H^1, but I insisted this was real H^1 and we had washed away analyticity. I was wrong. Because of a residual amount of analyticity,

$$B(f, [a, \infty)) \leq D\|f\|_{L^1}.$$

The sets defined by a L^1 bound are much better behaved because they are translation invariant, as are sets of spectral synthesis. Those satisfying a H^1 bound need not be because H^1, unlike L^1 is not invariant under multiplication by characters. But notice two paradoxical things: big sets are controlled by L^1, while small sets (points) require H^1 in the case of the origin and our proof requires H^1 for other singletons.

Second, while translation invariance does not follow from the theory because a character times a H^1 function need not be H^1, in practice the collection of all sets we have shown to be Q-sets are translation invariant. Another curious fact is that is that Bob and I had earlier proved [4] that the maximal ideal space of $H^1(R)$ is $\mathbb{R} \setminus \{0\}$, so we had treated $\{0\}$ like a third rail that we could not touch, yet it is a Q-set.

Not all S-sets are Q-sets. Any arithmetic progression, like $\mathbb{Z}$, is a C-set, but is not $\mathbb{Q}$.

I will mention two other questions. Is every compact S-set a Q-set? If that were true, you would have at least solved the boundary and union problems for compact S-sets, and I think it would be fairly easy to get the general result from the result for compact sets. I have not tried to prove this because I do not believe it to be true, but it might be.

I have tried (and failed) to prove that every compact C-set is a Q-set. If that were true it would not help with the boundary problem or the union problem because C-sets are already known to have both properties (and in fact are closed under countable unions). However, I think it would be possible to show that the Cantor set is not a Q-set; if you had proved the C-set, Q-set property, you would have resolved the C-set, S-set problem in the negative.

Acknowledgement Thanks to Kasso Okoudjou for his help with modern LaTeX.

References

1. Benedetto, J. J.: Spectral Synthesis, B. G. Teubner Stuttgart, (1975).
2. Beurling, A.: On the spectral synthesis of bounded functions, Acta Math. **81**, 225–238 (1949)
3. Graham, C. C. and McGehee, O. C.: Essays in Commutative Harmonic Analysis, Grundlehren 238, Springer-Verlag, (1979).
4. Johnson, R. and Warner, C. R.: $H^1(\mathbb{R})$ as a convolution algebra, J. Function Spaces, **8** no. 2 , 167-179 (2010).
5. Johnson, R., and C. R. Warner, A characterization of some sets of spectral synthesis. J. Fourier Anal. Appl. v. 27.
6. Pollard, H.: The harmonic analysis of bounded functions, Duke Math J., 499-512 (1953)
7. Reiter, H. and Stegeman, J.: Classical Harmonic Analysis and Locally Compact Groups, Oxford Science Publications, Clarendon Press, (2000)
8. Rudin, W.: Fourier Analysis on Groups, Interscience Publishers, John Wiley and Sons, Second Printing, (1967)
9. Stein, E.: Harmonic Analysis, Princeton University Press, Princeton, NJ, (1993)
10. French lecture notes, Synthese Harmonique, Faculté des Sciences, Nancy, D. E. A. de Mathématiques Pures,1966-67

Universal Upper Bound on the Blowup Rate of Nonlinear Schrödinger Equation with Rotation

Yi Hu, Christopher Leonard, and Shijun Zheng

Abstract In this chapter, we prove a universal upper bound on the blowup rate of a focusing nonlinear Schrödinger equation with an angular momentum under a trapping harmonic potential, assuming that the initial data is radially symmetric in the weighted Sobolev space. The nonlinearity is in the mass supercritical and energy subcritical regime. Numerical simulations are also presented.

1 Introduction

Consider the focusing nonlinear Schrödinger (NLS) equation with an angular momentum term in $\mathbb{R}^{1+n}$:

$$\begin{cases} iu_t = -\Delta u + Vu - \lambda |u|^{p-1}u + L_A u \\ u(0,x) = u_0 \in \mathscr{H}^1. \end{cases} \tag{1.1}$$

Here $u = u(t,x) : \mathbb{R} \times \mathbb{R}^n \to \mathbb{C}$ denotes the wave function, $V(x) := \gamma^2|x|^2$ ($\gamma > 0$) is a trapping harmonic potential that confines the movement of particles, and λ is a positive constant indicating that the self-interaction between particles is attractive. The nonlinearity has the exponent $1 \le p < 2^* - 1$, where by convention $2^* := \frac{2n}{n-2}$ if $n \ge 3$; ∞ if $n = 1, 2$. The operator $L_A u := iA \cdot \nabla u$ is the angular momentum term, where $A = Mx$ with $M = (M_{j,k})_{1\le j,k\le n}$ being an $n \times n$ real-valued skew-symmetric matrix, i.e., $M = -M^T$. It generates a rotation in $\mathbb{R}^n$ in the sense that $e^{-itL_A} f(x) = f(e^{tM}x)$ for $(t,x) \in \mathbb{R} \times \mathbb{R}^n$. The space $\mathscr{H}^1 = \mathscr{H}^{1,2}$ denotes the weighted Sobolev space

Y. Hu · S. Zheng (✉)
Department of Mathematical Sciences, Georgia Southern University, Statesboro, GA, USA
e-mail: yihu@GeorgiaSouthern.edu; szheng@GeorgiaSouthern.edu

C. Leonard
Department of Mathematics, North Carolina State University, Raleigh, NC, USA
e-mail: cleonar@ncsu.edu

M. Hirn et al. (eds.), *Excursions in Harmonic Analysis, Volume 6*, Applied and Numerical Harmonic Analysis, https://doi.org/10.1007/978-3-030-69637-5_4

$$\mathscr{H}^{1,r}(\mathbb{R}^n) := \{f \in L^r(\mathbb{R}^n) : \nabla f,\ xf \in L^r(\mathbb{R}^n)\}$$

for $r \in (1, \infty)$, and the endowed norm is given by

$$\|f\|_{\mathscr{H}^{1,r}} = \|\nabla f\|_r + \|xf\|_r + \|f\|_r ,$$

where $\|\cdot\|_r := \|\cdot\|_{L^r}$ is the usual L^r-norm. The linear Hamiltonian $H_{A,V} := -\Delta + V + iA \cdot \nabla$ is essentially self-adjoint in L^2, whose eigenvalues are associated to the Landau levels as quantum numbers.

When $n = 3$, Eq. (1.1) is also known as Gross–Pitaevskii equation, which models rotating Bose–Einstein condensation (BEC) with attractive particle interactions in a dilute gaseous ultra-cold superfluid. The operator L_A is usually denoted by $-\Omega \cdot L$, where $\Omega = (\Omega_1, \Omega_2, \Omega_3) \in \mathbb{R}^3$ is a given angular velocity vector and $L = -ix \wedge \nabla$. In this case the skew-symmetric matrix M is equal to $\begin{pmatrix} 0 & -\Omega_3 & \Omega_2 \\ \Omega_3 & 0 & -\Omega_1 \\ -\Omega_2 & \Omega_1 & 0 \end{pmatrix}$.

Such system as given in Eq. (1.1) describing rotating particles in a harmonic trap has acquired significance in connection with optics, plasma, quantized vortices, superfluids, and spinor BEC in theoretical and experimental physics [1, 4, 5, 15, 17, 21]. Meanwhile, mathematical study of the solutions to Eq. (1.1) has been conducted in order to provide insight and rigorous understanding for the dynamical behaviors of such wave matter. For $\lambda \in \mathbb{R}$ and $1 \leq p < 1 + \frac{4}{n-2}$, the local well-posedness results of Eq. (1.1) were obtained in, e.g., [3, 6, 16]; see also [9, 10, 13, 26] for the treatment in a general magnetic setting. In the focusing case $\lambda > 0$ and $p \geq 1 + \frac{4}{n}$, there exist solutions that blow up in finite time [7, 8, 22, 24, 25].

Let $Q \in H^1$ be the unique positive, nonincreasing, and radial ground state solution of the elliptic equation:

$$-\Delta Q + Q - Q^p = 0, \tag{1.2}$$

where H^1 denotes the usual Sobolev space. In the mass-critical case $p = 1 + \frac{4}{n}$, the paper [6] showed that $\|Q\|_2$ serves as the sharp threshold for blowup and global existence for Eq. (1.1). Moreover, if $p = 1 + \frac{4}{n}$ and $\|u_0\|_2$ is slightly greater than $\|Q\|_2$, the paper [7] obtained the exact blowup rate $\|\nabla u(t)\|_2 = (2\pi)^{-1/2}\|\nabla Q\|_2\sqrt{\frac{\log|\log(T-t)|}{T-t}}$ as $t \to T = T_{max}$. The analogous results for the standard NLS were initially proven in [22] and [18], where $A = V = 0$. In [7] we apply the so-called $\mathcal{R}$-transform method, which is a composite of the lens transform and a time-dependent rotation that allows to convert Eq. (1.1) into the standard NLS. We would like to mention that the case $A = 0$ and $V = \gamma^2|x|^2$ was considered in [24, 28]. Also if the harmonic potential is repulsive, i.e., $V = -\gamma^2|x|^2$, there are similar blowup results for Eq. (1.1) without angular momentum, see, e.g., [27].

The purpose of this chapter is to give a space-time universal upper bound on the blowup rate for the blowup solution to Eq. (1.1) with radial data in the mass-supercritical regime $p \in (1 + \frac{4}{n}, 1 + \frac{4}{n-2})$. Our main result is stated as follows.

Theorem 1.1 *Let $n \geq 3$ and $1 + \frac{4}{n} < p < 1 + \frac{4}{n-2}$, or $n = 2$ and $3 < p < 5$. Let $u_0 \in \mathscr{H}^1$ be radially symmetric, and assume that the corresponding solution $u \in C([0, T), \mathscr{H}^1)$ blows up in finite time T. Then,*

$$\int_t^T (T-s)\|\nabla u(s)\|_2^2 ds \leq C(T-t)^{\frac{2(5-p)}{5-p+(n-1)(p-1)}}. \tag{1.3}$$

Theorem 1.1 is motivated by a similar result by Merle et al. [20], where they proved such an upper bound on the blowup rate for the standard NLS without potential or angular momentum. The proof of Theorem 1.1 mainly follows the idea in [20] but relies on a refined version of the localized virial identity (Lemma 2, Section 3) in the magnetic setting. Note that the $\mathcal{R}$-transform introduced in [7] does not apply here for $p > 1 + \frac{4}{n}$. In Sect. 4 we shall give the proof of the main theorem. In Sect. 5, we include numerical figures to show the threshold of blowup for various cases of interest.

2 Preliminaries

In this section we recall the local well-posedness theory for Eq. (1.1) and a radial version of Gagliardo–Nirenberg inequality that we shall apply in the proof of our main theorem.

2.1 Local Well-Posedness for $p \in [1, 1 + 4/(n-2))$

For $u_0 \in \mathscr{H}^1$, the local well-posedness of Eq. (1.1) was obtained as a special case in, e.g., [10, 26] and [3, 7]. The paper [10, 26] dealt with a general class of magnetic potentials and quadratic electric potentials, where A is sublinear and V is subquadratic and essentially of positive sign. The case where V is subquadratic of both signs, e.g., $V = \pm \sum_{j=1}^n \gamma_j^2 x_j^2$, $\gamma_j > 0$, was treated in [3] when $n = 2, 3$, and in [7] for higher dimensions.

Let $H_{A,V} = -\Delta + V + iA\cdot\nabla = -(\nabla - \frac{i}{2}A)^2 + V_e$, where $V_e(x) = V(x) - \frac{|A|^2}{4}$ and $\operatorname{div} A = 0$. The proof for the local result relies on local in time-dispersive estimates for the propagator $U(t) = e^{-itH_{A,V}}$, the fundamental solution operator in $t \in [0, \delta)$, for some small constant $\delta > 0$, constructed in [23]. Alternatively, this can also be done by means of

$$e^{-itH_{A,V}}(x, y) = \left(\frac{\gamma}{2\pi i \sin(2\gamma t)}\right)^{\frac{n}{2}} e^{i\frac{\gamma}{2}(|x|^2+|y|^2)\cot(2\gamma t)} e^{-i\gamma\frac{(e^{tM}x)\cdot y}{\sin(2\gamma t)}}, \tag{2.1}$$

the fundamental solution to $iu_t = H_{A,V}u$ defined in $(0, \frac{\pi}{2\gamma}) \times \mathbb{R}^{2n}$ if $V(x) = \sum \gamma_j^2 x_j^2$; and replacing $\gamma \to i\gamma$ if $V(x) = -\sum \gamma_j^2 x_j^2$. The above formula (2.1) can be obtained via the $\mathcal{R}$-transform, a type of pseudo-conformal transform in the rotational setting; see [7].

Proposition 1 *For Eq.* (1.1)*, we have the following known results on well-posedness and conservation laws. Let* $r := p + 1$ *and* $q := \frac{4(p+1)}{n(p-1)}$.

(a) Well-posedness and blowup alternative:

- *(i) If* $1 \le p < 1 + \frac{4}{n}$*, then Eq.* (1.1) *has an* $\mathscr{H}^1$*-bounded global solution* $u \in C(\mathbb{R}; \mathscr{H}^1) \cap L^q_{\mathrm{loc}}(\mathbb{R}; \mathscr{H}^{1,r})$.
- *(ii) If* $1 + \frac{4}{n} \le p < 1 + \frac{4}{n-2}$*, then there exists* $T > 0$ *such that Eq.* (1.1) *has a unique maximal solution* $u \in C([0, T), \mathscr{H}^1) \cap L^q_{\mathrm{loc}}([0, T), \mathscr{H}^{1,r})$*. If* $T < \infty$*, then* u *blows up at* T *with a lower bound*

$$\|\nabla u(t)\|_2 \ge C(T - t)^{-(\frac{1}{p-1} - \frac{n-2}{4})}. \tag{2.2}$$

(b) The followings are conserved on the maximal lifespan $[0, T)$*:*

- *(i) Mass:* $M[u] = \int |u|^2$.
- *(ii) Energy:* $E[u] = \int \left(|\nabla u|^2 + V|u|^2 - \frac{2\lambda}{p+1}|u|^{p+1} + \overline{u}L_A u \right)$.
- *(iii) Angular momentum:* $\ell_A[u] = \int \overline{u}L_A u$.

Here we briefly outline the proof given in the literature [3, 7, 23, 26]. From Eq. (2.1), it follows the dispersive estimate for $|t| < \frac{\pi}{2\gamma}$

$$\|U(t)f\|_{L^\infty} \lesssim \frac{1}{|t|^{n/2}} \|f\|_{L^1} .$$

This, together with $\|U(t)f\|_2 = \|f\|_2$, yields the Strichartz type estimates on $I := [0, \frac{\pi}{2\gamma})$

$$\|U(t)f\|_{L^q_t \mathscr{H}^{1,r}_x} \lesssim \|f\|_{\mathscr{H}^1} \tag{2.3}$$

$$\| \int_0^t U(t-s)F(s,\cdot)ds \|_{L^q_t \mathscr{H}^{1,r}_x} \lesssim C\|F\|_{L^{\tilde{q}'}_t \mathscr{H}^{1,\tilde{r}'}_x} \tag{2.4}$$

where $(q, r) = (q, r, n)$, $(\widetilde{q}, \widetilde{r}) = (\widetilde{q}, \widetilde{r}, n)$ are admissible pairs satisfying $q, r \in [2, \infty]$, $(q, r, n) \neq (2, \infty, 2)$, and

$$\frac{2}{q} + \frac{n}{r} = \frac{n}{2},$$

with q' denoting the Hölder conjugate of q. Hence the local in time existence of Eq. (1.1) holds. If $p \in [1 + 4/n, 1 + 4/(n - 2))$, the blowup alternative and the lower bound for blowup rate of Eq. (1.1) follow from standard argument as in [7].

Remark 1 The Strichartz estimates (2.3)–(2.4) generalize those given in [10, 26], where $V_e(x) \approx \beta|x|^2$, $\beta > 0$, as $|x| \to \infty$. Here we allow V_e to be any quadratic function asymptotically $V_e(x) \approx \sum \beta_{ij} x_i x_j$ with $\beta_{ij} \in \mathbb{R}$. In the proof of (2.3)–(2.4), we directly study the action of $U(t - s)$ in the weighted space $\mathscr{H}^{1,r}$ based on [23, Lemma 3.1], an oscillatory integral operator (OIO) formula of Yajima in the magnetic setting. The OIO method was initially applied by Fujiwara in treating electric potentials. Our approach allows to technically deal with the commuting issue between x, ∇, and $U(t - s)$.

Let $Q = Q_0$ be the ground state solution of Eq. (1.2). In the L^2-critical case $p = 1 + 4/n$, from [6, 7] we know that $\|Q_0\|_2$ is the sharp threshold in the sense that:

(a) If $\|u_0\|_2 < \|Q_0\|_2$, then Eq. (1.1) has a unique global in time solution.
(b) For all $c \geq \|Q_0\|_2$, there exists u_0 with $\|u_0\|_2 = c$ so that u is a finite time blowup solution of Eq. (1.1).

According to [7, Proposition 4.5], if $p = 1 + 4/n$ and $\|u_0\|_2 = \|Q_0\|_2$, then all blowup solutions of Eq. (1.1) have the pseudo-conformal blowup rate

$$\|\nabla u\|_2 = O\left((T - t)^{-1}\right), \quad \text{as } t \to T. \tag{2.5}$$

The case $p > 1 + 4/n$ is technically more challenging. As far as we know, there have not been results on the characterization for the blowup profile or blowup rate. Theorem 1.1 provides an upper bound for the rotational NLS Eq. (1.1) under a harmonic potential. For the standard NLS, such upper bound is sharp, which is shown by constructing a ring-blowup solution in [20]. However, we do not know if the estimate (1.3) is sharp for Eq. (1.1), since the $\mathcal{R}$ transform does not apply for the L^2-supercritical case.

2.2 Radial Gagliardo–Nirenberg Inequality

The following is a radial version of Gagliardo–Nirenberg inequality due to W.A. Strauss.

Lemma 1 *Let $u \in H^1$ be a radial function. Then for $R > 0$, there is*

$$\|u\|_{L^\infty(|x| \geq R)} \leq C \frac{\|\nabla u\|_2^{1/2} \|u\|_2^{1/2}}{R^{\frac{n-1}{2}}}.$$

To prove Lemma 1, first note that when $n = 1$, the classical Gagliardo–Nirenberg inequality reads $\|u\|_\infty \le C\|u'\|_2^{1/2}\|u\|_2^{1/2}$. For general dimensions, since u is radial, we denote $u(x) = v(|x|) = v(r)$ and note that

$$\|u\|_2 \ge \|u\|_{L^2(|x|\ge R)} = C\left(\int_R^\infty |v(r)|^2 r^{n-1} dr\right)^{1/2} \ge CR^{\frac{n-1}{2}}\left(\int_R^\infty |v(r)|^2 dr\right)^{1/2} = CR^{\frac{n-1}{2}}\|v_R\|_2,$$

where $v_R = v|_{\{r\ge R\}}$. Similarly, we have $\|\nabla u\|_2 \ge CR^{\frac{n-1}{2}}\|v_R'\|_2$. Combining these with the above one-dimensional inequality, we obtain

$$\|u\|_{L^\infty(|x|\ge R)} = \|v_R\|_\infty \le C\|v_R'\|_2^{1/2}\|v_R\|_2^{1/2} \le C\frac{\|\nabla u\|_2^{1/2}\|u\|_2^{1/2}}{R^{\frac{n-1}{2}}}.$$

3 Localized Virial Identity

To prove Theorem 1.1, we derive certain localized virial identity associated to Eq. (1.1). This type of identities were shown in [19] in the case $A = V = 0$ and in [12, 14] for some general electromagnetic potentials. Here we present a direct proof for $A = Mx$ (M is skew symmetric) and general V, which is different than that in [12, 14]. Let $C_0^\infty = C_0^\infty(\mathbb{R}^n)$ denote the space of C^∞ functions with compact support.

Lemma 2 (Localized Virial Identity) *Assume that* $u \in C([0,T), \mathscr{H}^1)$ *is a solution to Eq. (1.1). Define* $J(t) := \int \varphi|u|^2$ *for any real-valued radial function* $\varphi \in C_0^\infty$. *Then,*

$$J'(t) = 2\Im\int \bar{u}\nabla\varphi\cdot\nabla u, \tag{3.1}$$

and

$$J''(t) = -\int \Delta^2\varphi|u|^2 - \frac{2\lambda(p-1)}{p+1}\int \Delta\varphi|u|^{p+1} + 4\int\left(\frac{\varphi''}{r^2} - \frac{\varphi'}{r^3}\right)|x\cdot\nabla u|^2 + 4\int\frac{\varphi'}{r}|\nabla u|^2 - 2\int\nabla\varphi\cdot\nabla V|u|^2. \tag{3.2}$$

Proof Note that

$$\begin{aligned}J'(t) &= \int \varphi u_t\bar{u} + \int \varphi u\bar{u}_t = 2\Re\int\varphi\bar{u}u_t\\ &= 2\Re\int\varphi\bar{u}(i\Delta u - iVu + i\lambda|u|^{p-1}u + A\cdot\nabla u) := 2(I_1 + I_2 + I_3 + I_4).\end{aligned}$$

The term I_1 is estimated as

$$I_1 = \Re\Big(i\int \varphi\overline{u}\Delta u\Big) = \Re\Big(-i\int \varphi|\nabla u|^2 - i\int \overline{u}\nabla\varphi\cdot\nabla u\Big) = \Im\int \overline{u}\nabla\varphi\cdot\nabla u.$$

Obviously $I_2 = \Re\Big(-i\int \varphi V|u|^2\Big) = 0$ and $I_3 = \Re\Big(i\lambda\int \varphi|u|^{p+1}\Big) = 0$. For I_4, we have

$$\begin{aligned}I_4 &= \Re\int \varphi\overline{u}A\cdot\nabla u = \Re\Big(-\int \nabla\varphi\cdot A|u|^2 - \int \varphi\nabla\overline{u}\cdot Au - \int \varphi(\nabla\cdot A)|u|^2\Big)\\ &= -\int \nabla\varphi\cdot A|u|^2 - I_4 - \int \varphi\nabla\cdot A|u|^2.\end{aligned}$$

Since φ is radial and M is skew symmetric, we know that (with $r = |x|$)

$$\nabla\varphi\cdot A = \varphi'(r)\frac{x}{r}\cdot A = \frac{\varphi'(r)}{r}x\cdot(Mx) = 0 \qquad \text{and} \qquad \nabla\cdot A = 0. \tag{3.3}$$

So $I_4 = -I_4$ and this implies $I_4 = 0$. Hence Eq. (3.1) follows.

Differentiating Eq. (3.1) again, we have

$$\begin{aligned}J''(t) &= 2\Big(\Im\int \overline{u}_t\nabla\varphi\cdot\nabla u + \Im\int \overline{u}\nabla\varphi\cdot\nabla u_t\Big) = 2\Big(\Im\int \overline{u}_t\nabla\varphi\cdot\nabla u - \Im\int \nabla\cdot(\overline{u}\nabla\varphi)u_t\Big)\\ &= 2\Big(-\Im\int \Delta\varphi\overline{u}u_t - 2\Im\int \nabla\varphi\cdot\nabla\overline{u}u_t\Big) := 2(-S-2T).\end{aligned}$$

To estimate S, first we write

$$S=\Im\int \Delta\varphi\,\overline{u}u_t=\Im\int \Delta\varphi\,\overline{u}(i\Delta u-iVu+i\lambda|u|^{p-1}u+A\cdot\nabla u) := S_1+S_2+S_3+S_4.$$

Since

$$\begin{aligned}S_1 &= \Im\Big(i\int \Delta\varphi\overline{u}\Delta u\Big)\\ &= \int \Delta^2\varphi|u|^2 + \Im\Big(-i\int \Delta\varphi\Delta\overline{u}u\Big) - 2\int \Delta\varphi|\nabla u|^2 = \int \Delta^2\varphi|u|^2 - S_1 - 2\int \Delta\varphi|\nabla u|^2,\end{aligned}$$

we have $S_1 = \dfrac{1}{2}\displaystyle\int \Delta^2\varphi|u|^2 - \int \Delta\varphi|\nabla u|^2$. Obviously, $S_2 = -\displaystyle\int \Delta\varphi V|u|^2$ and $S_3 = \lambda\displaystyle\int \Delta\varphi|u|^{p+1}$. In S_4, since $\Delta\varphi$ is also radial, by Eq. (3.3) we note that

$$\int \Delta\varphi \overline{u} A \cdot \nabla u = -\int \nabla(\Delta\varphi) \cdot A|u|^2 - \int \Delta\varphi \nabla\overline{u} \cdot Au - \int \Delta\varphi(\nabla \cdot A)|u|^2$$
$$= -\overline{\int \Delta\varphi \overline{u} A \cdot \nabla u}\,, \tag{3.4}$$

Indicating that $\int \Delta\varphi \overline{u} A \cdot \nabla u$ is imaginary. So $S_4 = \Im \int \Delta\varphi \overline{u} A \cdot \nabla u = -i \int \Delta\varphi \overline{u} A \cdot \nabla u$.

To estimate T, first we write

$$T = \Im \int \nabla\varphi \cdot \nabla\overline{u}(i\Delta u - iVu + i\lambda|u|^{p-1}u + A \cdot \nabla u) := T_1 + T_2 + T_3 + T_4.$$

For T_1, one has

$$T_1 = \Im\left(i\int \sum_{j,k} \varphi_{x_j}\overline{u}_{x_j}u_{x_k x_k}\right) = \Im\left(-i\int \sum_{j,k} \varphi_{x_j x_k}\overline{u}_{x_j}u_{x_k} - i\int \sum_{j,k} \varphi_{x_j}\overline{u}_{x_j x_k}u_{x_k}\right)$$
$$:= T_{1,1} + T_{1,2}.$$

Since φ is radial, we have $\varphi_{x_j} = \varphi'(r)\frac{x_j}{r}$ and $\varphi_{x_j x_k} = \varphi''(r)\frac{x_k x_j}{r^2} + \varphi'(r)\frac{\delta_{jk}}{r} - \varphi'(r)\frac{x_j x_k}{r^3}$, so

$$T_{1,1} = -\int \frac{\varphi''}{r^2}|x \cdot \nabla u|^2 - \int \frac{\varphi'}{r}|\nabla u|^2 + \int \frac{\varphi'}{r^3}|x \cdot \nabla u|^2.$$

Also,

$$T_{1,2} = \Im\left(i\int \Delta\varphi|\nabla u|^2 + i\int \sum_{j,k} \varphi_{x_j}\overline{u}_{x_k}u_{x_k x_j}\right) = \int \Delta\varphi|\nabla u|^2 - T_{1,2}\,,$$

which reveals $T_{1,2} = \frac{1}{2}\int \Delta\varphi|\nabla u|^2$. For T_2, one has

$$T_2 = \Im\left(-i\int \nabla\varphi \cdot \nabla\overline{u}Vu\right) = \int \Delta\varphi V|u|^2 + \int \nabla\varphi \cdot \nabla V|u|^2 - T_2\,,$$

so $T_2 = \frac{1}{2}\int \Delta\varphi V|u|^2 + \frac{1}{2}\int \nabla\varphi \cdot \nabla V|u|^2$. For T_3, there is

$$T_3 = \Im\left(i\lambda \int \nabla\varphi \cdot \nabla\overline{u}|u|^{p-1}u\right) = -\lambda \int \Delta\varphi|u|^{p+1} - pT_3\,,$$

and so $T_3 = -\frac{\lambda}{p+1}\int \Delta\varphi|u|^{p+1}$. For T_4, we have

$$T_4 = \Im\int (\nabla\varphi \cdot \nabla\overline{u})\,(A\cdot\nabla u) = \Im\left(-\int \overline{u}\Delta\varphi A\cdot\nabla u - \int \overline{u}\nabla\varphi\cdot\nabla(A\cdot\nabla u)\right) := T_{4,1} + T_{4,2}.$$

By Eq. (3.4), we obtain $T_{4,1} = i\int \overline{u}\Delta\varphi A\cdot\nabla u$. Also,

$$\begin{aligned}
T_{4,2} &= \Im\left(-\int \overline{u}\sum_j \varphi_{x_j}\left(\sum_{k,l} M_{k,l}x_l u_{x_k}\right)_{x_j}\right)\\
&= \Im\left(-\int \overline{u}\sum_{j,k,l}\varphi_{x_j}M_{k,l}\delta_{lj}u_{x_k} - \int \overline{u}\sum_{j,k,l}\varphi_{x_j}M_{k,l}x_l u_{x_k x_j}\right)\\
&= \Im\left(-\int \overline{u}\sum_{j,k}\varphi_{x_j}M_{k,j}u_{x_k} + \int \sum_{j,k,l}\overline{u}_{x_k}\varphi_{x_j}M_{k,l}x_l u_{x_j}\right.\\
&\qquad \left. + \int \overline{u}\sum_{j,k,l}\varphi_{x_j x_k}M_{k,l}x_l u_{x_j} + \int \overline{u}\sum_{j,k,l}\varphi_{x_j}M_{k,l}\delta_{lk}u_{x_j}\right)\\
&:= T_{4,2,1} + T_{4,2,2} + T_{4,2,3} + T_{4,2,4}.
\end{aligned}$$

Obviously $T_{4,2,2} = -T_4$, and the skew symmetry of M implies $T_{4,2,4} = 0$. Note that

$$\begin{aligned}
T_{4,2,3} &= \Im\left(-\int \sum_{j,k,l}\overline{u}_{x_j}\varphi_{x_k}M_{k,l}x_l u_{x_j} - \int \overline{u}\sum_{j,k,l}\varphi_{x_k}M_{k,l}\delta_{lj}u_{x_j} - \int \overline{u}\sum_{j,k,l}\varphi_{x_k}M_{k,l}x_l u_{x_j x_j}\right)\\
&= -\Im\int A\cdot\nabla\varphi|\nabla u|^2 - T_{4,2,1} - \Im\int \overline{u}A\cdot\nabla\varphi\Delta u = -T_{4,2,1}\,.
\end{aligned}$$

Hence $T_{4,2} = -T_4$ and so $T_4 = \frac{i}{2}\int \overline{u}\Delta\varphi A\cdot\nabla u$. Finally we obtain Eq. (3.2) by collecting all estimates on S's and T's. □

4 Proof of the Main Theorem

Now we are ready to prove Theorem 1.1.

Proof of Theorem 1.1 For a radial data u_0, let u be a corresponding radial solution that blows up in finite time $T < \infty$. Then $x \cdot \nabla u = ru'$ and $|\nabla u| = |u'|$, and the localized virial identity (3.2) can be written as, with $V = \gamma^2|x|^2$,

$$J''(t) = -\int \Delta^2\varphi|u|^2 - \frac{2\lambda(p-1)}{p+1}\int \Delta\varphi|u|^{p+1} + 4\int \varphi''|\nabla u|^2 - 4\gamma^2\int x\cdot\nabla\varphi|u|^2. \tag{4.1}$$

Choose a smooth radial function ψ such that $\psi(x) = \frac{|x|^2}{2}$ if $|x| \leq 2$ and $\psi(x) = 0$ if $|x| \geq 3$. Pick a time $0 < \tau < T$ and a radius $0 < R = R(\tau) \ll 1$ (to be determined later). Let $\varphi(x) = R^2\psi(\frac{x}{R})$. Then, with $r = |x|$,

$$\nabla\varphi(x) = R\psi'(\frac{r}{R})\frac{x}{r}, \quad \varphi''(r) = \psi''(\frac{r}{R}), \quad \Delta\varphi(x) = \Delta\psi(\frac{x}{R}), \quad \Delta^2\varphi(x) = \frac{1}{R^2}\Delta^2\psi(\frac{x}{R}),$$

so

$$\begin{aligned} J''(t) &= -\frac{1}{R^2}\int \Delta^2\psi(\frac{x}{R})|u|^2 - \frac{2(p-1)}{p+1}\lambda\int \Delta\psi(\frac{x}{R})|u|^{p+1} \\ &\quad + 4\int \psi''(\frac{r}{R})|\nabla u|^2 - 4\gamma^2 R\int \psi'(\frac{r}{R})r|u|^2 \\ &:= J_1 + J_2 + J_3 + J_4. \end{aligned}$$

Since $\Delta^2\psi$ is bounded, and $\Delta^2\psi(\frac{x}{R}) = 0$ if $|x| \leq 2R$ or $|x| \geq 3R$, we have $J_1 \leq \frac{C}{R^2}\int_{2R\leq|x|\leq 3R} |u|^2$. Also, since $\Delta\psi$ is bounded, and $\Delta\psi(\frac{x}{R}) = n$ when $|x| \leq 2R$, there is

$$\begin{aligned} J_2 &= -\frac{2n\lambda(p-1)}{p+1}\int_{|x|\leq 2R} |u|^{p+1} - \frac{2\lambda(p-1)}{p+1}\int_{|x|>2R} \Delta\psi(\frac{x}{R})|u|^{p+1} \\ &= -\frac{2n\lambda(p-1)}{p+1}\int |u|^{p+1} + \frac{2n\lambda(p-1)}{p+1}\int_{|x|>2R} |u|^{p+1} \\ &\quad - \frac{2\lambda(p-1)}{p+1}\int_{|x|\geq 2R} \Delta\psi(\frac{x}{R})|u|^{p+1} \\ &\leq -\frac{2n\lambda(p-1)}{p+1}\int |u|^{p+1} + C\int_{|x|\geq 2R} |u|^{p+1}. \end{aligned}$$

By choosing ψ such that $\psi'' \le 1$, we have $J_3 \le 4\int |\nabla u|^2$. And last, since $\psi'(\frac{r}{R}) = \frac{r}{R}$ when $|x| \le 2R$ and $\psi'(\frac{r}{R}) = 0$ when $|x| \ge 3R$, there is

$$
\begin{aligned}
J_4 &= -4\gamma^2 R \int_{|x|\le 2R} \frac{r}{R} r|u|^2 - 4\gamma^2 R \int_{2R<|x|\le 3R} \psi'(\frac{r}{R}) r|u|^2 \\
&\le -4\gamma^2 \int_{|x|\le 2R} |x|^2|u|^2 + CR \int_{2R<|x|\le 3R} |x||u|^2 \le CR \int_{2R<|x|\le 3R} |x||u|^2.
\end{aligned}
$$

Collecting all these terms, we have

$$
\begin{aligned}
J''(t) &\le 4\int |\nabla u|^2 - \frac{2n\lambda(p-1)}{p+1} \int |u|^{p+1} \\
&\quad + C\left(\frac{1}{R^2}\int_{2R\le|x|\le 3R} |u|^2 + \int_{|x|\ge 2R} |u|^{p+1} + R\int_{2R\le|x|\le 3R} |x||u|^2\right).
\end{aligned}
$$

Recall that the energy $E[u]$ is conserved, and $\ell_A[u] = 0$ since u is radial and $L_A u = 0$, we obtain

$$
\int |u|^{p+1} = \frac{p+1}{2\lambda}\int |\nabla u|^2 + \frac{p+1}{2\lambda}\int V|u|^2 - \frac{p+1}{2\lambda}E[u_0],
$$

so

$$
4\int |\nabla u|^2 - \frac{2n\lambda(p-1)}{p+1}\int |u|^{p+1} = n(p-1)E[u_0] - (n(p-1)-4)\int |\nabla u|^2 - n(p-1)\int V|u|^2.
$$

This yields

$$
\begin{aligned}
J''(t) &\le n(p-1)E[u_0] - (n(p-1)-4)\int |\nabla u|^2 - n(p-1)\int V|u|^2 \\
&\quad + C\left(\frac{1}{R^2}\int_{2R\le|x|\le 3R} |u|^2 + \int_{|x|\ge 2R} |u|^{p+1} + R\int_{2R\le|x|\le 3R} |x||u|^2\right).
\end{aligned}
$$

Since $R \ll 1$, we know $\frac{1}{R^2} \gg 3R^2$, so

$$
(n(p-1)-4)\int |\nabla u|^2 + J''(t) \le n(p-1)E[u_0] + C\left(\frac{1}{R^2}\int_{2R\le|x|\le 3R} |u|^2 + \int_{|x|\ge 2R} |u|^{p+1}\right).
$$

Also recall conservation of mass and $\frac{1}{R^2} \gg 1$, we have

$$
\begin{aligned}
(n(p-1)-4)\int |\nabla u|^2 + J''(t) &\le n(p-1)E[u_0] + C\left(\frac{1}{R^2} + \int_{|x|\ge 2R} |u|^{p+1}\right) \\
&\le C\left(\frac{1}{R^2} + \int_{|x|\ge 2R} |u|^{p+1}\right).
\end{aligned}
$$

To control the last term in the above inequality, we apply Lemma 1 and again the conservation of mass to obtain

$$\int_{|x|\ge 2R} |u|^{p+1} \le \|u\|^{p-1}_{L^\infty(|x|\ge 2R)} \int_{|x|\ge 2R} |u|^2 \le C\frac{\|\nabla u\|_2^{\frac{p-1}{2}}}{R^{\frac{(n-1)(p-1)}{2}}}$$

$$= \left[\left(\delta\frac{4n(p-1)-16}{p-1}\right)^{\frac{p-1}{4}} \|\nabla u\|_2^{\frac{p-1}{2}}\right]\frac{C}{R^{\frac{(n-1)(p-1)}{2}}}$$

$$\le \delta(n(p-1)-4)\|\nabla u\|_2^2 + \frac{C}{R^{\frac{2(n-1)(p-1)}{5-p}}},$$

where the last inequality is an application of Young's inequality with $\frac{1}{\frac{4}{p-1}}+\frac{1}{\frac{4}{5-p}} = 1$. By choosing $\delta > 0$ small enough and noting that $\frac{2(n-1)(p-1)}{5-p} > 2$, we obtain

$$\frac{n(p-1)-4}{2}\|\nabla u\|_2^2 + J''(t) \le \frac{C}{R^{\frac{2(n-1)(p-1)}{5-p}}}.$$

Integrate the above inequality over $[\tau, t]$ to obtain

$$\frac{n(p-1)-4}{2}\int_\tau^t \|\nabla u(s)\|_2^2 + J'(t) \le \frac{C(t-\tau)}{R^{\frac{2(n-1)(p-1)}{5-p}}} + J'(\tau).$$

Recalling Eq. (3.1) and integrating the inequality again with respect to t over $[\tau, t_0]$, we have

$$\frac{n(p-1)-4}{2}\int_\tau^{t_0} (t_0-s)\|\nabla u(s)\|_2^2 + \int \varphi|u(t_0)|^2$$

$$\le C\frac{(t_0-\tau)^2}{R^{\frac{2(n-1)(p-1)}{5-p}}} + 2(t_0-\tau)\Im\int \overline{u}(\tau)\nabla\varphi\cdot\nabla u(\tau) + \int \varphi|u(\tau)|^2.$$

Recall that $\varphi(x) = R^2\psi(\frac{x}{R})$, so

$$\int_\tau^{t_0} (t_0-s)\|\nabla u(s)\|_2^2 + \int \varphi|u(t_0)|^2$$

$$\le C\left(\frac{(t_0-\tau)^2}{R^{\frac{2(n-1)(p-1)}{5-p}}} + R(t_0-\tau)\left|\Im\int \overline{u}(\tau)\nabla\psi(\frac{x}{R})\cdot\nabla u(\tau)\right| + R^2\int |u(\tau)|^2\right)$$

$$\le C\left(\frac{(t_0-\tau)^2}{R^{\frac{2(n-1)(p-1)}{5-p}}} + R(t_0-\tau)\left\|\overline{u}(\tau)\nabla\psi(\frac{\cdot}{R})\right\|_2 \|\nabla u(\tau)\|_2 + R^2\|u_0\|_2\right)$$

$$\le C\left(\frac{(t_0-\tau)^2}{R^{\frac{2(n-1)(p-1)}{5-p}}} + R(t_0-\tau)\|\nabla u(\tau)\|_2 + R^2\right).$$

Letting $t_0 \to T$ and applying Young's inequality yield

$$\int_\tau^T (T-s)\|\nabla u(s)\|_2^2 \le C\left(\frac{(T-\tau)^2}{R^{\frac{2(n-1)(p-1)}{5-p}}} + R(T-\tau)\|\nabla u(\tau)\|_2 + R^2\right)$$
$$\le C\left(\frac{(T-\tau)^2}{R^{\frac{2(n-1)(p-1)}{5-p}}} + R^2\right) + (T-\tau)^2\|\nabla u(\tau)\|_2^2 .$$

By setting $\frac{(T-\tau)^2}{R^{\frac{2(n-1)(p-1)}{5-p}}} = R^2$, i.e., choosing $R = (T-\tau)^{\frac{5-p}{5-p+(n-1)(p-1)}}$, we have

$$\int_\tau^T (T-s)\|\nabla u(s)\|_2^2 \le C(T-\tau)^{\frac{2(5-p)}{5-p+(n-1)(p-1)}} + (T-\tau)^2\|\nabla u(\tau)\|_2^2.$$

To solve this inequality, let $g(\tau) = \int_\tau^T (T-s)\|\nabla u(s)\|_2^2$. Then the above inequality becomes

$$g(\tau) \le C(T-\tau)^{\frac{2(5-p)}{5-p+(n-1)(p-1)}} - (T-\tau)g'(\tau),$$

which is equivalent to

$$\frac{d}{d\tau}\left(\frac{g(\tau)}{T-\tau}\right) \le C(T-\tau)^{-\frac{2(n-1)(p-1)}{5-p+(n-1)(p-1)}}.$$

Integrating this with respect to τ over $[0, t]$ yields $g(t) \le C(T-t)^{\frac{2(5-p)}{5-p+(n-1)(p-1)}}$, the desired result in Eq. (1.3). □

Remark 2 Theorem 1.1 is still valid if we replace the positive constant λ in Eq. (1.1) by a C^1-function $\lambda(x)$ that satisfies the following three conditions:

(a) There exist $\lambda_1, \lambda_2 > 0$ such that $\lambda_1 \le \lambda(x) \le \lambda_2$;
(b) $x \cdot \nabla\lambda \le 0$;
(c) $\nabla\lambda$ is bounded.

The proof proceeds the same way as that given in this section but requires a version of Lemma 2 for the inhomogeneous NLS with rotation. We omit the details here.

Remark 3 Assume that $T = T_{max} < \infty$ is the blowup time for the solution u of Eq. (1.1). Then Eq. (1.3) implies that

$$\liminf_{t\to T} (T-t)^\delta \|\nabla u\|_2 < \infty ,$$

where we note that the function $\delta := \delta(p,n) = \frac{(n-1)(p-1)}{5-p+(n-1)(p-1)} \in \left(\frac{1}{2}, \frac{n-1}{2n-4}\right)$ is increasing in both p and n, given $p \in [1+4/n, 1+4/(n-2))$. From (2.2), we know that for any initial data in $\mathcal{H}^1$ one can derive a general lower bound for the collapse rate, namely, there exists $C = C_{p,n} > 0$ such that

$$\|\nabla u(t)\|_2 \geq C(T-t)^{-(\frac{1}{p-1}-\frac{n-2}{4})}.$$

In particular, if $p = 1+4/n$, the estimate (1.3) is only valid for the lower bound $(T-t)^{-1/2}$. Thus, comparing the mass-critical case, where the log–log law and pseudo-conformal blowup rate (2.5) can occur, the mass-supercritical case for larger data can be more subtle, see [7, Theorem 1.1] and [20].

5 Numerical Results for Mass-Critical and Mass-Supercritical RNLS in 2D

In this section we show numerical simulations for the blowup of Eq. (1.1) with $n = 2$ with the given initial data ψ_0 being a multiple of the ground state for the following nonlinear Schrödinger equation:

$$i\psi_t = -\frac{1}{2}\Delta\psi + \frac{1}{2}(\gamma_1^2 x^2 + \gamma_2^2 y^2)\psi - \lambda|\psi|^{p-1}\psi - i\Omega(y\partial_x - x\partial_y)\psi\,. \tag{5.1}$$

Let $Q = Q_{\Omega,V}$ be the ground state for Eq. (5.1) satisfying the associated Euler–Lagrange equation:

$$\omega Q = -\frac{1}{2}\Delta Q + \frac{1}{2}(\gamma_1^2 x^2 + \gamma_2^2 y^2)Q - \lambda|Q|^{p-1}Q - i\Omega(y\partial_x - x\partial_y)Q\,, \tag{5.2}$$

where ω is the chemical potential. The construction of the ground states can be found, e.g., in [6, 11] if $p \leq 1+4/n$. Here we use GPELab as introduced in [2] to do the computations and observe the blowup phenomenon for $\psi_0 = CQ_{\Omega,V}$ with appropriate constant C for $p = 3, 4$ and $p = 6$. Note that the case $p = 6$ is beyond the limit of exponents covered in Theorem 1.1. For certain convenience from the software, we compute the solution ψ of Eq. (5.1) rather than Eq. (1.1) on the (x, y)-domain $[-3, 3] \times [-3, 3]$ in the plane. There is an obvious scaling relation between ψ and u of these two equations. From Sect. 2.1 we know that when $p = 3$, the mass of $Q_{0,0}$ is the dichotomy that distinguishes the blow-up vs. global existence solutions. The main reason we use $Q_{\Omega,V}$ in place of $Q_{0,0}$ is that numerically the actual ground state $Q_{\Omega,V}$ is easier to compute and save as a stable profile under a trapping potential.

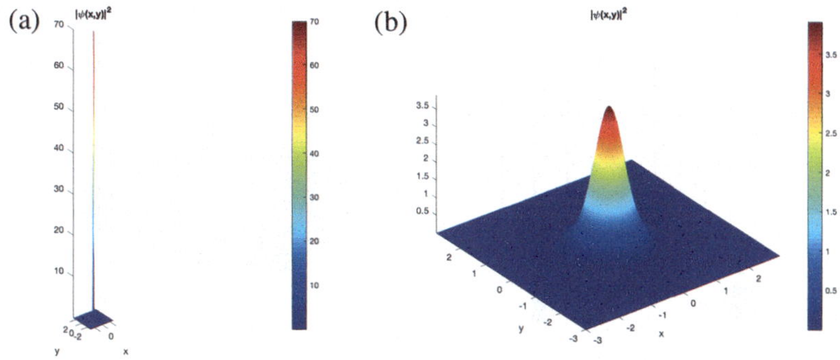

Fig. 1 $|\psi|^2$ when $p = 3$, $(\gamma_1, \gamma_2) = (1, 1)$, $\Omega = 0.5$. (**a**) $\psi_0 = 2.5 * Q_{\Omega,V}$ (max $|\psi|^2 \approx 1812$). (**b**) $\psi_0 = 2 * Q_{\Omega,V}$ (max $|\psi|^2 \approx 3.9$)

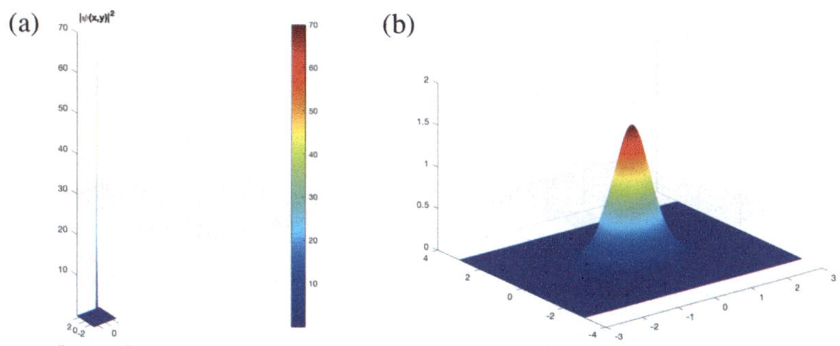

Fig. 2 $|\psi|^2$ when $p = 4$, $(\gamma_1, \gamma_2) = (1, 1)$, $\Omega = 0.5$. (**a**) $\psi_0 = 2 * Q_{\Omega,V}$ (max $|\psi|^2 \approx 102$). (**b**) $\psi_0 = 1.6 * Q_{\Omega,V}$ (max $|\psi|^2 \approx 1.63$)

1. Isotropic case: $\gamma_1 = \gamma_2 = 1$, $\lambda = 1$, $\Omega = 0.5$. Let $p = 3$. We see in Fig. 1 that the solution has energy concentration in short time and blows up with $\psi_0 = 2.5Q_{\Omega,V}$, but it shows stable smooth solution at the level $\psi_0 = 2Q_{\Omega,V}$. For $p = 4$, we observe in Fig. 2 that using $\psi_0 = 2Q_{\Omega,V}$ yields a blowup solution; but, there shows no blowup at $1.6Q_{\Omega,V}$.
2. Anisotropic case: $\gamma_1 = 1$, $\gamma_2 = 2$, $\lambda = 1$, $\Omega = 0.5$. Let $p = 4$. We observe that the anisotropic harmonic potential may yield blowup at a lower-level ground state. Figure 3 shows blowup when $\psi_0 = 1.8Q_{\Omega,V}$, while stable smooth solution at $\psi_0 = 1.5Q_{\Omega,V}$.
3. If turning off the rotation, i.e., $\Omega = 0$, then Fig. 4 shows that in the isotropic case $\gamma_1 = \gamma_2 = 1$, $p = 6$, $\lambda = 1$, then blowup threshold $\psi_0 = 1.565Q_{\Omega,V}$; and there exists a bounded solution in $\mathscr{H}^1$ if $\psi_0 = 1.56Q_{\Omega,V}$. However, in the anisotropic case for V, $\gamma_1 = 1$, $\gamma_2 = 2$, the blowup threshold is at level $\psi_0 = 1.395Q_{\Omega,V}$; and there exists a bounded solution in $\mathscr{H}^1$ if $\psi_0 = 1.39Q_{\Omega,V}$. The above results reveal that higher-order exponent p and anisotropic property for the potential contribute more to the wave collapse, which may make the system

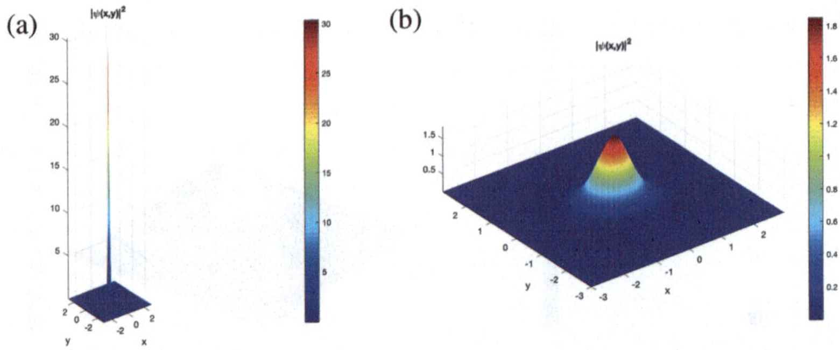

Fig. 3 $|\psi|^2$ when $p = 4$, $(\gamma_1, \gamma_2) = (1, 2)$, $\Omega = 0.5$. **(a)** $\psi_0 = 1.8 * Q_{\Omega,V}$ $(\max |\psi|^2 \approx 93)$. **(b)** $\psi_0 = 1.5 * Q_{\Omega,V}$ $(\max |\psi|^2 \approx 1.85)$

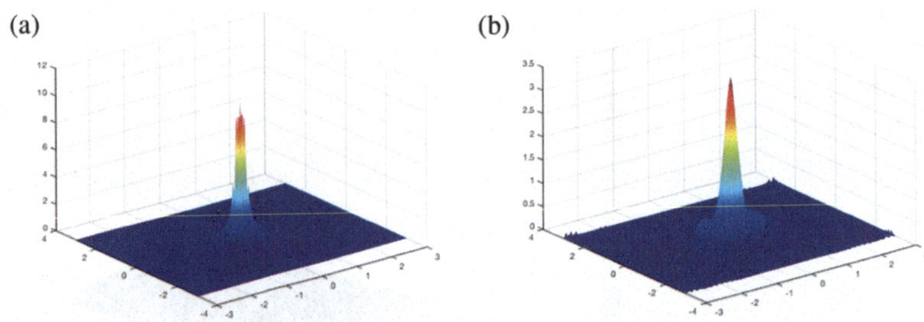

Fig. 4 $|\psi|^2$ when $p = 6$, $V = \frac{1}{2}(\gamma_1^2 x^2 + \gamma_2^2 y^2)$, $\Omega = 0$. **(a)** $(\gamma_1, \gamma_2) = (1, 1)$, $\psi_0 = 1.565 * Q_{\Omega,V}$. **(b)** $(\gamma_1, \gamma_2) = (1, 2)$, $\psi_0 = 1.395 * Q_{\Omega,V}$

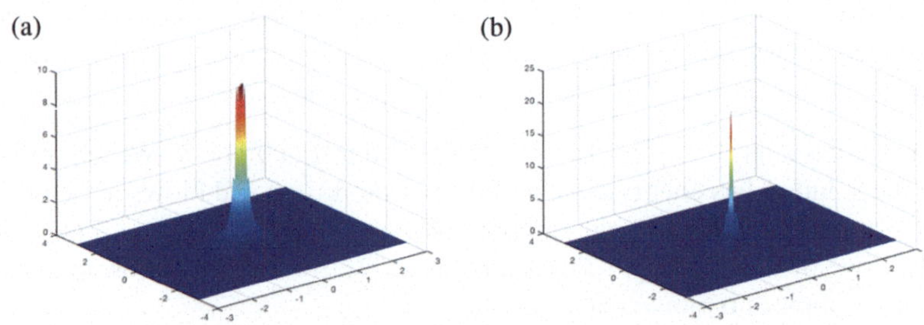

Fig. 5 $|\psi|^2$ when $p = 6$, $V = \frac{1}{2}(\gamma_1^2 x^2 + \gamma_2^2 y^2)$, $\Omega = 0.5$. **(a)** $(\gamma_1, \gamma_2) = (1, 1)$, $\psi_0 = 1.565 * Q_{\Omega,V}$. **(b)** $(\gamma_1, \gamma_2) = (1, 2)$, $\psi_0 = 1.395 * Q_{\Omega,V}$

unstable at a lower level of mass. It is of interest to observe that in the presence of rotation (Fig. 5), the threshold constants C remain the same in both isotropic and anisotropic cases, although $|\psi|^2$, the energies and chemical potentials grow at larger magnitude.

Notice that if $p = 6$, then the behavior of wave collapse is quite different than the case $p < 5$. The modulus square of $\psi(t, x)$ first forms growing singularity.

Then it quickly reduces to normal level but with large energy and $\|\nabla\psi\|_2$ after collapsing time although it does not seem to admit proper self-similar profile of energy concentration.

Acknowledgement The authors thank the anonymous referee for helpful comments that have helped improve the presentation of the chapter.

References

1. Aftalion, A.: Vortices in Bose-Einstein condensates. *Progress in Nonlinear Differential Equations and their Applications* **67**, Birkhäuser, 2006.
2. Antoine, X., Duboscq, R.: GPELab, a Matlab toolbox to solve Gross-Pitaevskii equations II: Dynamics and stochastic simulations. *Computer Physics Communications* **193** (2015), 95–117.
3. Antonelli, P., Marahrens, D., Sparber, C.: On the Cauchy problem for nonlinear Schrödinger equations with rotation. *Discrete Contin. Dyn. Syst.* **32** (2012), no. 3, 703–715.
4. Bao, W., Cai, Y.: Ground states and dynamics of spin-orbit-coupled Bose-Einstein condensates. *SIAM J. Appl. Math.* **75** (2015), no. 2, 492–517.
5. Bao, W., Wang, H., Markowich, P.: Ground, symmetric and central vortex states in rotating Bose-Einstein condensates. *Comm. Math. Sci.* **3** (2005), 57–88.
6. Basharat, N., Hajaiej, H., Hu, Y., Zheng, S.: Threshold for blowup and stability for nonlinear Schrödinger equation with rotation. Preprint.
7. Basharat, N., Hu, Y., Zheng, S.: Blowup rate for mass critical rotational nonlinear Schrödinger equations. *Nonlinear Dispersive Waves and Fluids. Contemp. Math.* **725** (2019), 1–12.
8. Carles, R., Remarks on nonlinear Schrödinger equations with harmonic potential. *Annales Henri Poincaré.* **3** (2002), no. 4, 757–772.
9. Cazenave, T., Esteban, M.: On the stability of stationary states for nonlinear Schrödinger equations with an external magnetic field. *Mat. Apl. Comput.* **7** (1988), 155–168.
10. De Bouard, A.: Nonlinear Schrödinger equations with magnetic fields. *Differential Integral Equations.* **4** (1991), no. 1, 73–88.
11. Esteban, M., Lions, P.: Stationary solutions of nonlinear Schrödinger equations with an external magnetic field. In: *Partial Differential Equations and the Calculus of Variations.* Progress in Nonlinear Differential Equations and Their Applications **1** (1989), 401–449, Birkhäuser.
12. Fanelli, L., Vega, L.: Magnetic virial identities, weak dispersion and Strichartz inequalities. *Math. Ann.* **344** (2009), 249–278.
13. Galati, L., Zheng, S.: Nonlinear Schrödinger equations for Bose-Einstein condensates. *Nonlinear and Modern Mathematical Physics.* AIP Conference Proceedings **1562** (1), (2013), 50–64.
14. Garcia, A.: Magnetic virial identities and applications to blow-up for Schrödinger and wave equations. *Journal of Physics.* A, Mathematical and Theoretical **45** (1), 015202.
15. Gross, E., Structure of a quantized vortex in boson systems. Nuovo Cimento **20** (1961), 454.
16. Hao, C., Hsiao, L., Li, H.: Global well posedness for the Gross-Pitaevskii equation with an angular momentum rotational term in three dimensions. *J. Math. Phys.* **48** (2007), no. 10, 102105.
17. Matthews, M., Anderson, B., Haljan, P., Hall, D., Wiemann, C., Cornell, E.: Vortices in a Bose-Einstein condensates. *Phys. Rev. Lett.* **83** (1999), 2498–2501.
18. Merle, F., Raphaël, P.: Profiles and quantization of the blow up mass for critical nonlinear Schrödinger equation. *Comm. Math. Phys.* **253** (2005), no. 3, 675–704.
19. Merle, F., Raphaël, P.: Blow up of the critical norm for some radial L^2 super critical nonlinear Schrödinger equations. *Amer. J. Math.* **130** (2008), 945–978.
20. Merle, F., Raphaël, P., Szeftel, J.: On collapsing ring blow-up solutions to the mass supercritical nonlinear Schrödinger equation. *Duke Math. J.* **163** (2014), no. 2, 369–431.

21. Recati, A., Zambelli, F., Stringari, S.: Overcritical rotation of a trapped Bose-Einstein condensate. *Phys. Rev. Lett.* **86** (2001), 377–380.
22. Weinstein, M.: Nonlinear Schrödinger equations and sharp interpolation estimates. *Comm. Math. Phys.* **87** (4):567–576, 1983.
23. Yajima, K.: Schrödinger evolution equations with magnetic fields. *J. Analyse Math.* **56** (1991), 29–76.
24. Zhang, J.: Stability of attractive Bose-Einstein condensates. *J. Statist. Phys.* **101**(3-4):731–746, 2000.
25. Zhang, J.: Sharp threshold for blowup and global existence in nonlinear Schrödinger equations under a harmonic potential. *Comm. Partial Differential Equations* **30** (2005), no. 10-12, 1429–1443.
26. Zheng, S.: Fractional regularity for nonlinear Schrödinger equations with magnetic fields. *Contemp. Math.* **581** (2012), 271–285.
27. Zhu, S., Li, X.: Sharp upper and lower bounds on the blow-up rate for nonlinear Schrödinger equation with potential. *Appl. Math. Comput.* **190** (2007), no. 2, 1267–1272.
28. Zhu, S., Zhang, J.: Profiles of blow-up solutions for the Gross-Pitaevskii equation. *Acta Math. Appl. Sin. Engl. Ser.* **26** (2010), no. 4, 597–606.

Almost Eigenvalues and Eigenvectors of Almost Mathieu Operators

Thomas Strohmer and Timothy Wertz

Dedicated to John Benedetto on the occasion of his eightieth birthday. In John's rich scientific oeuvre, the topics of spectrum, sampling, discretization, Fourier transform and time-frequency analysis play a central role. This book chapter is partly inspired by John's work and draws from all these topics.

Abstract The *almost Mathieu operator* is the discrete Schrödinger operator $H_{\alpha,\beta,\theta}$ on $\ell^2(\mathbb{Z})$ defined via $(H_{\alpha,\beta,\theta} f)(k) = f(k+1) + f(k-1) + \beta \cos(2\pi\alpha k + \theta) f(k)$. We derive explicit estimates for the eigenvalues at the edge of the spectrum of the finite-dimensional almost Mathieu operator $H^{(n)}_{\alpha,\beta,\theta}$. We furthermore show that the (properly rescaled) m-th Hermite function ϕ_m is an approximate eigenvector of $H^{(n)}_{\alpha,\beta,\theta}$, and that it satisfies the same properties that characterize the true eigenvector associated with the m-th largest eigenvalue of $H^{(n)}_{\alpha,\beta,\theta}$. Moreover, a properly translated and modulated version of ϕ_m is also an approximate eigenvector of $H^{(n)}_{\alpha,\beta,\theta}$, and it satisfies the properties that characterize the true eigenvector associated with the m-th largest (in modulus) negative eigenvalue. The results hold at the edge of the spectrum, for any choice of θ and under very mild conditions on α and β. We also give precise estimates for the size of the "edge," and extend some of our results to $H_{\alpha,\beta,\theta}$. The ingredients for our proofs comprise special recursion properties of Hermite functions, Taylor expansions, time-frequency analysis, Sturm sequences, and perturbation theory for eigenvalues and eigenvectors. Numerical simulations demonstrate the tight fit of the theoretical estimates.

T. Strohmer (✉)
University of California, Davis, CA, USA
e-mail: strohmer@math.ucdavis.edu

T. Wertz
Yale-NUS College, Singapore, Singapore
e-mail: tim.wertz@yale-nus.edu.sg

M. Hirn et al. (eds.), *Excursions in Harmonic Analysis, Volume 6*, Applied and Numerical Harmonic Analysis, https://doi.org/10.1007/978-3-030-69637-5_5

1 Introduction

We consider the *almost Mathieu operator* $H_{\alpha,\beta,\theta}$ on $\ell^2(\mathbb{Z})$, given by

$$(H_{\alpha,\beta,\theta}f)(x) = f(x+1) + f(x-1) + 2\beta\cos(2\pi\alpha x + \theta)f(x), \tag{1.1}$$

with $x \in \mathbb{Z}$, $\beta \in \mathbb{R}$, $\alpha \in [-\frac{1}{2}, \frac{1}{2})$, and $\theta \in [0, 2\pi)$. This operator is interesting both from a physical and a mathematical point of view [24, 26]. In physics, for instance, it serves as a model for Bloch electrons in a magnetic field [19]. In mathematics, it appears in connection with graph theory and random walks on the Heisenberg group [10, 17] and rotation algebras [13]. A major part of the mathematical fascination of almost Mathieu operators stems from their interesting spectral properties, obtained by varying the parameters α, β, θ, which has led to some deep and beautiful mathematics, see, e.g., [6, 7, 9, 11, 21, 23]. For example, it is known that the spectrum of the almost Mathieu operator is a Cantor set for all irrational α and for all $\beta \neq 0$, cf. [8]. Furthermore, if $\beta > 1$, then $H_{\alpha,\beta,\theta}$ exhibits Anderson localization, i.e., the spectrum is pure point with exponentially decaying eigenvectors [20].

A vast amount of literature exists devoted to the study of the bulk spectrum of $H_{\alpha,\beta,\theta}$ and its structural characteristics, but very little seems to be known about the edge of the spectrum. For instance, what is the size of the extreme eigenvalues of $H_{\alpha,\beta,\theta}$, how do they depend on α, β, θ, and what do the associated eigenvectors look like? These are exactly the questions we will address in this paper.

While in general the localization of the eigenvectors (when they exist) of $H_{\alpha,\beta,\theta}$ depends on the choice of β, it turns out that there exist approximate eigenvectors associated with the approximate eigenvalues of $H_{\alpha,\beta,\theta}$ at the edges of the spectrum which are *always* exponentially localized. Indeed, we will show that for small α the m-th Hermitian function ϕ_m as well as certain translations and modulations of ϕ_m form *almost* eigenvectors of $H_{\alpha,\beta,\theta}$ regardless whether α is rational or irrational and as long as the product $\alpha\sqrt{\beta}$ is small.

There is a natural heuristic explanation why Hermitian functions emerge in connection with almost Mathieu operators (however, as our results will show, this heuristics only holds at the edge of the spectrum). Consider the continuous-time version of $H_{\alpha,\beta,\theta}$ in (1.1) by letting $x \in \mathbb{R}$ and set $\alpha = 1, \beta = 1, \theta = 0$. Then $H_{\alpha,\beta,\theta}$ commutes with the Fourier transform on $L^2(\mathbb{R})$. It is well-known that Hermite functions are eigenfunctions of the Fourier transform, ergo Hermite functions are eigenvectors of the aforementioned continuous-time analog of the Mathieu operator. Of course, it is no longer true that the discrete $H_{\alpha,\beta,\theta}$ commutes with the corresponding Fourier transform (nor do we want to restrict ourselves to one specific choice of α and β). But nevertheless it may still be true that discretized (and perhaps truncated) Hermite functions are almost eigenvectors for $H_{\alpha,\beta,\theta}$. We will see that this is indeed the case under some mild conditions, but it only holds for the first m Hermite functions where the size of m is either $O(1)$ or $O(1/\sqrt{\gamma})$, where $\gamma = \pi\alpha\sqrt{\beta}$, depending on the desired accuracy (γ^2 and γ, respectively) of the approximation. We will also show a certain symmetry for the eigenvalues of $H_{\alpha,\beta,\theta}$ and use this fact to conclude that a properly translated and modulated Hermite function is an approximate eigenvector for the m-th largest (in modulus) negative eigenvalue.

The only other papers we are aware of that analyze the eigenvalues of the almost Mathieu operator at the edge of the spectrum are [14, 15, 28, 29]. In [28], the authors analyze a continuous-time model to obtain eigenvalue estimates of the discrete-time operator $H_{\alpha,\beta,\theta}$. They consider the case $\beta = 1, \theta = 0$, and small α and arrive at an estimate for the eigenvalues at the right edge of the spectrum that is not far from our expression for this particular case (after translating their notation into ours and correcting what seems to be a typo in [28]). But there are several differences to our work. First, [28] does not provide any results about the eigenvectors of $H_{\alpha,\beta,\theta}$. Second, [28] does not derive any error estimates for their approximation, and indeed, an analysis of their approach yields that their approximation is only accurate up to order γ and not γ^2. Third, [28] contains no quantitative characterization of the size of the edge of the spectrum. On the other hand, the scope of [28] is different from ours. The paper [14] derives quite accurate bounds for the largest eigenvalue for the case when $\alpha \in [\frac{1}{4}, \frac{1}{2}]$ (the translation parameter is also chosen to be α instead of 1, and $\theta = 0$). The authors use methods from rotation C^*-algebras. There is essentially no overlap with our paper, since [14] "only" focuses on estimates for the largest eigenvalue and contains no results on eigenvectors. Finally, in the very intriguing paper [15], which was written at about the same time as our manuscript, the authors provide three very different approaches to estimating the extreme eigenvalues and associated eigenvectors of Harper's Operator, i.e., of the almost Mathieu operator with $\theta = 0$, $\alpha = 1/n$, and $\beta = 1$. Some of the results in [15] are similar to some of our results, the proof techniques are very different from ours.

Harper's operator and discrete versions of the Hermite functions also appear in the context of attempts to derive the eigenvectors of the discrete Fourier transform analytically [5] and in definitions of the discrete fractional Fourier transform [4, 25] and related operators [16].

The remainder of the paper is organized as follows. In Sect. 1.1 we introduce some notation and definitions used throughout the paper. In Sect. 2 we derive eigenvector and eigenvalue estimates for the finite-dimensional model of the almost Mathieu operator. The ingredients for our proof comprise Taylor expansions, basic time-frequency analysis, Sturm sequences, and perturbation theory for eigenvalues and eigenvectors. The extension of our main results to the infinite dimensional almost Mathieu operator is carried out in Sect. 3. Finally, in Sect. 4 we complement our theoretical findings with numerical simulations.

1.1 Definitions and Notation

We define the unitary operators of translation and modulation, denoted T_a and M_b, respectively, by

$$(T_a f)(x) := f(x-a) \qquad \text{and} \qquad (M_b f)(x) := e^{2\pi i b x} f(x),$$

where the translation is understood in a periodic sense if f is a vector of finite length. It will be clear from the context if we are dealing with finite or infinite-dimensional versions of T_a and M_b. Recall the commutation relations (see, e.g., Section 1.2 in [18])

$$T_a M_b = e^{-2\pi i a b} M_b T_a. \tag{1.2}$$

The discrete and periodic Hermite functions we will be using are derived from the standard Hermite functions defined on $\mathbb{R}$ (see, e.g., [1]) by simple discretization and truncation. We do choose a slightly different normalization than in [1] by introducing the scaling terms $(\sqrt{2\gamma})^{2l}$ and $(\sqrt{2\gamma})^{2l+1}$, respectively.

Definition 1 The scaled Hermite functions φ_m with parameter $\gamma > 0$ are for

even m :

$$\varphi_m(x) = e^{-\gamma x^2} \sum_{l=0}^{\frac{m}{2}} (\sqrt{\gamma})^{2l} c_{m,l} x^{2l}, \quad \text{where} \quad c_{m,l} = \frac{m!(2\sqrt{2})^{2l}(-1)^{\frac{m}{2}-l}}{(2l)!(\frac{m}{2}-l)!}, \tag{1.3}$$

odd m :

$$\varphi_m(x) = e^{-\gamma x^2} \sum_{l=0}^{\frac{m-1}{2}} (\sqrt{\gamma})^{2l+1} c_{m,l} x^{2l+1}, \text{ where } c_{m,l} = \frac{m!(2\sqrt{2})^{2l+1}(-1)^{\frac{m-1}{2}-l}}{(2l+1)!(\frac{m-1}{2}-l)!}, \tag{1.4}$$

for $x \in \mathbb{Z}$ (or $x \in \mathbb{R}$) and $m = 0, 1, \ldots$. The discrete, periodic Hermite functions of period n, denoted by $\varphi_{m,n}$, are defined for $x \in \left\{k - \frac{n}{2} : k \in \{0, 1, \ldots, n-1\}\right\}$ with periodic boundary conditions by

$$\varphi_{m,n}(x) := \varphi_m(x), \quad m \in \{0, 1, \ldots, n-1\},$$

when n is even. For odd n, $\varphi_{m,n}$, are defined for $x \in \left\{k - \frac{n-1}{2} : k \in \{0, 1, \ldots, n-1\}\right\}$ with periodic boundary conditions by

$$\varphi_{m,n}(x) := \varphi_m(x), \quad m \in \{0, 1, \ldots, n-1\}.$$

We denote the finite almost Mathieu operator, acting on sequences of length n, by $H^{(n)}_{\alpha,\beta,\theta}$. It can be represented by an $n \times n$ tridiagonal matrix, which has ones on the two side diagonals and $\cos(x_k + \theta)$ as k-th entry on its main diagonal for $x_k = 2\pi\alpha k$, $k = -\frac{n}{2}, \ldots, \frac{n}{2} - 1$. Here, we have assumed for simplicity that n is even, the required modification for odd n is obvious. Sometimes it is convenient to replace the translation present in the infinite almost Mathieu operator by a periodic translation. In this case we obtain the $n \times n$ periodic almost Mathieu operator $P^{(n)}_{\alpha,\beta,\theta}$ which is almost a tridiagonal matrix; it is given by

$$P^{(n)}_{\alpha,\beta,\theta} = \begin{bmatrix} 2\beta\cos(x_{-\frac{n}{2}}+\theta) & 1 & \cdots & 0 & 1 \\ 1 & 2\beta\cos(x_{-\frac{n}{2}+1}+\theta) & \cdots & 0 & 0 \\ 0 & 1 & \cdots & 0 & 0 \\ \vdots & \vdots & \ddots & \vdots & \vdots \\ 0 & 0 & \cdots & 2\beta\cos(x_{\frac{n}{2}-2}+\theta) & 1 \\ 1 & 0 & \cdots & 1 & 2\beta\cos(x_{\frac{n}{2}-1}+\theta) \end{bmatrix} \tag{1.5}$$

where $x_k = 2\pi\alpha k$ for $k = -\frac{n}{2}, \ldots, \frac{n}{2} - 1$. If $\theta = 0$ we write $P^{(n)}_{\alpha,\beta}$ instead of $P^{(n)}_{\alpha,\beta,\theta}$.

2 Finite Hermite Functions as Approximate Eigenvectors of the Finite-Dimensional Almost Mathieu Operator

In this section we focus on eigenvector and eigenvalue estimates for the finite-dimensional almost Mathieu operator. Finite versions of $H_{\alpha,\beta,\theta}$ are interesting in their own right. On the one hand, numerical simulations are frequently based on truncated versions of $H_{\alpha,\beta,\theta}$, on the other hand certain problems, such as the study of random walks on the Heisenberg group, are often more naturally carried out in the finite setting. As we will be working with symmetric, tridiagonal matrices, it is useful to remember that if the entries of the first upper and lower sub-diagonals of such a matrix are all non-zero, then the eigenvalues of the matrix have multiplicity one. This is mostly easily seen from rank considerations, but can also be shown using determinants.

We first gather some properties of the true eigenvectors of the almost Mathieu operator, collected in the following proposition.

Proposition 1 *Consider the finite, non-periodic almost Mathieu operator $H^{(n)}_{\alpha,\beta,\theta}$. Let $\zeta_0 > \zeta_1 > \cdots > \zeta_{n-1}$ be its eigenvalues and $\psi_0, \psi_1, \ldots, \psi_{n-1}$ the associated eigenvectors. The following statements hold:*

1. $H^{(n)}_{\alpha,\beta,\theta}\psi_m = \zeta_m\psi_m$ *for all* $0 \le m \le n-1$;
2. *For each m, there exist constants $C_1(m)$, $C_2(m)$ (which are independent of n if $\beta > 1$) such that*

$$|\psi_m(i)| \le C_1(m)e^{-C_2(m)|i|};$$

3. *If ζ_m is the m-th largest eigenvalue, then ψ_m changes sign exactly m times.*

Proof The first statement is simply restating the definition of eigenvalues and eigenvectors. The second statement follows from the fact that the inverse (or pseudo-inverse in the non-invertible case) of a tridiagonal (or an almost tridiagonal operator) exhibits exponential off-diagonal decay and the relationship between the spectral projection and eigenvectors associated with isolated eigenvalues. See [12] for details. The independence from n in the case $\beta > 1$ is a consequence of Theorem 3.8 in [3]. The third statement is a result of Lemma 1 below. □

Lemma 1 *Let A be a symmetric tridiagonal $n \times n$ matrix with entries $a_1, \dots, a_n$ on its main diagonal and $b_1, \dots, b_{n-1}$ on its two non-zero side diagonals, with $b_j > 0$ for all $j \in \{1, \dots, n-1\}$. Let $\lambda_0 > \lambda_1 > \cdots > \lambda_{n-1}$ be the eigenvalues of A, and $v_0, \dots, v_{n-1}$ be the associated eigenvectors. Then, for each $0 \leq m \leq n-1$ the entries of the vector v_m change signs m times. That is, the entries of the vector v_0 all have the same sign, while the vector v_1 has only a single index where the sign of the entry at that index is different than the one before it, and so on.*

Proof This result follows directly from Theorem 6.1 in [2] which relates the Sturm sequence to the sequence of ratios of eigenvector elements. Using the assumption that $b_i > 0$ for all $i = 1, \dots, n$ yields the claim. □

The main results of this paper are summarized in the following theorem. In essence we show that the Hermite functions (approximately) satisfy all three eigenvector properties listed in Proposition 1. The theorem is stated in a somewhat heuristic fashion here. The technical details are presented later in this section.

Theorem 2.1 *Let A be either of the operators $P^{(n)}_{\alpha,\beta,\theta}$ or $H^{(n)}_{\alpha,\beta,\theta}$. Let $\varphi_{m,n}$ be as defined in Definition 1. Set $\gamma = \pi\alpha\sqrt{\beta}$, let $0 < \varepsilon < 1$, and assume $\frac{4}{n^{2-\varepsilon}} < \gamma < \frac{1}{n^{\varepsilon}}$. Then, for $m = 0, \dots, N$ where $N = O(1)$, the following statements hold:*

1. *$A\varphi_{m,n} \approx \lambda_m \varphi_{m,n}$, where $\lambda_m = 2\beta + 2e^{-\gamma} - 4m\gamma e^{-\gamma}$;*
2. *For each m, there exist constants C_1, C_2, independent of n such that*

$$|\varphi_{m,n}(i)| \leq C_1 e^{-C_2|i|};$$

3. *For each $0 \leq m \leq n-1$, the entries of $\varphi_{m,n}$ change signs exactly m times.*

The second property is an obvious consequence of the definition of $\varphi_{m,n}$, while the first and third are proved in Theorem 2.2 and Lemma 3, respectively. That the theorem applies to both $P^{(n)}_{\alpha,\beta,\theta}$ and $H^{(n)}_{\alpha,\beta,\theta}$ is a consequence of Corollary 1. In particular, the (truncated) Gaussian function is an approximate eigenvector associated with the largest eigenvalue of $P^{(n)}_{\alpha,\beta,\theta}$ (this was also proven by Persi Diaconis and brought to the attention of the authors via personal communication).

In fact, via the following symmetry property, Theorem 2.1 also applies to the m smallest eigenvalues and their associated eigenvectors of $P^{(n)}_{\alpha,\beta}$ (and with an additional argument, see the proof of Theorem 2.2 , also to $P^{(n)}_{\alpha,\beta,\theta}$).

Proposition 2 *If φ is an eigenvector of $H^{(n)}_{\alpha,\beta}$ with eigenvalue λ, then $M_{1/2}T_{1/2\alpha}\varphi$ is an eigenvector of $H^{(n)}_{\alpha,\beta}$ with eigenvalue $-\lambda$.*

Proof It is convenient to express $H_{\alpha,\beta}$ as

$$H_{\alpha,\beta} = T_1 + T_{-1} + \beta M_\alpha + \beta M_{-\alpha}.$$

Note that we may also consider this as an operator on $L^2(\mathbb{R})$, which we will refer to as $h_{\alpha,\beta}$. That is,

$$(H_{\alpha,\beta}f)(n) = \left[\left(h_{\alpha,\beta}f\right)(x)\right]\Big|_{\mathbb{Z}}.$$

Taking this view, we study the commutation relations between $h_{\alpha,\beta}$ and translation and modulation by considering

$$T_aM_bh_{\alpha,\beta}(T_aM_b)^* = T_aM_b(T_1 + T_{-1} + \beta M_\alpha + \beta M_{-\alpha})(T_aM_b)^*.$$

Using (1.2) we have

$$T_aM_bT_1M_{-b}T_{-a} = e^{2\pi ib}T_1, \qquad T_aM_bT_{-1}M_{-b}T_{-a} = e^{-2\pi ib}T_{-1},$$

and

$$T_aM_bM_\alpha M_{-b}T_{-a} = e^{-2\pi i\alpha a}M_\alpha, \qquad T_aM_bM_{-\alpha}M_{-b}T_{-a} = e^{2\pi i\alpha a}M_{-\alpha}.$$

Note that $e^{2\pi ib} = \pm 1$ if $b \in \frac{1}{2}\mathbb{Z}$ and $e^{-2\pi i\alpha a} = \pm 1$ if $a\alpha \in \frac{1}{2}\mathbb{Z}$. In particular, if $b = \frac{1}{2}$ and $a = \frac{1}{2\alpha}$ we obtain

$$T_{\frac{1}{2\alpha}}M_{\frac{1}{2}}h_{\alpha,\beta}(T_{\frac{1}{2\alpha}}M_{\frac{1}{2}})^* = -h_{\alpha,\beta}.$$

Then, it follows that if φ is an eigenvector of $h_{\alpha,\beta}$ with eigenvalue λ, then $T_{\frac{1}{2\alpha}}M_{\frac{1}{2}}\varphi$ is an eigenvector of $h_{\alpha,\beta}$ with eigenvalue $-\lambda$.

Now, note that if φ is an eigenvector of $h_{\alpha,\beta}$ with eigenvalue λ, then $\varphi\big|_{\mathbb{Z}}$ is an eigenvector of $H_{\alpha,\beta}$ with the same eigenvalue. The result follows. □

In the following lemma we establish an identity about the coefficients $c_{m,l}$ in (1.3) and (1.4), and the binomial coefficients $b_{2l,k}$, which we will need later in the proof of Theorem 2.2.

Lemma 2 *For $m = 0, \ldots, n-1$; $l = 0, \ldots, \frac{m}{2}$, there holds*

$$2c_{m,l}b_{2l,2} - 4c_{m,l-1}b_{2l-2,1} = -4mc_{m,l-1}, \tag{2.1}$$

where $b_{j,k} = \binom{j}{k} = \frac{j!}{k!(j-k)!}$.

Proof To verify the claim we first note that $b_{2l,2} = \frac{2l(2l-1)}{2}$ and $b_{2l-2,1} = 2l - 2$. Next, note that (2.1) is equivalent to

$$c_{m,l} = 4\frac{2l-2-m}{2l(2l-1)}c_{m,l-1}.$$

Now, for even m we calculate

$$c_{m,l} = m!\frac{(2\sqrt{2})^{2l}(-1)^{\frac{m}{2}-l}}{(2l)!(\frac{m}{2}-l)!} = \frac{-8\left(\frac{m}{2}-l+1\right)m!(2\sqrt{2})^{2l-2}(-1)^{\frac{m}{2}-l+1}}{2l(2l-1)(2l-2)!\left(\frac{m}{2}-l+1\right)!}$$

$$= \frac{-8\left(\frac{m}{2}-l+1\right)}{2l(2l-1)}c_{m,l-1} = 4\frac{2l-2-m}{2l(2l-1)}c_{m,l-1},$$

as desired. The calculation is almost identical for odd m, and is left to the reader. □

While the theorem below is stated for general α, β, θ (with some mild conditions on α, β), it is most instructive to first consider the statement for $\alpha = \frac{1}{n}, \beta = 1, \theta = 0$. In this case the parameter γ appearing below will take the value $\gamma = \frac{\pi}{n}$. The theorem then states that at the right edge of the spectrum of $P^{(n)}_{\alpha,\beta,\theta}$, $\varphi_{m,n}$ is an approximate eigenvector of $P^{(n)}_{\alpha,\beta,\theta}$ with approximate eigenvalue $\lambda_m = 2\beta + 2e^{-\gamma} - 4m\gamma e^{-\gamma}$, and a similar result holds for the left edge. The error is of the order $\frac{1}{n^2}$ and the edge of the spectrum is of size $O(1)$. If we allow the approximation error to increase to be of order $\frac{1}{n}$, then the size of the edge of the spectrum will increase to $O(\sqrt{n})$.

Theorem 2.2 *Let $P^{(n)}_{\alpha,\beta,\theta}$ be defined as in* (1.5) *and let $\alpha, \beta \in \mathbb{R}+$ and $\theta \in [0, 2\pi)$. Set $\gamma = \pi\alpha\sqrt{\beta}$ and assume that $\frac{4}{n^2} \le \gamma < 1$.*

(1) For $m = 0, 1, \ldots, N$, where $N = O(1)$, there holds for all $x = -\frac{n}{2}, \ldots, \frac{n}{2} - 1$

$$\left|(P^{(n)}_{\alpha,\beta,\theta}\varphi_{m,n})(x) - \lambda_m\varphi_{m,n}\left(x + \frac{\theta}{2\pi\alpha}\right)\right| \le O(\gamma^2), \tag{2.2}$$

where

$$\lambda_m = 2\beta + 2e^{-\gamma} - 4m\gamma e^{-\gamma}. \tag{2.3}$$

(2) For $m = -n+1, -n+2, \ldots, -N$, where $N = O(1)$, there holds for all $x = -\frac{n}{2}, \ldots, \frac{n}{2} - 1$

$$\left|(P^{(n)}_{\alpha,\beta,\theta}\varphi_{m,n})(x) - \lambda_m(-1)^x\varphi_{m,n}\left(x + \frac{\theta}{2\pi\alpha} - \frac{1}{2\alpha}\right)\right| \le O(\gamma^2), \tag{2.4}$$

where

$$\lambda_m = -(2\beta + 2e^{-\gamma} - 4m\gamma e^{-\gamma}). \tag{2.5}$$

Proof We prove the result for even m; the proof for odd m is similar and left to the reader.

We first assume that $\theta = 0$. We compute for $x = -\frac{n}{2}, \ldots, \frac{n}{2} - 1$ (recall that $\varphi_{m,n}(\frac{n}{2} - 1) = \varphi_{m,n}(-\frac{n}{2})$ due to our assumption of periodic boundary conditions)

$$
\begin{aligned}
(P^{(n)}_{\alpha,\beta}\varphi_{m,n})(x) &= \varphi_{m,n}(x+1) + \varphi_{m,n}(x-1) + 2\beta\cos(2\pi\alpha x)\big)\varphi_{m,n}(x) \\
&= e^{-\gamma x^2} e^{-\gamma}\Big(e^{-2\gamma x} \cdot \sum_l c_{m,l}\gamma^l (x+1)^{2l} + e^{2\gamma x} \cdot \sum_l c_{m,l}\gamma^l (x-1)^{2l}\Big) \\
&\quad + e^{-\gamma x^2} 2\beta \cos(2\pi\alpha x) \sum_l c_{m,l}\gamma^l x^{2l}.
\end{aligned}
$$

We expand each of the terms $(x \pm 1)^{2l}$ into its binomial series, i.e.,

$$
(x+1)^{2l} = \sum_{k=0}^{2l} b_{2l,k} x^{2l-k} 1^k \text{ and } (x-1)^{2l} = \sum_{k=0}^{2l} b_{2l,k} x^{2l-k} (-1)^k, \text{ where } b_{2l,k} = \binom{2l}{k},
$$

and obtain after some simple calculations

$$
\begin{aligned}
(P^{(n)}_{\alpha,\beta}\varphi_{m,n})(x) &= e^{-\gamma x^2}\Big[\sum_{l=0}^{m/2} c_{m,l}\gamma^l x^{2l}\Big(e^{-\gamma} b_{2l,0}(e^{-2\gamma x} + e^{2\gamma x}) + 2\beta\cos(2\pi\alpha x)\Big)\Big] \\
&\quad + e^{-\gamma x^2}\Big[\sum_{l=0}^{m/2} c_{m,l}\gamma^l \sum_{k=1}^{2l} b_{2l,k} x^{2l-k} e^{-\gamma}(e^{-2\gamma x} + (-1)^k e^{2\gamma x})\Big] \\
&= \text{(I)} + \text{(II)}.
\end{aligned}
\tag{2.6}
$$

We will now show that $\text{(I)} = \varphi_{m,n}(x) \cdot (2\beta + 2e^{-\gamma}) + O(\gamma^2)$ and $\text{(II)} = -\varphi_{m,n}(x) \cdot 4m\gamma e^{-\gamma} + O(\gamma^2)$, from which (2.2) and (2.3) will follow.

We first consider the term (I). We rewrite (I) as

$$
\text{(I)} = e^{-\gamma x^2}\Big(e^{-\gamma}(e^{-2\gamma x} + e^{2\gamma x}) + 2\beta\cos(2\pi\alpha x)\Big) \sum_{l=0}^{m/2} c_{m,l}\gamma^l x^{2l}.
$$

Using Taylor approximations for $e^{-2\gamma x}$, $e^{2\gamma x}$, and $\cos(2\pi\alpha x)$, respectively, we obtain after some rearrangements (which are justified due to the absolute summability of each of the involved infinite series)

$$
\begin{aligned}
&e^{-\gamma x^2}\Big(e^{-\gamma}(e^{-2\gamma x} + e^{2\gamma x}) + 2\beta\cos(2\pi\alpha x)\Big) = \\
&= e^{-\gamma x^2} e^{-\gamma}\Big[\Big(1 + (-2\gamma x) + \frac{(-2\gamma x)^2}{2!} + \frac{(-2\gamma x)^3}{3!} + R_1(x)\Big)
\end{aligned}
$$

$$+\left(1+(2\gamma x)+\frac{(2\gamma x)^2}{2!}+\frac{(2\gamma x)^3}{3!}+R_2(x)\right)\Big]$$

$$+e^{-\gamma x^2}2\beta\left(1-\frac{(2\pi\alpha x)^2}{2!}+R_3(x)\right)$$

$$=e^{-\gamma x^2}\Big(2e^{-\gamma}+2e^{-\gamma}\frac{(2\gamma x)^2}{2!}+2\beta-2\beta\frac{(2\pi\alpha x)^2}{2!}+e^{-\gamma}\big(R_1(x)+R_2(x)\big)+2\beta R_3(x)\Big), \tag{2.7}$$

where R_1, R_2, and R_3 are the remainder terms of the Taylor expansion for $e^{-2\gamma x}$, $e^{2\gamma x}$, and $\cos(2\pi\alpha x)$ (in this order), respectively, given by

$$R_1(x)=\frac{(-2\gamma)^4 e^{-2\gamma\xi_1}}{4!}x^4,\;\; R_2(x)=\frac{(2\gamma)^4 e^{2\gamma\xi_2}}{4!}x^4,\;\; R_3(x)=\frac{(2\pi\alpha)^4\cos(\xi_3)}{4!}x^4,$$

with real numbers ξ_1, ξ_2, ξ_3 between 0 and x. We use a second-order Taylor approximation for $e^{-\gamma}$ with corresponding remainder term $R_4(\gamma) = \frac{e^{-\xi_4}}{3!}\gamma^3$ (for some $\xi_4 \in (0, \gamma)$) in (2.7).

Hence (2.7) becomes

$$e^{-\gamma x^2}\left(2e^{-\gamma}+(2\gamma x)^2+\left(-\gamma+\frac{\gamma^2}{2!}+R_4(\gamma)\right)(2\gamma x)^2+2\beta-\beta(2\pi\alpha x)^2\right)+ \tag{2.8}$$

$$+e^{-\gamma x^2}\left(\left(1-\gamma+\frac{\gamma^2}{2!}+R_4(\gamma)\right)\big(R_1(x)+R_2(x)\big)+2\beta R_3(x)\right). \tag{2.9}$$

Since $\gamma = \pi\alpha\sqrt{\beta}$, the terms $(2\gamma x)^2$ and $-\beta(2\pi\alpha x)^2$ in (2.8) cancel. Clearly, $|R_1(x)| \le \frac{(2\gamma x)^4}{4!}$, $|R_2(x)| \le \frac{e^{2\gamma x}(2\gamma x)^4}{4!}$, and $|R_4(\gamma)| \le \frac{(\gamma)^3}{3!}$. Note that for R_1 and R_2, the preceding calculations only hold for non-negative x, but the same bounds can be similarly derived for negative x. It is convenient to substitute $\alpha = \frac{\gamma}{\pi\sqrt{\beta}}$ in $R_3(x)$, in which case we get $|R_3(x)| \le \frac{(2\gamma x)^4}{\beta^2 4!}$. Thus, we can bound the expression in (2.9) from above by

$$\left|e^{-\gamma x^2}\Big(\big(-\gamma+\frac{\gamma^2}{2!}+R_4(\gamma)\big)\big(R_1(x)+R_2(x)\big)+2\beta R_3(x)\Big)\right|$$

$$\le e^{-\gamma x^2}\Big(\big(\gamma+\frac{\gamma^2}{2!}+\frac{(\gamma)^3}{3!}\big)\big(\frac{(2\gamma x)^4}{4!}+\frac{e^{2\gamma x}(2\gamma x)^4}{4!}+2\beta\frac{(2\gamma x)^4}{\beta^2 4!}\big)\Big). \tag{2.10}$$

Assume now that $|x| \le \frac{1}{\sqrt{\gamma}}$ then we can further bound the expression in (2.10) from above by

$$e^{-\gamma x^2}\Big((\gamma+\frac{\gamma^2}{2!}+\frac{\gamma^3}{3!})(\frac{(2\gamma x)^4}{4!}+\frac{e^{2\gamma x}(2\gamma x)^4}{4!}+2\beta\frac{(2\gamma x)^4}{\beta^2 4!})\Big)$$
$$\le\Big((\gamma+\frac{\gamma^2}{2!}+\frac{\gamma^3}{3!})(\frac{2\gamma^2}{3}+\frac{2e^{2\sqrt{\gamma}}\gamma^2}{3}+\frac{4\gamma^2}{3\beta})\Big)\le O(\gamma^3)+O(\gamma^3/\beta).$$

Moreover, if $|x|\le\frac{1}{\sqrt{\gamma}}$, we can bound the term $\big(-\gamma+\frac{\gamma^2}{2!}+R_4(\gamma)\big)\frac{(2\gamma x)^2}{2!}$ in (2.8) by

$$\Big|\big(-\gamma+\frac{\gamma^2}{2!}+R_4(\gamma)\big)\frac{(2\gamma x)^2}{2!}\Big|\le 2\gamma^2+\gamma^3+\frac{\gamma^4}{3}\le O(\gamma^2).$$

Now suppose $|x|\ge\sqrt{\frac{1}{\gamma}}$. We set $|x|=\left(\sqrt{\frac{1}{\gamma}}\right)^c$ for some c with

$$1<c\le\frac{2\log(n/2)}{\log(1/\gamma)}. \tag{2.11}$$

The upper bound in (2.11) ensures that $|x|\le\frac{n}{2}$ and the assumption $\frac{4}{n^2}\le\gamma<1$ implies that condition (2.11) is not empty. Then, in (2.8), we use $R_4(\gamma)\le\frac{\gamma^3}{3!}$ and $|x|=\gamma^{-c/2}$ to compute

$$\left(-\gamma+\frac{\gamma^2}{2!}+R_4(\gamma)\right)\frac{(2\gamma x)^2}{2!}\le 2\gamma^{3-c}+O(\gamma^2),$$

so that

$$e^{-\gamma x^2}\left(-\gamma+\frac{\gamma^2}{2!}+R_4(\gamma)\right)\frac{(2\gamma x)^2}{2!}\le O(\gamma^2). \tag{2.12}$$

Moreover

$$e^{-\gamma x^2}\Big((\gamma+\frac{\gamma^2}{2!}+\frac{(\gamma)^3}{3!})(\frac{(2\gamma x)^4}{4!}+\frac{e^{2\gamma x}(2\gamma x)^4}{4!}+2\beta\frac{(2\gamma x)^4}{\beta^2 4!})\Big)= \tag{2.13}$$
$$=e^{-\gamma^{1-c}}\Big((\gamma+\frac{\gamma^2}{2!}+\frac{(\gamma)^3}{3!})(\frac{2^4\gamma^{4-2c}}{4!}+\frac{e^{2\gamma^{1-c}}2^4\gamma^{4-2c}}{4!}+\frac{4\gamma^{4-2c}}{3\beta})\Big)\le O(\gamma^2), \tag{2.14}$$

where (2.12) and (2.14) follow from basic inequalities like $e^{-\gamma^{1-c}}\gamma^{4-2c}\le\gamma^2$ (which in turn follows from $e^y\ge y^2$ for all $y\ge 0$).

Thus we have shown that

$$e^{-\gamma x^2}\Big(e^{-\gamma}(e^{-2\gamma x}+e^{2\gamma x})+2\beta\cos(2\pi\alpha x)\Big)=e^{-\gamma x^2}(2e^{-\gamma}+2\beta)+O(\gamma^2). \tag{2.15}$$

Returning to the term (I) in (2.6), we obtain, using (2.15),

$$\text{(I)} = \varphi_{m,n}(x)(2e^{-\gamma} + 2\beta) + C_\gamma \sum_{l=0}^{m/2} c_{m,l}\gamma^l x^{2l}, \tag{2.16}$$

where $C_\gamma = O(\gamma^2)$.

Let us analyze the error term $C_\gamma \sum_{l=0}^{m/2} c_{m,l}\gamma^l x^{2l}$. Using Stirling's Formula [1, Page 257] we note that $c_{m,l}$ grows at least as fast as $\left(m/e\right)^{m/2}$, but not faster than $(m/e)^m$. Hence, the error term will remain of size $O(\gamma^2)$, as long as we ensure that m does not exceed $O(1)$.

We now proceed to showing that (II) $= -\varphi_{m,n}(x) \cdot \left(4m\gamma e^{-\gamma}\right) + O(\gamma^2)$. The key to this part of the proof is the observation that $P^{(n)}_{\alpha,\beta}$ acts "locally" on the powers x^{2l} that appear in the definition of $\varphi_{m,n}$. Recall that the term (II) has the form

$$e^{-\gamma x^2}\Big[\sum_{l=0}^{m/2} c_{m,l}\gamma^l \sum_{k=1}^{2l} b_{2l,k}x^{2l-k}e^{-\gamma}(e^{-2\gamma x} + (-1)^k e^{2\gamma x})\Big]. \tag{2.17}$$

Analogous to the calculations leading up to (2.15) we can show that for odd k there holds

$$e^{-\gamma x^2}e^{-\gamma}(e^{-2\gamma x} + (-1)^k e^{2\gamma x}) = (-4\gamma x)e^{-\gamma x^2}e^{-\gamma} + O(\gamma^2), \tag{2.18}$$

and for even k

$$e^{-\gamma x^2}e^{-\gamma}(e^{-2\gamma x} + (-1)^k e^{2\gamma x}) = 2e^{-\gamma x^2}e^{-\gamma} + O(\gamma^2). \tag{2.19}$$

Using (2.18) and (2.19) we can express (2.17) as

$$\left(e^{-\gamma x^2}e^{-\gamma} + O(\gamma^2)\right)\Big[\sum_{l=0}^{m/2} c_{m,l}\gamma^l\Big(\sum_{\text{odd } k}(-4\gamma x)b_{2l,k}x^{2l-k} + \sum_{\text{even } k} 2b_{2l,k}x^{2l-k}\Big)\Big]. \tag{2.20}$$

Furthermore, using estimates similar to the ones used in deriving the bounds for (I), one easily verifies that

$$e^{-\gamma x^2}e^{-\gamma}\gamma^l(-4\gamma x)x^{2l-k} \le O(\gamma^2), \qquad \text{for } k \ge 3,$$

and

$$e^{-\gamma x^2}e^{-\gamma}c_{m,l}\gamma^l x^{2l-2} \le O(\gamma^2), \qquad \text{for } k \ge 4.$$

Therefore,

$$e^{-\gamma x^2}e^{-\gamma}\Big[\sum_{l=0}^{m/2} c_{m,l}\gamma^l\big(\sum_{\text{odd } k}(-4\gamma x)b_{2l,k}x^{2l-k}+\sum_{\text{even } k}2b_{2l,k}x^{2l-k}\big)\Big]$$

$$=\big(e^{-\gamma x^2}e^{-\gamma}+O(\gamma^2)\big)\Big[\sum_{l=0}^{m/2} c_{m,l}\gamma^l\big((-4\gamma x)b_{2l,1}x^{2l-1}+2b_{2l,2}x^{2l-2}\big)\Big].$$

The expression above implies that $P^{(n)}_{\alpha,\beta}$ acts "locally" on the powers x^{2l} of $\varphi_{m,n}$.

Moreover,

$$e^{-\gamma x^2}e^{-\gamma}\Big[\sum_{l=0}^{m/2} c_{m,l}\gamma^l\big((-4\gamma x)b_{2l,1}x^{2l-1}+2b_{2l,2}x^{2l-2}\big)\Big]$$

$$=e^{-\gamma x^2}e^{-\gamma}\Big[(-4\gamma m)c_{m,\frac{m}{2}}\gamma^{m/2}x^m+2c_{m,\frac{m}{2}}\gamma^{m/2}b_{m,2}x^{m-2}$$

$$+\sum_{l=0}^{m/2-1}(-4\gamma)c_{m,l}\gamma^l b_{2l,1}x^{2l}+2c_{m,l}\gamma^l b_{2l,2}x^{2l-2}\big)\Big],$$

and

$$(-4m\gamma)c_{m,\frac{m}{2}}\gamma^{m/2}x^m+2c_{m,\frac{m}{2}}\gamma^{m/2}b_{m,2}x^{m-2}$$

$$+\sum_{l=0}^{m/2-1}(-4\gamma)c_{m,l}\gamma^l b_{2l,1}x^{2l}+2c_{m,l}\gamma^l b_{2l,2}x^{2l-2}\big)$$

$$=(-4m\gamma)c_{m,\frac{m}{2}}\gamma^{m/2}x^m+\sum_{l=0}^{m/2}\big(2c_{m,l}\gamma^l b_{2l,2}x^{2l-2}-4\gamma c_{m,l-1}\gamma^{l-1}b_{2l-2,1}x^{2l-2}\big)$$

$$=(-4m\gamma)c_{m,\frac{m}{2}}\gamma^{m/2}x^m+\gamma\sum_{l=0}^{m/2}\gamma^{l-1}x^{2l-2}\big(2c_{m,l}b_{2l,2}-4c_{m,l-1}b_{2l-2,1}\big)$$

$$=(-4m\gamma)c_{m,\frac{m}{2}}\gamma^{m/2}x^m-4m\gamma\sum_{l=0}^{m/2}c_{m,l-1}\gamma^{l-1}x^{2l-2} \tag{2.21}$$

$$=4m\gamma\sum_{l=0}^{m/2}c_{m,l}\gamma^l x^{2l},$$

where we have used Lemma 2 in (2.21). Hence, up to an error of size $O(\gamma^2)$ we have

$$e^{-\gamma x^2}e^{-\gamma}\Big[\sum_{l=0}^{m/2} c_{m,l}\gamma^l\Big(\sum_{\text{odd } k}(-4\gamma x)b_{2l,k}x^{2l-k}+\sum_{\text{even } k}2b_{2l,k}x^{2l-k}\Big)\Big]$$
$$= -4m\gamma e^{-\gamma}e^{-\gamma x^2}\sum_{l=0}^{m/2} c_{m,l}\gamma^l x^{2l}. \tag{2.22}$$

Analogous to the estimate of (I), we need to choose m to be not larger than $O(1)$, in order to keep the error term

$$O(\gamma^2)\Big[\sum_{l=0}^{m/2} c_{m,l}\gamma^l\big((-4\gamma x)b_{2l,1}x^{2l-1}+2b_{2l,2}x^{2l-2}\big)\Big] \tag{2.23}$$

of size $O(\gamma^2)$.

Therefore, for such an m, by invoking (1.3), (2.20), and (2.22) we can express the term (II) in (2.17) as

$$\text{(II)} \;=\; -4m\gamma e^{-\gamma}\varphi_{m,n}(x)+O(\gamma^2).$$

Hence we have shown that

$$(P^{(n)}_{\alpha,\beta}\varphi_{m,n})(x) = (2\beta+2e^{-\gamma}-4m\gamma e^{-\gamma})\varphi_{m,n}(x)+O(\gamma^2), \tag{2.24}$$

which establishes claims (2.2) and (2.3) for $\theta = 0$.

Now we consider the case $\theta \neq 0$. We define $y = -\frac{\theta}{2\pi\alpha}$ so that $2\pi\alpha y = -\theta$ and compute

$$\begin{aligned}
P^{(n)}_{\alpha,\beta,\theta}T_y\varphi_{m,n}(x) &= (T_1T_y+T_{-1}T_y)\varphi_{m,n}(x)+\beta\Big((e^{i\theta}M_\alpha+e^{-i\theta}M_{-\alpha}T_y)\varphi_{m,n}(x)\Big)\\
&= T_yT_1\varphi_{m,n}(x)+T_yT_{-1}\varphi_{m,n}(x)\\
&\quad+\beta\Big(e^{i\theta}e^{2\pi i\alpha y}T_yM_\alpha\varphi_{m,n}(x)+e^{-i\theta}e^{-2\pi i\alpha y}T_yM_{-\alpha}\varphi_{m,n}(x)\Big)\\
&= T_yT_1\varphi_{m,n}(x)+T_yT_{-1}\varphi_{m,n}(x)+\beta\left(T_yM_\alpha\varphi_{m,n}(x)+T_yM_{-\alpha}\varphi_{m,n}(x)\right)\\
&= T_y\left[P^{(n)}_{\alpha,\beta,\theta}\varphi_{m,n}(x)\right] = (2\beta+2e^{-\gamma}-4m\gamma e^{-\gamma})T_y\varphi_{m,n}(x)+O(\gamma^2),
\end{aligned}$$

(where we have used (2.24) in the last step), which establishes (2.2) and (2.3) for arbitrary θ.

Claims (2.4) and (2.5) of Theorem 2.2 follow now from Proposition 2 and Corollary 1 below. □

Remark 1 If we allow the error bound in (2.2) and (2.4) to increase from $O(\gamma^2)$ to $O(\gamma)$, then the range for which the $\varphi_{m,n}$ serve as approximate eigenfunctions for

$P^{(n)}_{\alpha,\beta,\theta}$ extends from $m \sim O(1)$ to $m \sim O(\sqrt{1/\gamma})$. This can be seen by a suitable modification of the proof of Theorem 2.2 in the derivations in (2.13), (2.14) and in the error term (2.16), as well as in the calculations leading up to (2.22) and in the error term (2.23). (We leave the details to the reader.) See also the numerical simulations depicted in Fig. 2, which illustrate this fact.

Corollary 1 *Theorem 2.2 also holds if we replace* $P^{(n)}_{\alpha,\beta,\theta}$ *with* $H^{(n)}_{\alpha,\beta,\theta}$.

Proof A simple application of the triangle inequality yields

$$\left\| H^{(n)}_{\alpha,\beta,\theta}\varphi_{m,n} - \lambda_m\varphi_{m,n}\right\|_2 \leq \left\| H^{(n)}_{\alpha,\beta,\theta}\varphi_{m,n} - P^{(n)}_{\alpha,\beta,\theta}\varphi_{m,n}\right\|_2 + \left\| P^{(n)}_{\alpha,\beta,\theta}\varphi_{m,n} - \lambda_m\varphi_{m,n}\right\|_2.$$

We compute

$$\left\| H^{(n)}_{\alpha,\beta,\theta}\varphi_{m,n} - P^{(n)}_{\alpha,\beta,\theta}\varphi_{m,n}\right\|_2^2 = \left|\varphi_{m,n}\left(\frac{n}{2}-1\right)\right|^2 + \left|\varphi_{m,n}\left(-\frac{n}{2}\right)\right|^2$$

$$= \left| e^{-\gamma(\frac{n}{2}-1)^2}\sum_{l=0}^{m/2}\gamma^l c_{m,l}\left(\frac{n}{2}-1\right)^{2l}\right|^2 + \left| e^{-\gamma(-\frac{n}{2})^2}\sum_{l=0}^{m/2}\gamma^l c_{m,l}\left(-\frac{n}{2}\right)^{2l}\right|^2.$$

Now, from (2.16) we know that the sums are of size $O(\gamma^2)$, so we conclude (in a large overestimate when n is large) that

$$\left\| H^{(n)}_{\alpha,\beta,\theta}\varphi_{m,n} - P^{(n)}_{\alpha,\beta,\theta}\varphi_{m,n}\right\|_2 = O(\gamma^2).$$

Meanwhile, from Theorem 2.2, we know that

$$\left\| P^{(n)}_{\alpha,\beta,\theta}\varphi_{m,n} - \lambda_m\varphi_{m,n}\right\|_2 = O(\gamma^2).$$

□

Lemma 3 *Let* $\varphi_{m,n}$ *be as before. Let* $0 < \varepsilon < 1$ *and assume* $\frac{4}{n^{2-\varepsilon}} \leq \gamma < \frac{1}{n^{\varepsilon}}$. *Then, as long as* $m < n^{\varepsilon} - 1$, *the entries of* $\varphi_{m,n}$ *change signs* m *times.*

Proof Recall that the Hermite polynomial of order m has m distinct real roots. From [22], we know that the roots of the mth Hermite polynomial lie in the interval $[-\sqrt{2m+2}, \sqrt{2m+2}]$. Now, in our definition of $\varphi_{m,n}$, we have applied the transformation $x \mapsto \sqrt{2\gamma}x$ to the Hermite polynomials. Thus, in our case, we see that the zeroes of $\varphi_{m,n}$ lie in the interval

$$\left[-\sqrt{\frac{m+1}{\gamma}}, \sqrt{\frac{m+1}{\gamma}}\right].$$

Now, the condition $\frac{4}{n^{2-\varepsilon}} \leq \gamma$ implies that the zeroes of $\varphi_{m,n}$ lie in the interval

$$\left[-\frac{n^{1-\frac{\varepsilon}{2}}}{2}\sqrt{m+1}, \frac{n^{1-\frac{\varepsilon}{2}}}{2}\sqrt{m+1}\right],$$

from which we see that so long as $m \leq n^{\varepsilon} - 1$, all of the zeroes of the mth Hermite polynomial lie in the interval

$$\left[-\frac{n}{2}, \frac{n}{2}\right].$$

Then, if the distance between consecutive zeroes is always larger than 1, the lemma will follow. Now, from [27] we know that the distance between consecutive roots of the mth Hermite polynomial is at least $\frac{\pi}{\sqrt{2m+1}}$. Applying our scaling, we see that the distance between consecutive zeroes of $\varphi_{m,n}$ is at least $\frac{\pi}{\sqrt{2\gamma(2m+1)}}$. The condition $\gamma < \frac{1}{n^{\varepsilon}}$ leads to

$$\frac{\pi}{\sqrt{2\gamma(2m+1)}} > \pi\sqrt{\frac{n^{\varepsilon}}{4m+2}},$$

and so we see that whenever $m < \frac{\pi^2 n^{\varepsilon}-2}{4}$, the minimum distance between consecutive zeroes of $\varphi_{m,n}$ will be greater than 1. Finally, note that $n^{\varepsilon} - 1 < \frac{\pi^2 n^{\varepsilon}-2}{4}$. $\square$

In particular, by choosing $\varepsilon = \frac{1}{2}$ in Lemma 3, and thus $\frac{4}{\sqrt{n^3}} \leq \gamma \leq \frac{1}{\sqrt{n}}$, we obtain that for $m < \sqrt{n} + 1$, $\varphi_{m,n}$ changes sign m times.

Remark 2 Let $\alpha = \frac{r}{n}$, where $r \in \mathbb{N}_+$. Then it is easy to see that each eigenvalue of $P^{(n)}_{\alpha,\beta}$ has multiplicity (at least) $\frac{n}{r}$ and if f is an approximate eigenvector of $P^{(n)}_{\alpha,\beta}$ with eigenvalue λ, then so is $T_{kn/r} f$, for $k = 1, \ldots, r-1$.

3 Eigenanalysis of the Infinite Almost Mathieu Operator $H_{\alpha,\beta,\theta}$

In the preceding sections we have seen that $\varphi_{m,n}$ are approximate eigenfunctions with associated eigenvalues λ_m for both $P^{(n)}_{\alpha,\beta,\theta}$ and $H^{(n)}_{\alpha,\beta,\theta}$. We now consider the infinite-dimensional operator $H_{\alpha,\beta,\theta}$ and its relationship with the non-periodic version of the Hermite functions, denoted by φ_m and defined in (1.3) and (1.4). Note that the claims in Proposition 1 no longer make sense for the infinite-dimensional operator $H_{\alpha,\beta,\theta}$. Nevertheless, we see that λ_m and φ_m still function as approximate eigenpairs for $H_{\alpha,\beta,\theta}$, and the statements of Theorem 2.1 still hold.

Theorem 3.1 *Let $H_{\alpha,\beta,\theta}, \varphi_m, \lambda_m$ be as before. Set $\gamma = \pi\alpha\sqrt{\beta}$ and assume $0 < \gamma < 1$. Then, for $m = 0, \dots, N$, where $N = O(1)$, the following statements hold:*

1. *$H_{\alpha,\beta,\theta}\varphi_m \approx \lambda_m\varphi_m$;*
2. *For each m, there exist constants $C_1, C_2 > 0$ such that*

$$|\varphi_m(i)| \leq C_1 e^{-C_2|i|};$$

3. *For all $0 \leq m < \frac{\pi^2}{4\gamma} - \frac{1}{2}$, the entries of φ_m change signs exactly m times.*

Proof The first statement is proved in Corollary 2 to Theorem 2.2 below. The second follows from the definition of the Hermite functions. The third statement is an easy consequence of Lemma 3. □

Corollary 2 *Let $H_{\alpha,\beta,\theta}$ be as defined in (1.1) and let $\alpha, \beta \in \mathbb{R}^+$ and $\theta \in [0, 2\pi)$. Set $\gamma = \pi\alpha\sqrt{\beta}$ and assume that $0 < \gamma < 1$.*

(1) For $m = 0, 1, \dots, N$, where $N = O(1)$, $T_{\frac{-\theta}{2\pi\alpha}}\phi_m$ is an approximate eigenvector for $H_{\alpha,\beta,\theta}$ with approximate eigenvalue $\lambda_m = 2\beta + 2e^{-\gamma} - 4m\gamma e^{-\gamma}$. Moreover, there holds for all $x \in \mathbb{Z}$

$$\left|(H_{\alpha,\beta,\theta}\varphi_m)(x) - \lambda_m T_{\frac{-\theta}{2\pi\alpha}}\phi_m(x)\right| \leq O(\gamma^2). \tag{3.1}$$

(2) For $m = -n + 1, -n + 2, \dots, -N$, where $N = O(1)$, $T_{\frac{-\theta}{2\pi\alpha}}M_{1/2}\phi_m$ is an approximate eigenvector and $\lambda_m = -(2\beta + 2e^{-\gamma} - 4m\gamma e^{-\gamma})$ is an approximate eigenvalue associated with the m-th largest negative eigenvalue of $H_{\alpha,\beta}$. Moreover, there holds for all $x \in \mathbb{Z}$

$$\left|(H_{\alpha,\beta,\theta}\varphi_m)(x) - \lambda_m T_{\frac{\pi-\theta}{2\pi\alpha}}M_{\frac{1}{2}}\phi_m(x)\right| \leq O(\gamma^2). \tag{3.2}$$

If we replace $O(\gamma^2)$ in (3.1) and (3.2) by $O(\gamma)$, then the range for m increases from $N = O(1)$ to $N = O(\sqrt{1/\gamma})$.

Proof As in the proof of Theorem 2.2, we prove the result for even m and $\theta = 0$. Indeed, the proof proceeds exactly as before, except that in this case we no longer need the assumption (2.11). Thus, (3.1) holds for $0 < \gamma < 1$ and for all $x \in \mathbb{Z}$. We apply the infinite-dimensional version of Proposition 2 to obtain (3.2). □

4 Numerical Simulations

In this section we illustrate our theoretical findings with numerical examples. We consider $P^{(n)}_{\alpha,\beta,\theta}$ with the parameters $n = 10000$, $\alpha = \frac{1}{n}$, $\beta = 1$, $\theta = 0$, hence in this case $\gamma = \frac{\pi}{n}$.

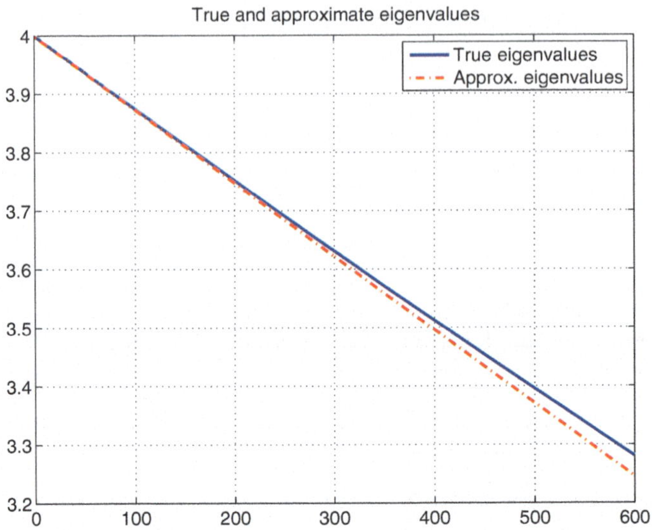

Fig. 1 True eigenvalues of $P^{(n)}_{\alpha,\beta,\theta}$ and their approximation via (2.2). The values for $P^{(n)}_{\alpha,\beta,\theta}$ are $n = 10000$, $\alpha = \frac{1}{n}$, $\beta = 1$, $\theta = 0$

Figure 1 shows the true eigenvalues of $P^{(n)}_{\alpha,\beta}$ and the approximate eigenvalues given by formula (2.3). Depicted are the 600 algebraically largest eigenvalues and their approximations.

Theorem 2.2 predicts that the approximation error will be of size γ^2 for the first few approximate eigenvalues, while about $\sqrt{n}$ many approximate eigenvalues deviate from the true eigenvalues by an error of less than about $1/n$. Figure 2 confirms this prediction. In Fig. 2 we show the number of approximate eigenvalues that approximate the true eigenvalues of $P^{(n)}_{\alpha,\beta,\theta}$ within an error of γ (red solid line) as well as within an error of γ^2 (green dash-dotted line). Also depicted is the graph of $1/\sqrt{\gamma}$ (blue dashed line). The picture confirms clearly our theoretical findings, which imply that the number of approximate eigenvalues that deviates from the true eigenvalues by an error less than γ^2 is of size $O(1)$, i.e., independent of n, while the number of approximate eigenvalues that deviates from the true eigenvalues by an error less than γ scales like $1/\sqrt{\gamma}$, thus in this example like $\sqrt{\frac{n}{\pi}}$.

Acknowledgments T.S. wants to thank Persi Diaconis for introducing him to the topic of this paper. This research was partially supported by the National Science Foundation via grant DTRA-DMS 1322393.

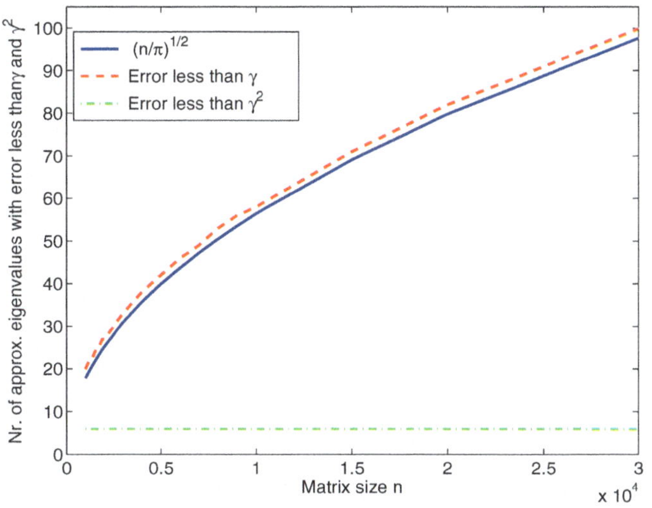

Fig. 2 Shown are the number of approximate eigenvalues that approximate the true eigenvalues of $P^{(n)}_{\alpha,\beta,\theta}$ within an error of γ (red solid line) as well as within an error of γ^2 (green dash-dotted line). Also depicted is the graph of $1/\sqrt{\gamma}$ (blue dashed line). The values for $P^{(n)}_{\alpha,\beta,\theta}$ are $n = 1000, \ldots, 30000$, $\alpha = \frac{1}{n}$, $\beta = 1$, $\theta = 0$, hence in this example $\gamma = \frac{\pi}{n}$

References

1. Abramowitz, M. and Stegun, I.A.: *Handbook of mathematical functions with formulas, graphs, and mathematical tables*, ninth Dover printing, tenth GPO printing ed., Dover, New York, 1964.
2. Albrecht, J.T., Chan, C.P., and Edelman, A.: *Sturm sequences and random eigenvalue distributions*, Foundations of Computational Mathematics **9** (2009), no. 4, 461–483 (English).
3. Arveson, W.: *C*-algebras and numerical linear algebra*, Journal of Functional Analysis **122** (1994), no. 2, 333–360.
4. Atakishiyev, N.M., Vicent, L.E., and Wolf, K.B.: *Continuous vs. discrete fractional Fourier transforms*, J. Comput. Appl. Math. **107** (1999), no. 1, 73–95.
5. Atakishiyeva, M.K. and Atakishiyev, N.M.: *On algebraic properties of the discrete raising and lowering operators, associated with the N-dimensional discrete Fourier transform*, Adv. Dyn. Syst. Appl. **11** (2016), no. 2, 81–92.
6. Avila, A. and Damanik, D.: *Absolute continuity of the integrated density of states for the almost Mathieu operator with non-critical coupling*, Invent. Math. **172** (2008), no. 2, 439–453.
7. Avila, A. and Jitomirskaya, S.: *Solving the ten martini problem*, Mathematical Physics of Quantum Mechanics, Springer, 2006, pp. 5–16.
8. Avila, A. and Jitomirskaya, S.Y.: *The Ten Martini Problem*, Ann. of Math. (2) **170** (2009), no. 1, 303–342.
9. Avron, J., v. Mouche, P.H.M., and Simon, B.: *On the measure of the spectrum for the almost Mathieu operator*, Communications in Mathematical Physics **132** (1990), no. 1, 103–118 (English).
10. Bèguin, C., Valette, A., and Zuk, A.: *On the spectrum of a random walk on the discrete Heisenberg group and the norm of Harper's operator*, Journal of Geometry and Physics **21** (1997), no. 4, 337–356.

11. Bellissard, J. and Simon, B.: *Cantor spectrum for the almost Mathieu equation*, Journal of Functional Analysis **48** (1982), no. 3, 408–419.
12. Benzi, M., Boito, P., and Razouk, N.: *Decay properties of spectral projectors with applications to electronic structure*, SIAM Review **55** (2013), no. 1, 3–64.
13. Boca, F.P.: *Rotation C^*-algebras and almost Mathieu operators*, The Theta Foundation, Bucharest, 2001.
14. Boca, F.P. and Zaharescua, A.: *Norm estimates of almost Mathieu operators*, Journal of Functional Analysis **220** (2008), 76–96.
15. Bump, D., Diaconis, Hicks, P.A., Miclo, L., and Widom, H.: *Useful bounds on the extreme eigenvalues and vectors of matrices for Harper's operators*, Preprint, arxiv (2015).
16. Campos, R.G.: *The XFT quadrature in discrete Fourier analysis*, Applied and Numerical Harmonic Analysis, Birkhäuser/Springer, Cham, 2019.
17. Diaconis, P. and Saloff-Coste, L.: *Moderate growth and random walk on finite groups*, Geom. Funct. Anal. **4** (1994), 1–36.
18. Gröchenig, K.: *Foundations of time-frequency analysis*, Birkhäuser, Boston, 2001.
19. Hofstadter, D.R.: *Energy levels and wave functions of Bloch electrons in rational and irrational magnetic fields*, Physical Review B **14** (1976), 2239–2249.
20. Jitomirskaya, S.: *Anderson localization for the almost mathieu equation: II. point spectrum for $\lambda > 2$*, Communications in mathematical physics **168** (1995), no. 3, 563–570.
21. Jitomirskaya, S. and Zhang. S.: *Quantitative continuity of singular continuous spectral measures and arithmetic criteria for quasiperiodic Schrödinger operators.* arXiv preprint arXiv:1510.07086. 2015
22. Krasikov, I.: *Nonnegative quadratic forms and bounds on orthogonal polynomials*, Journal of Approximation Theory **111** (2001), no. 1, 31–49.
23. Lamoureux, M.P. and Mingo, J.A.: *On the characteristic polynomial of the almost Mathieu operator*, Proc. Amer. Math. Soc. **135** (2007), no. 10, 3205–3215.
24. Last, Y.: *Spectral theory of Sturm-Liouville operators on infinite intervals: a review of recent developments*, Sturm-Liouville Theory, Springer, 2005, pp. 99–120.
25. de Oliveira Neto, J.R. and Lima, J.B.: *Discrete fractional Fourier transforms based on closed-form Hermite-Gaussian-like DFT eigenvectors*, IEEE Trans. Signal Process. **65** (2017), no. 23, 6171–6184.
26. Shubin, M.A.: *Discrete Magnetic Laplacian*, Communications in Mathematical Physics **164** (1994), no. 2, 259–275 (English).
27. Szegö, G.: *Orthogonal polynomials*, American Mathematical Society colloquium publications, no. v. 23, American Mathematical Society, 1967.
28. Wang, Y.Y., Pannetier, B., and Rammal, R.: *Quasiclassical approximations for almost-Mathieu equations*, J. Physique **48** (1987), 2067–2079.
29. Zhang, Y., Bulmash, D., Maharaj, A.V., Jian, C., and Kivelson, S.A.: *The almost mobility edge in the almost Mathieu equation*, Preprint arXiv:1504.05205 (2015).

Spatio–Spectral Limiting on Redundant Cubes: A Case Study

Jeffrey A. Hogan and Joseph D. Lakey

Abstract The operator that first truncates to a neighborhood of the origin in the spatial domain then truncates to a neighborhood of zero in the spectral domain is investigated in the case of redundant cubes—Boolean cubes with added generators. This operator is self-adjoint on a space of spectrum-limited signals. Certain invariant subspaces of this iterated projection operator, in which eigenspaces lie, are studied for a specific example. These observations suggest a general structure of eigenspaces of spatio–spectral limiting on redundant cubes.

1 Overview

This work is part of a study of spatio–spectral limiting on finite graphs through specific structured graphs amenable to some level of combinatorial analysis. By *spatial limiting* we mean restriction to a vertex neighborhood in a connected graph, and by *spectral limiting* we mean, specifically, restriction to the span of eigenvectors of small eigenvalues of the graph Laplacian. Spatio–spectral limiting then refers to composition of a spatial-limiting operator and a spectral-limiting operator. These operators are analogues on graphs of time- and bandlimiting operators on $\mathbb{R}$ studied by Landau, Slepian, and Pollack beginning in the 1960s, e.g.,[8, 9, 18, 19] and [10]. The eigenfunctions of time and bandlimiting were identified in [19] and the (asymptotic) distribution of eigenvalues was quantified in [10]. Eigen-decompositions of spatio–spectral are related to general uncertainty principles on graphs, e.g., [1] and other aspects of *graph signal processing* [2–4, 11–13, 15–17, 21–23]. Our view is that nuanced—as opposed to general—results about

---

J. A. Hogan
School of Mathematical and Physical Sciences, University of Newcastle, Callaghan, NSW, Australia
e-mail: jeff.hogan@newcastle.edu.au

J. D. Lakey (✉)
New Mexico State University, Las Cruces, NM, USA
e-mail: jlakey@nmsu.edu

M. Hirn et al. (eds.), *Excursions in Harmonic Analysis, Volume 6*, Applied and Numerical Harmonic Analysis, https://doi.org/10.1007/978-3-030-69637-5_6

important operators in harmonic analysis defined on spaces of vertex functions on graphs will depend on specific structure that particular graphs possess. For example, basic spectral properties of Cartesian product graphs are inherited from the factors, e.g., [7]. Our recent work [6] takes advantage of the product structure of Boolean cubes, or simply *cubes* to establish their basic spatio–spectral limiting properties. The current work presents through a few explicit examples some preliminary results pertaining to *redundant cubes* which are special cases of *generalized cubes*, e.g., [20]. The latter are not typically products, but do possess structure: because $\mathbb{Z}_2^N$ is an abelian group, the spectral domain is isomorphic $\mathbb{Z}_2^N$ and inherits a graph structure from the spatial domain. This provides leverage for quantifying eigen-decompositions of spatio–spectral limiting.

The rest of the paper is outlined as follows. In Sect. 2 we review definitions of graph Fourier transforms (GFTs) appropriate for the present context, and define cubes and their Fourier transforms. In Sect. 3 we define redundant cubes as Cayley graphs, observe that they have the same GFTs as cubes but with different Laplacian eigenvalues, and define spatio–spectral limiting (SSL) operators on redundant cubes, emphasizing distinctions from the case of standard cubes. In Sect. 4 we summarize prior results of [6] that identified eigenspaces of spatio–spectral limiting on cubes and outline the present approach relative to the case of standard cubes. In Sect. 5 we proceed to analyze two concrete examples of redundant cubes, first a very simple example to illustrate differences from the case of standard cubes, and then a moderate-sized example to illustrate the core problem of identifying spaces that are defined in terms of Laplacian eigenvectors of given spectral order that are invariant under a suitable submatrix of the adjacency matrix (acting in the spectral domain). A complete set of eigenvectors of a corresponding SSL operator is described for this example and plotted according to an indexing of Laplacian eigenvectors that is also described. Proposition 2 serves as a partial explanation of what is observed in the main example, and a conjecture, whose proof would provide a more complete explanation, is formulated. Finally, Sect. 6 provides further details of the adjacency structure of the moderate-sized redundant cube studied here.

2 Cubes, Space Limiting and Spectrum Limiting

2.1 Graph Laplacian and Graph Fourier Transform

A finite unweighted, symmetric graph $\Gamma = (V, E)$ consists of a finite set V of vertices and a symmetric binary relation $E : V \times V \to \{0, 1\}$. For $v, w \in V$ we write $v \sim w$ if $E(v, w) = 1$ and say that v and w are adjacent. If V has cardinality $|V| = n$ and $V = \{v_1, \ldots, v_n\}$ is an ordering of the vertices, we define the $n \times n$ *adjacency matrix* A by $A_{ij} = E(v_i, v_j)$. The (unnormalized) graph Laplacian $\mathbf{L}$ is the map $\mathbf{L}(f)(v) = \sum_{u \sim v} f(u) - f(v)$ for a vertex function $f : V \to \mathbb{R}$. Given

$v \in V, d(v) = \sum_{u \in V} E(v, u)$ is the degree of the vertex v, i.e., the number of edges that terminates at v, or equivalently the number of vertices that is adjacent to v. The Laplacian is represented by the matrix $L = D - A$ where D is the diagonal matrix with diagonal entries $D(v, v) = d(v)$. The graph Fourier transform $\mathbf{F}$ is the unitary operator defined by the eigenvectors of $\mathbf{L}$. It can be represented by a matrix F whose columns are the eigenvectors of L, usually ordered by increasing eigenvalue of L.

2.2 Cubes and Their Fourier Transforms

The Cayley graph Γ of an abelian group G with symmetric generating set $S \subseteq G$ ($S = -S$) is the graph whose vertices are the elements $v \in G$ and whose edges correspond to pairs v and w in G such that $vw^{-1} \in S$. We also assume the identity of G is not in S so Γ is loopless. The Boolean cube $\mathcal{B}_N$ is the Cayley graph of the group of $\mathbb{Z}_2^N$ with symmetric generators e_i having entry 1 in the ith coordinate and zeros in the other $N - 1$ coordinates. With componentwise addition modulo two, vertices corresponding to elements of $\mathbb{Z}_2^N$ are adjacent precisely when their difference in $\mathbb{Z}_2^N$ is equal to e_i for some i. It is convenient to index vertices v of $\mathcal{B}_N$ by subsets $R_v \subseteq \{1, \ldots, N\}$ defined by $R_v = \{i : \varepsilon_i = 1\}$ when $v = (\varepsilon_1, \ldots, \varepsilon_N) \in \{0, 1\}^N$. Distance between vertices is defined by Hamming distance—the number of differing coordinates—which is equal to the path distance. The (unnormalized) graph Laplacian of $\mathcal{B}_N$ is represented by the matrix L, thought of as a function on $\mathcal{B}_N \times \mathcal{B}_N$, with $L_{vv} = N$ and $L_{vw} = -1$ if $v \sim w$, that is, v and w are nearest neighbors—they differ in a single coordinate—and $L_{vw} = 0$ otherwise. Thus $L = NI - A$ where A is the adjacency matrix $A_{vw} = 1$ if $v \sim w$ and $A_{vw} = 0$ otherwise. We denote by $|R_v|$ the number of elements of R_v—the number of unit coordinates in the vertex $v \in \mathcal{B}_N$. The eigenvalues of L are $0, 2, \ldots, 2N$ where $2K$ has multiplicity $\binom{N}{K}$. The corresponding eigenvectors, which together provide a matrix representation of the graph (and group) Fourier transform, can be defined by $H_v(w) = (-1)^{|R_v \cap R_w|}$. H_v is an eigenvector of L with eigenvalue $2|R_v|$. This is a special case of Lemma 1 below. The normalized vectors $\bar{H}_v = H_v/2^{N/2}$ together make the Fourier transform unitary.

We denote by H the matrix with columns H_v, $v \in \mathbb{Z}_2^N$ and $\bar{H} = 2^{-N/2}H$. Up to an indexing of the columns and a factor $2^{N/2}$, H is the (real symmetric) *Hadamard matrix* of order N obtained by taking the N-fold tensor product of the matrix $\begin{pmatrix} 1 & 1 \\ 1 & -1 \end{pmatrix}$. It should be mentioned that the Fourier (Hadamard) transform on $\mathcal{B}_N$ corresponds to the *group Fourier transform*, that is, the collection of linear homomorphisms or *characters* from $\mathbb{Z}_2^N$ to $\{-1, 1\}$ (for a general abelian group, the image is the unit circle). These characters also form a group isomorphic to $\mathbb{Z}_2^N$ (e.g., [14]) and therefore also can be equipped with the same graph structure as $\mathcal{B}_N$, a fact that we will use implicitly in what follows.

3 Redundant Cubes

3.1 Definition of Redundant Cubes and Laplacian Eigenvalues

By a redundant cube Γ_S we simply mean the Cayley graph on $\mathbb{Z}_2^N$ with generating set S that properly contains the generators of $\mathcal{B}_N$. If $|S| = M > N$, then Γ_S is said to be M-regular. It is a simple exercise to show that the Hadamard vectors H_v, $v \in \mathbb{Z}_2^N$, also form a complete set of eigenvectors of the Laplacian L_{Γ_S} of Γ_S.

Lemma 1 *The eigenvalue λ_v in $L_{\Gamma_S} H_v = \lambda_v H_v$ is*

$$\lambda_v = \sum_{s \in S} (1 - (-1)^{|R_s \cap R_v|}). \tag{3.1}$$

Proof $L = MI - A$ so it is equivalent to show that

$$(AH_v)(w) = \Big(\sum_{s \in S} (-1)^{|R_v \cap R_s|}\Big)(-1)^{|R_v \cap R_w|} = \sum_{s \in S} (-1)^{|R_v \cap R_s| + |R_v \cap R_w|}.$$

In general, $(Af)(w) = \sum_{v \sim w} f(v) = \sum_{s \in S} f(w - s)$ so $(AH_v)(w) = \sum_{s \in S} H_v(w - s) = \sum_{s \in S} (-1)^{|R_v \cap R_{w-s}|}$. We claim that the parity of $|R_v \cap R_{w-s}|$ is the same as that of $|R_v \cap R_w| + |R_v \cap R_s|$. To do so, it is enough to show that each $i \in \{1, \dots, N\}$ contributes the same parity to $|R_v \cap R_{w-s}|$ and $|R_v \cap R_w| + |R_v \cap R_s|$. Fix such i.

Case 1 $i \notin R_v$. Then i does not contribute to either $|R_v \cap R_{w-s}|$ or $|R_v \cap R_w| + |R_v \cap R_s|$, so its parity contribution is the same.

Case 2 $i \in R_v$. Then it suffices to show that the parity contributions of i to $|R_{w-s}|$ and $|R_w| + |R_s|$ are the same. If $i \in R_{w-s}$, then $i \in R_w \Delta R_s$ so contributes one also to $|R_w| + |R_s|$. Otherwise $i \notin R_{w-s}$. Then either $i \in R_w \cap R_s$ or $i \notin R_w \cup R_s$ so its contribution to $|R_w| + |R_s|$ is two or zero. In both cases it provides an even parity contribution to both. This establishes the parity claim, hence the lemma. □

As above, R_s is the *bit index set* of s: if $s = (\varepsilon_1, \dots, \varepsilon_N)$, then $R_s = \{i : \varepsilon_i = 1\}$. We define the *spectral order* of H_v to be half the Laplacian eigenvalue. It is the number of generators whose bit set has odd intersection with R_v. We refer to such generators as *boost generators* of v. Clearly $0 \le \lambda_v \le 2M$.

3.2 Neighborhood Limiting and Spectrum Limiting on Redundant Cubes

Neighborhoods of the origin (zero element) in Γ_S are defined in terms of edge path distance. Distance to the origin is determined by the number of factors $v = s_1 +$

$\cdots + s_\ell$ in a *minimal length* factorization of v into a sum of generators. We define the closed ball $B_K = B_K(\Gamma_S)$ of radius K to consist of all vertices of path distance at most K from the origin. The (spatio–spectral) *order* of $v \in \Gamma_S$ is the pair whose first entry is the spectral order and whose second is the distance to the origin (spatial order).

One denotes by $Q = Q_K$ the cutoff of a vertex function onto B_K, $(Qf)(v) = \mathbf{1}_{B_K} f(v)$. We define a spectrum cutoff similarly. Because Laplacian eigenvalues have multiplicity, we consider a spectral function g to be a function of eigenvectors. Since, for redundant cubes, the eigenvectors H_v are indexed again by vertices v, we can express g as a vertex function. A spectral cutoff Q_R is then defined by $(Q_R)(g)(v) = \mathbf{1}_{\lambda_v \leq 2R}\, g(v)$ with λ_v as in (3.1). That is, Q_R restricts to spectral order at most R. In the classical setting of the real line $\mathbb{R}$, it is typical to denote the Ω-bandlimiting operator $(P_\Omega f)(t) = (\widehat{f}(\xi)\mathbf{1}_{[-\Omega/2,\Omega/2]}(\xi))^\vee$ where $\widehat{f}$ denotes the Fourier transform of f on $\mathbb{R}$. By analogy one can define a spectrum-limiting operator $P = P_R$ by $P = \bar{H}^T Q_R \bar{H}$ (recall that H has real entries). Equivalently, P_R is the projection onto the span of those H_v with $\lambda_v \leq 2R$. By analogy with Paley–Wiener spaces on $\mathbb{R}$, we refer to the range of P_R as the *Paley–Wiener* space of order R on Γ_S and refer to its elements as being R-spectrum limited. Its dimension is evidently the number of v such that $\lambda_v \leq 2R$. For standard cubes, this number is $\sum_{k=0}^{R} \binom{N}{k}$. We will also consider the *dual* operator $\mathcal{P}_K = \bar{H}^T Q_K \bar{H}$, defined on spectral vertex functions.

As observed, the character group of $\mathbb{Z}_2^N$ is also (isomorphic to) $\mathbb{Z}_2^N$ and so inherits the graph structure of $\mathcal{B}_N$ as well. We will use the symbol $\mathcal{A}$ to denote the adjacency matrix on this *character* or *spectral graph*, where two characters, indexed by H_v, H_w are adjacent precisely when $v \sim w$ (in the spatial graph).

The fundamental question that we begin to address here for the case of redundant cubes Γ_S in general is: what are the eigenspaces of PQ on Γ_S? For standard cubes, we showed in [6] that these eigenspaces can be characterized in the spectral domain in terms of certain invariant subspaces of $\mathcal{A}$. Dependence of the distribution of eigenvalues of PQ on redundancy and on structure of the generators S will be studied elsewhere.

4 SSL Results on $\mathcal{B}_N$ and Outline of Current Approach

4.1 Summary of Prior Results

Our prior work [5, 6] characterized eigenspaces of PQP operators on $\mathcal{B}_N$ where P and Q are spectral-limiting and spatial-limiting operators on $\mathcal{B}_N$. In the classical case of time and bandlimiting on $\mathbb{R}$ developed initially at Bell Labs, e.g., [8, 9, 18, 19] and [10], a certain second-order self-adjoint *prolate* differential operator commutes with the time- and bandlimiting operator. The eigenvectors of the latter are then those of the former. In [5] it was shown that the natural matrix analogue on $\mathcal{B}_N$ of the prolate differential operator does not commute with PQP. The commutator matrix was expressed in terms of values of hypergeometric functions in [5]. While explicit norm bounds for the commutator were not derived there, numerical estimates indicated that the matrix 2-norm of the commutator is a small fraction of the corresponding norm of the prolate operator matrix analogue. This suggests that eigenvectors of the Boolean matrix analogue of the prolate differential operator, which turn out to be relatively simple to compute, can be used effectively as seed vectors in a numerical algorithm (essentially the power method) to compute eigenvectors of corresponding Boolean SSL operators. Eigenspaces of $Q_R\mathcal{P}_K Q_R$ were identified in the spectral domain in [6] first by identifying certain spaces invariant under the adjacency matrix. Denote by Σ_r the (spectral) r-Hamming sphere of those $v \in \mathbb{Z}_2^N$ having r coordinates equal to one so ($\lambda_v = 2r$). We express the (spectral) adjacency matrix $\mathcal{A}$ as $\mathcal{A} = \mathcal{A}_+ + \mathcal{A}_-$ where the *outer adjacency* matrix $\mathcal{A}_+$ maps vertex functions supported on Σ_r to vertex functions supported at adjacent vertices in Σ_{r+1} and the *inner adjacency* matrix $\mathcal{A}_-$ maps vertex functions supported on Σ_r to vertex functions supported on Σ_{r-1}. The invariant spaces have the form

$$\mathcal{V}_r = \left\{ \sum_{k=0}^{N-r} c_k \mathcal{A}_+^k g, \quad g \in \mathcal{W}_r \right\},$$

where $\mathcal{W}_r$ consists of those vertex functions in the kernel of $\mathcal{A}_-$ supported in Σ_r. The space $\mathcal{V}_r$ is then linearly isomorphic to a product $\mathcal{W}_r \times \mathbb{R}^{N+1-r}$. The invariance property follows from identities of the form $\mathcal{A}_- \mathcal{A}_+^{k+1} g = m(r,k)\mathcal{A}_+^k g$, $g \in \mathcal{W}_r$ where $m(r,k) = (k+1)(N-2r-k)$. This structure allows the action of $\mathcal{A}$ on $\mathcal{V}_r$ to factor across $\mathcal{W}_r$ and be expressed solely in terms of a size $N+1-r$ *factor matrix* on $\mathbb{R}^{N+1-r}$. One then shows that the operator $\mathcal{P}_K = \bar{H}^T Q_K \bar{H}$ can be expressed as a polynomial in $\mathcal{A}$ and, on $\mathcal{V}_r$, can likewise be expressed in terms of the same polynomial of the factor matrix of $\mathcal{A}$ acting on $\mathbb{R}^{N+1-r}$. This matrix is not self-adjoint on $\mathbb{R}^{N+1-r}$ with standard inner product, but is so when expressed in terms of a certain weighted inner product. Even so, numerical computation of the matrix is problematic because magnitudes of its entries prohibit its computation in floating point even for moderate N (say $N \geq 20$). Corresponding analysis for the Boolean analogue of the prolate differential operator leads to a tridiagonal matrix whose eigenvectors, as indicated above, are *nearly* eigenvectors of SSL and can be used to seed a weighted version of the power method, leading to accurate numerical approximations of the eigenvectors of $Q\mathcal{P}Q$. Eigenvalues can be estimated directly then by applying $Q\mathcal{P}Q$.

4.2 Summary of Present Approach

Fundamental to our results on standard cubes was identification of suitable spaces invariant under $\mathcal{A}$. The main question we begin to address here is whether a parallel approach can be used to identify and compute eigenspaces of spatio–spectral limiting (in the spectral domain) on redundant cubes. Specifically, we seek an understanding of eigenspaces of $\mathcal{QPQ}$ in terms of invariant subspaces of $\mathcal{A}$, from which a simpler characterization of the eigenspaces can be recovered. Our approach to begin to address this question here will be to analyze systematically a small scale example (an $N = 11$ cube with 2048 vertices, taking five redundant generators in addition to the 11 standard generators of $\mathbb{Z}_2^{11}$, resulting in a 16-regular redundant cube with 2^{14} edges). We choose limiting operators that restrict to vertices of *spectral order* at most four and path distance at most four to the origin. In our example there are just 79 such vertices. We identify a complete set of eigenvectors of $\mathcal{QPQ}$ numerically via singular value decomposition. Our goal is to describe these vectors in terms of adjacency-invariant subspaces. Such a description depends fundamentally on quantifying certain *characteristic sets* or cells that play the role of Hamming spheres in the case of standard cubes, namely sets Σ of vertices such that certain $\mathcal{A}$-invariant spaces can be expressed in the form $\sum_{k=0}^{K} c_k \mathcal{A}_+^k g$ where g is supported in Σ and in the kernel of an analogue of $\mathcal{A}_-$. Decomposition of spectral adjacency $\mathcal{A}$ into inner and outer components in our example is described further below.

In what follows we will use decimal indexing separately to reference vertices in cubes (in a compact way) and to index vertices according to an ordering that facilitates analysis of spatio–spectral limiting. Addition will always be in $\mathbb{Z}_2^N$. Thus, the vertices in $\mathcal{B}_6$ with binary coordinates $(\varepsilon_0, \dots, \varepsilon_5)$, respectively, equal to $(1, 1, 1, 0, 0, 0)$ and $(1, 0, 1, 0, 1, 0)$ have respective decimal representations $(\sum_{k=0}^{n-1} \varepsilon_k 2^k)$ equal to 7 and 21, but their sum $(0, 1, 0, 0, 1, 0)$ has decimal representation 18. We also partition the set S of generators by their spectral orders determined by (3.1), writing $s \in S_j$ if $LH_s = 2j$ where H_s is the Hadamard vector of the element $s \in \mathbb{Z}_2^N$. Finally, if $v \in \mathbb{Z}_2^N$ can be expressed as a sum $v = s_1 + \cdots + s_k$ of generators $s_i \in S$, then we will write $v \sim [n_1, \dots, n_k]$ where n_i is the decimal representation of s_i. For example, we will write $(1, 0, 1, 0, 1, 0 \dots)$ $\sim [1, 4, 16]$.

5 Two Concrete Examples

5.1 A Tiny Example: $\mathbb{Z}_2^4$ with Generators 1, 2, 4, 8, 3, 6, 12

We consider the very simple redundant cube on $\mathbb{Z}_2^4$, whose extra generators have binaries 1100, 0110, and 0011 corresponding to decimal indices 3, 6, and 12, in

Table 1 Spatio–spectral orders of elements of Γ

Count	Binary	Order	Factors	Count	Binary	Order	Factors
0	0000	(0,0)	0	8	0001	(2,1)	8
1	1000	(2,1)	1	9	1001	(4,2)	[1,8]
2	0100	(3,1)	2	10	0101	(5,2)	[2,8];[6,12]
3	1100	(3,1)	3;[1,2]	11	1101	(5,2)	[3,8];[1,2,8]
4	0010	(3,1)	4	12	0011	(3,1)	12,[4,8]
5	1010	(5,2)	[1,4];[3,6]	13	1011	(5,2)	[1,12];[1,4,8]
6	0110	(4,1)	6;[2,4]	14	0111	(4,2)	[2,12];[2,4,8];[6,8]
7	1110	(4,2)	[3,4];[1,6]; [1,2,4]	15	1111	(4,2)	[3,12];[1,2,12], [3,4,8],[1,2,4,8]

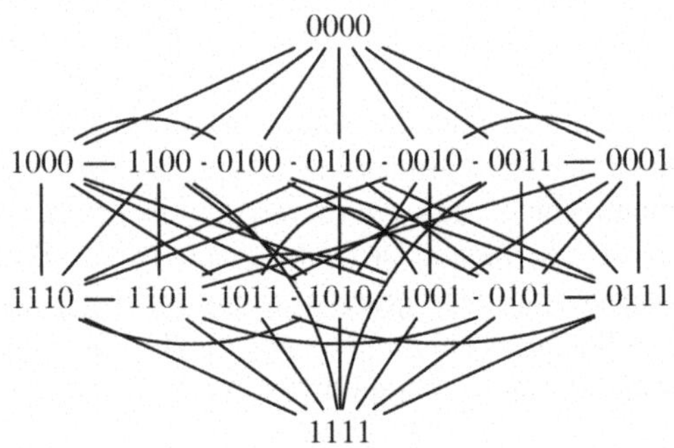

Fig. 1 Redundant cube on 16 vertices with redundant generators 1100, 0110, and 0011. The row below the origin shows all generators. The next row shows all elements that are distance two from the origin. The element 1111 also has distance two, but is drawn below to simplify the illustration. In contrast, the vertices in the r-th row of a corresponding representation of the standard Boolean cube on $\mathbb{Z}_2^4$ would be those binaries having r bit values equal to one

order to illustrate spatio–spectral groupings. These groupings play an important role in our subsequent analysis of a moderately more complex example, which is too large to represent graphically with any visual appeal. The orders of the vertices in this smaller example are recorded in Table 1.

Figure 1 draws this redundant cube representing the vertices according to their path distance to the origin (the vertex 1111 has distance two but is drawn below to simplify the drawing). Figure 2, on the other hand, lays out the vertices differentiating vertices with different spectral order by placing them in corresponding rows. The first three rows below the origin contain the vertices of unit spatial distance from the origin, laid out in increasing spectral orders 2 through 4. The next two rows contain vertices of spatial distance two and respective spectral order four and five.

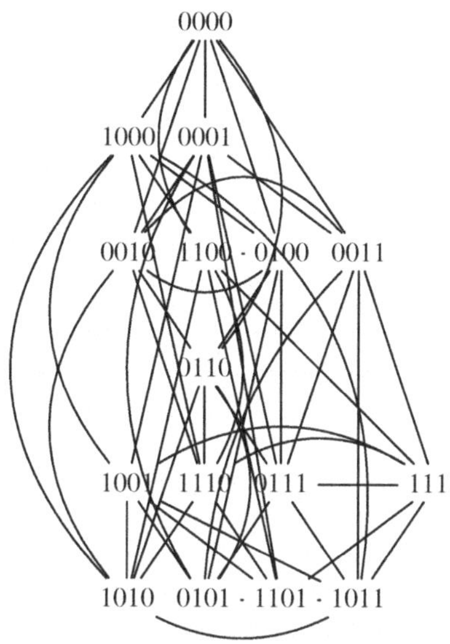

Fig. 2 Redundant cube on 16 vertices with redundant generators 1100, 0110, and 0011. Rows are drawn first with increasing spatial order, then by increasing spectral order, according to values in Table 1

Table 2 Generators of Γ_S and their spectral orders

Generator binary	Decimal	Spectral order	Boost generators (decimal)
(1,0,0,0,0,0,0,0,0,0,0)	1	2	1,7
(0,1,0,0,0,0,0,0,0,0,0)	3	3	2,7,14
(0,0,1,0,0,0,0,0,0,0,0)	4	4	4,7,14,28
(0,0,0,1,0,0,0,0,0,0,0)	8	4	8,14,28,56
(0,0,0,0,1,0,0,0,0,0,0)	16	4	16,28,56,112
(0,0,0,0,0,1,0,0,0,0,0)	32	3	32,56,112
(0,0,0,0,0,0,1,0,0,0,0)	64	2	64,112
(0,0,0,0,0,0,0,1,0,0,0)	128	1	128
(0,0,0,0,0,0,0,0,1,0,0)	256	1	256
(0,0,0,0,0,0,0,0,0,1,0)	512	1	512
(0,0,0,0,0,0,0,0,0,0,1)	1024	1	1024
(1,1,1,0,0,0,0,0,0,0,0)	7	5	1,2,4,7,28
(0,1,1,1,0,0,0,0,0,0,0)	14	5	2,4,8,14,56
(0,0,1,1,1,0,0,0,0,0,0)	28	6	4,8,16,7,28,112
(0,0,0,1,1,1,0,0,0,0,0)	56	5	8,16,32,14,56
(0,0,0,0,1,1,1,0,0,0,0)	112	5	16,32,64,28,112

5.2 A Moderate Example

In what follows, Γ_S will always refer to the specific redundant cube where $N = 11$ and S is a set of $M = 15$ generators, consisting of the standard generators e_i along with generators $e_i + e_{i+1} + e_{i+2}$, $i = 0, 1, 2, 3, 4$. The generators are tabulated in Table 2.

The graph Γ_S can be expressed as the Cartesian product $\Gamma_{S^{(1)}} \times \Gamma_{S^{(2)}}$ having respective generators $S^{(1)} = \{1, 2, 4, 8, 16, 32, 64, 7, 14, 28, 56, 112\}$, which separately generate a redundant cube Γ_1 on the underlying group $\mathbb{Z}_2^6$, and the generators $S^{(2)} = \{128, 256, 512, 1024\}$ that separately generate a copy of $\Gamma_2 = \mathcal{B}_4$. In a Cartesian product, two product vertices are adjacent if there is an edge in one of the coordinates. Any vertex $v = \sum s_i$ can be expressed as a product via $v = \sum s_i^{(1)} + \sum s_i^{(2)}$ quantifying the components in $\Gamma_{S^{(1)}}$ and $\Gamma_{S^{(2)}}$. The challenge in analyzing SSL operators on redundant—as opposed to standard—cubes is that spatial and spectral cutoffs affect the different factors of the product differently.

In this example we will consider spatial limiting to vertices having path distance at most four to the origin and spectrum limiting to Fourier components having spectral order at most four (note that $\lambda_v \leq 2 \times 4$ is not the same as v having distance four from the origin). Table 3 lists all 79 vertices that are associated with a Fourier vector of spectral (and spatial) order at most four. The spatial order of a vertex cannot exceed its spectral order. Thus, the space of vertex functions on Γ_S that is spectrum limited to order at most four is also 79-dimensional. On the other hand, a product of generators can have spectral order smaller than those of its factors. For example, $00001100\ldots$ has spectral order three but its factors $0000100\ldots$ and $00000100\ldots$ have spectral orders 4 and 3, respectively.

Vertex Ordering, Spatio–Spectral Cells, and Decomposition of Spectral Adjacency *Dyadic lexicographic order* on $\mathcal{B}_N$ is consistent with spectral order in that the Laplacian eigenvalues λ_v satisfy $\lambda_v \leq \lambda_u$ if $|R_v| \leq |R_u|$. We order vertices of Γ_S initially according to spectral then spatial order. In $\mathcal{B}_N$, adjacent vertices necessarily have different spectral orders, but in redundant cubes, adjacent vertices can have equal spectral orders. We say that two vertices v, w are *spectral neutral neighbors* (SNNs) if $v \sim w$ and v, w have the same spectral order. For reasons outlined below, we revise our initial ordering to order SNNs consecutively. We call a set of vertices that have the same spectral and spatial ordering, together with any of their SNNs, a (spatio–spectral) *cell*. We do not otherwise specify ordering within cells.

In [6], decomposition of the spectral adjacency matrix $\mathcal{A}$ into its respective lower and upper triangular components $\mathcal{A}_+$ (outer adjacency) and $\mathcal{A}_-$ (inner adjacency) played a critical role. We consider a parallel decomposition here, but add a third component $\mathcal{A}_0$ that accounts for SNN-adjacencies. We write $\mathcal{A} = \mathcal{A}_+ + \mathcal{A}_0 + \mathcal{A}_-$ where $\mathcal{A}_+$ is the lower triangular part of the matrix of $\mathcal{A} - \mathcal{A}_0$ in the indexing just outlined. That is, in any nonzero entry of the matrix, the spectral order of the row index vertex exceeds that of the column index vertex.

5.3 *Eigenvectors of Spatio–Spectral Limiting*

As previously outlined, the spatial-limiting operator $Q = Q_K$ is equal to the identity on those vertices in $B_K(\Gamma_S)$ and zeros out other vertices. It can be expressed as a diagonal matrix with ones on the diagonal indices of vertices in B_K and zeros

Table 3 Factorizations of elements of spectral order at most four

Count	Integer	Order	Factors		Integer	Order	Factors
1	0	(0,0)	0				
2	1024	(1,1)	1024	38	1120	(4,3)	[16,112,1024]
	512	(1,1)	512		608	(4,3)	[16,112,512]
	256	(1,1)	256		352	(4,3)	[16,112,256]
	128	(1,1)	128		224	(4,3)	[16,112,128]
					1027	(4,3)	[1,2,1024]
6	64	(2,1)	64		515	(4,3)	[1,2,512]
	1	(2,1)	1		259	(4,3)	[1,2,256]
					131	(4,3)	[1,2,128]
8	1536	(2,2)	[512,1024]		1056	(4,2)	[32,1024]
	1280	(2,2)	[256,1024]		544	(4,2)	[32,512]
	768	(2,2)	[256,512]		288	(4,2)	[32,256]
	1152	(2,2)	[128,1024]		160	(4,2)	[32,128]
	640	(2,2)	[128,512]		1026	(4,2)	[2,1024]
	384	(2,2)	[128,256]		514	(4,2)	[2,512]
					258	(4,2)	[2,256]
14	96	(3,2)	[16,112]=[32,64]		130	(4,2)	[2,128]
	3	(3,2)	[1,2]=[4,7]	54	88	(4,3)	[4,64,28]
	32	(3,1)	32		13	(4,3)	[1,2,14]
	2	(3,1)	2		24	(4,2)	[4,28]
					12	(4,2)	[2,14]
18	1088	(3,2)	[64,1024]	58	54	(4,2)	[14,56]
	576	(3,2)	[64,512]				
	320	(3,2)	[64,256]	59	1600	(4,3)	[64,512,1024]
	192	(3,2)	[64,128]		1344	(4,3)	[64,256,1024]
	1025	(3,2)	[1,1024]		832	(4,3)	[64,256,512]
	513	(3,2)	[1,512]		1216	(4,3)	[64,128,1024]
	257	(3,2)	[1,256]		704	(4,3)	[64,128,512]
	129	(3,2)	[1,128]		448	(4,3)	[64,128,256]
					1537	(4,3)	[1,512,1024]
26	48	(3,2)	[8,56]=[16,32]		1281	(4,3)	[1,256,1024]
27	6	(3,2)	[1,7]=[2,4]		769	(4,3)	[1,256,512]
					1153	(4,3)	[1,128,1024]
28	1792	(3,3)	[256,512,1024]		641	(4,3)	[1,128,512]
	1664	(3,3)	[128,512,1024]		385	(4,3)	[1,128,256]
	1408	(3,3)	[128,256,1024]				
	896	(3,3)	[128,256,512]	71	1072	(4,3)	[8,56,1024]
					560	(4,3)	[8,56,512]
32	80	(4,2)	[16,64]		304	(4,3)	[8,56,256]
	5	(4,2)	[1,4]		176	(4,3)	[8,56,128]
	16	(4,1)	16		1030	(4,3)	[1,7,1024]
	4	(4,1)	4		518	(4,3)	[1,7,512]
					262	(4,3)	[1,7,256]
36	8	(4,1)	8		134	(4,3)	[1,7,128]
37	65	(4,2)	[1,64]	79	1920	(4,4)	$[2^7, 2^8, 2^9, 2^{10}]$

elsewhere. Similarly, spectral limiting can be defined by $P = P_{\Gamma,K} = \bar{H}_\Gamma Q_K \bar{H}_\Gamma^T$ where $\bar{H}_\Gamma$ is the unitary Fourier matrix on Γ, again, ordered so as to restrict to elements of spectral order at most K. We will use the same value of K for spatial and spectral order, but only for notational convenience. For small values of K, as in the example at hand, the eigenvectors of PQ can be found by singular value decomposition.

As in [6] we study the eigenspaces of PQ in the spectral domain. If f is a λ-eigenvector of PQ, then (using $P = P^T$) $\bar{H}_\Gamma f$ satisfies

$$\mathcal{QP}\bar{H}_\Gamma f = \mathcal{Q}\bar{H}_\Gamma Q\bar{H}_\Gamma^T \bar{H}_\Gamma f = \mathcal{Q}\bar{H}_\Gamma Qf = \bar{H}_\Gamma \bar{H}_\Gamma^T \mathcal{Q}\bar{H}_\Gamma Qf = \bar{H}_\Gamma P^T Qf = \lambda \bar{H}_\Gamma f \,.$$

Thus, to find eigenvectors of PQ it suffices instead to find eigenvectors of $\mathcal{QP}$, which are Fourier transforms of those of PQ.

Eigenvectors of $\mathcal{QP}$ on Γ_S for $K = 4$ are plotted in Figs. 3 and 4. Figure 3 plots those eigenvectors that are nonzero at the origin. In each case the corresponding vector is constant on each of the spectral cells indicated in Table 3. There are 17 such cells including the origin, and the Paley–Wiener space of 4-spectrum-limited vectors that are cell-wise constant is 17-dimensional. The remaining eigenvectors are plotted in Fig. 4. They are sorted by base support, meaning the lowest spectral order cell in the support. The other groups of supporting vertices in each case correspond to certain (not necessarily all) outer adjacent cells of the base, as indicated in Tables 3 and 5. In the order described above and catalogued in Tables 3, 4, and 5, $\mathcal{A}_+$ is a cell-wise lower triangular part of the adjacency matrix. The dimensions of the corresponding eigenspaces of $\mathcal{QP}$ are listed in Table 6. The plots at least suggest the possibility that eigenvectors have the form $\sum_{k=0}^{K} c_k \mathcal{A}_+^k g$ for some g supported in the base support cell of the vector (Fig. 5).

Proposition 1 *Let $C = [\mathcal{A}_-, \mathcal{A}_+] = \mathcal{A}_-\mathcal{A}_+ - \mathcal{A}_+\mathcal{A}_-$ denote the commutator of $\mathcal{A}_-$ and $\mathcal{A}_+$. Suppose that C acts as multiplication by a constant on vectors*

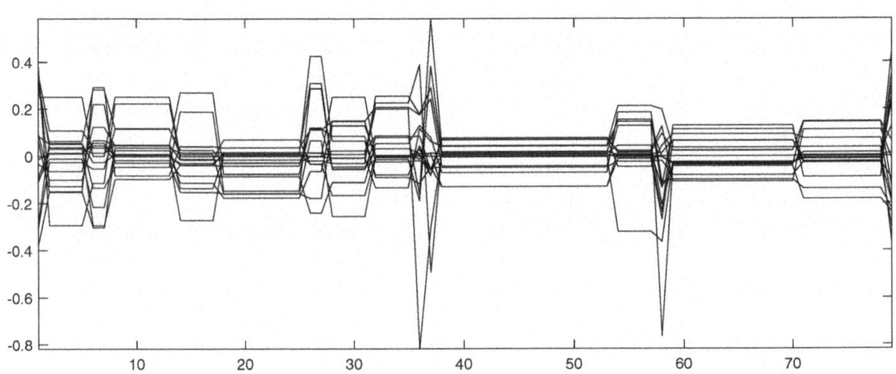

Fig. 3 Eigenvectors of $\mathcal{QPQ}$, $K = 4$, that are constant on spatio–spectral cells. There are 17 such eigenvectors

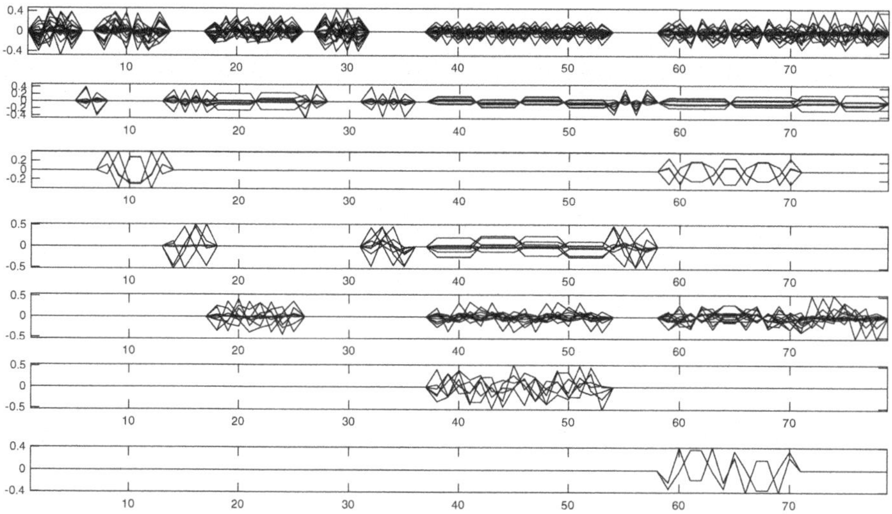

Fig. 4 Remaining eigenvectors of QPQ $K = 4$, grouped by lowest supported (base) spatio–spectral bin in the support of the vectors, see Table 4. Remaining cells in the support of each vector are determined by outer adjacencies

Table 4 Spatio–spectral cells

Index	Spectral order	Spatial order	Description	UF/NUF
1	0	0		
2–5	1	1	2^p, $p = 7, \ldots, 10$	UF
6–7	2	1	2^p, $p = 0, 6$	UF
8–13	2	2	$(1, 1)^2$	UF
14–17	3	1 (2)	2^p, $p = 2, 5$ and SNN	NUF
18–25	3	2	$(2, 1) \times (1, 1)$	UF
26–27	3	2	01100... & 0000110...	NUF
28–31	3	3	$(1, 1)^3$	UF
32–35	4	1 (2)	2^p, $p = 2, 4$ &SNN	NUF
36–37	4	1 (2)	00010... & 10000010...	UF
38–53	4	2 (3)	$14 - 17 \times (1, 1)$	NUF
54–57	4	2 (3)	$32 - 35 \times 00010\ldots$	NUF
58	4	2	$01110\cdots \times 0001110\ldots$	UF
59–70	4	3	$(2, 1) \times (1, 1)^2$	UF
71–78	4	3	$26 - 27 \times (1, 1)$	UF
79	4	4	$(1, 1)^4$	UF

supported in any fixed spatio–spectral cell that are also eigenvectors of $\mathcal{A}_0$, and that powers of $\mathcal{A}_+$ map eigenvectors of $\mathcal{A}_0$ to eigenvectors of $\mathcal{A}_0$. Let $\mathcal{V} = \{\sum_{k=0}^{N} c_k \mathcal{A}_+^k g,\ g \in \mathcal{W}_{\Sigma,\lambda}\}$ where $\mathcal{W}_{\Sigma,\lambda} = \{g : \operatorname{supp} g \subseteq \Sigma,\ \mathcal{A}_- g = 0,\ \mathcal{A}_0 g = \lambda g\}$. Then $\mathcal{V}$ is invariant under $\mathcal{A} = \mathcal{A}_+ + \mathcal{A}_- + \mathcal{A}_0$.

Table 5 Multiplicative relations between spatio–spectral cells

Cell indices	1 (0,0)	2–5 (1,1)	6–7 (2,1)	8–13 (2,2)	14–17 (3,1) (3,2) SNN	18–25 (3,2) UF	26–27 (3,2) NUF	28–31 (3,3) UF	32–25 (4,1) (4,2) SNN	36–37 (4,1) (4,2)	38–53 (4,2) (4,3)
1											
2–5	(1,1)										
6–7	(2,1)										
8–13		(1,1)									
		(1,0)									
14–15			(1,1)		(1,0)						
					(0,0)						
16–17	(3,1)										
18–25		(2,1)	(1,1)								
26–27					(1,0)						
					(0,0)						
28–31				(1,1)							
				(1,0)							
32–35	(4,1)		(2,1)				(1,1)				
							(1,0)		(0,1)		
36	(4,1)										
37			(2,1)								
38–53		(3,1)		(1,1)		(1,1)					(0,1)
		(3,0)									(1,0)
		(2,0)									
54–57									(0,1)	(0,1)	
									(0,0)		
58	$(2,1)^2$										
59–70		(2,1)		(2,1)		(1,1)					
		(1,1)		(2,0)							
71–78							(1,1)				(0,1)
											(1,0)
79								(1,1)			

Proof It suffices to show that $\mathcal{V}$ is invariant under each of $\mathcal{A}_+$, $\mathcal{A}_0$, and $\mathcal{A}_-$. If $f \in \mathcal{V}$, $f = \sum_{k=0}^{N} c_k \mathcal{A}_+^k g$, then $\mathcal{A}_+ f = \sum_{k=0}^{N} c_k \mathcal{A}_+^{k+1} g = \sum_{k=1}^{N} c_{k-1} \mathcal{A}_+^k g$ so $\mathcal{A}_+ f \in \mathcal{V}$. Note that $\mathcal{A}_+^{N+1} g = 0$ because vectors have spectral order at most N. Since $\mathcal{A}_+^\ell$ maps eigenvectors of $\mathcal{A}_0$ to eigenvectors of $\mathcal{A}_0$, $\mathcal{V}$ is also invariant under $\mathcal{A}_0$. For each cell Σ let $m(\Sigma, \lambda)$ be such that $Cg = m(\Sigma, \lambda) g$ when g is supported in Σ and $\mathcal{A}_0 g = \lambda g$. If g is such a vector also in the kernel of $\mathcal{A}_-$, then $\mathcal{A}_- \mathcal{A}_+ g = Cg = m(\Sigma, \lambda) g$. We claim that for each ℓ there is a fixed number $m(\Sigma, \lambda, \ell)$ such that $\mathcal{A}_- \mathcal{A}_+^{\ell+1} g = m(\Sigma, \lambda, \ell) \mathcal{A}_- \mathcal{A}_+^\ell g$. Suppose such an identify holds for $\ell - 1$. Then

Table 6 Eigenspaces of $Q_K P_K Q_K$, $K = 4$

Base cell index	Base cell order	Dimension per eigenspace	Number of eigenspaces	Net dimension
1	(0,0)	1	17	17
2–5	(1,1)	3	7	21
6–7	(2,1)	1	9	9
8–13	(2,2)	2	2	4
14–17	(3,1) & SNN	2	4	8
18–25	(3,2) $((2,1) \times (1,1))$	3	4	12
38–53	(4,2) $((3,1) \times (1,1))$ & SNN	6	1	6
59–70	$(2,1) \times (2,2)$	1	2	2

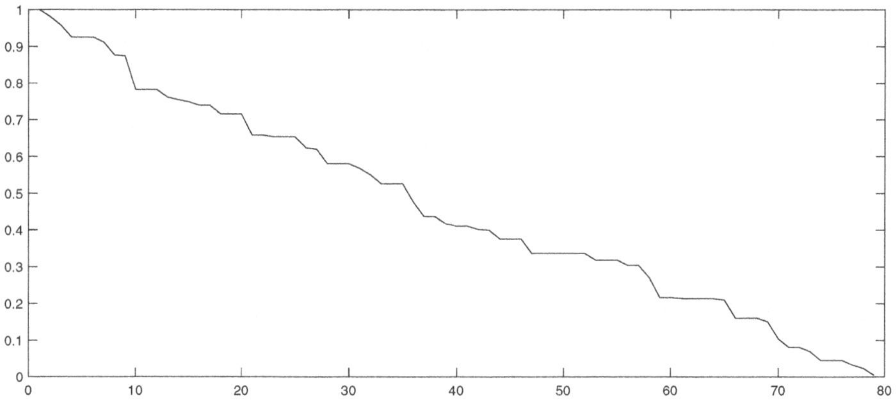

Fig. 5 Eigenvalues of PQP with multiplicity. Observe that the multiplicities agree with dimensions of eigenspaces observed in Table 6. For example, the six dimensional eigenspace is associated with the eigenvalue 0.33593750 and the (4,2)-base cell with indices 38–53 in the sixth plot in Fig. 4

$$\begin{aligned}
\mathcal{A}_-\mathcal{A}_+^{\ell+1}g &= (\mathcal{A}_-\mathcal{A}_+ - \mathcal{A}_+\mathcal{A}_-)\mathcal{A}_+^{\ell}g + \mathcal{A}_+\mathcal{A}_-\mathcal{A}_+^{\ell}g \\
&= C\mathcal{A}_+^{\ell}g + \mathcal{A}_+ m(\Sigma, \lambda, \ell-1)\mathcal{A}_+^{\ell-1}g \\
&= C(\Sigma_\ell, \lambda_\ell)\mathcal{A}_+^{\ell}g + m(\Sigma, \lambda, \ell-1)\mathcal{A}_+^{\ell}g \equiv m(\Sigma, \lambda, \ell)\mathcal{A}_+^{\ell}g,
\end{aligned}$$

where Σ_ℓ supports images under $\mathcal{A}_+^{\ell}$ of vectors supported in Σ and λ_ℓ is such that $\mathcal{A}_0\mathcal{A}_+^{\ell}g = \lambda_\ell\mathcal{A}_+^{\ell}g$. Here we used the hypothesis that the property of being an eigenvector of $\mathcal{A}_0$ is preserved under powers of $\mathcal{A}_+$ and the fact that $\mathcal{A}_+$ maps one spatio–spectral cell to another. The calculation shows that any space $\mathcal{V}$ defined as in the statement of the proposition is also invariant under $\mathcal{A}_-$ and therefore under $\mathcal{A}$.

□

In [6] it was proved that $[\mathcal{A}_-, \mathcal{A}_+]$ acts as multiplication by $N-2r$ on the Hamming sphere Σ_r of radius r. Since $\mathcal{A}_0$ is trivial on $\mathcal{B}_N$, the result was that spaces of the form $\mathcal{V} = \{\sum_{k=0}^{N} c_k\mathcal{A}_+^k g,\ g \in \mathcal{W}_r\}$ where $\mathcal{W}_r$ are those vectors supported in

Σ_r in the kernel of $\mathcal{A}_-$, is invariant under the (spectral) adjacency map on $\mathcal{B}_N$. Eigenspaces of SSL on $\mathcal{B}_N$ were then identified as spans of Fourier transforms of certain vectors in spaces $\mathcal{V}$ whose coefficients are eigenvectors of a certain *reduced* matrix that acts of coefficients of elements of $\mathcal{V}$.

The goal here is much more modest: to provide evidence that the approach suggested in Proposition 1 is feasible for identifying invariant subspaces of $\mathcal{A}$ from which eigenspaces of SSL can be extracted. Specifically, the following proposition suggests that the inner adjacency matrix $\mathcal{A}_-$ acts like a cell-wise multiplier on terms of the sums defining the spaces in Proposition 1. If so, the action of the adjacency operator $\mathcal{A}$ on such sums can be expressed as a linear mapping on the coefficients of such sums, and determining eigenvectors of $\mathcal{QP}$ can be reduced to computing eigenvectors of a polynomial in the corresponding matrix. In the example at hand, such a matrix would have size 17, versus the size-79 matrix from which the eigenvectors plotted below were estimated. This suggests a means to compute eigen-decompositions of $\mathcal{QP}$ in much larger cases by such reduction.

Proposition 2 *Let f be supported in Σ, the set of all v expressible as a product of k unique generators, where $k = k_1 + \cdots + k_t$ and k_j is the number of factors in S_j, the generators of spectral order j, and such that no two vertices in Σ are adjacent. Suppose also that (i) each $(k+1)$-generator extension also has a unique factorization, either with $k_j + 1$-terms in S_j for some j, or with a generator in $\widetilde{S} = S \setminus \cup_{j=1}^t S_j$, and (ii) f lies in the kernel of $\mathcal{A}_-$. Then $\mathcal{A}_-\mathcal{A}_+ f(v) = m(k,\Sigma) f(v)$ where $m(k,\Sigma) = |\widetilde{S}| + \sum_{j=1}^t m_j(k, S_j)$ and $m_j(k, S_j) = |S_j| - 2k_j$.*

Proof Let $v \in \Sigma$. Contributions to $(\mathcal{A}_-\mathcal{A}_+ f)(v)$ involving vertices with a factor in $\widetilde{S}$ can only arise via a path from v to vertices of the form $v + \widetilde{s}$, $\widetilde{s} \in \widetilde{S}$, and back, giving a net contribution of $|\widetilde{S}| f(v)$. The condition $\mathcal{A}_- f = 0$ is the same as $\sum f(w) = 0$ where the sum extends over those $w \sim v$ obtained by eliminating a single generating factor from the unique product defining v. Let $(\mathcal{A}_-\mathcal{A}_+ f)_j(v)$ denote the terms associated with adjacencies that pass through vertices having an additional factor in S_j. One can write

$$(\mathcal{A}_-\mathcal{A}_+ f)_j(v) = (|S_j| - k_j) f(v) + \sum_{s_j | v, s_j \in S_j} \sum_{\sigma_j \nmid v, \sigma_j \in S_j} f(v + \sigma_j - s_j)$$

($s|v$ means s is a factor in a minimal factorization of v) accounting for those terms that pass from v to a $k+1$-term product with $k_j + 1$ factors in S_j then directly back to v (there are $|S_j| - k_j$ choices for the extra factor), and those terms that pass from a different element of Σ to v via a one-term substitution of an element σ of S_j that is not a factor of v by an element of S_j that is a factor of v. As in the standard cube argument (see [6]), we add and subtract $f(v)$ once for each $s_j | v$ to rewrite as

$$(\mathcal{A}_-\mathcal{A}_+ f)_j(v) = (|S_j| - k_j) f(v) + \sum_{s_j | v, s_j \in S_j} \Big[\Big(\sum_{w \in \Sigma, w \sim v - s_j} f(w)\Big) - f(v)\Big].$$

Summing now over the different S_j represented in v and using the condition $\mathcal{A}_- f(v) = 0$ in the form $\sum_{w\in\Sigma, w\sim v-s_j} f(w) = 0$ gives the result. □

As an example, let Σ consist of all order (2, 2) products corresponding to indices 8–13 in Table 3. The one-factor extensions of these consist of the order (3, 3) products (indices 28–31), and (4,3) products (indices 59–71). There are four (3,3)-extensions, one for each generator in S_1 so $|S_1| = 4$, whereas each (2,2)-product has $k_1 = 2$. On the other hand, $|S_2| = 2$ corresponding to indices 6–7 in Table 3, whereas each (2,2)-product has $k_2 = 0$. Therefore

$$(\mathcal{A}_-\mathcal{A}_+)f(v) = [(|S_1|-2k_1)+(|S_2|-2k_2)]f(v) = [(4-4)+(2-0)]f(v) = 2f(v)$$

in this example.

The proposition does not address potential multiplier relations in the case of non-unique factorizations. Consider the cell Σ that consists of the (2,1)-generators with decimal expressions 1 and 64. As in Table 3, this cell has outer neighbors $S_2 + S_1$ (indices 18–25) and other unique product extensions $[16, 64]$ and $[1, 4]$ (indices 32–33) and $[1, 64]$ (index 37), along with non-unique extensions $[1, 2] = [4, 7]$ and $[32, 64] = [16, 112]$ (indices 14–15) and $[16, 32] = [8, 56] = [64, 112]$ and $[1, 7] = [2, 4]$ (indices 26–27). Proposition 2 *suggests* (as is verified numerically) that if g is a vector supported in $\Sigma = S_2$ in the kernel of $\mathcal{A}_-$ (in this case, the sum of the two values at indices 6 and 7 is zero), then $\mathcal{A}_-\mathcal{A}_+g = 7g$. The proposition also does not consider neutral adjacencies. In Proposition 1 we added a hypothesis that the vector g supported in Σ is also an eigenvector of $\mathcal{A}_0$. Typically an identity of the form $(C + \mathcal{A}_0)g(v) = m(\Sigma)g(v)$ applies to any g supported in a cell Σ containing neutral neighbors (as before, $C = [\mathcal{A}_-, \mathcal{A}_+]$).

We conjecture that Proposition 2 extends to non-unique products, provided one *reduces* the adjacency matrix to account for non-uniqueness to a suitable matrix $\mathcal{A}_{\mathrm{red}}$.

Conjecture Fix a set Σ as in Proposition 2. The space $\sum_{k=0}^{N} c_k\mathcal{A}_{\mathrm{red},+}^k g$, $g \in \mathcal{W}_\Sigma$ where g is supported in Σ and lies in the kernel of $\mathcal{A}_{\mathrm{red},-}$ forms an invariant subspace of $\mathcal{QP}$ that is also invariant under $\mathcal{A}_{\mathrm{red}}$.

The fourth plot in Fig. 4 shows eight eigenvectors of $\mathcal{QP}$ with base support in indices 14–17 and other supports in groups 32–35, 38–53, and 54–57. The indices 16–17 are SNNs of 14–15. The values in indices 14–17 are such that the values at 14 and 16 and at 15 and 17 are additive inverses. As per Table 3, there are adjacencies in index pairs (16,26) and (17,27) but not (14,26) or (15,27). This means that the vanishing of the given eigenvectors at indices 26–27 does not occur because of cancellation of inputs of the outer adjacency matrix. A similar situation occurs for the eigenvectors in the sixth plot in Fig. 4. In all other plots, it can be verified numerically that the eigenvectors of $\mathcal{QP}$ that are plotted all have the form $\sum c_k\mathcal{A}_+^k g$ where g is supported in the base support set of the vector and $\mathcal{A}_- g = 0$. Conditions on a reduced adjacency matrix and the neutral matrix $\mathcal{A}_0$ in the decomposition

$\mathcal{A} = \mathcal{A}_+ + \mathcal{A}_0 + \mathcal{A}_-$ to verify the connection will be taken up in subsequent work.

6 Adjacency on Γ_S

This section catalogues the adjacency properties on Γ_S referenced in the previous section. Because our example considers spatio–spectral limiting to orders at most four, we catalog only the adjacencies in the size 79 submatrix of the full size 2048 adjacency matrix corresponding to vertices that are order-limited to at most four in both space and spectrum. Table 3 lists those vertices in Γ_S that have spatial and spectral orders at most four. The grouping of vertices by spatial and spectral cell order is summarized in Table 4, and multiplicative relations between cells are summarized in Table 5.

Table 4 further lists the elements of spectral order at most four in terms of properties that determine their roles as components of eigenvectors of spatio–spectral limiting. This includes identifying which cells contain SNNs and which cells allow non-unique minimal length factorizations (NUF) of at least some of the elements in the cell.

Groups of elements that have non-unique minimal length factorizations include the elements with indices 14–15 with respective non-unique factorizations $0000100\cdots + 000011100\cdots = 0000010\cdots + 00000010\ldots$ and $1110\cdots + 0010\cdots = 100\cdots + 0100\ldots$. The factorizations on the right express these elements as SNNs of the respective (3, 1)-elements $0000010\ldots$ and $0100\ldots$. In contrast, elements with indices 18–25 uniquely factor (UF) into pairs having decimal representation $2^p + 2^q$ where $p = 1, 6$ and $q = 7, \ldots, 10$.

Table 5 lists adjacency cells and types of factorization that link different cells. Factorization types can be inferred starting from the bottom and working upwards. For example, elements with indices 59–70 can be expressed as (1,1)-extensions of elements of the (3,2) UF-bin with indices 18–25 which in turn can be expressed as (1,1)-extensions of elements of the (2,1)-bin. Obviously the order of these factors does not matter.

Table 6 summarizes the structure of the eigenspaces of $Q_K P_K Q_K$, $K = 4$.

Acknowledgement The authors would like to thank the anonymous referee for several constructive comments to make the presentation more palatable.

References

1. Benedetto, J.J. and Koprowski, P. J.: Graph theoretic uncertainty principles. In: 2015 International Conference on Sampling Theory and Applications (SampTA), pp. 357-361. (2015) doi: 10.1109/SAMPTA.2015.7148912

2. Chen, L., Cheng, S., Stankovic, V., and Stankovic, L.: Shift-Enabled Graphs: Graphs Where Shift-Invariant Filters are Representable as Polynomials of Shift Operations. IEEE Signal Processing Letters **25**, 1305–1309 (2018)
3. Gavili, A. and Zhang, X.: On the Shift Operator, Graph Frequency, and Optimal Filtering in Graph Signal Processing. IEEE Trans. Signal Process. **65**, 6303–6318 (2017)
4. Han, M., Shi, J., Deng, Y., and Song, W.: On Sampling of Bandlimited Graph Signals. In: Gu, Xi, Liu, G., Li, B. (eds) Machine Learning and Intelligent Communications, pp. 577-584. Springer International Publishing, Cham (2018)
5. Hogan, J.A. and Lakey, J.: An analogue of Slepian vectors on Boolean hypercubes. J. Fourier Anal. Appl. **25**, 2004–2020 (2019)
6. Hogan, J.A. and Lakey, J.: Spatio-spectral limiting on hypercubes: eigenspaces. (2018) arXiv:1812.08905
7. Kurokawa, T., Oki, Y., and Nagao, H.: Multi-dimensional graph Fourier transform. (2017) arXiv:1712.0781
8. Landau, H.J. and Pollak, H.O.: Prolate spheroidal wave functions, Fourier analysis and uncertainty. II. Bell System Tech. J. **40**, 5–84 (1961)
9. Landau, H.J. and Pollak, H.O.: Prolate spheroidal wave functions, Fourier analysis and uncertainty. III. The dimension of the space of essentially time- and band-limited signals. Bell System Tech. J. **41**, 1295–1336 (1962)
10. Landau, H.J. and Widom, H.: Eigenvalue distribution of time and frequency limiting. J. Math. Anal. Appl. **77**, 469–481 (1980)
11. Pesenson, I.: Sampling in Paley-Wiener spaces on combinatorial graphs. Trans. Amer. Math. Soc. **360**, 5603–5627 (2008)
12. Pesenson, I.Z. and Pesenson, M.Z.: Sampling, Filtering and Sparse Approximations on Combinatorial Graphs. J. Fourier Anal. Appl. **169**, 321–354 (1995)
13. Puy, G., Tremblay, N., Gribonval, R., and Vandergheynst, P.: Random sampling of bandlimited signals on graphs. Appl. Comput. Harmon. Anal. **44**, 446–475 (2018)
14. Rudin, W.: Fourier Analysis on Groups. Wiley-Interscience, New York (1962)
15. Sandryhaila, A. and Moura, J.M.F. : Big Data Analysis with Signal Processing on Graphs: Representation and processing of massive data sets with irregular structure. IEEE Signal Processing Magazine **31**, 80–90 (2014)
16. Sandryhaila, A. and Moura, J.M.F.: Discrete signal processing on graphs: Graph filters. In: 2013 IEEE International Conference on Acoustics, Speech and Signal Processing, pp. 6163-6166. (2013)
17. Segarra, S., Marques, A.G., Leus, G., and Ribeiro, A.: Interpolation of graph signals using shift-invariant graph filters. In: 2015 23rd European Signal Processing Conference (EUSIPCO), pp. 210-214. (2015)
18. Slepian, D.: Prolate spheroidal wave functions, Fourier analysis and uncertainty. IV. Extensions to many dimensions; generalized prolate spheroidal functions. Bell System Tech. J. **43**, 3009–3057 (1964)
19. Slepian, D. and Pollak, H.O.: Prolate spheroidal wave functions, Fourier analysis and uncertainty. I. Bell System Tech. J. **40**, 43–63 (1961)
20. Spielman, D.A.: Spectral graph theory http://www.cs.yale.edu/homes/spielman/561/ Cited 20 Oct 2018
21. Strichartz, R.S.: Half Sampling on Bipartite Graphs. J. Fourier Anal. Appl. **22**, 157–1173 (2016)
22. Tremblay, N., Gonçalves, P., and Borgnat, P.: Design of graph filters and filterbanks. In: Djurić, P.M., Richard, C. (eds.) Cooperative and Graph Signal Processing, pp. 299-324. Academic Press (2018)
23. Tsitsvero, M., Barbarossa, and S., Di Lorenzo, P.: Signals on Graphs: Uncertainty Principle and Sampling. IEEE Trans. Signal Process. **169**, 4845–4860 (2016)

Part III
Wavelets and Frames

This part is dedicated to wavelets and frames, topics that include countless generalizations and extensions to the most diverse contexts. Here we present some of the most recent trend developments, characterized by mathematical rigor combined with the constructive aspect, and modern variations of classic topics.

We are sure that they will capture the attention of the interested reader both in the purely mathematical aspect and in the search for solutions to be applied.

In particular, the first chapter written by P. G. Casazza, J. I. Haas, J. Stueck, and T. T. Tran introduces (and solves) the problem of packing subspaces of various dimensions in a finite-dimensional Hilbert space (mixed-rank packing problem).

In the second chapter, D.-C. Chang, Y. Han, and X. Wu use the Calderón–Zygmund operator theory to show how the Frazier–Jawert construction of frames can be carry out without the use of the Fourier transform.

In the third chapter, S. Datta takes us back to the equiangular tight frames (ETFs), glimpsed in the first chapter of this part, studies equiangular frames that are not necessarily tight, and investigates whether they can have an equiangular dual.

In the fourth chapter, K. D. Merrill focuses on wavelets obtained by an action of a 2-dimensional crystallographic group on $L^2(\mathbb{R}^2)$ and an integer dilation δ (which replace the usual integer translations and dilations) and shows that single wavelet sets, and thus single wavelets, exist for all the two dimensional crystallographic groups and for all the integer dilations.

Finally, in the last chapter of this part, A. Olevskii and A. Ulanovskii discuss many fundamental results related to completeness of systems of translates in several function spaces over $\mathbb{R}$, including general Banach function spaces not necessarily translation-invariant.

A Notion of Optimal Packings of Subspaces with Mixed-Rank and Solutions

Peter G. Casazza, Joshua Stueck, and Tin T. Tran

Abstract We resolve a longstanding open problem by reformulating the Grassmannian fusion frames to the case of mixed dimensions and show that this satisfies the proper properties for the problem. In order to compare elements of mixed dimension, we use a classical embedding to send all fusion frame elements to points on a higher dimensional Euclidean sphere, where they are given "equal footing." Over the embedded images—a compact subset in the higher dimensional embedded sphere—we define optimality in terms of the corresponding restricted coding problem. We then construct infinite families of solutions to the problem by using maximal sets of mutually unbiased bases and block designs. Finally, we show that using Hadamard 3-designs in this construction leads to infinitely many examples of maximal orthoplectic fusion frames of constant-rank. Moreover, any such fusion frames constructed by this method must come from Hadamard 3-designs.

1 Introduction

Let $\mathcal{X}$ be a compact metric space endowed with a distance function $d_{\mathcal{X}}$. The packing problem is the problem of finding a finite subset of $\mathcal{X}$ so that the minimum pairwise distance between points of this set is maximized. When $\mathcal{X}$ is the Grassmannian manifold and $d_{\mathcal{X}}$ is the chordal distance, the problem has received considerable attention over the last century. This has been motivated in part by emerging applications such as in quantum state tomography [31, 38], compressed sensing [2], and coding theory [1, 33, 37]. Much previous work also focuses on the special

In memory of John I. Haas.

P. G. Casazza (✉) · J. Stueck · T. T. Tran
Department of Mathematics, University of Missouri, Columbia, MO, USA
e-mail: casazzap@missouri.edu; jss4yf@mail.missouri.edu; tttrz9@mail.missouri.edu

M. Hirn et al. (eds.), *Excursions in Harmonic Analysis, Volume 6*, Applied and Numerical Harmonic Analysis, https://doi.org/10.1007/978-3-030-69637-5_7

case where the Grassmannian manifold contains subspaces of dimension one. This direction has become an active area of research in the context of frame theory, see [3–6, 19–26, 29, 33, 34], for instance.

In this work, we will consider the problem of optimal packings of subspaces with multiple dimensions. We expect that this generalized notion will similarly be useful in the signal processing realm.

The paper is organized as follows. After preliminaries in Sect. 2, we will take a look at the constant-rank packing problem in Sect. 3. In particular, after dissecting the well-known proof of the (chordal) simplex and orthoplex bounds, we can see that solutions for this problem can be found by solving a corresponding restricted coding problem in a higher dimensional space. This motivates us to give a definition for optimal packings of subspaces with multiple dimensions in Sect. 4. Next, we present some properties of solutions to the mixed-rank packing problem in Sect. 5. In particular, as in the constant-rank case, we show that the solutions to the problem always form fusion frames. Finally, we will construct some infinite families of solutions in Sect. 6. We also focus on constructing maximal orthoplectic fusion frames of constant-rank in this section.

2 Preliminaries

Fix $\mathbb{F}$ as $\mathbb{R}$ or $\mathbb{C}$; furthermore, prescribe $d, l, m, n \in \mathbb{N}$, where the last three numbers satisfy $l \leq m \leq n$. We also denote $[[m]]$ for the set $[[m]] := \{1, 2, \dots, m\}$. A **code** is a sequence of n points on the real unit sphere, $\mathcal{S}^{d-1} \subseteq \mathbb{R}^d$. A **packing** is a sequence of $m \times m$ orthogonal projections onto subspaces of $\mathbb{F}^m$, which we often identify with their associated subspaces; that is, we frequently identify a given packing $\mathcal{P} := \{P_i\}_{i=1}^n$ with its subspace images, $\mathcal{P} := \{W_i := \text{Im}(P_i)\}_{i=1}^n$. The **mixture** of a packing is the number of different ranks occurring among the packing's elements. If a packing has a mixture of one, meaning the ranks of a packing's elements equate, then the packing is **constant-rank**; otherwise, it is **mixed-rank**. We define the **Grassmannian manifold**, $\mathcal{G}(l, \mathbb{F}^m)$, as the set of all orthogonal projections over $\mathbb{F}^m$ with rank l. Thus, a constant-rank packing, $\mathcal{P}$, is a sequence of points in the n-fold Cartesian product of Grassmannian manifolds:

$$\mathcal{P} = \{P_i\}_{i=1}^n \in \prod_{i=1}^n \mathcal{G}(l, \mathbb{F}^m).$$

Conventionalizing the preceding notation, we reserve d for the ambient dimension of a given code, m for the ambient dimension of a given packing, and n for the number of points in an arbitrary code or packing.

The existence of a mixed-rank packing, $\mathcal{P}$, implicates a sequence of pairs, $\{(n_i, l_i)\}_{i=1}^s$, where each n_i denotes the number of packing elements of rank l_i, meaning $n = \sum_{i=1}^s n_i$, and $1 \leq l_i \leq m$, for $i \in [[s]]$. This leads to a dictionary-type ordering on the elements, which—as in the constant-rank case—respects the interpretation of mixed-rank packings as sequences in a Cartesian product of Grassmannian manifolds:

$$\mathcal{P} = \{P_j^{(l_i)}\}_{i=1,j=1}^{s,\ n_i} \in \prod_{i=1}^{s} \prod_{j=1}^{n_i} \mathcal{G}(l_i, \mathbb{F}^m).$$

A **fusion frame**, $\mathcal{P} = \{P_i\}_{i=1}^n$, is a packing such that its **fusion frame operator**,

$$F_P := \sum_{i=1}^{n} P_i,$$

is positive-definite. This is equivalent to the family of associated subspaces, $\{W_i\}_{i=1}^n$ spans $\mathbb{F}^m$. If the F_P is a multiple of the $m \times m$ identity operator, henceforth denoted I_m, then $\mathcal{P}$ is **tight.** A constant-rank fusion frame for which all elements are rank one is a **frame**. For more information on (fusion) frames, we recommend [12, 14, 35].

3 Constant-Rank Packings Revisited

In order to generalize the notion of optimal packings of subspaces with different dimensions, we will take a look at the constant-rank packing problem as discussed, for example, in [7, 13, 16].

Definition 1 Given two l-dimensional subspaces of $\mathbb{F}^m$ with corresponding orthogonal projections P and Q, the chordal distance between them is

$$d_c(P, Q) := \frac{1}{\sqrt{2}} \|P - Q\|_{H.S} = (l - \text{tr}(PQ))^{1/2}.$$

Definition 2 Given a constant-rank packing, $\mathcal{P} = \{P_i\}_{i=1}^n \subseteq \mathcal{G}(l, \mathbb{F}^m)$, its **(chordal) coherence** is

$$\mu(\mathcal{P}) := \max_{\substack{1 \le i,j \le n \\ i \ne j}} \text{tr}(P_i P_j).$$

The constant-rank packing $\mathcal{P}$ is **optimally spread** if:

$$\mu(\mathcal{P}) = \min_{\substack{\mathcal{P}' = \{P_i'\}_{i=1}^n \\ \mathcal{P}' \subseteq \mathcal{G}(l, \mathbb{F}^m)}} \mu(\mathcal{P}').$$

$\mathcal{P}$ is said to be **equiangular** if there is a number $C \ge 0$ such that $\text{tr}(P_i P_j) = C$ for each pair $i \ne j$ in the index set $[[n]]$.

For reasons clarified below, we denote the dimension, $d_{\mathbb{F}^m}$, as a function of m and $\mathbb{F}$ as follows

$$d_{\mathbb{F}^m} := \begin{cases} \frac{m(m+1)}{2} - 1, & \mathbb{F} = \mathbb{R} \\ m^2 - 1, & \mathbb{F} = \mathbb{C} \end{cases}. \tag{3.1}$$

We recall two well-known bounds for the coherence $\mu(\mathcal{P})$, the simplex and orthoplex bounds, as derived by Conway, Hardin, and Sloane [16] from the Rankin bound [30]:

Theorem 3.1 ([16], See Also in [7, 28])

1. ***Simplex bound***: *If $\mathcal{P} = \{P_i\}_{i=1}^n \subseteq \mathcal{G}(l, \mathbb{F}^m)$ is a constant-rank packing for $\mathbb{F}^m$, then*

$$\mu(\mathcal{P}) \geq \frac{nl^2 - ml}{m(n-1)},$$

and equality is achieved if and only if $\mathcal{P}$ is an equiangular, tight fusion frame.
2. ***Orthoplex bound***: *If $\mathcal{P} = \{P_i\}_{i=1}^n \subseteq \mathcal{G}(l, \mathbb{F}^m)$ is a constant-rank packing for $\mathbb{F}^m$ and $n > d_{\mathbb{F}^m} + 1$, then*

$$\mu(\mathcal{P}) \geq \frac{l^2}{m},$$

and if equality is achieved, then $\mathcal{P}$ is optimally spread and $n \leq 2d_{\mathbb{F}^m}$.

Among constant-rank packings, it seems the most commonly known solutions to this problem arise as equiangular tight fusion frames (ETFFs) [8, 18, 27, 28], including the special case where all projections of packings are of rank one as mentioned. ETFFs are characterized by several special properties, including (1) equiangular, (2) tightness, and, of course, (3) coherence minimized to the simplex bounds in the previous theorem. When the cardinality of a constant-rank packing is sufficiently high, the orthoplex bound characterizes several other infinite families of optimally spread constant-rank solutions, including maximal sets of mutually unbiased bases [36] among others [5–8, 13, 32]. Although unnecessary for solutions characterized by the orthoplex bound (see [11] for a discussion of this phenomenon), many solutions arising from this bound are also tight.

One way to prove Theorem 3.1 is using a classical embedding to send all fusion frame elements to points on a higher dimensional Euclidean sphere. The desired claims then follow from a result of Rankin:

Theorem 3.2 ([16, 30]) *Let d be a positive integer and $\{v_1, v_2, \ldots, v_n\}$ be n vectors on the unit sphere in $\mathbb{R}^d$. Then*

$$\min_{1\le i\ne j\le n} \|v_i - v_j\| \le \sqrt{\frac{2n}{n-1}},$$

and if equality is achieved, then $n \le d+1$ and the vectors form a simplex. Additionally, if $n > d+1$, then the minimum Euclidean distance improves to:

$$\min_{1\le i\ne j\le n} \|v_i - v_j\| \le \sqrt{2},$$

and if equality holds in this case, then $n \le 2d$. Moreover, if $n = 2d$, then equality holds if and only if the vectors form an orthoplex, the union of an orthonormal basis with the negatives of its basis vectors.

In terms of the inner products between n unit vectors in $\mathbb{R}^d$, the Rankin bound is $\max_{i\ne j}\langle v_i, v_j\rangle \ge -\frac{1}{n-1}$, and if $n > d+1$, then $\max_{i\ne j}\langle v_i, v_j\rangle \ge 0$.

To see Theorem 3.2 implies Theorem 3.1, recall the well-known, isometric embedding, which maps constant-rank projections to points on a real, higher dimensional sphere, see [7, 13, 16]. Given any $m \times m$ rank l orthogonal projection P, the **l-traceless map** T_l is defined as:

$$T_l(P) = P - \frac{l}{m} I_m.$$

As proven in [7, 16], T_l isometrically injects P into the $d_{\mathbb{F}^m}$-dimensional subspace of the $m \times m$ symmetric/Hermitian matrices,

$$\mathbb{H} = \{A \in \mathbb{F}^{m\times m} : \mathrm{tr}(A) = 0\},$$

endowed with the standard Hilbert–Schmidt inner product. Dimension counting yields the value of $d_{\mathbb{F}^m}$ as function of m, as described in (3.1). With an appropriate choice of isomorphism, which we record as $\mathcal{V}$, we interpret the inner product between elements of this subspace as the standard Euclidean inner product between vectors in $\mathbb{R}^{d_{\mathbb{F}^m}}$.

Denote by Ω_l the image of $\mathcal{G}(l, \mathbb{F}^m)$ under the l-traceless map T_l and $\mathcal{V}$ as:

$$\Omega_l := \mathcal{V}\left(T_l\left(\mathcal{G}\left(l, \mathbb{F}^m\right)\right)\right) \subseteq \mathbb{R}^{d_{\mathbb{F}^m}}.$$

It is clear that the image Ω_l lies on the sphere in $\mathbb{R}^{d_{\mathbb{F}^m}}$, with squared radius

$$r_l^2 = \mathrm{tr}\left(\left(P - \frac{l}{m} I_m\right)^2\right) = \frac{l(m-l)}{m}.$$

Normalizing the image, Ω_l, to lie on the surface of the unit sphere

$$\mathcal{K}_l := \left\{ \frac{A}{r_l} : A \in \Omega_l \right\} \subseteq \mathcal{S}^{d_{\mathbb{F}^m}-1} \subseteq \mathbb{R}^{d_{\mathbb{F}^m}},$$

thereby converting any constant-rank packing into a code. Because $\mathcal{V}$ and the l-traceless map are all continuous actions, and because the Grassmannian manifold is well-known to be compact, elementary topological theory implies that $\mathcal{K}_l$ is compact.

For any orthogonal projection $P \in \mathcal{G}(l, \mathbb{F}^m)$, set $v_P = \dfrac{\mathcal{V}(T_l(P))}{r_l}$, then $v \in \mathcal{K}_l$. We will say that v_P is the **embedded vector** corresponding to P. A simple computation yields the identity:

$$\mathrm{tr}\,(PQ) = \frac{l^2}{m} + \frac{l(m-l)}{m} \langle v_P, v_Q \rangle, \tag{3.2}$$

for any $P, Q \in \mathcal{G}(l, \mathbb{F}^m)$, where v_P and v_Q are the corresponding embedded vectors of P and Q, respectively. Theorem 3.1 then follows by using (3.2) and Rankin's result.

Thus, the optimality of the packing $\mathcal{P}$ has a close connection with the restricted coding problem. More precisely, for any compact set $\mathcal{K}$ in the unit sphere in $\mathbb{R}^d$, we can consider the following restricted coding problem:

Problem 1 A code $C = \{v_i\}_{i=1}^n \subseteq \mathcal{K}$ is said to be a solution to the restricted coding problem respective to $\mathcal{K}$ if it satisfies

$$\max_{1 \le i \ne j \le n} \langle v_i, v_j \rangle = \min_{\{u_i\}_{i=1}^n \subseteq \mathcal{K}} \max_{1 \le i \ne j \le n} \langle u_i, u_j \rangle.$$

By what we have discussed, the following proposition is obvious.

Proposition 1 *A packing $\mathcal{P} = \{P_i\}_{i=1}^n \subseteq \mathcal{G}(l, \mathbb{F}^m)$ is optimally spread if and only if the corresponding embedded unit vectors $\{v_i\}_{i=1}^n \subseteq \mathbb{R}^{d_{\mathbb{F}^m}}$ are a solution to the restricted coding problem respective to $\mathcal{K}_l$.*

4 Generalized to the Mixed-Rank Packing Problem

In the previous section, we have seen that solutions to the constant-rank packing problem can be found by solving the corresponding restricted coding problem. We will use this idea to define a notion of optimal packings of subspaces with various dimensions.

Given a sequence of pairs $\{(n_i, l_i)\}_{i=1}^{s}$. By the previous section, for each i, $\mathcal{K}_{l_i} := \frac{1}{r_{l_i}}\mathcal{V}(T_{l_i}(\mathcal{G}(l_i, \mathbb{F}^m)))$ is a compact set in the unit sphere of $\mathbb{R}^{d_{\mathbb{F}^m}}$. Note that $d_{\mathbb{F}^m}$ is independent of l_i. Moreover, they are disjoint sets.

Proposition 2 *For any* $1 \le l_i, l_j \le m$, $i \ne j$, *we have*

$$\mathcal{K}_{l_i} \cap \mathcal{K}_{l_j} = \emptyset.$$

Proof Suppose by way of contradiction that $\mathcal{K}_{l_i} \cap \mathcal{K}_{l_j} \ne \emptyset$. Then there exists $v \in \mathcal{K}_{l_i} \cap \mathcal{K}_{l_j}$. We can write v in two ways as follows:

$$v = \frac{1}{r_{l_i}}\mathcal{V}\left(P - \frac{l_i}{m}I_m\right) = \frac{1}{r_{l_j}}\mathcal{V}\left(Q - \frac{l_j}{m}I_m\right)$$

for some $P \in \mathcal{G}(l_i, \mathbb{F}^m)$ and $Q \in \mathcal{G}\left(l_j, \mathbb{F}^m\right)$.

Hence,

$$\frac{1}{r_{l_i}}\left(P - \frac{l_i}{m}I_m\right) = \frac{1}{r_{l_j}}\left(Q - \frac{l_j}{m}I_m\right).$$

Since P has eigenvalues 1 and 0 with corresponding multiplicities l_i and $m - l_i$. It follows that $\frac{1}{r_{l_i}}\left(P - \frac{l_i}{m}I_m\right)$ has eigenvalues $\frac{1}{r_{l_i}}\left(1 - \frac{l_i}{m}\right)$ and $-\frac{l_i}{mr_{l_i}}$ with multiplicities l_i and $m - l_i$, respectively. Likewise, $\frac{1}{r_{l_j}}\left(Q - \frac{l_j}{m}I_m\right)$ has eigenvalues $\frac{1}{r_{l_j}}\left(1 - \frac{l_j}{m}\right)$ and $-\frac{l_j}{mr_{l_j}}$ with respective multiplicities l_j and $m - l_j$. This contradicts the fact that $l_i \ne l_j$. □

Similar to (3.2), the following identity gives a relation between the Hilbert–Schmidt inner products of orthogonal projections and the inner products of their embedded vectors.

Proposition 3 *Let P and Q be orthogonal projections of rank l_P and l_Q, respectively. Let v_P and v_Q be the corresponding embedded vectors. Then*

$$\langle v_P, v_Q\rangle = \sqrt{\frac{m^2}{l_P l_Q(m - l_P)(m - l_Q)}}\left(\operatorname{tr}(PQ) - \frac{l_P l_Q}{m}\right). \tag{4.1}$$

For any mixed-rank packing $\mathcal{P} = \{P_j^{(l_i)}\}_{i=1,j=1}^{s,\ n_i} \in \prod_{i=1}^{s}\prod_{j=1}^{n_i}\mathcal{G}(l_i, \mathbb{F}^m)$, if $v_j^{(l_i)}$ is the embedded vector corresponding to $P_j^{(l_i)}$, then $C = \{v_j^{(l_i)}\}_{i=1,j=1}^{s,\ n_i}$ is a code in $\mathbb{R}^{d_{\mathbb{F}^m}}$. Note that for each $i \in [[s]]$, $\{v_j^{(l_i)}\}_{j=1}^{n_i}$ is a sequence of unit vectors lying on the compact set $\mathcal{K}_{l_i}$ of the unit sphere. In other words, $\{v_j^{(l_i)}\}_{i=1,j=1}^{s,\ n_i} \in$

$\prod_{i=1}^{s}\prod_{j=1}^{n_i}\mathcal{K}_{l_i}$. We will call the value $\mu(\mathcal{P}) := \max_{(j,l_i)\neq(j',l_{i'})}\langle v_j^{(l_i)}, v_{j'}^{(l_{i'})}\rangle$ the **coherence** of $\mathcal{P}$.

Motivated by Proposition 1, we will now give a definition for optimally spread of mixed-rank packings:

Definition 3 Let $\mathcal{P} = \{P_j^{(l_i)}\}_{i=1,j=1}^{s,\ n_i}$ be a mixed-rank packing in $\mathbb{F}^m$. $\mathcal{P}$ is said to be **optimally spread** if the corresponding embedded vectors $\{v_j^{(l_i)}\}_{i=1,j=1}^{s,\ n_i}$ satisfy

$$\max_{(j,l_i)\neq(j',l_{i'})}\langle v_j^{(l_i)}, v_{j'}^{(l_{i'})}\rangle = \min_{\{u_j^{(l_i)}\}_{i=1,j=1}^{s,\ n_i}\in\prod_{i=1}^{s}\prod_{j=1}^{n_i}\mathcal{K}_{l_i}}\ \max_{(j,l_i)\neq(j',l_{i'})}\langle u_j^{(l_i)}, u_{j'}^{(l_{i'})}\rangle.$$

The value

$$\mu := \min_{\{u_j^{(l_i)}\}_{i=1,j=1}^{s,\ n_i}\in\prod_{i=1}^{s}\prod_{j=1}^{n_i}\mathcal{K}_{l_i}}\ \max_{(j,l_i)\neq(j',l_{i'})}\langle u_j^{(l_i)}, u_{j'}^{(l_{i'})}\rangle$$

is referred to as the **packing constant**.

We also say that their associated subspaces $\mathcal{W} = \{W_j^{(l_i)}\}_{i=1,j=1}^{s,\ n_i}$ are an optimally spread mixed-rank packing.

Remark 1 This definition is well-posed because each $\mathcal{K}_{l_i}$ is compact and the objective functions are continuous.

5 Properties of Optimally Spread Mixed-Rank Packings

This section presents some properties of solutions to the mixed-rank packing problem. In particular, as in the constant-rank case [13], we will see that solutions to the mixed-rank packing problem are always fusion frames.

Suppose $\mathcal{P} = \{P_j^{(l_i)}\}_{i=1,j=1}^{s,\ n_i}$ is an optimally spread mixed-rank packing for $\mathbb{F}^m$. Let $\mathcal{W} = \{W_j^{(l_i)}\}_{i=1,j=1}^{s,\ n_i}$ be their associated subspaces. To simplify notation, in the following theorems, we will enumerate $\mathcal{P}$ and $\mathcal{W}$ as $\mathcal{P} = \{P_i^{(l_i)}\}_{i=1}^{n}$ and $\mathcal{W} = \{W_i^{(l_i)}\}_{i=1}^{n}$, where $n = \sum_{i=1}^{s} n_i$ is the number of subspaces of $\mathcal{W}$, $\dim W_i = l_i$, $1 \leq l_i \leq m$. Note that $l_i's$ here need not to be distinct as before.

Let μ be the packing constant and let $\mathcal{W} = \{W_i^{(l_i)}\}_{i=1}^{n}$ be an optimally spread mixed-rank packing. We say that an element $W_j^{(l_j)} \in \mathcal{W}$ **achieves the packing constant** if there exists some $i \in [[n]]$ such that the inner product between the corresponding embedded vectors of $W_j^{(l_j)}$ and $W_i^{(l_i)}$ equals μ. We call each element $W_i^{(l_i)}$ satisfying this condition a **packing neighbor** of $W_j^{(l_j)}$.

Theorem 5.1 *Let* $\mathcal{W} = \{W_i^{(l_i)}\}_{i=1}^n, n \geq m$ *be an optimally spread mixed-rank packing. Denote*

$$\mathcal{I} := \{i : W_i^{(l_i)} \text{ achieves the packing constant}\}.$$

Then

$$\text{span}\{W_i^{(l_i)} : i \in \mathcal{I}\} = \mathbb{F}^m.$$

In order to prove Theorem 5.1, we need some lemmas. The first one can be proved similarly to Lemma 3.1 in [13].

Lemma 1 *Let* $0 < \varepsilon < \alpha$ *and let* $\{x_i\}_{i=1}^l$ *and* $\{y_i\}_{i=1}^k$ *be unit vectors in* $\mathbb{F}^m$ *satisfying:*

$$\sum_{i=1}^{l}\sum_{j=1}^{k} \left|\langle x_i, y_j\rangle\right|^2 < \alpha - \varepsilon.$$

Let δ *be such that*

$$2\sqrt{l\delta(\alpha - \varepsilon)} + l\delta < \frac{\varepsilon}{2}.$$

If $\{z_i\}_{i=1}^k$ *is a sequence of unit vectors in* $\mathbb{F}^m$ *satisfying*

$$\sum_{i=1}^{k} \|y_i - z_i\|^2 < \delta,$$

then

$$\sum_{i=1}^{l}\sum_{j=1}^{k} \left|\langle x_i, z_j\rangle\right|^2 < \alpha - \frac{\varepsilon}{2}.$$

Lemma 2 *Let* $\mathcal{W} = \{W_i^{(l_i)}\}_{i=1}^n, n \geq m$ *be an optimally spread mixed-rank packing. Suppose that* $\mathcal{W}$ *does not contain a pair of orthogonal lines for which the inner product of their corresponding embedded vectors equals the packing constant. If* $W_k^{(l_k)}$ *is an element of* $\mathcal{W}$ *that achieves the packing constant, then it contains a unit vector which is not orthogonal to any of its packing neighbors.*

Proof Denote

$$\mathcal{I}_k = \{i : W_i^{(l_i)} \text{ is a packing neighbor of } W_k^{(l_k)}\}.$$

Let μ be the packing constant. By the definition of packing neighbor and (4.1), we have that

$$\operatorname{tr}(P_i^{(l_i)} P_k^{(l_k)}) = \mu\sqrt{\frac{l_i l_k (m-l_i)(m-l_k)}{m^2}} + \frac{l_i l_k}{m}, \text{ for all } i \in \mathcal{I}_k.$$

First, we will show that $\operatorname{tr}(P_i^{(l_i)} P_k^{(l_k)}) > 0$ for all $i \in \mathcal{I}_k$.

If $n > d_{\mathbb{F}^m} + 1$, then this is obvious since $\mu \geq 0$.

Consider the case $m \leq n \leq d_{\mathbb{F}^m} + 1$. In this case, the Rankin bound gives $\mu \geq -\frac{1}{n-1}$.

Therefore

$$\operatorname{tr}(P_i^{(l_i)} P_k^{(l_k)}) \geq \left(-\frac{1}{n-1}\right)\sqrt{\frac{l_i l_k (m-l_i)(m-l_k)}{m^2}} + \frac{l_i l_k}{m}, \text{ for all } i \in \mathcal{I}_k.$$

Note that

$$\left(-\frac{1}{n-1}\right)\sqrt{\frac{l_i l_k (m-l_i)(m-l_k)}{m^2}} + \frac{l_i l_k}{m} > 0$$

is equivalent to

$$(n-1)^2 l_i l_k > (m-l_i)(m-l_k).$$

We estimate

$$\begin{aligned}(n-1)^2 l_i l_k - (m-l_i)(m-l_k) &\geq (m-1)^2 l_i l_k - (m-l_i)(m-l_k)\\ &= (m^2-2m+1) l_i l_k - m^2 + m l_i + m l_k - l_i l_k\\ &= (m^2-2m) l_i l_k - m^2 + m l_i + m l_k.\end{aligned}$$

If $l_i \neq l_k$, then we can assume that $l_i > l_k$, so $l_i \geq l_k + 1$. This implies

$$\begin{aligned}(n-1)^2 l_i l_k - (m-l_i)(m-l_k) &\geq (m^2-2m) l_i l_k - m^2 + m l_i + m l_k\\ &\geq (m^2-2m)(l_k+1) l_k - m^2 + m(l_k+1) + m l_k\\ &= m l_k^2 (m-2) + m^2 (l_k - 1) + m > 0.\end{aligned}$$

If $l_i = l_k \geq 2$, then

$$\begin{aligned}(n-1)^2 l_i l_k - (m-l_i)(m-l_k) &\geq (m^2-2m) l_i l_k - m^2 + m l_i + m l_k\\ &= (m^2-2m) l_k^2 - m^2 + 2m l_k\\ &= (m-1)^2 l_k^2 - (m-l_k)^2 > 0.\end{aligned}$$

Thus, we have shown that $\mathrm{tr}(P_i^{(l_i)}P_k^{(l_k)}) > 0$ for all $i \in \mathcal{I}_k$.

Now for each $i \in \mathcal{I}_k$, let $V_i := W_k^{(l_k)} \cap [W_i^{(l_i)}]^{\perp}$. Since $\mathrm{tr}(P_i^{(l_i)}P_k^{(l_k)}) > 0$, every V_i is a proper subspace of $W_k^{(l_k)}$, and since a linear space cannot be written as a finite union of proper subspaces, it follows that $W_k^{(l_k)} \setminus \cup_{i\in\mathcal{I}_k} V_i$ is nonempty, so the claim follows. □

Proof of Theorem 5.1 We first consider the case for which $\mathcal{W}$ satisfies the assumption in Lemma 2. Let μ be the packing constant. We now proceed by way of contradiction. Iteratively replacing elements of $\mathcal{W}$ that achieve the packing constant in such a way that we eventually obtain a new packing in $\mathbb{F}^m$ with coherence strictly less than μ, which cannot exist. With the contradictory approach in mind, fix a unit vector $z \in \mathbb{F}^m$ so that z is orthogonal to all $W_i^{(l_i)}$, $i \in \mathcal{I}$.

For a fixed $k \in \mathcal{I}$, denote

$$\mathcal{I}_k = \left\{1 \le i \le n : i \ne k, \mathrm{tr}(P_i^{(l_i)}P_k^{(l_k)}) = \mu\sqrt{\frac{l_i l_k (m-l_i)(m-l_k)}{m^2}} + \frac{l_i l_k}{m}\right\}.$$

Then

$$\mathrm{tr}(P_i^{(l_i)}P_k^{(l_k)}) < \mu\sqrt{\frac{l_i l_k (m-l_i)(m-l_k)}{m^2}} + \frac{l_i l_k}{m}, \text{ for all } i \in \mathcal{I}_k^c \setminus \{k\}.$$

This implies there exists $\varepsilon > 0$ such that

$$\mathrm{tr}(P_i^{(l_i)}P_k^{(l_k)}) < \mu\sqrt{\frac{l_i l_k (m-l_i)(m-l_k)}{m^2}} + \frac{l_i l_k}{m} - \varepsilon, \text{ for all } i \in \mathcal{I}_k^c \setminus \{k\}.$$

By Lemma 2, there exists a unit vector $x_{k,1} \in W_k^{(l_k)}$ which is not orthogonal to any $W_i^{(l_i)}$, $i \in \mathcal{I}_k$. Extend it to an orthonormal basis $\{x_{k,j}\}_{j=1}^{l_k}$ and let $\{x_{i,j}\}_{j=1}^{l_i}$ be an orthonormal basis for $W_i^{(l_i)}$, for $i \ne k$.

For each $i \in \mathcal{I}_k^c \setminus \{k\}$, denote

$$\alpha(i) := \mu\sqrt{\frac{l_i l_k (m-l_i)(m-l_k)}{m^2}} + \frac{l_i l_k}{m}$$

and let δ be such that

$$0 < \delta < \frac{1}{2} \text{ and } 2\sqrt{l_i\delta(\alpha(i)-\varepsilon)} + l_i\delta < \frac{\varepsilon}{2}, \text{ for all } i \in \mathcal{I}_k^c \setminus \{k\}.$$

Define

$$y_{k,1} := \sqrt{1-\delta^2}\, x_{k,1} + \delta z \text{ and } y_{k,j} := x_{k,j} \text{ for } 2 \le j \le l_k,$$

and define $V_k^{(l_k)} := \text{span}\{y_{k,j}\}_{j=1}^{l_k}$ with the corresponding orthogonal projection, $Q_k^{(l_k)}$. Because z is orthogonal to $W_k^{(l_k)}$, it follows that $\{y_{k,j}\}_{j=1}^{l_k}$ is an orthonormal basis for $V_k^{(l_k)}$.

Noting that $1-\delta^2 < \sqrt{1-\delta^2}$ implies $2(1-\sqrt{1-\delta^2}) < 2\delta^2$, we estimate

$$\sum_{j=1}^{l_k} \|x_{k,j} - y_{k,j}\|^2 = \|x_{k,1} - y_{k,1}\|^2 = 2(1-\sqrt{1-\delta^2}) < 2\delta^2 < \delta.$$

Hence, using Lemma 1, where $\alpha = \alpha(i) = \mu\sqrt{\frac{l_i l_k(m-l_i)(m-l_k)}{m^2}} + \frac{l_i l_k}{m}$ we have that for each $i \in \mathcal{I}_k^c \setminus \{k\}$,

$$\text{tr}(P_i^{(l_i)} Q_k^{(l_k)}) = \sum_{j=1}^{l_i}\sum_{j'=1}^{l_k} \left|\langle x_{i,j}, y_{k,j'}\rangle\right|^2 < \mu\sqrt{\frac{l_i l_k(m-l_i)(m-l_k)}{m^2}} + \frac{l_i l_k}{m} - \frac{\varepsilon}{2}.$$

Furthermore, the nonorthogonality of $x_{k,1}$ with W_i for every $i \in \mathcal{I}_k$ implies

$$0 < \sum_{j=1}^{l_i} \left|\langle x_{i,j}, y_{k,1}\rangle\right|^2 = (1-\delta^2)\sum_{j=1}^{l_i} \left|\langle x_{i,j}, x_{k,1}\rangle\right|^2 < \sum_{j=1}^{l_i} \left|\langle x_{i,j}, x_{k,1}\rangle\right|^2 \text{ for all } i \in \mathcal{I}_k.$$

It follows that

$$\begin{aligned}
\text{tr}(P_i^{(l_i)} Q_k^{(l_k)}) &= \sum_{j=1}^{l_i}\sum_{j'=1}^{l_k} \left|\langle x_{i,j}, y_{k,j'}\rangle\right|^2 \\
&< \sum_{j=1}^{l_i}\sum_{j'=1}^{l_k} \left|\langle x_{i,j}, x_{k,j'}\rangle\right|^2 \\
&= \text{tr}(P_i^{(l_i)} P_k^{(l_k)}) = \mu\sqrt{\frac{l_i l_k(m-l_i)(m-l_k)}{m^2}} + \frac{l_i l_k}{m}, \text{ for all } i \in \mathcal{I}_k.
\end{aligned}$$

Thus, replacing $W_k^{(l_k)}$ by $V_k^{(l_k)}$, we have that

$$\text{tr}(P_i^{(l_i)} Q_k^{(l_k)}) < \mu\sqrt{\frac{l_i l_k(m-l_i)(m-l_k)}{m^2}} + \frac{l_i l_k}{m}, \text{ for all } i \neq k,$$

i.e.,

$$\mu > \sqrt{\frac{m^2}{l_i l_k(m-l_i)(m-l_k)}}\left(\text{tr}(P_i^{(l_i)} Q_k^{(l_k)}) - \frac{l_i l_k}{m}\right), \text{ for all } i \neq k.$$

Now, pick another $k \in \mathcal{I}$ and iterate this replacement procedure. Note that we do not need to change the fixed vector z for each step since $\mathcal{I}_k \subseteq \mathcal{I}$. After finite repetitions of this process, we obtain a final packing, with its coherence strictly less than μ, the desired contradiction.

Now we consider the case where $\mathcal{W}$ contains a pair of orthogonal lines, W_i, W_j, such that the inner product of their corresponding embedded vectors, v_i, v_j, equals the packing constant μ. Then by (4.1) and the Rankin bound, we have

$$\mu = \langle v_i, v_j \rangle = -\frac{1}{m-1} \geq -\frac{1}{n-1}.$$

This implies that $n = m$ and hence the corresponding embedded vectors of the packing form a simplex in $\mathbb{R}^{d_{\mathbb{F}^m}}$. We claim that $\mathcal{W}$ spans $\mathbb{F}^m$.

Indeed, the claim is clear if all elements of $\mathcal{W}$ are one-dimensional subspaces since in this case, $\mathcal{W}$ must be a set of m pairwise orthogonal lines in $\mathbb{F}^m$. Now suppose that $\mathcal{W}$ contains a subspace $W_k^{(l_k)}$ of dimension $l_k > 1$. Note that every element of $\mathcal{W}$ is a packing neighbor of $W_k^{(l_k)}$ and as in the proof of Lemma 2, $\text{tr}(P_i^{(l_i)} P_k^{(l_k)}) > 0$ for all $i \in [[m]], i \neq k$. If $\mathcal{W}$ does not span the space, then we can replace $W_k^{(l_k)}$ by $V_k^{(l_k)}$ so that the embedded vector of $V_k^{(l_k)}$ has inner product with all other embedded vectors of the packing less than μ. This contradicts to the fact that these embedded vectors (of this new packing) are a simplex. The proof of the theorem is completed. □

For any packing $\mathcal{P} = \{P_i\}_{i=1}^n$, its spatial complement is the family $\mathcal{P}^\perp = \{I_m - P_i\}_{i=1}^n$, where I_m is the identity operator in $\mathbb{F}^m$. The next property demonstrates that optimality of a packing $\mathcal{P}$ is preserved when we switch to its spatial complement.

Proposition 4 *If $\mathcal{P}$ is an optimally spread mixed-rank packing with mixture s, then so is its spatial complement.*

Proof Suppose $\mathcal{P}$ is an optimally spread mixed-rank packing for $\mathbb{F}^m$. Let any $P \in \mathcal{P}$, then $Q := I_m - P \in \mathcal{P}^\perp$. It is easy to see that $v_Q = -v_P$, where v_P and v_Q are the embedded vectors corresponding to P and Q. The conclusion then follows by our definition of optimally spread mixed-rank packings. □

Given a mixed-rank packing $\mathcal{P}$ in $\mathbb{F}^m$, if the corresponding embedded vectors form an orthoplex in $\mathbb{R}^{d_{\mathbb{F}^m}}$ (so $|\mathcal{P}| = 2d_{\mathbb{F}^m}$), then we will say that $\mathcal{P}$ is a **maximal orthoplectic fusion frame**. Note that such fusion frames exist, for example, see [7]. The following is a nice property of these optimal packings.

Theorem 5.2 *All maximal orthoplectic fusion frames are tight.*

Proof Suppose $\mathcal{P} = \{P_i^{(l_i)}\}_{i=1}^{2d_{\mathbb{F}^m}}$ is a maximal orthoplectic fusion frame. By definition, the corresponding embedded vectors of $\mathcal{P}$ form an orthoplex in $\mathbb{R}^{2d_{\mathbb{F}^m}}$. This implies that for every i, there exists j such that

$$\sqrt{\frac{m}{l_i(m-l_i)}}\left(P_i^{(l_i)}-\frac{l_i}{m}I_m\right)=-\sqrt{\frac{m}{l_j(m-l_j)}}\left(P_j^{(l_j)}-\frac{l_j}{m}I_m\right). \tag{5.1}$$

Let W_i and W_j be the associated subspaces to $P_i^{(l_i)}$ and $P_j^{(l_j)}$, respectively. It is sufficient to show that $W_i \oplus W_j = \mathbb{F}^m$.

We proceed by way of contradiction. If $\text{span}\{W_i, W_j\} \neq \mathbb{F}^m$, then let x be orthogonal to both W_i and W_j. Using (5.1), we get

$$-\frac{l_i}{m}\sqrt{\frac{m}{l_i(m-l_i)}}=\frac{l_j}{m}\sqrt{\frac{m}{l_j(m-l_j)}},$$

which is impossible.

Likewise, if $W_i \cap W_j \neq \{0\}$, let a non-zero vector x be in the intersection, then we get

$$\sqrt{\frac{m}{l_i(m-l_i)}}\left(1-\frac{l_i}{m}\right)=-\sqrt{\frac{m}{l_j(m-l_j)}}\left(1-\frac{l_j}{m}\right),$$

which is again impossible. The conclusion then follows. □

6 Constructing Solutions

In this section, we will construct some infinite families of solutions to the mixed-rank packing problem in $\mathbb{F}^m$ when the number of projections n exceeds $d_{\mathbb{F}^m}+1$. By Identity (4.1), we will construct packings $\mathcal{P} = \{P_j^{(l_i)}\}_{i=1,j=1}^{s,\ n_i} \in \prod_{i=1}^{s}\prod_{j=1}^{n_i}\mathcal{G}(l_i,\mathbb{F}^m)$ such that $n=\sum_{i=1}^{s} n_i > d_{\mathbb{F}^m}+1$ and $\text{tr}(P_j^{(l_i)}P_{j'}^{(l_{i'})}) \le \frac{l_i l_{i'}}{m}$, for any $(l_i, j) \neq (l_{i'}, j')$. This implies that the corresponding embedded vectors $\{v_j^{(l_i)}\}_{i=1,j=1}^{s,\ n_i}$ are a solution for the usual (unrestricted) coding problem by Theorem 3.2. Thus, by Definition 3, $\mathcal{P}$ is a solution to the mixed-rank packing problem.

Similar to the constructions in [7, 13], we will construct mixed-rank packings containing coordinate projections as defined below:

Definition 4 Given an orthonormal basis $\mathcal{B}=\{b_j\}_{j=1}^m$ for $\mathbb{F}^m$ and a subset $\mathcal{J} \subseteq [[m]]$, the $\mathcal{J}$-coordinate projection with respect to $\mathcal{B}$ is $P_{\mathcal{J}}^{\mathcal{B}}=\sum_{j\in\mathcal{J}} b_j b_j^*$.

Our main tools for the constructions are mutually unbiased bases (MUBs) and block designs.

Definition 5 If $\mathcal{B} = \{b_j\}_{j=1}^m$ and $\mathcal{B}' = \{b'_j\}_{j=1}^m$ are a pair of orthonormal bases for $\mathbb{F}^m$, then they are mutually unbiased if $|\langle b_j, b'_{j'}\rangle|^2 = \frac{1}{m}$ for every $j, j' \in [[m]]$. A collection of orthonormal bases $\{\mathcal{B}_k\}_{k\in K}$ is called a set of mutually unbiased bases (MUBs) if the pair $\mathcal{B}_k$ and $\mathcal{B}_{k'}$ is mutually unbiased for every $k \neq k'$.

As in [7, 13], maximal sets of MUBs play an important role in our constructions. The following theorems give an upper bound for the cardinality of the set of MUBs in terms of m, and sufficient conditions to attain this bound.

Theorem 6.1 ([17]) *If $\{B_k\}_{k\in K}$ is a set of MUBs for $\mathbb{F}^m$, then $|K| \le m/2 + 1$ if $\mathbb{F} = \mathbb{R}$, and $|K| \le m + 1$ if $\mathbb{F} = \mathbb{C}$.*

Theorem 6.2 ([10, 36]) *If m is a prime power, then a family of $m + 1$ MUBs for $\mathbb{C}^m$ exists. If m is a power of 4, then a family of $m/2 + 1$ MUBs for $\mathbb{R}^m$ exists.*

Henceforth, we abbreviate $k_{\mathbb{R}^m} = m/2 + 1$ and $k_{\mathbb{C}^m} = m + 1$. In order to construct the desired packings, the following simple proposition is very useful.

Proposition 5

1. *Let $\mathcal{B} = \{b_j\}_{j=1}^m$ be an orthonormal basis for $\mathbb{F}^m$. Then for any subsets $\mathcal{J}, \mathcal{J}' \subseteq [[m]]$, we have*

$$\mathrm{tr}(P_{\mathcal{J}}^{\mathcal{B}} P_{\mathcal{J}'}^{\mathcal{B}}) = |\mathcal{J} \cap \mathcal{J}'|.$$

2. *Let $\mathcal{B}_1 = \{x_j\}_{j=1}^m$ and $\mathcal{B}_2 = \{y_j\}_{j=1}^m$ be a pair of MUBs for $\mathbb{F}^m$. Then for any subsets $\mathcal{J}, \mathcal{J}' \subseteq [[m]]$, we have*

$$\mathrm{tr}(P_{\mathcal{J}}^{\mathcal{B}_1} P_{\mathcal{J}'}^{\mathcal{B}_2}) = \frac{|\mathcal{J}||\mathcal{J}'|}{m}.$$

Proof We compute

$$\mathrm{tr}(P_{\mathcal{J}}^{\mathcal{B}} P_{\mathcal{J}'}^{\mathcal{B}}) = \sum_{j\in\mathcal{J}} \sum_{j'\in\mathcal{J}'} \mathrm{tr}(b_j b_j^* b_{j'} b_{j'}^*) = \sum_{j\in\mathcal{J}} \sum_{j'\in\mathcal{J}'} |\langle b_j, b_{j'}\rangle|^2 = |\mathcal{J} \cap \mathcal{J}'|,$$

which is (1).

For (2), since $\mathcal{B}_1$ and $\mathcal{B}_2$ are mutually unbiased, it follows that $|\langle x_j, y_{j'}\rangle|^2 = 1/m$, for every $j \in \mathcal{J}$, $j' \in \mathcal{J}'$. Therefore,

$$\mathrm{tr}(P_{\mathcal{J}}^{\mathcal{B}_1} P_{\mathcal{J}'}^{\mathcal{B}_2}) = \sum_{j\in\mathcal{J}} \sum_{j'\in\mathcal{J}'} \mathrm{tr}(x_j x_j^* y_{j'} y_{j'}^*) = \sum_{j\in\mathcal{J}} \sum_{j'\in\mathcal{J}'} |\langle x_j, y_{j'}\rangle|^2 = \frac{|\mathcal{J}||\mathcal{J}'|}{m},$$

which is the claim. □

To control the trace between coordinate projections generated by the same orthonormal basis, we need the following concept stated in [7].

Definition 6 Let $\mathbb{S}$ be a collection of subsets of $[[m]]$ such that each $\mathcal{J}$ of $\mathbb{S}$ has the same cardinality. We say that $\mathbb{S}$ is c-cohesive if there exists $c > 0$ such that $\max_{\mathcal{J},\mathcal{J}'\in\mathbb{S},\mathcal{J}\neq\mathcal{J}'} |\mathcal{J}\cap\mathcal{J}'| \leq c$.

Another ingredient for our constructions is block designs.

Definition 7 A t-(m, l, λ) block design $\mathbb{S}$ is a collection of subsets of $[[m]]$, called blocks, where each block $\mathcal{J} \in \mathbb{S}$ has cardinality l, such that every subset of $[[m]]$ with cardinality t is contained in exactly λ blocks and each element of $[[m]]$ occurs in exactly r blocks. We also denote $b = |\mathbb{S}|$, the cardinality of $\mathbb{S}$. If $t = 2$, then the design is called a **balanced incomplete block design or BIBD**. When the parameters are not important or implied by the context, then $\mathbb{S}$ is also referred to as a t-block design.

The following proposition gives a few simple facts about block designs.

Proposition 6 *For any t-(m, l, λ) block design, the following conditions hold:*

1. $mr = bl$,
2. $r(l-1) = \lambda(m-1)$ *if* $t = 2$.

Furthermore, a t-block design is also a $(t-1)$-block design for $t \geq 1$.

Tight fusion frames can be formed by coordinate projections from block designs.

Proposition 7 ([7, 13]) *Let $\mathcal{B} = \{b_j\}_{j=1}^m$ be an orthonormal basis for $\mathbb{F}^m$ and $\mathbb{S} = \{\mathcal{J}_1, \mathcal{J}_2, \ldots, \mathcal{J}_b\}$ be a t-(m, l, λ)-block design. Then the set of coordinate projections with respect to $\mathcal{B}$, $\{P_{\mathcal{J}_1}^{\mathcal{B}}, P_{\mathcal{J}_2}^{\mathcal{B}}, \ldots, P_{\mathcal{J}_b}^{\mathcal{B}}\}$, forms a tight fusion frame for $\mathbb{F}^m$.*

Proof Since $\mathbb{S}$ is a t-block design, every element $j \in [[m]]$ occurs in exactly r blocks. It follows that

$$\sum_{j=1}^{b} P_{\mathcal{J}_j}^{\mathcal{B}} = r \sum_{j=1}^{m} b_j b_j^* = r I_m.$$

This means that $\{P_{\mathcal{J}_j}^{\mathcal{B}}\}_{j=1}^b$ forms a tight fusion frame. □

The main idea of our constructions is using a set of MUBs and a collection of block designs to form a packing of coordinate projections. We are now ready for the main theorem of the constructions.

Theorem 6.3 *Let $\{\mathcal{B}_k\}_{k\in K}$ be a family of MUBs for $\mathbb{F}^m$, and let $\mathbb{S}_i$ be a 1-(m, l_i, λ_i) block design, $i = 1, \ldots, s$, where $s \leq |K|$. Let $\{A_1, A_2, \ldots, A_s\}$ be any partition of $[[K]]$ and for each $i \in [[s]]$, let $\mathcal{P}_i = \{P_{\mathcal{J}}^{\mathcal{B}_k} : k \in A_i, \mathcal{J} \in \mathbb{S}_i\}$ be the family of coordinate projections of rank l_i. If for each i, $\mathbb{S}_i$ is l_i^2/m-cohesive and $\sum_{i=1}^s |\mathbb{S}_i||A_i| > d_{\mathbb{F}^m} + 1$, then the union of families $\{\mathcal{P}_i\}_{i=1}^s$ forms an optimally spread mixed-rank tight fusion frame with mixture s for $\mathbb{F}^m$.*

Proof For every fixed $i \in [[s]]$, consider any coordinate projections $P_{\mathcal{J}}^{\mathcal{B}_k}, P_{\mathcal{J}'}^{\mathcal{B}_{k'}}$ in $\mathcal{P}_i$. If $k = k'$ then by (1) of Proposition 5 and the assumption that $\mathbb{S}_i$ is l_i^2/m-cohesive, we get

$$\mathrm{tr}(P_{\mathcal{J}}^{(k)} P_{\mathcal{J}'}^{(k')}) = |\mathcal{J} \cap \mathcal{J}'| \leq l_i^2/m$$

for all $\mathcal{J}, \mathcal{J}' \in \mathbb{S}_i, \mathcal{J} \neq \mathcal{J}'$. If $k \neq k'$, then by (2) of Proposition 5, we have

$$\mathrm{tr}(P_{\mathcal{J}}^{(k)} P_{\mathcal{J}'}^{(k')}) = l_i^2/m$$

for all $\mathcal{J}, \mathcal{J}' \in \mathbb{S}_i$. Thus, for each i, the family of coordinate projections of rank l_i, $\mathcal{P}_i = \{P_{\mathcal{J}}^{\mathcal{B}_k} : k \in A_i, \mathcal{J} \in \mathbb{S}_i\}$ forms a packing of $|\mathbb{S}_i||A_i|$ elements and the trace of the product of any two projections is at most l_i^2/m.

Now consider any $P \in \mathcal{P}_i$ and $Q \in \mathcal{P}_j$ where $i \neq j$. Then again, by (2) of Proposition 5, $\mathrm{tr}(PQ) = \dfrac{l_i l_j}{m}$. Moreover, by assumption, the number of elements of the union is greater than $d_{\mathbb{F}^m} + 1$. Therefore, this family forms an optimally spread mixed-rank packing with a mixture of s. It is a tight fusion frame by Proposition 7. □

Symmetric block designs are a special type of block designs and are well-studied objects in design theory, for example, see [9, 15]. In [13], the authors exploited their nice properties to construct optimally spread packings of constant-rank. In this paper, we will continue to use them for our constructions.

Definition 8 A 2-(m, l, λ) block design is symmetric if $m = b$, or equivalently $l = r$.

A very useful property of symmetric block designs is that the pairwise block intersections have the same cardinality.

Theorem 6.4 ([9]) *For a symmetric (m, l, λ) block design $\mathbb{S}$, every $\mathcal{J}, \mathcal{J}' \in \mathbb{S}, \mathcal{J} \neq \mathcal{J}'$ satisfies*

$$|\mathcal{J} \cap \mathcal{J}'| = \lambda.$$

Moreover, it has been shown in [13] that every symmetric block design has cohesive property.

Proposition 8 *Every symmetric (m, l, λ) block design is l^2/m-cohesive.*

We also need another property of symmetric block design.

Proposition 9 ([15]) *The complement of a symmetric (m, l, λ) block design $\mathbb{S}$ is a symmetric $(m, m-l, m-2l+\lambda)$ block design $\mathbb{S}^c$, whose blocks are the complements of blocks of $\mathbb{S}$.*

The following theorem will consider a special case where all block designs are symmetric.

Theorem 6.5 *Let $\{\mathcal{B}_k\}_{k\in K}$ be a family of MUBs and let $\mathbb{S}_i$ be a symmetric (m, l_i, λ_i) block design, $i = 1, \ldots, s$, $s \leq |K|$. Let $\{A_1, A_2, \ldots, A_s\}$ be any partition of $[[K]]$ and for each $i \in [[s]]$, let $\mathcal{P}_i = \{P_{\mathcal{J}}^{\mathcal{B}_k} : k \in A_i, \mathcal{J} \in \mathbb{S}_i\}$ be the family of coordinate projections of rank l_i. If $m \sum_{i=1}^{s} |A_i| > d_{\mathbb{F}^m} + 1$, then the union of families $\{\mathcal{P}_i\}_{i=1}^{s}$ forms an optimally spread mixed-rank tight fusion frame with mixture s for $\mathbb{F}^m$.*

Proof For each $i \in [[s]]$, by Proposition 8, $\mathbb{S}_i$ is l_i^2/m-cohesive. Note that all the sets $\mathbb{S}_i$ have the same cardinality, m. The conclusion then follows by Theorem 6.3. □

Example

1. Let $\{\mathcal{B}_1, \ldots, \mathcal{B}_9\}$ be 9 MUBs in $\mathbb{R}^{16}$ and let $\mathbb{S}$ be a symmetric $(16, 6, 2)$ block design, which exists, see [15]. Let $\{P_i^{\mathcal{B}_1}\}_{i=1}^{16}$ be 16 projections on lines, each spanned by a vector in $\mathcal{B}_1$ and let $\{P_{\mathcal{J}}^{\mathcal{B}_k} : \mathcal{J} \in \mathbb{S}\}$ be the collection of coordinate projections respective to $\mathcal{B}_k$, for $k = 2, \ldots, 9$. Then the family
$$\{P_i^{\mathcal{B}_1}\}_{i=1}^{16} \cup \{P_{\mathcal{J}}^{\mathcal{B}_k} : \mathcal{J} \in \mathbb{S}\}_{k=2}^{9}$$
forms an optimally spread mixed-rank tight fusion frame for $\mathbb{R}^{16}$ with a mixture of 2. This family has 16 rank 1 projections and 128 rank 6 projections.
2. Let $\mathbb{S}_1$ and $\mathbb{S}_2$ be symmetric block designs with parameters $(71, 15, 3)$ and $(71, 21, 6)$, respectively, see [15]. Since 71 is a prime, there are 72 MUBs in $\mathbb{C}^{71}$. Let $\{A_1, A_2\}$ be any partition of $[[72]]$. Then the family
$$\cup_{i=1}^{2}\{P_{\mathcal{J}}^{\mathcal{B}_k} : k \in A_i, \mathcal{J} \in \mathbb{S}_i\}$$
forms an optimally spread mixed-rank tight fusion frame for $\mathbb{C}^{71}$. It has 71×72 elements, namely $71|A_1|$ elements of rank 15 and $71|A_2|$ elements of rank 21.

As a consequence of Theorem 6.5 and noting that maximal sets of MUBs exist by Theorem 6.2, we get:

Corollary 1 *Let $\mathbb{S}$ be a symmetric block design and $\mathbb{S}^c$ be its complement.*

1. *If m is a power of 4, then for any partition $\{A_1, A_2\}$ of $\big[[m/2+1]\big]$, the family*

$$\{P_{\mathcal{J}}^{\mathcal{B}_k} : k \in A_1, \mathcal{J} \in \mathbb{S}\} \cup \{P_{\mathcal{J}}^{\mathcal{B}_k} : k \in A_2, \mathcal{J} \in \mathbb{S}^c\}$$

   forms an optimally spread mixed-rank tight fusion frame with a mixture of 2 for $\mathbb{R}^m$, where $\{\mathcal{B}_k\}_{k=1}^{m/2+1}$ is a maximal set of MUBs in $\mathbb{R}^m$.
2. *If m is a prime power, then for any partition $\{A_1, A_2\}$ of $\big[[m+1]\big]$, the family*

$$\{P_{\mathcal{J}}^{\mathcal{B}_k} : k \in A_1, \mathcal{J} \in \mathbb{S}\} \cup \{P_{\mathcal{J}}^{\mathcal{B}_k} : k \in A_2, \mathcal{J} \in \mathbb{S}^c\}$$

   forms an optimally spread mixed-rank tight fusion frame with a mixture of 2 for $\mathbb{C}^m$, where $\{\mathcal{B}_k\}_{k=1}^{m+1}$ is a maximal set of MUBs in $\mathbb{C}^m$.

Besides symmetric block designs, affine designs are also very useful for our constructions.

Definition 9 A 2-(m, l, λ) block design $\mathbb{S}$ is resolvable if its collection of blocks $\mathbb{S}$ can be partitioned into subsets, called parallel classes, such that:

1. the blocks within each class are disjoint, and
2. for each parallel class, every element of $\big[[m]\big]$ is contained in a block.

Moreover, if the number of elements occurring in the intersection between blocks from different parallel classes is constant, then it is called an affine design.

For any resolvable 2-(m, l, λ) block design, Bose's condition gives a lower bound for the number of blocks b.

Theorem 6.6 ([15]) *Given any resolvable* 2-(m, l, λ) *block design, the number of blocks is bounded by the other parameters:*

$$b \geq m + r - 1,$$

and this lower bound is achieved if and only if the design is a l^2/m-cohesive affine design.

Proposition 10 *If $\mathbb{S}$ is a l^2/m-cohesive affine (m, l, λ) design, then its complement $\mathbb{S}^c$ is $(m-l)^2/m$-cohesive.*

Proof Let any $\mathcal{J}_1, \mathcal{J}_2 \in \mathbb{S}$. If they are in the same parallel class, then $\mathcal{J}_1 \cap \mathcal{J}_2 = \emptyset$. Hence

$$|\mathcal{J}_1^c \cap \mathcal{J}_2^c| = |[[m]] \setminus (\mathcal{J}_1 \cup \mathcal{J}_2)| = m - 2l < \frac{(m-l)^2}{m}.$$

If $\mathcal{J}_1$ and $\mathcal{J}_2$ are in different parallel classes, then by definition, $|\mathcal{J}_1 \cap \mathcal{J}_2| = \frac{l^2}{m}$. Therefore, $|\mathcal{J}_1 \cup \mathcal{J}_2| = 2l - l^2/m$ and so

$$|\mathcal{J}_1^c \cap \mathcal{J}_2^c| = m - 2l + \frac{l^2}{m} = \frac{(m-l)^2}{m}.$$

This completes the proof. □

In the following, we will give some infinite families of optimally spread mixed-rank packings. Before giving such examples, we recall some block designs from [15].

- Affine designs with parameters

$$(m, l, \lambda) = \left(q^{t+1}, q^t, \frac{q^t - 1}{q - 1}\right), \ t \geq 1$$

  exist if q is a prime power.
- Menon symmetric designs with parameters $(m, l, \lambda) = (4t^2, 2t^2 - t, t^2 - t), t \geq 1$. These designs exist, for instance, when Hadamard matrices of order $2t$ exist, and it is conjectured that they exist for all values of t.
- Hadamard symmetric designs with parameters $(m, l, \lambda) = (4t - 1, 2t - 1, t - 1)$, $t \geq 1$. These designs exist if and only if Hadamard matrices of order $4t$ exist.

Example

Let $\mathbb{S}$ be an affine design with parameters $\left(q^{t+1}, q^t, \frac{q^t-1}{q-1}\right)$, where q is a prime power. Note that we can verify the values of the remaining parameters

$$r = \sum_{i=0}^{t} q^i, \text{ and } b = \sum_{i=1}^{t+1} q^i.$$

According to Bose's condition, this design is q^{t-1}-cohesive. Let $\mathbb{S}^c$ be its complement. By Proposition 10, it is $q^{t-1}(q-1)^2$-cohesive. Since q^{t+1} is a prime power, a maximal set of MUBs, $\{\mathcal{B}_k\}_{k=1}^{q^{t+1}+1}$, for $\mathbb{C}^{q^{t+1}}$ exists. By Theorem 6.3, for any partition $\{A_1, A_2\}$ of $\left[[q^{t+1} + 1]\right]$, the family

$$\{P_{\mathcal{J}}^{\mathcal{B}_k} : k \in A_1, \mathcal{J} \in \mathbb{S}\} \cup \{P_{\mathcal{J}}^{\mathcal{B}_k} : k \in A_2, \mathcal{J} \in \mathbb{S}^c\}$$

forms an optimally spread mixed-rank tight fusion frame with a mixture of 2 for $\mathbb{C}^{q^{t+1}}$.

Example
Let $\mathbb{S}_1$ be an affine design with parameters $\left(4^{t+1}, 4^t, \frac{4^t-1}{4-1}\right)$, $t \geq 1$ and $\mathbb{S}_2$ be its complement. Let $\mathbb{S}_3$ be the Menon symmetric design with parameters $(4^{t+1}, 2^{2t+1} - 2^t, 2^{2t} - 2^t)$. Denote $\mathbb{S}_4$ the complement of $\mathbb{S}_3$.

Let $\{\mathcal{B}_k\}_{k=1}^{4^{t+1}/2+1}$ be a maximal set of MUBs which exists in $\mathbb{R}^{4^{t+1}}$. By Theorem 6.3, for any partition $\{A_1, A_2, A_3, A_4\}$ of $\left[[4^{t+1}/2+1]\right]$, the family

$$\mathcal{P} = \cup_{i=1}^{4}\{P_{\mathcal{J}}^{\mathcal{B}_k} : k \in A_i, \mathcal{J} \in \mathbb{S}_i\}$$

forms an optimally spread mixed-rank tight fusion frame with a mixture of 4 for $\mathbb{R}^{4^{t+1}}$.

Finally, we will construct optimally spread mixed-rank packings for $\mathbb{F}^m$ whose corresponding embedded vectors are vertices of an orthoplex in $\mathbb{R}^{d_{\mathbb{F}^m}}$.

Theorem 6.7 *Let $\mathbb{S}$ be a collection of subsets of $\left[[m]\right]$, each of size l. Suppose that $|\mathcal{J} \cap \mathcal{J}'| = l^2/m$ for any $\mathcal{J}, \mathcal{J}' \in \mathbb{S}$. Let $\mathbb{S}^c$ be the complement of $\mathbb{S}$, i.e.,*

$$\mathbb{S}^c := \{\mathcal{J}^c = \left[[m]\right] \setminus \mathcal{J} : \mathcal{J} \in \mathbb{S}\}.$$

Let $\{\mathcal{B}_k\}_{k\in K}$ be a set of MUBs for $\mathbb{F}^m$ and let $\mathcal{P} = \{P_{\mathcal{J}}^{\mathcal{B}_k} : k \in K, \mathcal{J} \in \mathbb{S} \cup \mathbb{S}^c\}$. If $2|\mathbb{S}||K| > d_{\mathbb{F}^m} + 1$, then $\mathcal{P}$ is an optimally spread mixed-rank packing for which the embedded vectors corresponding to the coordinate projections in $\mathcal{P}$ occupy the vertices of an orthoplex in $\mathbb{R}^{d_{\mathbb{F}^m}}$. Furthermore, if $|\mathbb{S}| = m - 1$ and a maximal set of MUBs, $\{\mathcal{B}_k\}_{k=1}^{k_{\mathbb{F}^m}}$, exists in $\mathbb{F}^m$, then the collection of projections $\mathcal{P} = \{P_{\mathcal{J}}^{\mathcal{B}_k} : k \in \left[[k_{\mathbb{F}^m}]\right], \mathcal{J} \in \mathbb{S} \cup \mathbb{S}^c\}$ is a maximal orthoplectic fusion frame for $\mathbb{F}^m$.

Proof Since $|\mathcal{J} \cap \mathcal{J}'| = l^2/m$, for every $\mathcal{J}, \mathcal{J}' \in \mathbb{S}$, it is easy to see that $|\mathcal{J} \cap \mathcal{J}'| = \frac{(m-l)^2}{m}$, for every $\mathcal{J}, \mathcal{J}' \in \mathbb{S}^c$.

Moreover, for every $\mathcal{J} \in \mathbb{S}, \mathcal{J}' \in \mathbb{S}^c, \mathcal{J}' \neq \mathcal{J}^c$, we have

$$|\mathcal{J} \cap \mathcal{J}'| = |\mathcal{J} \setminus (\mathcal{J}')^c| = l - \frac{l^2}{m}.$$

Hence, for each $k \in K$, by (1) of Proposition 5, it follows that

$$\mathrm{tr}(P_{\mathcal{J}}^{\mathcal{B}_k} P_{\mathcal{J}'}^{\mathcal{B}_k}) - \frac{\mathrm{rank}(P_{\mathcal{J}}^{\mathcal{B}_k})\,\mathrm{rank}(P_{\mathcal{J}'}^{\mathcal{B}_k})}{m} = 0, \text{ for every } \mathcal{J}, \mathcal{J}' \in \mathbb{S} \cup \mathbb{S}^c, \mathcal{J}' \neq \mathcal{J}, \mathcal{J}^c.$$

Note also that for every $\mathcal{J} \in \mathbb{S}$, the corresponding embedded vectors of $P_{\mathcal{J}}^{\mathcal{B}_k}$ and $P_{\mathcal{J}^c}^{\mathcal{B}_k}$ are antipodal points in $\mathbb{R}^{d_{\mathbb{F}^m}}$. Moreover, by (2) of Proposition 5, for any $k, k' \in K$, $k \neq k'$, we have

$$\operatorname{tr}(P_{\mathcal{J}}^{\mathcal{B}_k} P_{\mathcal{J}'}^{\mathcal{B}_{k'}}) - \frac{\operatorname{rank}(P_{\mathcal{J}}^{\mathcal{B}_k}) \operatorname{rank}(P_{\mathcal{J}'}^{\mathcal{B}_{k'}})}{m} = 0, \text{ for every } \mathcal{J}, \mathcal{J}' \in \mathbb{S} \cup \mathbb{S}^c.$$

Note that $|\mathbb{S} \cup \mathbb{S}^c| = 2|\mathbb{S}|$ and so there are $2|\mathbb{S}|$ coordinate projections respective to each $\mathcal{B}_k$, $k \in K$. Hence, the total number of projections is $|\mathcal{P}| = 2|\mathbb{S}||K| > d_{\mathbb{F}^m} + 1$ by assumption. The conclusion of the first part of the theorem then follows by using Identity (4.1).

The "furthermore part" is obvious since in this case,

$$|\mathcal{P}| = 2(m-1)k_{\mathbb{F}^m} = 2d_{\mathbb{F}^m}.$$

This completes the proof. □

Recall that for any symmetric (m, l, λ) block design, the intersection between any of its blocks has exactly λ elements. Note that $\lambda = \frac{l(l-1)}{m-1} < \frac{l^2}{m}$. However, if $\frac{m-l}{l-1}$ is an integer, then we can view these blocks as subsets of a bigger set, $[[m']]$ so that $\lambda = \frac{l^2}{m'}$. In other words, these blocks, viewed as subsets of $[[m']]$, satisfy the assumption of Theorem 6.7. We will record this by the following proposition.

Proposition 11 *Let $\mathbb{S}$ be a symmetric (m, l, λ) block design. Suppose that $\frac{m-l}{l-1}$ is an integer. Let $m' = m + \frac{m-l}{l-1}$. Then for any $\mathcal{J}, \mathcal{J}' \in \mathbb{S}$, we have $|\mathcal{J} \cap \mathcal{J}'| = \frac{l^2}{m'}$.*

Example

Let q be a prime power and let $\mathbb{S}$ be a symmetric block design with parameters

$$m = q^2 + q + 1,\ l = q + 1,\ \lambda = 1.$$

Such designs exist, called Projective planes, see [15]. We have $\frac{m-l}{l-1} = q$, which is an integer. Let $m' = m + q = (q+1)^2$. If m' is a prime power, then by Theorem 6.2, a maximal set of MUBs, $\{\mathcal{B}_k\}_{k=1}^{m'+1}$ exists in $\mathbb{C}^{m'}$. By Theorem 6.7, the family

$$\mathcal{P} = \{P_{\mathcal{J}}^{\mathcal{B}_k} : k \in [[m'+1]], \mathcal{J} \in \mathbb{S} \cup \mathbb{S}^c\},$$

(continued)

where $\mathbb{S}^c = \{[[m']] \setminus \mathcal{J} : \mathcal{J} \in \mathbb{S}\}$ is an optimally spread mixed-rank packing for $\mathbb{C}^{m'}$ with a mixture of 2. A similar result is obtained for $\mathbb{R}^{m'}$ when m' is a power of 4. However, in both cases, the projections cannot embed exhaustively into every vertex of the orthoplex since $|\mathcal{P}| = 2mk_{\mathbb{F}^{m'}} < 2d_{\mathbb{F}^{m'}}$.

It is known that an extension of a Hadamard symmetric 2-$(4t-1, 2t-1, t-1)$ block design is a Hadamard 3-$(4t, 2t, t-1)$ block design. This design exists if and only if a Hadamard matrix of order $4t$ exists. Moreover, it can be constructed from a Hadamard matrix as follows.

Let $H = (h_{ij})$ be a Hadamard matrix of order $4t$. Normalize H so that all elements in the last row are $+1$. For each row i other than the last, we define a pair of subsets of $[[4t]]$ by

$$\mathcal{J}_i = \{j \in [[4t]] : h_{ij} = +1\}, \text{ and } \mathcal{J}_i' = \{j \in [[4t]] : h_{ij} = -1\}.$$

Then the collection $\mathbb{S}$ of all these subsets forms a Hadamard 3-design. Note that $\mathcal{J}_i' = \mathcal{J}_i^c$, for all i. Hence if we let $\mathbb{S}_1 = \{\mathcal{J}_i : i = 1, 2, \ldots, 4t-1\}$, then the set of blocks of a Hadamard 3-design has the form $\mathbb{S} = \mathbb{S}_1 \cup \mathbb{S}_1^c$. One of the useful properties of a Hadamard 3-$(4t, 2t, t-1)$ design is that the cardinality of the intersection of any of its two blocks $\mathcal{J}, \mathcal{J}'$, where $\mathcal{J}' \neq \mathcal{J}^c$ is the same, namely $|\mathcal{J} \cap \mathcal{J}'| = t$. In other words, they are affine designs. For more properties of these designs, see, for example, [9, 15].

It turns out that we can use Hadamard 3-designs and maximal sets of mutually unbiased bases to construct maximal orthoplectic fusion frames of constant-rank. Moreover, any such fusion frames constructed in this way must come from Hadamard 3-designs. This does not seem to be mentioned in previous papers [7, 13]. Note that in [7], the authors give a construction of a family of block designs and then use them to construct a family of maximal orthoplectic fusion frames of constant-rank. They also claim that these block designs are 2-designs. However, as a consequence of the following theorem, they are actually Hadamard 3-designs.

Theorem 6.8 *Let $m \in \mathbb{N}$ be such that a maximal set of MUBs, $\{\mathcal{B}_k\}_{k\in K}$, for $\mathbb{F}^m$ exists. Let $\mathbb{S}$ be a collection of subsets of $[[m]]$, each of size l, and let $\mathcal{P} = \{P_{\mathcal{J}}^{\mathcal{B}_k} : k \in K, \mathcal{J} \in \mathbb{S}\}$. If $\mathbb{S}$ is a Hadamard 3-design, then $\mathcal{P}$ is a maximal orthoplectic fusion frame of constant-rank for $\mathbb{F}^m$. In particular,*

1. *if m is a power of 2, then $\mathcal{P}$ is a maximal orthoplectic fusion frame for $\mathbb{C}^m$, and*
2. *if m is a power of 4, then $\mathcal{P}$ is a maximal orthoplectic fusion frame for $\mathbb{R}^m$.*

Conversely, if $\mathcal{P}$ is a maximal orthoplectic fusion frame of constant-rank for $\mathbb{F}^m$, then $\mathbb{S}$ is a Hadamard 3-design, and we can construct a Hadamard matrix via $\mathcal{P}$.

Proof Suppose $\mathbb{S}$ is a Hadamard 3-design of parameters $(m, l, \lambda) = (4t, 2t, t-1)$. Then it is an affine 2-$(4t, 2t, 2t-1)$ design, and the remaining parameters are $r = 4t-1, b = 8t-2$. Hence $|\mathbb{S}| = 8t-2 = 2(m-1)$. Note that $l = 2t = m/2$, so every projection of $\mathcal{P}$ has the same rank, $m/2$. Moreover, $|\mathcal{J} \cap \mathcal{J}'| = t = l^2/m$, for every distinct blocks $\mathcal{J}, \mathcal{J}' \in \mathbb{S}, \mathcal{J}' \neq \mathcal{J}^c$. The conclusions then follow from Theorem 6.7.

Conversely, suppose $\mathcal{P}$ is a maximal orthoplectic fusion frame of constant-rank for $\mathbb{F}^m$. Then each projection of $\mathcal{P}$ must have rank $l = m/2$, (see Corollary 2.7 in [7]), and hence every $\mathcal{J} \in \mathbb{S}$ has size $m/2$. Moreover, $\mathbb{S}$ is of size $2d_{\mathbb{F}^m}/k_{\mathbb{F}^m} = 2(m-1)$. Since $\mathcal{P}$ is maximal, it follows from (3.2) that for each $\mathcal{J} \in \mathbb{S}$, its complement $\mathcal{J}^c$ is also in $\mathbb{S}$. Thus, $\mathbb{S}$ is of the form $\mathbb{S} = \mathbb{S}_1 \cup \mathbb{S}_1^c$. Furthermore, by (3.2), $\mathrm{tr}(P_{\mathcal{J}}^{\mathcal{B}_k} P_{\mathcal{J}'}^{\mathcal{B}_{k'}}) = l^2/m = m/4$, for any $k, k' \in K$, and $\mathcal{J}, \mathcal{J}' \in \mathbb{S}, \mathcal{J}' \neq \mathcal{J}^c$.

Denote the subsets of $\mathbb{S}_1$ by $\mathcal{J}_1, \mathcal{J}_2, \ldots, \mathcal{J}_{m-1}$. Let H be an $m \times m$-matrix whose the last row is of all +1's, and the entries of row i is defined by $H_{ij} = +1$ if $j \in \mathcal{J}_i$, and $H_{ij} = -1$, otherwise. We will show that H is a Hadamard matrix and therefore $\mathbb{S}$ is a Hadamard 3-design. By the construction, each row other than the last has precisely $m/2$ entries that are +1 and $m/2$ entries that are -1. Therefore, it is enough to show that the inner product of any two of them is zero.

Let R_i and $R_{i'}$ be any two distinct rows of H, $i, i' \leq m-1$. Since

$$|\mathcal{J}_i \cap \mathcal{J}_{i'}| = \mathrm{tr}(P_{\mathcal{J}_i}^{\mathcal{B}_k} P_{\mathcal{J}_{i'}}^{\mathcal{B}_k}) = m/4, \text{ for all } k \in K,$$

it follows that row R_i and row $R_{i'}$ have exactly $m/4$ entries that are +1 in the same column. This implies $\langle R_i, R_{i'} \rangle = 0$, which completes the proof. □

Acknowledgement The authors were supported by NSF DMS 1609760, 1906725, and NSF ATD 1321779.

References

1. Bachoc, C., Bannai, E., and Coulangeon, R.: Codes and designs in Grassmannian spaces. Discrete Math. **277**(1–3), 15–28 (2004)
2. Bandeira, A.S., Fickus, M., Mixon, D.G., and Wong, P.: The road to deterministic matrices with the restricted isometry property. J. Fourier Anal. Appl. **19**(6), 1123–1149 (2013)
3. Benedetto, J.J. and Kolesar, J.D.: Geometric properties of Grassmannian frames for $\mathbb{R}^2$ and $\mathbb{R}^3$. EURASIP J. Appl. Signal Process. 1–17 (2006)
4. Bodmann, B.G. and Elwood, H.J.: Complex equiangular Parseval frames and Seidel matrices containing pth roots of unity. Proc. Amer. Math. Soc. **138**(12), 4387–4404 (2010)
5. Bodmann, B.G. and Haas, J.: Frame potentials and the geometry of frames. J. Fourier Anal. Appl. **21**(6), 1344–1383 (2015)
6. Bodmann, B.G. and Haas, J.: Achieving the orthoplex bound and constructing weighted complex projective 2-designs with Singer sets. Linear Algebra Appl. **511**, 54–71 (2016)
7. Bodmann, B.G. and Haas, J.: Maximal orthoplectic fusion frames from mutually unbiased bases and block designs. Proc. Amer. Math. Soc. **146**(6), 2601–2616 (2018)

8. Calderbank, A.R., Hardin, R.H., Rains, E.M., Shor, P.W., and Sloane, N.J.A.: A group theoretic framework for the construction of packings in Grassmannian spaces. J. Algebraic Combin. **9**(2), 129–140 (1999)
9. Cameron, P.J. and van Lint, J.H.: Designs, Graphs, Codes and their Links. London Mathematical Society Student Texts. Cambridge University Press. (1991)
10. Cameron, P.J. and Seidel, J.J.: Quadratic forms over GF(2). Nederl. Akad. Wetensch. Proc. Ser. A 76=Indag. Math. **35**, 1–8 (1973)
11. Casazza, P.G. and Haas, J.I.: On the rigidity of geometric and spectral properties of Grassmannian frames. ArXiv e-prints (May 2016)
12. Casazza, P.G. and Kutyniok, G. (editors).: Finite Frames. Applied and Numerical Harmonic Analysis. Birkhäuser/Springer, New York (2013)
13. Casazza, P.G., Haas IV, I.J., Stueck, J., and Tran, T.T.: Constructions and properties of optimally spread subspace packings via symmetric and affine block designs and mutually unbiased bases. ArXiv e-prints (June 2018)
14. Christensen, O.: An Introduction to Frames and Riesz Bases. An introduction to frames and Riesz bases. Second expanded edition, Birkhäuser, Boston, Basel, Berlin (2016)
15. Colbourn, C.J. and Dinitz, J.H. (editors): Handbook of Combinatorial Designs. Discrete Mathematics and its Applications (Boca Raton). Chapman Hall/CRC, Boca Raton, FL, second edition (2007)
16. Conway, J.H., Hardin, R.H., and Sloane, N.J.A.: Packing lines, planes, etc.: packings in Grassmannian spaces. Exp. Math. **5**(2), 139–159 (1996)
17. Delsarte, P., Goethals, J.M., and Seidel, J.J.: Bounds for systems of lines, and Jacobi polynomials. Philips Res. Rep. **30**, 91–105 (1975)
18. Dhillon, I.S., Heath Jr., R.W., Strohmer, T., and Tropp, J.A.: Constructing packings in Grassmannian manifolds via alternating projection. Exp. Math. **17**(1), 9–35 (2008)
19. Et-Taoui, B.: Complex conference matrices, complex Hadamard matrices and equiangular tight frames. In: Convexity and Discrete Geometry including Graph Theory, 181–191, Springer Proc. Math. Stat., 148, Springer, [Cham] (2016)
20. Fickus, M., Mixon, D.G., and Tremain, J.C.: Steiner equiangular tight frames. Linear Algebra Appl. **436**(5), 1014—1027 (2012)
21. Fickus, M., Jasper, J., Mixon, D.G., and Peterson, J.: Tremain equiangular tight frames. J. Combin. Theory Ser. A **153**, 54–66 (2018)
22. Fickus, M., Jasper, J., and Mixon, D.G.: Packings in real projective spaces. SIAM J. Appl. Algebra Geom. **2**(3), 377–409 (2018)
23. Hoffman, T.R. and Solazzo, J.P.: Complex equiangular tight frames and erasures. Linear Algebra Appl. **437**(2), 549–558 (2012)
24. Holmes, R.B. and Paulsen, V.I.: Optimal frames for erasures. Linear Algebra Appl. **377**, 31–51 (2004)
25. Jasper, J., Mixon, D.G., and Fickus, M.: Kirkman equiangular tight frames and codes. IEEE Trans. Inf. Theory. **60**(1), 170–181 (2014)
26. Kalra, D.: Complex equiangular cyclic frames and erasures. Linear Algebra Appl. **419**(2–3), 373–399 (2006)
27. King, E.J.: New constructions and characterizations of flat and almost flat Grassmannian fusion frames. ArXiv e-prints, arXiv:1612.05784 (February 2019)
28. Kutyniok, G., Pezeshki, A., Calderbank, R., and Liu, T.: Robust dimension reduction, fusion frames, and Grassmannian packings. Appl. Comput. Harmon. Anal. **26**(1), 64–76 (2009)
29. Oktay, O.: Frame quantization theory and equiangular tight frames. ProQuest LLC, Ann Arbor, MI. Dissertation (Ph.D.)-University of Maryland, College Park, MD (2007)
30. Rankin, R.A.: The closest packing of spherical caps in n dimensions. Proc. Glasgow Math. Assoc. **2**, 139–144 (1955)
31. Renes, J.M., Blume-Kohout, R., Scott, A.J., and Caves, C.M.: Symmetric informationally complete quantum measurements. J. Math. Phys. **45**(6), 2171–2180 (2004)
32. Shor, P.W. and Sloane, N.J.A.: A family of optimal packings in Grassmannian manifolds. J. Algebraic Combin. **7**(2), 157–163 (1998)

33. Strohmer, T. and Heath Jr., R.W.: Grassmannian frames with applications to coding and communication. Appl. Comput. Harmon. Anal. **14**(3), 257–275 (2003)
34. Sustik, M.A., Tropp, J.A., Dhillon, I.S., and Heath Jr., R.W.: On the existence of equiangular tight frames. Linear Algebra Appl. **426**(2–3), 619–635 (2007)
35. Waldron, S.F.D.: An Introduction to Finite Tight Frames. Applied and Numerical Harmonic Analysis. Birkhäuser/Springer, New York (2018)
36. Wootters, W.K. and Fields, B.D.: Optimal state-determination by mutually unbiased measurements. Ann. Phys. **191**(2), 363–381 (1989)
37. Xia, P., Zhou, S., and Giannakis, G.B.: Achieving the Welch bound with difference sets. IEEE Trans. Inf. Theory. **51**(5), 1900–1907 (2005)
38. Zauner, G.: Quantendesigns - Grundz¨uge einer nichtkommutativen Designtheorie. University Wien (Austria). Dissertation (Ph.D.) (1999). English translation in International Journal of Quantum Information (IJQI). **9**(1), 445–507 (2011)

Construction of Frames Using Calderón–Zygmund Operator Theory

Der-Chen Chang, Yongsheng Han, and Xinfeng Wu

Dedicated to Professor John Benedetto on his eightieth birthday.

Abstract In this paper, we present a construction of frame without using the Fourier transform. Our methods are based on the Calderón–Zygmund operator theory and Coifman's decomposition of the identity operator, which also work on homogeneous spaces in the sense of Coifman and Weiss.

1 Introduction

The theory of wavelet analysis has played an important role in many different branches of science and technology; see, for instance, [1, 2, 5, 6, 9] and the references therein. Wavelet analysis provides a simpler and more efficient way to analyze those functions and distributions that have been studied through Fourier series and integrals. R. Coifman and G. Weiss invented the atoms and molecules (cf. [4, 14]) which were to form the basic building blocks of various function spaces. The atom decomposition can be obtained by using a discrete version of a well-known identity, due to A. Calderón ([3]), in which wavelets were implicitly

D.-C. Chang
Department of Mathematics and Statistics, Georgetown University, Washington, DC, USA

Graduate Institute of Business Administration, College of Management, Fu Jen Catholic University, New Taipei City, Taiwan
e-mail: chang@georgetown.edu

Y. Han
Department of Mathematics and Statistics, Auburn University, Auburn, AL, USA
e-mail: hanyong@auburn.edu

X. Wu (✉)
Department of Mathematics, China University of Mining and Technology, Beijing, P.R. China
e-mail: wuxf@cumtb.edu.cn

M. Hirn et al. (eds.), *Excursions in Harmonic Analysis, Volume 6*, Applied and Numerical Harmonic Analysis, https://doi.org/10.1007/978-3-030-69637-5_8

involved. The wavelet series decompositions are nowadays effective expansion by unconditional bases in various function spaces arising from the theory of harmonic analysis.

Let us now recall the frame constructed by Frazier and Jawerth in [11] using the Fourier transform. Let $\psi \in \mathcal{S}(\mathbb{R}^n)$ with

1. $\operatorname{supp} \widehat{\psi} \subseteq \{1/2 \leq |\xi| \leq 2\}$,
2. $\sum_{j\in\mathbb{Z}} |\widehat{\psi}(2^{-j}\xi)|^2 = 1, \qquad$ for all $\xi \neq 0$.

Then

$$f(x) = \sum_{j\in\mathbb{Z}} \sum_{Q:\ell(Q)=2^{-j}} 2^{-jn} \psi_j(x - x_Q)\psi'_j * f(x_Q), \qquad \forall f \in L^2(\mathbb{R}^n),$$

where Q represents dyadic cubes with $\ell(Q) = 2^{-j}$, x_Q represents the lower-left corners of Q, $\psi_j(x) = 2^{jn}\psi(2^j x)$, $\psi'_j(x) = 2^{jn}\psi(-2^j x)$, and the series converges in $L^2(\mathbb{R}^n)$.

Note that Frazier–Jawerth approach relies on the Fourier transform on Euclidean spaces. However, there are many non-Euclidean situations (for instance, the Ahlfors-type spaces and Carnot-Carathédory-type spaces) in which the Fourier transform is not available. The purpose of this article is to present a new approach for constructing a frame without using the Fourier transform. To be more precise, let $0 < \varepsilon \leq 1$, and let φ be a function defined on $\mathbb{R}^n$ with

1. $|\varphi(x)| \leq \frac{C}{(1+|x|)^{n+\varepsilon}}$,
2. $|\varphi(x) - \varphi(x')| \leq C\left(\frac{|x-x'|}{1+|x|}\right)^{\varepsilon} \frac{1}{(1+|x|)^{n+\varepsilon}}$ for $|x - x'| \leq (1/2)(1 + |x|)$,
3. $\int_{\mathbb{R}^n} \varphi(x)dx = 1$.

For $j \in \mathbb{Z}$, let $\varphi_j(x) = 2^{jn}\varphi(2^j x)$.

The following is our main result, whose proof is based on the Calderón–Zygmund operator theory.

Theorem 1.1 *Suppose that φ is a function defined on $\mathbb{R}^n$ satisfying the above conditions. Then there exist families of functions $\widetilde{\varphi}_j(x, y)$ and $\widetilde{\varphi}^*_j(x, y)$ such that for all $f \in L^2(\mathbb{R}^n)$,*

$$\begin{aligned} f(x) &= \sum_{j\in\mathbb{Z}} \sum_{\substack{Q \\ \ell(Q)=2^{-j-N}}} |Q|\widetilde{\varphi}_j(x, x_Q)\varphi_j * f(x_Q) \\ &= \sum_{j\in\mathbb{Z}} \sum_{\substack{Q \\ \ell(Q)=2^{-j-N}}} |Q|\varphi_j(x - x_Q)\widetilde{\varphi}^*_j(f)(x_Q), \end{aligned} \tag{1}$$

where Q represents dyadic cubes of side length $\ell(Q) = 2^{-j-N}$ with N being a large fixed positive integer, x_Q is any fixed point in Q,

$$\widetilde{\varphi}_j^*(f)(x_Q) = \int_{\mathbb{R}^n} \widetilde{\varphi}_j^*(x_Q, y) f(y) dy,$$

and $\widetilde{\varphi}_j(x, x_Q)$ and $\widetilde{\varphi}_j^*(x_Q, x)$ satisfy the following conditions: For any fixed $0 < \varepsilon' < \varepsilon$,

(i) $|\widetilde{\varphi}_j(x, x_Q)| \le C \frac{2^{-j\varepsilon'}}{(2^{-j}+|x-x_Q|)^{n+\varepsilon'}}$,
(ii) $|\widetilde{\varphi}_j(x, x_Q) - \widetilde{\varphi}_j(x', x_Q)| \le C \left(\frac{|x-x'|}{2^{-j}+|x-x_Q|}\right)^{\varepsilon'} \frac{2^{-j\varepsilon'}}{(2^{-j}+|x-x_Q|)^{n+\varepsilon'}}$ *for all* $|x - x'| \le 1/2(2^{-j} + |x - x_Q|)$,
(iii) $\int_{\mathbb{R}^n} \widetilde{\varphi}_j(x, x_Q) dx = 0$ *for all j and x_Q,*
(i′) $|\widetilde{\varphi}_j^*(x_Q, x)| \le C \frac{2^{-j\varepsilon'}}{(2^{-j}+|x-x_Q|)^{n+\varepsilon'}}$,
(ii′) $|\widetilde{\varphi}_j^*(x_Q, x) - \widetilde{\varphi}_j^*(x_Q, x')| \le C \left(\frac{|x-x'|}{2^{-j}+|x-x_Q|}\right)^{\varepsilon'} \frac{2^{-j\varepsilon'}}{(2^{-j}+|x-x_Q|)^{n+\varepsilon'}}$ *for all* $|x - x'| \le 1/2(2^{-j} + |x - x_Q|)$,
(iii′) $\int_{\mathbb{R}^n} \widetilde{\varphi}_j^*(x_Q, x) dx = 0$ *for all j and x_Q.*

For clarification purposes, we point out that the number N is needed to be taken sufficiently large depending on p, φ, and the dimension.

As mentioned before, the key idea behind the proof for Frazier–Jawerth's result is the use of the Fourier transform. In contrast, Theorem 1.1 is proven through the Calderón–Zygmund operator theory.

Definition 1 Let $\mathcal{D}(\mathbb{R}^n)$ be the space of smooth functions on $\mathbb{R}^n$ with compact support, and let $\mathcal{D}'(\mathbb{R}^n)$ denote its dual space. An operator T is said to be a *Calderón–Zygmund singular integral operator* with the kernel K if $T : \mathcal{D}(\mathbb{R}^n) \to \mathcal{D}'(\mathbb{R}^n)$ with

$$\langle Tf, g\rangle = \int_{\mathbb{R}^n} \int_{\mathbb{R}^n} K(x, y) f(y) g(x)\, dy\, dx,$$

for $f, g \in \mathcal{D}(\mathbb{R}^n)$ with supp $f \cap$ supp $g = \varnothing$, where the kernel $K(x, y)$ is a complex-valued continuous function on $\mathbb{R}^n \times \mathbb{R}^n \backslash \{x = y\}$ and satisfies the following estimates: For some $0 < \varepsilon \le 1$,

(i) $|K(x, y)| \le \frac{C}{|x-y|^n}$ for $x \ne y$,
(ii) $|K(x, y) - K(x', y)| \le C \frac{|x-x'|^{\varepsilon}}{|x-y|^{n+\varepsilon}}$ for $|x - x'| \le (1/2)|x - y|$,
(iii) $|K(x, y) - K(x, y')| \le C \frac{|y-y'|^{\varepsilon}}{|x-y|^{n+\varepsilon}}$ for $|y - y'| \le (1/2)|x - y|$.

Definition 2 A function f defined on $\mathbb{R}^n$ is said to be a test function if there exist β, γ, r, x_0 with $0 < \beta \le 1$, $\gamma, r > 0$ and $x_0 \in \mathbb{R}^n$ such that

(i) $|f(x)| \le C\frac{r^\gamma}{(r+|x-x_0|)^{n+\gamma}}$,

(ii) $|f(x)-f(x')| \le C\left(\frac{|x-x'|}{r+|x-x_0|}\right)^\beta \frac{r^\gamma}{(r+|x-x_0|)^{n+\gamma}}$ for $|x-x'| \le (1/2)(r+|x-x_0|)$,

(iii) $\int_{\mathbb{R}^n} f(x)dx = 0$.

If f is a test function as above, we denote $f \in \mathcal{M}(\beta,\gamma,r,x_0)$ and the norm of $f \in \mathcal{M}(\beta,\gamma,r,x_0)$ is defined to be the smallest constant given in (i) and (ii) above.

Now fix $x_0 \in \mathbb{R}^n$ and denote $\mathcal{M}(\beta,\gamma) = \mathcal{M}(\beta,\gamma,1,x_0)$. It is easy to see that $\mathcal{M}(\beta,\gamma,r,x_1) = \mathcal{M}(\beta,\gamma)$ with equivalent norms for all $x_1 \in \mathbb{R}^n$ and $r > 0$. Furthermore, it is also easy to check that $\mathcal{M}(\beta,\gamma)$ is a Banach space with respect to the norm in $\mathcal{M}(\beta,\gamma)$.

The main tool we will use to show Theorem 1.1 is the following result in [12].

Theorem 1.2 *Suppose that T is a Calderón–Zygmund singular integral operator and extends to be a bounded operator on $L^2(\mathbb{R}^n)$. Furthermore, if $T(1) = T^*(1) = 0$ and the kernel $K(x,y)$ satisfies the following double difference condition*

$$\big|[K(x,y)-K(x',y)]-[K(x,y')-K(x',y')]\big| \le C\frac{|x-x'|^\varepsilon|y-y'|^\varepsilon}{|x-y|^{n+2\varepsilon}} \tag{2}$$

for $|x-x'| \le \frac{1}{4}|x-y|$ and $|y-y'| \le \frac{1}{4}|x-y|$, then T maps $\mathcal{M}(\beta,\gamma,r,x_0)$ to $\mathcal{M}(\beta,\gamma,r,x_0)$ for $0 < \beta,\gamma < \varepsilon$ and all $r > 0$, $x_0 \in \mathbb{R}^n$. Moreover, there exists a constant C such that

$$\|Tf\|_{\mathcal{M}(\beta,\gamma,r,x_0)} \le C\|T\|\cdot\|f\|_{\mathcal{M}(\beta,\gamma,r,x_0)},$$

where $\|T\| = \|T\|_{L^2\to L^2} + \|T\|_{CZ}$, the $\|T\|_{CZ}$ being the smallest constants in the definition of Calderón–Zygmund kernels and (2).

We would also like to mention a result of Meyer. To study the property of the Besov space $\dot{B}_1^{0,1}(\mathbb{R}^n)$, Meyer introduced the following definition of smooth atoms.

Definition 3 [13] A function $f(x)$ is said to be a *smooth atom* if there exist $0 < \beta \le 1$, $\gamma, r > 0$, $x_0 \in \mathbb{R}^n$, and a constant C such that

(i) $|f(x)| \le C\frac{r^\gamma}{(r+|x-x_0|)^{n+\gamma}}$,

(ii) $|f(x)-f(x')| \le C\left(\frac{|x-x'|}{r}\right)^\beta\left[\frac{r^\gamma}{(r+|x-x_0|)^{n+\gamma}} + \frac{r^\gamma}{(r+|x'-x_0|)^{n+\gamma}}\right]$,

(iii) $\int_{\mathbb{R}^n} f(x)dx = 0$.

If f is a smooth atom as above, then the norm of f is defined by the smallest constant given in (i) and (ii) above and is denoted by $\|f\|_{M(\beta,\gamma,r,x_0)}$. We would like to point out that if $f \in \mathcal{M}(\beta,\gamma,r,x_0)$, then $f \in M(\beta,\gamma,r,x_0)$.

Meyer in [13] proved the following theorem.

Theorem 1.3 *[13] If T is a Calderón–Zygmund singular integral operator and extends to be a bounded operator on $L^2(\mathbb{R}^n)$ and $T(1) = T^*(1) = 0$, then there exists a constant C such that*

$$\|Tf\|_{M(\beta',\gamma',r,x_0)} \le C\|f\|_{M(\beta,\gamma,r,x_0)}$$

for $0 < \beta' < \beta < \varepsilon$, $0 < \gamma' < \gamma < \varepsilon$ and all $r > 0$, $x_0 \in \mathbb{R}^n$.

Note that the indices β, γ describe the smoothness (regularity) of functions in the definitions of $\mathcal{M}(\beta', \gamma', r, x_0)$ and $M(\beta', \gamma', r, x_0)$. Thus, Theorem 1.3 indicates that under certain cancellation conditions, a Calderón–Zygmund singular integral operator maps $M(\beta, \gamma, r, x_0)$ into another larger atom space $M(\beta', \gamma', r, x_0)$ boundedly, with certain loss of regularity. Theorem 1.2 shows that, under certain additional regularity assumptions of K, T maps $\mathcal{M}(\beta, \gamma, r, x_0)$ into itself boundedly, without loss of regularity of functions.

We conclude this section with the following remark. Our approach for constructing the frame actually works in a more general setting, namely homogeneous spaces in the sense of Coifman and Weiss [7], which cover Ahlfors-type spaces, homogeneous groups (in particular, Heisenberg groups), and Carnot-Carathédory-type spaces. We prefer to present our result in the Euclidean setting for the sake of simplicity.

2 Preliminaries and Some Lemmas

The following is the definition of generalized approximation to the identity, whose kernels have only Lipschitz smoothness.

Definition 4 Let $0 < \varepsilon \le 1$. A sequence $\{S_k\}_{k\in\mathbb{Z}}$ of operators is said to be an *ε-approximation to the identity* (or an approximation to the identity) if there exists $0 < C < \infty$ such that for all x, x', y, and $y' \in \mathbb{R}^n$, $S_k(x, y)$, the kernel of S_k, are functions from $\mathbb{R}^n \times \mathbb{R}^n$ into $\mathbb{C}$ satisfying

$$|S_k(x, y)| \le C\frac{2^{-k\varepsilon}}{(2^{-k} + |x - y|)^{1+\varepsilon}}; \tag{3}$$

$$|S_k(x, y) - S_k(x', y)| \le C\Big(\frac{|x - x'|}{2^{-k} + |x - y|}\Big)^{\varepsilon}\frac{2^{-k\varepsilon}}{(2^{-k} + |x - y|)^{1+\varepsilon}} \tag{4}$$

for $|x - x'| \le \frac{1}{2}(2^{-k} + |x - y|)$;

$$|S_k(x, y) - S_k(x, y')| \le C\Big(\frac{|y - y'|}{2^{-k} + |x - y|}\Big)^{\varepsilon}\frac{2^{-k\varepsilon}}{(2^{-k} + |x - y|)^{1+\varepsilon}}$$

for $|y-y'| \le \frac{1}{2}(2^{-k}+|x-y|)$;

$$\int_{\mathbb{R}^n} S_k(x,y)dy = 1$$

for all $x \in \mathbb{R}^n$;

$$\int_{\mathbb{R}^n} S_k(x,y)dx = 1$$

for all $y \in \mathbb{R}^n$.

Note that $\varphi_k(x-y)$ is an ε-approximation to the identity. Let $D_k = S_k - S_{k-1}$ and $D_k(x,y)$ be the kernel for $k \in \mathbb{Z}$. Then $D_k(\cdot,y) \in \mathcal{M}(\varepsilon,\varepsilon,2^{-k},y)$ for any fixed y and k, and similarly, $D_k(x,\cdot) \in \mathcal{M}(\varepsilon,\varepsilon,2^{-k},x)$ for fixed x and k.

Lemma 1 *[9, Lemmas 3.7 and 3.11] Let* $0 < \varepsilon \le 1$. *Suppose that* $\{S_k\}_{k\in\mathbb{Z}}$ *is an* ε*-approximation to the identity. Set* $D_k = S_k - S_{k-1}$ *for all* $k \in \mathbb{Z}$. *Then for* $0 < \varepsilon' < \varepsilon$, *there exists a constant* C *which depends on* ε' *and* ε, *but not on* k, l *such that*

$$|D_k D_l(x,y)| \le C 2^{-|k-l|\varepsilon'} \frac{2^{-(k\wedge l)\varepsilon}}{(2^{-(k\wedge l)}+|x-y|)^{n+\varepsilon}};$$

for $k \ge l$ *and* $|x-x'| \le \frac{1}{2}(2^{-l}+|x-y|)$

$$|D_k D_l(x,y) - D_k D_l(x',y)| \le C\Big(\frac{|x-x'|}{2^{-l}+|x-y|}\Big)^{\varepsilon} \frac{2^{-l\varepsilon}}{(2^{-l}+|x-y|)^{n+\varepsilon}};$$

for $k \le l$ *and* $|x-x'| \le \frac{1}{2}(2^{-k}+|x-y|)$

$$|D_k D_l(x,y) - D_k D_l(x',y)| \le C\Big(\frac{|x-x'|}{2^{-k}+|x-y|}\Big)^{\varepsilon'} \frac{2^{-k(\varepsilon-\varepsilon')}}{(2^{-k}+|x-y|)^{n+\varepsilon-\varepsilon'}};$$

for $k \le l$ *and* $|y-y'| \le \frac{1}{2}(2^{-k}+|x-y|)$

$$|D_k D_l(x,y) - D_k D_l(x,y')| \le C\Big(\frac{|y-y'|}{2^{-k}+|x-y|}\Big)^{\varepsilon} \frac{2^{-k\varepsilon}}{(2^{-k}+|x-y|)^{n+\varepsilon}};$$

for $l \le k$ *and* $|y-y'| \le \frac{1}{2}(2^{-l}+|x-y|)$

$$|D_k D_l(x,y) - D_k D_l(x,y')| \le C\Big(\frac{|y-y'|}{2^{-l}+|x-y|}\Big)^{\varepsilon'} \frac{2^{-l(\varepsilon-\varepsilon')}}{(2^{-l}+|x-y|)^{n+\varepsilon-\varepsilon'}}.$$

Definition 5 An approximation to the identity $\{S_k\}_{k\in\mathbb{Z}}$ is said to satisfy the double Lipschitz condition if

$$\begin{aligned}|S_k(x,y)-S_k(x',y)-S_k(x,y')+S_k(x',y')| \\ \leq C\Big(\frac{|x-x'|}{2^{-k}+|x-y|}\Big)^{\varepsilon}\Big(\frac{|y-y'|}{2^{-k}+|x-y|}\Big)^{\varepsilon}\frac{2^{-k\varepsilon}}{(2^{-k}+|x-y|)^{n+\varepsilon}}\end{aligned} \tag{5}$$

for $|x-x'|\leq\frac{1}{2}(2^{-k}+|x-y|)$ and $|y-y'|\leq\frac{1}{2}(2^{-k}+|x-y|)$.

Lemma 2 *[9, Lemma 3.12] Suppose that $\{S_k\}_{k\in\mathbb{Z}}$ is an approximation to the identity and $S_k(x,y)$, the kernels of S_k, satisfy the condition (5). Set $D_k=S_k-S_{k-1}$ for all $k\in\mathbb{Z}$. Then for any $0<\varepsilon'<\varepsilon$, there exists a constant C which depends on ε' and ε, but not on k or l, such that*

$$\begin{aligned}|D_kD_l(x,y)-D_kD_l(x',y)-D_kD_l(x,y')+D_kD_l(x',y')| \\ \leq C\Big(\frac{|x-x'|}{2^{-(k\wedge l)}+|x-y|}\Big)^{\varepsilon'}\Big(\frac{|y-y'|}{2^{-(k\wedge l)}+|x-y|}\Big)^{\varepsilon'}\frac{2^{-(k\wedge l)(\varepsilon-\varepsilon')}}{(2^{-(k\wedge l)}+|x-y|)^{n+\varepsilon-\varepsilon'}}\end{aligned}$$

for $|x-x'|\leq\frac{1}{2}(2^{-(k\wedge l)}+|x-y|)$ and $|y-y'|\leq\frac{1}{2}(2^{-(k\wedge l)}+|x-y|)$.

3 Proof of Theorem 1.1

Let $\varphi_j(x)=2^{jn}\varphi(2^jx)$, and $S_j(f)(x)=\varphi_j*f(x)$. Then $\{S_j\}$ is an ε-approximation to the identity satisfying the double Lipschitz condition (5), and the following properties

$$\lim_{j\to\infty}S_j=I,\qquad \text{the identity operator on } L^2(\mathbb{R}^n)$$
$$\lim_{j\to-\infty}S_j=0,$$

both in the strong operator topology on $\mathcal{B}(L^2(\mathbb{R}^n))$. Set $D_j=S_{j+1}-S_j$ and denote $D_j(f)(x)=\psi_j*f(x)$, where $\psi_j(x)=2^{jn}\psi(2^jx)$ and $\psi(x)=\varphi(x)-\varphi_1(x)$. Then we will apply the Coifman's decomposition of the identity operator as follows.

$$f(x)=\lim_{j\to\infty}S_j(f)(x)=\sum_{j\in\mathbb{Z}}D_j(f)(x)=\sum_{j\in\mathbb{Z}}\sum_{k\in\mathbb{Z}}D_kD_j(f)(x).$$

For any large positive integer N, we write

$$f(x)=\sum_{|j-k|\leq N}D_kD_j(f)(x)+\sum_{|j-k|>N}D_kD_j(f)(x).$$

We mention that David–Journé–Semmes used in [8] the Coifman's decomposition for the identity as above to show the $T1$ and Tb theorem on spaces of homogeneous type and the following orthogonality estimate was proved: For any $0 < \varepsilon' < \varepsilon$,

$$|D_k D_j(x, y)| \leq C 2^{-|j-k|\varepsilon'} \frac{2^{-(j\wedge k)\varepsilon}}{(2^{-(j\wedge k)} + |x - y|)^{n+\varepsilon}}, \tag{6}$$

where $D_k D_j(x, y) = \psi_k * \psi_j(x - y)$ is the kernel of the operator $D_k D_j$ and $j \wedge k = \min(j, k)$. This estimate allowed them to apply the Cotlar–Stein lemma and provide the Littlewood–Paley theory on L^2 :

$$C'\|f\|_2 \leq \left\| \left\{ \sum_j |D_j(f)|^2 \right\}^{1/2} \right\|_2 \leq C\|f\|_2.$$

To prove Theorem 1.1, we write f as follows.

$$f(x) = \sum_{|j-k|\leq N} D_k D_j(f)(x) + R_N^1(f)(x),$$

where $R_N^1(f)(x) = \sum_{|j-k|>N} D_k D_j(f)(x)$. Denoting $D_j^N = \sum_{|k|\leq N} D_{j+k}$, we write

$$\begin{aligned}
&\sum_{|j-k|\leq N} D_k D_j(f)(x)\\
&= \sum_j D_j^N D_j(f)(x) = \sum_j \int_{\mathbb{R}^n} D_j^N(x - y) D_j(f)(y) dy\\
&= \sum_j \sum_{\substack{Q\\ \ell(Q)=2^{-j-N}}} \int_Q D_j^N(x - y) D_j(f)(y) dy\\
&= \sum_j \sum_{\substack{Q\\ \ell(Q)=2^{-j-N}}} |Q| D_j^N(x - x_Q) D_j(f)(x_Q)\\
&\quad + \sum_j \sum_{\substack{Q\\ \ell(Q)=2^{-j-N}}} \int_Q [D_j^N(x - y) D_k(f)(y) - D_j^N(x - x_Q) D_j(f)(x_Q)] dy\\
&= T_N(f)(x) + R_N^2(f)(x),
\end{aligned}$$

where

$$T_N(f)(x)=\sum_{j\in\mathbb{Z}}\sum_{\substack{Q\\ \ell(Q)=2^{-j-N}}}|Q|D_j^N(x-x_Q)D_j(f)(x_Q),$$

$$R_N^2(f)(x)=\sum_{j\in\mathbb{Z}}\sum_{\substack{Q\\ \ell(Q)=2^{-j-N}}}\int_Q[D_j^N(x-y)D_j(f)(y)-D_j^N(x-x_Q)D_j(f)(x_Q)]\,dy,$$

and each x_Q denotes *any* fixed point in Q.

The key estimates are the following

$$\|R_N^1(f)\|_{\mathcal{M}(\beta,\gamma,r,x_0)}\le C2^{-N\delta}\|f\|_{\mathcal{M}(\beta,\gamma,r,x_0)}, \tag{7}$$

$$\|R_N^2(f)\|_{\mathcal{M}(\beta,\gamma,r,x_0)}\le C2^{-N\delta}\|f\|_{\mathcal{M}(\beta,\gamma,r,x_0)}, \tag{8}$$

for some $\delta>0$, $0<\beta,\gamma<\varepsilon$, $r>0$, $x_0\in\mathbb{R}^n$, and $f\in\mathcal{M}(\beta,\gamma,r,x_0)$. Assuming these estimates hold for the moment, let us finish the proof of Theorem 1.1. The estimates (7) and (8) imply that $T_N^{-1}=\sum_{l=0}^{\infty}(R_N^1+R_N^2)^l$ and hence T_N^{-1} maps the test functions in $\mathcal{M}(\beta,\gamma,r,x_0)$ to $\mathcal{M}(\beta,\gamma,r,x_0)$ for $0<\beta,\gamma<\varepsilon, r>0$ and $x_0\in\mathbb{R}^n$. Moreover,

$$\|T_N^{-1}(f)\|_{\mathcal{M}(\beta,\gamma,r,x_0)}\le C\|f\|_{\mathcal{M}(\beta,\gamma,r,x_0)}.$$

Finally, we obtain

$$\begin{aligned} f(x)=T_N^{-1}T_N(f)(x)&=T_N^{-1}\left(\sum_j\sum_{\substack{Q\\ \ell(Q)=2^{-j-N}}}|Q|D_j^N(\cdot-x_Q)D_j(f)(x_Q)\right)(x)\\ &= \\ &=\sum_j\sum_{\substack{Q\\ \ell(Q)=2^{-j-N}}}|Q|T_N^{-1}\left(D_j^N(\cdot-x_Q)\right)(x)D_j(f)(x_Q). \end{aligned} \tag{9}$$

Note that $D_j^N(\cdot-x_Q)\in\mathcal{M}(\varepsilon,\varepsilon,2^{-j},x_Q)$ and hence for $0<\varepsilon'<\varepsilon$, $T_N^{-1}(D_j^N(\cdot-x_Q))(x)\in\mathcal{M}(\varepsilon',\varepsilon',2^{-j},x_Q)$. Denote $T_N^{-1}(D_j^N(\cdot-x_Q))(x)=\widetilde{\varphi}_j(x,x_Q)$. Then properties (i), (ii), and (iii) in Theorem 1.1 are fulfilled and

$$f(x)=\sum_{j\in\mathbb{Z}}\sum_{\substack{Q\\ \ell(Q)=2^{-j-N}}}|Q|\widetilde{\varphi}_j(x,x_Q)\varphi_j*f(x_Q),$$

which establishes the first equality in (1). The second equality in (1) can be obtained similarly, by interchanging the order of T_N^{-1} and T_N in (9). We omit the details.

To finish the proof, it remains to prove (7) and (8).

Proof of (8). We write

$$\begin{aligned} R_N^2(f)(x) &= \sum_{j\in\mathbb{Z}} \sum_{\substack{Q \\ \ell(Q)=2^{-j-N}}} \int_Q D_j^N(x-y)D_j(f)(y)\,dy \\ &\quad -\sum_{j\in\mathbb{Z}} \sum_{\substack{Q \\ \ell(Q)=2^{-j-N}}} \int_Q D_j^N(x-x_Q)D_j(f)(x_Q)\,dy \\ &=: K_1(f)(x) + K_2(f)(x), \end{aligned}$$

where K_1 and K_2 are linear operators with kernels

$$K_1(x,y) = \sum_{j\in\mathbb{Z}} \sum_{\substack{Q \\ \ell(Q)=2^{-j-N}}} \int_Q D_j^N(x-z)D_j(z-y)\,dz$$

and

$$K_2(x,y) = \sum_{j\in\mathbb{Z}} \sum_{\substack{Q \\ \ell(Q)=2^{-j-N}}} \int_Q D_j^N(x-x_Q)D_j(x_Q-y)\,dz.$$

By Theorem 1.2, it suffices to show that K_1 and K_2 are Calderón–Zygmund kernels satisfying the double difference condition (2), with constants at most $C2^{-\delta N}$, and that K_1 and K_2 are bounded on L^2 with bound at most $C2^{-\delta N}$ for some $\delta > 0$. The latter can be shown by arguments similar to that given in [10, Proof of Lemma 3.2]. We omit the details.

Now we begin to show that K_1 and K_2 are Calderón–Zygmund kernels satisfying the double difference condition (2), with constants at most $C2^{-\delta N}$. We only treat K_1, since K_2 can be handled in the same way. Notice that D_j^N satisfies the same size, regularity, and double difference conditions as D_j with the constant C replaced by CN. It then follows from (3) and (4) that

$$\begin{aligned} &|K_1(x,y)| \\ &\leq \sum_{j\in\mathbb{Z}} \sum_{\substack{Q \\ \ell(Q)=2^{-j-N}}} \int_Q |D_j^N(x-z)| \cdot |D_j(z-y) - D_j(x_Q-y)|\,dz \\ &\lesssim N \sum_{j\in\mathbb{Z}} \sum_{\substack{Q \\ \ell(Q)=2^{-j-N}}} \int_Q \frac{2^{-j\varepsilon}}{(2^{-j}+|x-z|)^{n+\varepsilon}} \left(\frac{|z-x_Q|}{2^{-j}+|z-y|}\right)^{\varepsilon} \frac{2^{-j\varepsilon}}{(2^{-j}+|z-y|)^{n+\varepsilon}}\,dz \end{aligned}$$

$$\lesssim N2^{-N\varepsilon}\sum_{j\in\mathbb{Z}}\int_{\mathbb{R}^n}\frac{2^{-j\varepsilon}}{(2^{-j}+|x-z|)^{n+\varepsilon}}\frac{2^{-j\varepsilon}}{(2^{-j}+|z-y|)^{n+\varepsilon}}dz$$

$$\lesssim 2^{-N\delta}\sum_{j\in\mathbb{Z}}\frac{2^{-j\varepsilon}}{(2^{-j}+|x-y|)^{n+\varepsilon}}$$

$$\lesssim \frac{2^{-N\delta}}{|x-y|^n}\qquad \text{for } 0<\delta<\varepsilon,$$

establishing the size condition of K_1.

Next, let us verify the regularity conditions for K_1. We only prove the regularity condition in the first variable as the other one can be treated similarly. We claim that, for $|x-x'|\le\frac{1}{2}|x-y|$ and any $0<\varepsilon'<\varepsilon$,

$$\Big|\sum_{\substack{Q\\ \ell(Q)=2^{-j-N}}}\int_Q[D_j^N(x-z)-D_j^N(x'-z)]D_j(z-y)\,dz\Big| \lesssim N\Big(\frac{|x-x'|}{2^{-j}+|x-y|}\Big)^{\varepsilon}\frac{2^{-j\varepsilon}}{(2^{-j}+|x-y|)^{1+\varepsilon}} \tag{10}$$

and

$$\Big|\sum_{\substack{Q\\ \ell(Q)=2^{-j-N}}}\int_Q[D_j^N(x-z)-D_j^N(x'-z)]D_j(z-y)\,dz\Big|\lesssim 2^{-N\varepsilon'}\frac{2^{-j\varepsilon}}{(2^{-j}+|x-y|)^{1+\varepsilon}}. \tag{11}$$

Assuming that (10) and (11) hold for the moment, by interpolation it follows that for any $0<\varepsilon'<\varepsilon$ and $0<\delta<\varepsilon-\varepsilon'$,

$$\Big|K_1(x,y)-K_1(x',y)\Big|\lesssim\sum_j 2^{-N\delta}\Big(\frac{|x-x'|}{2^{-j}+|x-y|}\Big)^{\varepsilon'}\frac{2^{-j\varepsilon}}{(2^{-j}+|x-y|)^{n+\varepsilon}}$$

$$\lesssim 2^{-N\delta}|x-x'|^{\varepsilon'}\sum_j\frac{2^{-j\varepsilon}}{(2^{-j}+|x-y|)^{n+\varepsilon+\varepsilon'}}$$

$$\lesssim 2^{-N\delta}\frac{|x-x'|^{\varepsilon'}}{|x-y|^{n+\varepsilon'}},$$

which proves the regularity condition for K_1.

Let us now prove (10) and (11). To show (11), using the cancellation condition of $D_j^N(x-\cdot)-D_j^N(x'-\cdot)$, we have

$$
\begin{aligned}
&\Big|\sum_{\substack{Q\\ \ell(Q)=2^{-j-N}}}\int_Q [D_j^N(x-z)-D_j^N(x'-z)]D_j(z-y)\,dz\Big|\\
&\le \sum_{\substack{Q\\ \ell(Q)=2^{-j-N}}}\int_Q [|D_j^N(x-z)|+|D_j^N(x'-z)|]\cdot|D_j(z-y)-D_j(x_Q-y)|\,dz\\
&\lesssim_N \sum_{\substack{Q\\ \ell(Q)=2^{-j-N}}}\int_Q \left[\frac{2^{-j\varepsilon}}{(2^{-j}+|x-z|)^{n+\varepsilon}}+\frac{2^{-j\varepsilon}}{(2^{-j}+|x'-z|)^{n+\varepsilon}}\right]\\
&\quad\times\left(\frac{|z-x_Q|}{2^{-j}+|z-y|}\right)^{\varepsilon}\frac{2^{-j\varepsilon}}{(2^{-j}+|z-y|)^{n+\varepsilon}}dz\\
&\lesssim 2^{-N\varepsilon'}\int_{W_1}\left[\frac{2^{-j\varepsilon}}{(2^{-j}+|x-z|)^{n+\varepsilon}}+\frac{2^{-j\varepsilon}}{(2^{-j}+|x'-z|)^{n+\varepsilon}}\right]\frac{2^{-j\varepsilon}}{(2^{-j}+|z-y|)^{n+\varepsilon}}dz\\
&\quad+2^{-N\varepsilon'}\int_{W_2}\left[\frac{2^{-j\varepsilon}}{(2^{-j}+|x-z|)^{n+\varepsilon}}+\frac{2^{-j\varepsilon}}{(2^{-j}+|x'-z|)^{n+\varepsilon}}\right]\frac{2^{-j\varepsilon}}{(2^{-j}+|z-y|)^{n+\varepsilon}}dz\\
&= I_1+I_2,
\end{aligned}
$$

where

$$
W_1=\left\{z\in\mathbb{R}^n: |y-z|>\frac{1}{3}|x-y|\right\},\quad W_2=\left\{z\in\mathbb{R}^n: |y-z|\le\frac{1}{3}|x-y|\right\}.
$$

For I_1, we have

$$
\begin{aligned}
I_1 &\lesssim 2^{-N\varepsilon'}\frac{2^{-j\varepsilon}}{(2^{-j}+|x-y|)^{n+\varepsilon}}\int_{W_1}\left[\frac{2^{-j\varepsilon}}{(2^{-j}+|x-z|)^{n+\varepsilon}}+\frac{2^{-j\varepsilon}}{(2^{-j}+|x'-z|)^{n+\varepsilon}}\right]dz\\
&\lesssim 2^{-N\varepsilon'}\frac{2^{-j\varepsilon}}{(2^{-j}+|x-y|)^{n+\varepsilon}}.
\end{aligned}
$$

As for I_2, noticing that for $z\in W_2$ and $|x-x'|\le(1/2)|x-y|$. Then by the triangle inequality,

$$
\begin{aligned}
|x-z| &\ge |x-y|-|y-z|\ge(2/3)|x-y|,\\
|x'-z| &\ge |x-z|-|x-x'|\ge(1/6)|x-y|.
\end{aligned}
$$

This implies

$$
I_2\lesssim N2^{-N\varepsilon'}\frac{2^{-j\varepsilon}}{(2^{-j}+|x-y|)^{n+\varepsilon}}\int_{W_2}\frac{2^{-j\varepsilon}}{(2^{-j}+|z-y|)^{n+\varepsilon}}dz
$$

$$\lesssim 2^{-N\varepsilon'} \frac{2^{-j\varepsilon}}{(2^{-j}+|x-y|)^{n+\varepsilon}},$$

establishing (11).

To prove (10), using the cancellation condition of $D_j^N(x-\cdot) - D_j^N(x'-\cdot)$, we write

$$\begin{aligned}
\text{LHS of (10)} &= \Big|\int_{\mathbb{R}^n} [D_j^N(x-z) - D_j^N(x'-z)][D_j(z-y) - D_j(x-y)]\,dz\Big| \\
&= \Big|\int_{U_1} [D_j^N(x-z) - D_j^N(x'-z)][D_j(z-y) - D_j(x-y)]\,dz\Big| \\
&\quad + \Big|\int_{U_2} [D_j^N(x-z) - D_j^N(x'-z)][D_j(z-y) - D_j(x-y)]\,dz\Big| \\
&\quad + \Big|\int_{U_3} [D_j^N(x-z) - D_j^N(x'-z)][D_j(z-y) - D_j(x-y)]\,dz\Big| \\
&=: II_1 + II_2 + II_3,
\end{aligned}$$

where

$$U_1 = \Big\{z \in \mathbb{R}^n : |x-x'| \le \frac{1}{4}(2^{-j}+|x-y|) \le \frac{1}{2}(2^{-j}+|x-z|)\Big\},$$

$$U_2 = \Big\{z \in \mathbb{R}^n : |x-x'| \le \frac{1}{2}(2^{-j}+|x-z|) \le \frac{1}{4}(2^{-j}+|x-y|)\Big\},$$

$$U_3 = \Big\{z \in \mathbb{R}^n : |x-x'| > \frac{1}{2}(2^{-j}+|x-z|)\Big\}.$$

Note that if $z \in U_2$, then $|x-z| \le (1/2)(2^{-j}+|x-y|)$. Using the estimate (4),

$$\begin{aligned}
II_2 &\lesssim N\int_{U_2} \Big(\frac{|x-x'|}{2^{-j}+|x-z|}\Big)^{\varepsilon} \frac{2^{-j\varepsilon}}{(2^{-j}+|x-z|)^{n+\varepsilon}} \Big(\frac{|z-x|}{2^{-j}+|x-y|}\Big)^{\varepsilon} \frac{2^{-j\varepsilon}}{(2^{-j}+|x-y|)^{n+\varepsilon}}\,dz \\
&\lesssim N\Big(\frac{|x-x'|}{2^{-j}+|x-y|}\Big)^{\varepsilon} \frac{2^{-j\varepsilon}}{(2^{-j}+|x-y|)^{n+\varepsilon}}.
\end{aligned}$$

For II_1, using the estimates (3) and (4), we get

$$\begin{aligned}
II_1 &\lesssim N\int_{U_1} \Big(\frac{|x-x'|}{2^{-j}+|x-z|}\Big)^{\varepsilon} \frac{2^{-k\varepsilon}}{(2^{-j}+|x-z|)^{n+\varepsilon}} \\
&\quad \cdot\Big[\frac{2^{-j\varepsilon}}{(2^{-j}+|z-y|)^{n+\varepsilon}} + \frac{2^{-j\varepsilon}}{(2^{-j}+|x-y|)^{n+\varepsilon}}\Big]dz
\end{aligned}$$

$$\begin{aligned}
&\lesssim N\Big(\frac{|x-x'|}{2^{-j}+|x-y|}\Big)^{\varepsilon}\frac{2^{-j\varepsilon}}{(2^{-j}+|x-y|)^{n+\varepsilon}}\int_{U_1}\frac{2^{-j\varepsilon}}{(2^{-j}+|z-y|)^{n+\varepsilon}}dz\\
&\quad+N\Big(\frac{|x-x'|}{2^{-j}+|x-y|}\Big)^{\varepsilon}\frac{2^{-j\varepsilon}}{(2^{-j}+|x-y|)^{n+\varepsilon}}\int_{U_1}\frac{2^{-j\varepsilon}}{(2^{-j}+|x-z|)^{n+\varepsilon}}dz\\
&\lesssim N\Big(\frac{|x-x'|}{2^{-j}+|x-y|}\Big)^{\varepsilon}\frac{2^{-j\varepsilon}}{(2^{-j}+|x-y|)^{n+\varepsilon}}.
\end{aligned}$$

To estimate II_3, note that if $z \in U_3$, then $|x-z| \le (1/2)(2^{-j}+|x-y|)$ and $|x-z| \le 2|x-x'|$. By (4) and (3),

$$\begin{aligned}
II_3 &\lesssim N\int_{U_3}\Big[\frac{2^{-j\varepsilon}}{(2^{-j}+|x-z|)^{n+\varepsilon}}+\frac{2^{-j\varepsilon}}{(2^{-j}+|x'-z))^{n+\varepsilon}}\Big]\\
&\qquad\times\Big(\frac{|z-x|}{2^{-j}+|x-y|}\Big)^{\varepsilon}\frac{2^{-j\varepsilon}}{(2^{-j}+|x-y|)^{n+\varepsilon}}dz\\
&\lesssim N\Big(\frac{|x-x'|}{2^{-j}+|x-y|}\Big)^{\varepsilon}\frac{2^{-j\varepsilon}}{(2^{-j}+|x-y|)^{n+\varepsilon}}.
\end{aligned}$$

Putting together the estimates obtained above for II_1, II_2, and II_3 yields (10), and hence the regularity condition of K_1.

Finally, let us verify the double difference condition for K_1. By Lemma 2, for any $\varepsilon' < \varepsilon_1 < \varepsilon$,

$$\begin{aligned}
&|D_j^N D_j(x,y)-D_j^N D_j(x',y)-D_j^N D_j(x,y')+D_j^N D_j(x',y')|\\
&\lesssim N\Big(\frac{|x-x'|}{2^{-j}+|x-y|}\Big)^{\varepsilon_1}\Big(\frac{|y-y'|}{2^{-j}+|x-y|}\Big)^{\varepsilon_1}\frac{2^{-j(\varepsilon-\varepsilon_1)}}{(2^{-j}+|x-y|)^{n+(\varepsilon-\varepsilon_1)}}.
\end{aligned}\tag{12}$$

On the other hand, using the cancellation condition of D_j, we have for $|x-x'| < (1/4)|x-y|$ and $|y-y'| < (1/4)|x-y|$,

$$\begin{aligned}
&|D_j^N D_j(x,y)-D_j^N D_j(x',y)-D_j^N D_j(x,y')+D_j^N D_j(x',y')|\\
&\lesssim \int_{\mathbb{R}^n}|D_j^N(x,z)[D_j(z,y)-D_j(x_Q,y)]|dz\\
&\quad+\int_{\mathbb{R}^n}|D_j^N(x',z)[D_j(z,y)-D_j(x_Q,y)]|dz\\
&\quad+\int_{\mathbb{R}^n}|D_j^N(x,z)[D_j(z,y')-D_j(x_Q,y')]|dz\\
&\quad+\int_{\mathbb{R}^n}|D_j^N(x',z)[D_j(z,y')-D_j(x_Q,y')]|dz.
\end{aligned}$$

Since the estimate for these terms are similar, we just treat the last one. Note that for $|x-x'|<(1/4)|x-y|$ and $|y-y'|<(1/4)|x-y|$, by the triangle inequality,

$$|x'-y'|\geq |x-y|-|x-x'|-|y-y'|\geq (1/2)|x-y|.$$

Hence, for any $0<\varepsilon''<\varepsilon$,

$$\begin{aligned}
&\int_{\mathbb{R}^n} |D_j^N(x',z)[D_j(z,y')-D_j(x_Q,y')]|dz\\
&\lesssim N\int_{\mathbb{R}^n}\frac{2^{-j\varepsilon}}{(2^{-j}+|x'-z|)^{n+\varepsilon}}\left(\frac{|z-x_Q|}{2^{-j}+|y'-z|}\right)^{\varepsilon}\frac{2^{-j\varepsilon}}{(2^{-j}+|y'-z|)^{n+\varepsilon}}dz\\
&\lesssim 2^{-N\varepsilon''}\int_{\mathbb{R}^n}\frac{2^{-j\varepsilon}}{(2^{-j}+|x'-z|)^{n+\varepsilon}}\frac{2^{-j\varepsilon}}{(2^{-j}+|y'-z|)^{n+\varepsilon}}dz\\
&\lesssim 2^{-N\varepsilon''}\frac{2^{-j\varepsilon}}{(2^{-j}+|x'-y'|)^{n+\varepsilon}}\\
&\lesssim 2^{-N\varepsilon''}\frac{2^{-j(\varepsilon-\varepsilon_1)}}{(2^{-j}+|x-y|)^{n+(\varepsilon-\varepsilon_1)}}.
\end{aligned}$$

Similarly, we can treat the other terms and obtain the same bound. So

$$\begin{aligned}
&|D_j^N D_j(x,y)-D_j^N D_j(x',y)-D_j^N D_j(x,y')+D_j^N D_j(x',y')|\\
&\lesssim 2^{-N\varepsilon''}\frac{2^{-j(\varepsilon-\varepsilon_1)}}{(2^{-j}+|x-y|)^{n+\varepsilon-\varepsilon_1}}.
\end{aligned}$$

Let $\theta=\frac{\varepsilon'}{\varepsilon_1}\in(0,1)$. Interpolating this with (12) yields

$$\begin{aligned}
&\big|[K(x,y)-K(x',y)]-[K(x,y')-K(x',y')]\big|\\
&\lesssim \sum_j |D_j^N D_j(x,y)-D_j^N D_j(x',y)-D_j^N D_j(x,y')+D_j^N D_j(x',y')|\\
&\lesssim 2^{-N\varepsilon''(1-\theta)}\sum_{j\in\mathbb{Z}}\left(\frac{|x-x'|}{2^{-j}+|x-y|}\right)^{\theta\varepsilon_1}\left(\frac{|y-y'|}{2^{-j}+|x-y|}\right)^{\theta\varepsilon_1}\frac{2^{-j(\varepsilon-\varepsilon_1)}}{(2^{-j}+|x-y|)^{n+\varepsilon-\varepsilon_1}}\\
&\lesssim 2^{-N\varepsilon''(1-\theta)}\frac{|x-x'|^{\varepsilon}|y-y'|^{\varepsilon}}{|x-y|^{n+2\varepsilon}}.
\end{aligned}$$

This concludes the proof of (8).

Proof of (7). We write

$$|R_N^1(x,y)| = \left|\sum_{|l|>N}\sum_k D_{k+l}D_k(x-y)\right| \le \sum_{|l|>N}\sum_k |D_{k+l}D_k(x-y)|.$$

By (6), we have

$$|R_N^1(x,y)| \lesssim \sum_{|l|>N}\sum_k 2^{-|l|\varepsilon'} \frac{2^{-((k+l)\wedge k)\varepsilon}}{(2^{-((k+l)\wedge k)} + |x-y|)^{n+\varepsilon}} \lesssim 2^{-N\varepsilon'}|x-y|^{-n},$$

which gives the size condition.

To show the regularity condition, taking the geometric mean of (6) and Lemma 1, for $|x-x'| \le \frac{1}{2}|x-y|$ we get

$$\begin{aligned} &|D_{k+l}D_k(x,y) - D_{k+l}D_k(x',y)| \\ &\lesssim 2^{-|l|\delta}\Big(\frac{|x-x'|}{2^{-((k+l)\wedge k)} + |x-y|}\Big)^{\varepsilon'} \frac{2^{-((k+l)\wedge k)\varepsilon'}}{(2^{-((k+l)\wedge k)} + |x-y|)^{n+\varepsilon'}}, \end{aligned} \tag{13}$$

where $0 < \varepsilon' < \varepsilon$ and $\delta > 0$. Thus, for $|x-x'| \le \frac{1}{2}|x-y|$,

$$\begin{aligned} |R_N^1(x,y) - R_N^1(x',y)| &= \left|\sum_{|l|>N}\sum_k [D_{k+l}D_k(x-y) - D_{k+l}D_k(x'-y)]\right| \\ &\le \sum_{|l|>N}\sum_k |D_{k+l}D_k(x-y) - D_{k+l}D_k(x'-y)| \\ &\lesssim \sum_{|l|>N}\sum_k 2^{-|l|\delta}\Big(\frac{|x-x'|}{2^{-(k\wedge l)} + |x-y|}\Big)^{\varepsilon'} \frac{2^{-(k\wedge l)\varepsilon'}}{(2^{-(k\wedge l)} + |x-y|)^{n+\varepsilon'}} \\ &\lesssim 2^{-N\delta}|x-x'|^{\varepsilon'}|x-y|^{-(n+\varepsilon')}. \end{aligned}$$

The proof of the smoothness condition in the second variable is the same. By the geometric mean of estimates in Lemmas 1 and (6), we obtain that for $|y-y'| \le \frac{1}{2}|x-y|$,

$$\begin{aligned} &|D_{k+l}D_k(x-y) - D_{k+l}D_k(x-y')| \\ &\lesssim 2^{-|l|\delta}\Big(\frac{|y-y'|}{2^{-((k+l)\wedge k)} + |x-y|}\Big)^{\varepsilon'} \frac{2^{-((k+l)\wedge k)\varepsilon'}}{(2^{-((k+l)\wedge k)} + |x-y|)^{n+\varepsilon'}}, \end{aligned} \tag{14}$$

where $0 < \varepsilon' < \varepsilon$ and $\delta > 0$. Thus, for $|y - y'| \le \frac{1}{2}|x - y|$,

$$\begin{aligned}|R_N^1(x, y) - R_N^1(x, y')| &= \Big|\sum_{|l|>N} \sum_k [D_{k+l}D_k(x - y) - D_{k+l}D_k(x - y')\Big| \\ &\le \sum_{|l|>N} \sum_k |D_{k+l}D_k(x - y) - D_{k+l}D_k(x - y')| \\ &\lesssim \sum_{|l|>N} \sum_k 2^{-|l|\delta}\Big(\frac{|y - y'|}{2^{-(k\wedge l)} + |x - y|}\Big)^{\varepsilon'} \frac{2^{-(k\wedge l)\varepsilon'}}{(2^{-(k\wedge l)} + |x - y|)^{n+\varepsilon'}} \\ &\lesssim 2^{-N\delta}|y - y'|^{\varepsilon'}|x - y|^{-(n+\varepsilon')}.\end{aligned}$$

To verify the double difference condition (2), by the geometric mean of the estimates in Lemma 2, (13) and (14), we get that for $|x-x'| \le \frac{1}{2}|x-y|$ and $|y-y'| \le \frac{1}{2}|x-y|$,

$$\begin{aligned}&|[D_{k+l}D_k(x - y) - D_{k+l}D_k(x' - y)] - [D_{k+l}D_k(x - y') - D_{k+l}D_k(x' - y')]| \\ &\lesssim 2^{-|l|\delta}\Big(\frac{|x - x'|}{2^{-((k+l)\wedge k)} + |x - y|}\Big)^{\varepsilon'}\Big(\frac{|y - y'|}{2^{-((k+l)\wedge k)} + |x - y|}\Big)^{\varepsilon'} \\ &\quad\times\frac{2^{-((k+l)\wedge k)(\varepsilon-\varepsilon')}}{(2^{-((k+l)\wedge k)} + |x - y|)^{n+(\varepsilon-\varepsilon')}}.\end{aligned}$$

Thus, for $|x - x'| \le \frac{1}{2}|x - y|$ and $|y - y'| \le \frac{1}{2}|x - y|$,

$$\begin{aligned}&|R_N^1(x, y) - R_N^1(x', y) - R_N^1(x, y') + R_N^1(x', y')| \\ &\le \sum_{|l|>N} \sum_k |[D_{k+l}D_k(x - y) - D_{k+l}D_k(x' - y)] \\ &\qquad\qquad -[D_{k+l}D_k(x - y') - D_{k+l}D_k(x' - y')]| \\ &\lesssim \sum_{|l|>N} \sum_k 2^{-|l|\delta}\Big(\frac{|x - x'|}{2^{-((k+l)\wedge k)} + |x - y|}\Big)^{\varepsilon'}\Big(\frac{|y - y'|}{2^{-((k+l)\wedge k)} + |x - y|}\Big)^{\varepsilon'} \\ &\qquad\qquad\times\frac{2^{-((k+l)\wedge k)(\varepsilon-\varepsilon')}}{(2^{-((k+l)\wedge k)} + |x - y|)^{n+(\varepsilon-\varepsilon')}} \\ &\lesssim 2^{-N\delta}|x - x'|^{\varepsilon'}|y - y'|^{\varepsilon'}|x - y|^{-(n+2\varepsilon')}.\end{aligned}$$

This completes the proof of (7) and hence Theorem 1.1.

Acknowledgments The first author is partially supported by an NSF grant DMS-1408839 and a McDevitt Endowment Fund at Georgetown University, and the third author partially supported by NNSF of China (Grant No. 11671397) and Special Funds for Yueqi Young Scholars of CUMTB (2017QN29).

References

1. Bremer, J., Coifman, R., Maggioni, M., and Szlam, A.: Diffusion wavelet packets. Appl. Comput. Harmon. Anal. **21** 95–112 (2006)
2. Bui, H. and Laugesen R.S.: Wavelet frame bijectivity on Lebesgue and Hardy spaces. J. Fourier Anal. Appl. **19** 376–409 (2013)
3. Calderón, A.: Intermediate spaces and interpolation, the complex method. Studia Math. **24** 113–190 (1964)
4. Coifman, R.: A real variable characterization of H^p. Studia Math. **51** 269–274 (1974)
5. Coifman, R.: Wavelet analysis and signal processing. In: Signal Processing, Part I, in: IMA Vol. Math. Appl., vol. 22, pp. 59–68. Springer, New York (1990)
6. Coifman, R. and Maggioni, M.: Diffusion wavelets. Appl. Comput. Harmon. Anal. **21** (1) 53–94 (2006)
7. Coifman, R. and Weiss, G.: Extensions of Hardy spaces and their use in analysis. Bull. Amer. Math. Soc. **83** 569–645 (1977)
8. David, G., Journé, J.-L., and Semmes, S.: Calderón-Zygmund operators, para-accretive functions and interpolation. Rev. Mat. Iberoamericana **1** 1–56 (1985)
9. Deng, D. and Han, Y.: Harmonic analysis on spaces of homogeneous type. Lecture Notes in Math., vol. 1966. Springer-Verlag, Berlin (2009)
10. Ding, Y., Han, Y., Lu, G., and Wu, X.: Boundedness of singular integrals on multiparameter weighted Hardy spaces $H^p_w(\mathbb{R}^n \times \mathbb{R}^m)$. Potential Anal. **37** 31–56 (2012)
11. Frazier, M. and Jawerth, B.: A discrete transform and decomposition of distribution spaces. J. Funct. Anal. **93** 34–170 (1990)
12. Han, Y.: Calderón-type reproducing formula and the Tb theorem. Rev. Mat. Iberoamericana **10** 51–91 (1994)
13. Meyer, Y.: Wavelets and operators. Cambridge studies in advanced mathematics 37. Cambridge University Press (1992)
14. Taibleson, M. and Weiss, G.: The molecular characterization of certain Hardy spaces. Astérisque **77** 67–149 (1980)

Equiangular Frames and Their Duals

Somantika Datta

Abstract Systems of m equiangular lines spanning $\mathbb{R}^d$ or $\mathbb{C}^d$ that satisfy the so-called Welch bound have recently gained a lot of attention due to various applications in signal processing. Such sets are called *equiangular tight frames* (ETFs). One of the geometrically appealing aspects of an ETF is that any vector can be represented in terms of an ETF by using a dual frame that is also an equiangular set. However, for a given m and d, with $m > d + 1$, ETFs are rare. Here we study some properties of equiangular lines spanning $\mathbb{R}^d$ when the Welch bound is not met. Such equiangular sets are more common than ETFs. In this case, the properties of the canonical dual, in particular, the angle set of the canonical dual are studied. We determine conditions on equiangular lines spanning $\mathbb{R}^d$ whose canonical dual has few distinct angles.

1 Introduction

Given a set $\{f_i\}_{i=1}^m$ of m unit vectors in $\mathbb{C}^d$, with $m > d$, the lower bound on the maximum cross correlation between distinct vectors is given by

$$\max_{i \neq j} |\langle f_i, f_j \rangle|^2 \geq \frac{m-d}{d(m-1)}. \tag{1}$$

The quantity on the right of (1) is known as the *Welch bound* after L. R. Welch who gave a family of bounds [31] parametrized by integers $k \geq 1$ as follows:

$$\max_{i \neq j} |\langle f_i, f_j \rangle|^{2k} \geq \frac{1}{m-1} \left[\frac{m}{\binom{d+k-1}{k}} - 1 \right]. \tag{2}$$

S. Datta (✉)
Department of Mathematics, University of Idaho, Moscow, ID, USA
e-mail: sdatta@uidaho.edu

M. Hirn et al. (eds.), *Excursions in Harmonic Analysis, Volume 6*, Applied and Numerical Harmonic Analysis, https://doi.org/10.1007/978-3-030-69637-5_9

Welch obtained the bounds in (2) as a consequence of the following inequality:

$$\sum_{i=1}^{m}\sum_{j=1}^{m}|\langle f_i, f_j\rangle|^{2k} \geq \frac{m^2}{\binom{d+k-1}{k}},$$

and often this is referred to as the Welch bound. The special case of $k = 1$ which is given in (1) has gained a lot of attention among researchers mainly in regard to the study of sets that attain the lower bound. Sets that attain the lower bound in (1), often called *Welch bound equality* sets, arise in various application areas such as communication systems, quantum information processing, and coding theory [14, 16, 18–23, 32]. In a purely mathematical setting, sets that attain the lower bound in (1) are objects called *equiangular tight frames* (ETFs). Consequently, the problem of constructing ETFs and determining conditions under which they exist has gained substantial attention [1, 2, 11, 15, 23, 24, 27, 30].

The definition of a frame originates from work by Duffin and Schaeffer on nonharmonic Fourier series [10]. In general, frames can be thought of as redundant sets that generalize orthonormal bases. A frame $\{f_i\}_{i=1}^m$ for a finite d-dimensional space can be used to represent any element f in the underlying space as

$$f = \sum_{i=1}^{m}\langle f, f_i\rangle g_i = \sum_{i=1}^{m}\langle f, g_i\rangle f_i.$$

The set $\{g_i\}_{i=1}^m$ is called a *dual frame* of $\{f_i\}_{i=1}^m$. If $\{f_i\}_{i=1}^m$ is an ETF, then it has a dual that is also equiangular and tight. Roughly speaking, equiangular means that the vectors are equally spaced, and being tight means that any given f can be represented in a form that is similar to an orthogonal expansion. One can say that for ETFs, both the frame and its dual have a nice geometric structure. Unfortunately, for many pairs (m, d), ETFs either do not exist or it is unknown whether or not they exist [24]. Here we study equiangular frames that are not necessarily tight and investigate whether they can have an equiangular dual. If an equiangular dual cannot be found, the desire is to be able to classify conditions under which an equiangular frame has a dual such that the number of distinct angles among the dual frame vectors is *small*.

Some definitions and known results that will be used are collected next. We will be concerned with d-dimensional Hilbert spaces of the form $\mathbb{F}^d$, where $\mathbb{F} = \mathbb{R}$ or $\mathbb{C}$. In $\mathbb{F}^d$, a *frame* is the same as a spanning set. Given a set $\Phi = \{f_1, \ldots, f_m\}$ in $\mathbb{F}^d$, let T be a matrix whose columns are the vectors $f_1, \ldots, f_m$. T will be called the *synthesis operator* of Φ. If Φ is a frame, then the $d \times d$ matrix $S = TT^*$ is called the *frame operator* of Φ. The set Φ is said to be a *tight frame* if S is a constant multiple of the identity. The set $\{\widetilde{f_i} = S^{-1}f_i\}$ is a dual frame for Φ, and is called the *canonical dual*. The frame operator of the canonical dual is S^{-1}. If for all $1 \leq i \leq m$ there is a constant c such that $\|f_i\| = c$, then Φ is called an *equal norm frame* or a *unit norm frame* if $c = 1$. For basics on frame theory, the reader is referred to [4]. A frame of m vectors in a d-dimensional space will be referred to as an (m, d) *frame*. Unless otherwise stated, it will be assumed that $m > d$.

Definition 1 (Equiangular Tight Frame [23, 24]) An *equiangular tight frame* (ETF) is a set $\{f_i\}_{i=1}^m$ in $\mathbb{F}^d$ satisfying

(i) $TT^* = \frac{m}{d}I$, i.e., the set is a tight frame.
(ii) $\|f_i\| = 1$, for $i = 1, \ldots, m$, i.e., the set is unit norm.
(iii) $|\langle f_i, f_j\rangle| = \sqrt{\frac{m-d}{d(m-1)}}$, $1 \le i \ne j \le m$.

The quantity $\sqrt{\frac{m-d}{d(m-1)}}$ appearing in (iii) in Definition 1 will be referred to here as the *Welch bound*. Relaxing the condition of being tight in Definition 1 gives an equiangular frame as is defined next.

Definition 2 (Equiangular Frame [23, 30]) An *equiangular frame* (EF) is a set $\{f_i\}_{i=1}^m$ that spans $\mathbb{F}^d$, and satisfies

(i) $\|f_i\| = c$, for $i = 1, \ldots, m$, and some $c > 0$, i.e., the set is equal norm.
(ii) $|\langle f_i, f_j\rangle| = \alpha$, $1 \le i \ne j \le m$.

Note that in Definition 2, α equals the Welch bound only when the set is also unit norm and tight [23].

The matrix T^*T is the *Gram matrix* G of the set Φ. The (i, j)th entry of G is the inner product $\langle f_j, f_i\rangle$. If Φ is an equiangular frame (EF), then for all $i \ne j$

$$\alpha := |\langle f_i, f_j\rangle|.$$

If m vectors in $\mathbb{F}^d$ form an EF, then its Gram matrix G can be written as

$$G = cI + \alpha Q, \tag{3}$$

where Q is an $m \times m$ Hermitian matrix with zero diagonal and unimodular entries elsewhere, called the *signature matrix*. For a real EF, the off-diagonal entries of Q are ± 1. By $[\lambda]^n$ it will be meant that the eigenvalue λ has multiplicity n. The Gram matrix of a frame of m vectors in $\mathbb{F}^d$ will have eigenvalues that can be written as

$$[0]^{m-d} < \lambda_1 \le \cdots \le \lambda_d,$$

and this implies that the eigenvalues of Q are [15, 17]

$$[-c/\alpha]^{m-d}, (\lambda_1 - c)/\alpha, \ldots, (\lambda_d - c)/\alpha.$$

Due to (3), the study of EFs reduces to the study of properties of Gram matrices or the corresponding signature matrices. Note that G and the corresponding frame operator S have the same nonzero eigenvalues. Thus the Gram matrix G of a tight frame has only one distinct nonzero eigenvalue.

Lemma 1 *A signature matrix Q always has at least one negative eigenvalue.*

Proof This follows from the fact that the trace of Q is zero, and the sum of the eigenvalues of a matrix is equal to the trace. □

Definition 3 ([25]) Two signature matrices, Q_1 and Q_2, are *equivalent* if there exists a signed permutation matrix P such that $Q_2 = PQ_1P^{\mathrm{T}}$. Two signature matrices Q_1 and Q_2 are called *cospectral* if they have the same set of eigenvalues.

Equivalent matrices are cospectral but starting from $m = 8$ there exist examples of cospectral matrices of size $m \times m$ that are not equivalent [25].

For a given set $\{f_i\}_{i=1}^m$ in $\mathbb{F}^d$, the *angle set* is defined to be the set

$$\left\{ \left| \left\langle \frac{f_i}{\|f_i\|}, \frac{f_j}{\|f_j\|} \right\rangle \right|, 1 \leq i \neq j \leq m \right\}.$$

The number of distinct values in the angle set will be referred to as the number of angles in the set. For an equiangular frame, the angle set has only one distinct value α, for some $\alpha \in \mathbb{R}$. Generalizing this gives the following.

Definition 4 (k-Angle Frame) A frame $\{f_i\}_{i=1}^m$ for $\mathbb{F}^d$ is called a *k-angle frame* if the angle set has k distinct values $\{\alpha_1, \ldots, \alpha_k\}$.

The focus here is mainly on real equiangular frames, i.e., we take $\mathbb{F} = \mathbb{R}$. Nevertheless, where possible, the results have been presented in the setting of $\mathbb{F}^d$.

2 Construction and Existence of Equiangular Frames

It is well known that there always exists a $(d+1, d)$ ETF that can be obtained from the vertices of a regular simplex [24]. Other constructions can be found in [17] for $(d+1, d)$ real ETFs, and, more recently, in [8] for both real and complex $(d+1, d)$ ETFs. However, as already mentioned in Sect. 1, an ETF does not exist for many pairs (m, d), and one might wish to relax the conditions of an ETF and instead look at equiangular frames that are not necessarily tight. By relaxing the requirement of being tight, one can expect EFs to be more common than ETFs. For an arbitrary size m, the existence of an EF is restricted by *Gerzon's bound*, i.e., the maximum number of equiangular lines is bounded above by $d(d+1)/2$ in $\mathbb{R}^d$, and d^2 in $\mathbb{C}^d$ [9].

As already discussed in Sect. 1, for any equiangular frame $\{f_i\}_{i=1}^m$ in $\mathbb{F}^d$ there is a corresponding signature matrix. See (3) and the discussion that follows. Conversely, for a given signature matrix, an EF can be constructed as discussed below. For convenience, we consider EFs that are unit norm. The method can be adapted to the construction of EFs with any given norm.

Construction of Unit Norm Equiangular Frames from Signature Matrices
To construct an EF of size m, start with an $m \times m$ signature matrix Q. By Lemma 1, Q has at least one negative eigenvalue. Let the minimum eigenvalue of Q be $-\mu$, $\mu > 0$. Then

$$G = I + \frac{1}{\mu} Q$$

is the Gram matrix of a unit norm EF of m vectors in $\mathbb{R}^d$ where $m - d$ is the multiplicity of $-\mu$. Due to (1), we have

$$\frac{1}{\mu} \geq \sqrt{\frac{m-d}{d(m-1)}}$$

with equality attained if and only if the frame is also tight. From G, the actual frame vectors can be constructed via a diagonalization of G. Since G is Hermitian, it can be diagonalized by means of a unitary matrix. This means that there exists an $m \times m$ unitary matrix V and a diagonal matrix $D = \text{diag}(0, \ldots, 0, \lambda_1, \ldots, \lambda_d)$, such that

$$G = VDV^*.$$

Take the last d columns of $V : \{v_{m-d+1}, \ldots, v_m\}$, and consider the $m \times d$ matrix F whose ith column is $\sqrt{\lambda_i} v_{m-d+i}$.

$$F = \begin{bmatrix} | & | & & | \\ \sqrt{\lambda_1} v_{m-d+1} & \sqrt{\lambda_2} v_{m-d+2} & \cdots & \sqrt{\lambda_d} v_m \\ | & | & & | \end{bmatrix}. \tag{4}$$

Then the rows of F are the frame vectors of a unit norm (m, d) EF whose synthesis operator is F^*. This leads to the following.

Proposition 1 *Given $m, d \in \mathbb{N}$, with $m > d > 1$. An equiangular frame of m vectors in $\mathbb{F}^d$ exists if and only if there is an $m \times m$ signature matrix Q whose minimum eigenvalue $-\mu$, $\mu > 0$, has multiplicity $m - d$. Further, if the frame is unit norm, then the α in Definition 2 is $\alpha = \frac{1}{\mu} \geq \sqrt{\frac{m-d}{d(m-1)}}$ with equality holding if and only if the frame is also tight.*

For a given m, there can be $M = 2^{m(m-1)/2}$ different real $m \times m$ signature matrices, some of which might be equivalent. From these M signature matrices one can get EFs of m unit vectors in $\mathbb{R}^d$ for $2 \leq d \leq m - 1$, according to Proposition 1. As already mentioned, in $\mathbb{R}^d$, $m \leq \frac{d(d+1)}{2}$ [9]. The possibilities for $3 \leq m \leq 7$, found using MATLAB, are shown in Table 1. Beyond $m = 7$, the total number of $m \times m$ signature matrices becomes too large to store and process at once.

The method of construction discussed above can be used to construct EFs whose frame operator is diagonal. This is shown below in Proposition 2. See also

Table 1 Table showing number of $m \times m$ signature matrices resulting in equiangular frames in $\mathbb{R}^d$

m	d	Total no. of $m \times m$ signature matrices	No. of signature matrices giving an (m, d) real EF
3	2	8	4
4	3	64	56
5	3	1024	192
5	4	1024	816
6	3	32768	384
6	4	32768	480
6	5	32768	31872
7	5	2097152	106528
7	6	2097152	1990560

Examples 3.1 and 3.2. It is worthwhile to note that the result in Proposition 2 is not specific to EFs.

Proposition 2 *Let G be an $m \times m$ Hermitian matrix of rank d. Then there exists a frame of m vectors in $\mathbb{F}^d$ whose Gram matrix is G such that the frame operator is diagonal.*

Proof Following a diagonalization of G, one can use the exact same notation as above, and (4), to get the frame operator to be

$$S = F^*F = \begin{bmatrix} - \sqrt{\lambda_1} v_{m-d+1} - \\ - \sqrt{\lambda_2} v_{m-d+2} - \\ \vdots \\ - \quad \sqrt{\lambda_d} v_m \quad - \end{bmatrix} \begin{bmatrix} | & | & & | \\ \sqrt{\lambda_1} v_{m-d+1} & \sqrt{\lambda_2} v_{m-d+2} & \cdots & \sqrt{\lambda_d} v_m \\ | & | & & | \end{bmatrix}.$$

Since V is a unitary matrix

$$\langle v_i, v_j \rangle = \delta_{ij},$$

and this gives

$$S = \begin{bmatrix} \lambda_1 & 0 & \cdots & 0 \\ 0 & \lambda_2 & \cdots & 0 \\ \vdots & \vdots & \ddots & \vdots \\ 0 & 0 & \cdots & \lambda_d \end{bmatrix}$$

which is diagonal. □

Two frames $\Phi = \{f_j\}_{j \in J}$ and $\Psi = \{g_j\}_{j \in J}$ are *unitarily equivalent* if there is a unitary transformation U such that for all $j \in J$, $g_j = U f_j$ [30].

Corollary 1 *Let $\Phi = \{f_i\}_{i=1}^m$ be a frame of $\mathbb{F}^d$. Then there exists a frame $\Psi = \{g_i\}_{i=1}^m$ of $\mathbb{F}^d$ such that the frame operator of Ψ is diagonal, and Φ is unitarily equivalent to Ψ.*

Proof Let G be the Gram matrix of Φ. By Proposition 2, one can construct a frame $\Psi = \{g_i\}_{i=1}^m$ such that the Gram matrix of Ψ is G, and the frame operator of Ψ is diagonal. As shown in [28], two frames are unitarily equivalent if and only if their Gram matrices are equal. Thus, Φ and Ψ are unitarily equivalent. □

3 Nontight Equiangular Frames and Their Duals

From the definition of a tight frame it is obvious that an ETF will have its canonical dual to be both equiangular and tight. The canonical dual vectors are just scaled versions of the frame vectors, and the angle set of the canonical dual will also have only one distinct value. This means that both the frame and its canonical dual have a nice structure. Phrased in terms of Definition 4, an ETF is an 1-angle frame such that the canonical dual is also an 1-angle frame. However, it is well known that for many pairs (m, d), ETFs do not exist [24]. Here, in Sect. 3.2, we study some properties of the canonical dual for a nontight equiangular frame. The angle set of the canonical dual is investigated in Sect. 3.3.

We first give some examples of nontight equiangular frames and their canonical duals.

3.1 Examples of Nontight Equiangular Frames

For a given d, we first investigate the number of equiangular frames of size $d + 1$ that one can have in $\mathbb{R}^d$ aside from the $(d+1, d)$ ETF that is already known to exist. For a $(d + 1, d)$ ETF, by the Welch bound,

$$|\langle f_i, f_j\rangle| = 1/d, \quad i \neq j.$$

Thus for a $(d + 1, d)$ nontight EF, since the Welch bound is not attained,

$$|\langle f_i, f_j\rangle| > 1/d, \quad i \neq j.$$

Example 3.1 (($(d+1, d)$) Equiangular Frames)

1. All (3, 2) real unit norm equiangular frames are ETFs. This is determined from the minimum eigenvalue of each of the $2^3 = 8$ possible signature matrices of size 3×3, by checking when the multiplicity of this eigenvalue is $3 - 2 = 1$. In each of the 4 feasible cases, see Table 1, the minimum eigenvalue equals -2. The corresponding unit norm EF has common angle $\alpha = \frac{1}{2}$, which matches with the Welch bound. Thus each EF is an ETF.
2. For $d \geq 3$, there can be $(d+1, d)$ *nontight* equiangular frames. Take $d = 3$. There are 64 possible real signature matrices Q of size 4×4. Out of these, 56 give equiangular frames in $\mathbb{R}^3$. The resulting unit norm equiangular frames have the common angle α to be either $\alpha = \frac{1}{3}$ or $\alpha = \frac{1}{\sqrt{5}}$. These values come from the minimum eigenvalue of the corresponding Q. The value $\alpha = \frac{1}{3}$ is the corresponding Welch bound, and corresponds to a (4, 3) ETF. The value $\alpha = \frac{1}{\sqrt{5}}$ comes from a (4, 3) nontight EF. One signature matrix Q giving rise to such a frame, and the corresponding Gram matrix G are

$$Q = \begin{bmatrix} 0 & 1 & -1 & -1 \\ 1 & 0 & -1 & -1 \\ -1 & -1 & 0 & -1 \\ -1 & -1 & -1 & 0 \end{bmatrix}, \quad G = \begin{bmatrix} 1 & \frac{1}{\sqrt{5}} & -\frac{1}{\sqrt{5}} & -\frac{1}{\sqrt{5}} \\ \frac{1}{\sqrt{5}} & 1 & -\frac{1}{\sqrt{5}} & -\frac{1}{\sqrt{5}} \\ -\frac{1}{\sqrt{5}} & -\frac{1}{\sqrt{5}} & 1 & -\frac{1}{\sqrt{5}} \\ -\frac{1}{\sqrt{5}} & -\frac{1}{\sqrt{5}} & -\frac{1}{\sqrt{5}} & 1 \end{bmatrix}.$$

As shown in Sect. 2, a diagonalization of G will yield a frame corresponding to this G coming from the rows of the matrix F. In this case, the rounded values of F are

$$\begin{bmatrix} 0.5257 & 0 & 0.8506 \\ -0.5257 & 0 & 0.8506 \\ 0 & -0.8506 & -0.5257 \\ 0 & 0.8506 & -0.5257 \end{bmatrix}.$$

Here F^* is the synthesis operator. The rounded values of the frame operator $S = F^*F$ are

$$\begin{bmatrix} 0.5528 & 0 & 0 \\ 0 & 1.4472 & 0 \\ 0 & 0 & 2 \end{bmatrix}$$

(continued)

which confirms that the frame is not tight. Note that the frame operator is diagonal, see Proposition 2. The canonical dual comes from the columns of $S^{-1}F^*$ which is the synthesis operator of the canonical dual. The rounded values of $S^{-1}F^*$ are

$$\begin{bmatrix} 0.9510 & -0.9510 & 0 & 0 \\ 0 & 0 & -0.5878 & 0.5878 \\ 0.4253 & 0.4253 & -0.2629 & -0.2629 \end{bmatrix}.$$

The matrix $[\langle \frac{\tilde{f}_i}{\|\tilde{f}_i\|}, \frac{\tilde{f}_j}{\|\tilde{f}_j\|} \rangle]$ is

$$\begin{bmatrix} 1 & -\frac{2}{3} & -\frac{1}{6} & -\frac{1}{6} \\ -\frac{2}{3} & 1 & -\frac{1}{6} & -\frac{1}{6} \\ -\frac{1}{6} & -\frac{1}{6} & 1 & -\frac{2}{3} \\ -\frac{1}{6} & -\frac{1}{6} & -\frac{2}{3} & 1 \end{bmatrix}.$$

By inspection, the angle set of the canonical dual is $\{\frac{1}{6}, \frac{2}{3}\}$, and so the canonical dual of this frame has 2 angles.

For $2 \leq d \leq 6$, Table 2 gives the number of feasible signature matrices of size $(d+1) \times (d+1)$ that can result in equiangular frames in $\mathbb{R}^d$. The table also gives how many different values of the common angle α are possible for each d. One of these values will correspond to the Welch bound of the corresponding $(d+1, d)$ ETF. Others will correspond to the common angles of nontight equiangular frames of $d+1$ vectors in $\mathbb{R}^d$.

Table 2 Table showing number of $(d+1) \times (d+1)$ signature matrices resulting in equiangular frames in $\mathbb{R}^d$

d	No. of $(d+1) \times (d+1)$ signature matrices giving an EF in $\mathbb{R}^d$	No. of different possibilities of the common angle α
2	4	1
3	56	2
4	816	5
5	31872	11
6	1990560	40

Example 3.2 ((n, d) **Equiangular Frames and Their Duals;** $n > d + 1$)
There does not exist a (5, 3) real ETF [23]. The Welch bound for $m = 5$ and $d = 3$ is $\frac{1}{\sqrt{6}}$ which cannot be attained by any set of 5 vectors in $\mathbb{R}^3$. However, there exists a (5, 3) real EF with signature matrix

$$Q = \begin{bmatrix} 0 & -1 & 1 & 1 & -1 \\ -1 & 0 & -1 & 1 & 1 \\ 1 & -1 & 0 & -1 & 1 \\ 1 & 1 & -1 & 0 & -1 \\ -1 & 1 & 1 & -1 & 0 \end{bmatrix}.$$

The eigenvalues of Q are $-\sqrt{5}, -\sqrt{5}, 0, \sqrt{5}, \sqrt{5}$. The Gram matrix is

$$G = I + \frac{1}{\sqrt{5}}Q = \begin{bmatrix} 1 & -\frac{1}{\sqrt{5}} & \frac{1}{\sqrt{5}} & \frac{1}{\sqrt{5}} & -\frac{1}{\sqrt{5}} \\ -\frac{1}{\sqrt{5}} & 1 & -\frac{1}{\sqrt{5}} & \frac{1}{\sqrt{5}} & \frac{1}{\sqrt{5}} \\ \frac{1}{\sqrt{5}} & -\frac{1}{\sqrt{5}} & 1 & -\frac{1}{\sqrt{5}} & \frac{1}{\sqrt{5}} \\ \frac{1}{\sqrt{5}} & \frac{1}{\sqrt{5}} & -\frac{1}{\sqrt{5}} & 1 & -\frac{1}{\sqrt{5}} \\ -\frac{1}{\sqrt{5}} & \frac{1}{\sqrt{5}} & \frac{1}{\sqrt{5}} & -\frac{1}{\sqrt{5}} & 1 \end{bmatrix}.$$

G has eigenvalues $\{0, 0, 1, 2, 2\}$. Thus G has two distinct nonzero eigenvalues, and the corresponding frame is nontight. For this (5, 3) equiangular frame, the absolute value of the inner product between any two distinct vectors is $\frac{1}{\sqrt{5}}$ as can be seen from G, and this minimizes the maximum cross correlation among all 5 vectors in $\mathbb{R}^3$ [23, 26]. To get the frame vectors of a (5, 3) equiangular frame, consider the SVD of G:

$$G = PDP',$$

where

$$D = \begin{bmatrix} 0 & 0 & 0 & 0 & 0 \\ 0 & 0 & 0 & 0 & 0 \\ 0 & 0 & 1 & 0 & 0 \\ 0 & 0 & 0 & 2 & 0 \\ 0 & 0 & 0 & 0 & 2 \end{bmatrix},$$

and the rounded values of P are

(continued)

$$\begin{bmatrix} 0.1995 & -0.6002 & 0.4472 & -0.3820 & -0.5041 \\ -0.5091 & -0.3752 & 0.4472 & 0.0127 & 0.6323 \\ -0.5142 & 0.3683 & 0.4472 & 0.3614 & -0.5190 \\ 0.1913 & 0.6028 & 0.4472 & -0.5974 & 0.2075 \\ 0.6324 & 0.0043 & 0.4472 & 0.6053 & 0.1833 \end{bmatrix}.$$

Following Sect. 2, the frame vectors of a frame corresponding to the above G are the rows of the matrix F whose rounded values are

$$\begin{bmatrix} 0.4472 & -0.5402 & -0.7129 \\ 0.4472 & 0.0180 & 0.8942 \\ 0.4472 & 0.5111 & -0.7340 \\ 0.4472 & -0.8449 & 0.2935 \\ 0.4472 & 0.8560 & 0.2592 \end{bmatrix}.$$

It can be checked that the frame operator is

$$S = F^*F = \begin{bmatrix} 1 & 0 & 0 \\ 0 & 2 & 0 \\ 0 & 0 & 2 \end{bmatrix}$$

and that $FF^* = G$. Note that the frame operator is diagonal, see Proposition 2. The inverse of the frame operator is

$$S^{-1} = \begin{bmatrix} 1 & 0 & 0 \\ 0 & 1/2 & 0 \\ 0 & 0 & 1/2 \end{bmatrix}.$$

The canonical dual can be obtained from the columns of $S^{-1}F^*$ whose rounded values are

$$\begin{bmatrix} 0.4472 & 0.4472 & 0.4472 & 0.4472 & 0.4472 \\ -0.2701 & 0.0092 & 0.2555 & -0.4225 & 0.4280 \\ -0.3564 & 0.4471 & -0.3670 & 0.1467 & 0.1296 \end{bmatrix}.$$

Upon calculating the Gram matrix of the canonical dual it can be seen that the canonical dual of this frame has two angles [30].

3.2 Frame Properties of Canonical Duals of Equiangular Frames

In what follows (see Corollary 2), we give a necessary and sufficient condition when the canonical dual of certain equiangular frames will also be equiangular.

Proposition 3 *[7] Given $\{\alpha_i\}_{i=1}^m \subseteq \mathbb{F}$, the following are equivalent:*

(a) *There exist dual frames $\{f_i\}_{i=1}^m$ and $\{g_i\}_{i=1}^m$ for $\mathbb{F}^d$ such that $\alpha_i = \langle f_i, g_i\rangle$ for all $1 \le i \le m$;*
(b) $d = \sum_{i=1}^m \alpha_i$.

Theorem 3.1 *[6] Let $\{f_i\}_{i=1}^m$ be an equal norm frame in $\mathbb{F}^d$. Let $\{\widetilde{f_i}\}_{i=1}^m$ denote the canonical dual of $\{f_i\}_{i=1}^m$. Then $\{f_i\}_{i=1}^m$ is a tight frame if and only if $\|f_i\|\,\|\widetilde{f_i}\| = \frac{d}{m}$ for $i = 1, \dots, m$.*

Proof Suppose that $\{f_i\}_{i=1}^m$ is a tight frame in $\mathbb{F}^d$ with frame bound A. Note that $\widetilde{f_i} = \frac{1}{A} f_i$ for $i = 1, \dots, m$. Since $\{f_i\}_{i=1}^m$ is equal norm, Proposition 3 implies that for $i = 1, \dots, m$,

$$d = \sum_{j=1}^m \langle f_j, \widetilde{f_j}\rangle = \frac{m}{A}\|f_i\|^2.$$

Thus we have

$$\|f_i\| = \sqrt{\frac{Ad}{m}},\ \|\widetilde{f_i}\| = \sqrt{\frac{d}{Am}}.$$

This implies that $\|f_i\|\,\|\widetilde{f_i}\| = \frac{d}{m}$ for $i = 1, \dots, m$.

Conversely, suppose that $\|f_i\|\,\|\widetilde{f_i}\| = \frac{d}{m}$ for $i = 1, \dots, m$. Using Proposition 3 again, we have

$$d = \sum_{i=1}^m \langle f_i, \widetilde{f_i}\rangle \le \sum_{i=1}^m |\langle f_i, \widetilde{f_i}\rangle| \le \sum_{i=1}^m \|f_i\|\,\|\widetilde{f_i}\| = d.$$

This implies that $|\langle f_i, \widetilde{f_i}\rangle| = \|f_i\|\,\|\widetilde{f_i}\|$ for $i = 1, \dots, m$. Then for $i = 1, \dots, m$, there exists a constant λ_i such that $\widetilde{f_i} = \lambda_i f_i$. Since $\{f_i\}_{i=1}^m$ is equal norm, and $\|f_i\|\,\|\widetilde{f_i}\| = \frac{d}{m}$ for $i = 1, \dots, m$, $|\lambda_i|$ is a constant for $i = 1, \dots, m$. Note that

$$0 \le \sum_{j=1}^m |\langle f_i, \widetilde{f_j}\rangle|^2 = \langle f_i, \widetilde{f_i}\rangle = \overline{\lambda_i}\|f_i\|^2$$

for $i = 1, \ldots, m$. Then λ_i is a positive constant for $i = 1, \ldots, m$. This implies that $\{f_i\}_{i=1}^m$ is a tight frame. □

Corollary 2 *Let $\{f_i\}_{i=1}^m$ be an equiangular frame in $\mathbb{F}^d$. Then the canonical dual $\{\widetilde{f}_i\}_{i=1}^m$ is equiangular with $\|f_i\| \, \|\widetilde{f}_i\| = \frac{d}{m}$ for $i = 1, \ldots, m$ if and only if $\{f_i\}_{i=1}^m$ is a tight frame.*

Remark 1 Even if the canonical dual is not equiangular, due to Proposition 2, one can construct EFs such that the frame operator of the canonical dual is diagonal.

As mentioned in Sect. 1, equivalent signature matrices are cospectral. Thus it is obvious from (3), and the discussion that follows, that equivalent signature matrices give rise to EFs having the same common angle α, where $-\frac{1}{\alpha}$ is the minimum eigenvalue of the equivalent signature matrices. Coming to canonical duals of EFs, it is shown below in Proposition 4 that the canonical duals of certain equiangular frames corresponding to equivalent signature matrices have the same angle set. The following lemma is needed.

Lemma 2 *Let Q_1 and Q_2 be two $m \times m$ equivalent signature matrices. There exist equiangular frames $\{f_i\}_{i=1}^m$ and $\{\psi_i\}_{i=1}^m$ corresponding to Q_1 and Q_2, respectively, such that their respective frame operators S_f and S_ψ are the same.*

Proof Obtain equiangular frames $\{f_i\}_{i=1}^m$ and $\{\psi_i\}_{i=1}^m$ corresponding to Q_1 and Q_2, respectively, by following the construction of Sect. 2. By Proposition 2, these frames will have diagonal frame operators S_f and S_ψ, respectively. Since Q_1 and Q_2 are cospectral, the corresponding Gram matrices have the same eigenvalues. Recall that frame operators have the same nonzero eigenvalues as the corresponding Gram matrices, and so S_f and S_ψ are the same. □

Let T and F be the synthesis operators of two frames. These frames are said to have *equivalent* Gram matrices if there exists a unitary matrix U and a signed (when $\mathbb{F} = \mathbb{R}$) or phased (when $\mathbb{F} = \mathbb{C}$) permutation matrix P such that $F = UTP$. Note that frames with equivalent Gram matrices have the same angle set.

Proposition 4 *Let Q_1 and Q_2 be two $m \times m$ equivalent signature matrices. Let $\{f_i\}_{i=1}^m$ and $\{\psi_i\}_{i=1}^m$ denote equiangular frames corresponding to Q_1 and Q_2, respectively, that also satisfy Proposition 2. If the canonical dual $\{\widetilde{f}_i\}_{i=1}^m$ has angle set equal to $A = \{\alpha_1, \ldots, \alpha_k\}$, then the canonical dual $\{\widetilde{\psi}_i\}_{i=1}^m$ also has the same angle set A. Moreover, if $\{\widetilde{f}_i\}_{i=1}^m$ is equal norm, then $\{\widetilde{\psi}_i\}_{i=1}^m$ is also equal norm.*

Proof By the proof of Lemma 2, $\{f_i\}_{i=1}^m$ and $\{\psi_i\}_{i=1}^m$ have the same frame operator S. Let F and T denote the synthesis operators of $\{f_i\}_{i=1}^m$ and $\{\psi_i\}_{i=1}^m$, respectively. Since the frames have equivalent signature matrices Q_1 and Q_2, they have equivalent Gram matrices, and there exists a unitary matrix U and a signed or phased permutation matrix P such that $F = UTP$. Then

$$\begin{aligned} S &= FF^* = (UTP)(UTP)^* = (UTP)P^*T^*U^* = UTT^*U^* \\ &= USU^*, \end{aligned}$$

and thus S must commute with U. This further implies that $\widetilde{U} := S^{-1}US$ is unitary. The synthesis operators of the canonical duals $\{\widetilde{f}_i\}_{i=1}^m$ and $\{\widetilde{\psi}_i\}_{i=1}^m$ are $S^{-1}F$ and $S^{-1}T$, respectively.

$$\widetilde{U}(S^{-1}T)P = S^{-1}US(S^{-1}T)P = S^{-1}UTP = S^{-1}F.$$

Thus the canonical duals yield equivalent Gram matrices, and the result follows. □

3.3 Angle Sets of Canonical Duals of Equiangular Frames

In this section we investigate conditions under which an equiangular frame that is not tight can have a dual that is a k-angle frame, where it is desired that k is a small positive integer. Gram matrices of ETFs have one nonzero eigenvalue. This means that signature matrices corresponding to ETFs have two distinct eigenvalues. Examples of such signature matrices are somewhat rare [24, 25], as are ETFs. It is known that various large sets of equiangular lines have corresponding signature matrices with three distinct eigenvalues [13]. This has motivated extensive study of signature matrices with exactly three eigenvalues in [13, 25]. The results in this section and some other related results can be found in [5]. In Theorem 3.2 and Theorem 3.3 below, we analyze signature matrices with three distinct eigenvalues and study the number of possible angles in the canonical dual of any corresponding equiangular frame. It is worth noting that a result similar to Theorem 3.2 using strongly regular graphs is given in [30]. The following Lemma 3 will be used.

Lemma 3 *If $-\lambda_1, \lambda_2, \lambda_3$ are the three distinct eigenvalues of a signature matrix Q, ordered such that $-\lambda_1 < \lambda_2 < \lambda_3$, $\lambda_1 > 0$, then*

$$-\lambda_1 \neq \frac{\lambda_2 + \lambda_3}{2}.$$

Proof Since Q has zero trace, it must have at least one positive eigenvalue, so $\lambda_3 > 0$. Now, if $-\lambda_1 = \frac{\lambda_2+\lambda_3}{2}$, then $\lambda_2+\lambda_3 < 0$; thus $-\lambda_1 < \lambda_2 < \lambda_2+\lambda_3 < \frac{\lambda_2+\lambda_3}{2} < 0$, which is a contradiction. □

The following lemma can be proved by a direct calculation of the characteristic polynomial of Q.

Lemma 4 *If Q is an $m \times m$ signature matrix whose off-diagonal entries are all* 1, *then Q has two distinct eigenvalues: $m - 1$ with multiplicity* 1, *and -1 with multiplicity $m - 1$.*

An eigenvector is said to be *regular* if its entries are ± 1. In what follows, $\mathbf{1}$ denotes the vector whose each entry is 1, and J is the matrix whose entries are all 1. The (i, j)th entry of a matrix A will be denoted by $A(i, j)$.

Theorem 3.2 *Let Q be an $m \times m$ signature matrix with three distinct eigenvalues $-\lambda_1, \lambda_2, \lambda_3$, ordered such that $-\lambda_1 < \lambda_2 < \lambda_3$, with $\lambda_1 > 0$. Let Φ denote any corresponding equiangular frame of m vectors in $\mathbb{R}^d$. Then the following hold.*

(a) Suppose that λ_2 or λ_3 is a simple eigenvalue with a regular eigenvector, and the multiplicity of $-\lambda_1$ is r. If the sum of this simple eigenvalue and λ_1 is not m, then the canonical dual is an equal norm 2-angle nontight frame, and $d = m - r$.

(b) Suppose that the minimum eigenvalue $-\lambda_1$ is simple with a regular eigenvector. Then the canonical dual is an equal norm 2-angle nontight frame. In this case, $d = m - 1$.

Proof The Gram matrix of a tight frame can have only one nonzero eigenvalue. Since Q has three distinct eigenvalues, G must have two distinct nonzero eigenvalues and so Φ is a not a tight frame. Thus the dual is also not tight in both (a) and (b).

Let P_1, P_2, and P_3 denote the orthogonal projections onto the eigenspaces of $-\lambda_1, \lambda_2$, and λ_3, respectively. By the Spectral Theorem $Q = -\lambda_1 P_1 + \lambda_2 P_2 + \lambda_3 P_3$, where $P_1 + P_2 + P_3 = I$, and for $i \neq j$, $P_i P_j = 0$. The Gram matrix of Φ is

$$G = I + \frac{1}{\lambda_1} Q = \frac{\lambda_1 + \lambda_2}{\lambda_1} P_2 + \frac{\lambda_1 + \lambda_3}{\lambda_1} P_3. \tag{5}$$

The Gram matrix of the canonical dual is the pseudo inverse of G [3], and given by

$$G^{\dagger} = \frac{\lambda_1}{\lambda_1 + \lambda_2} P_2 + \frac{\lambda_1}{\lambda_1 + \lambda_3} P_3. \tag{6}$$

(a) Since the multiplicity of the minimum eigenvalue of Q is $r = m - d$, the value of d is obvious. Without loss of generality assume that λ_3 is simple with a regular eigenvector v. Then $P_3 = \frac{1}{\|v\|^2} v v^{\mathrm{T}} = \frac{1}{m} v v^{\mathrm{T}}$. Note that the diagonal entries of P_3 are all equal to $\frac{1}{m}$. This implies, from (5), that P_2 also has constant diagonal. Therefore, in (6), $G^{\dagger}$ must have constant diagonal too, implying that the canonical dual is equal norm.

Equating the off-diagonal entries of G in (5) gives

$$\pm \frac{1}{\lambda_1} = \frac{\lambda_1 + \lambda_2}{\lambda_1} P_2(i, j) + \frac{\lambda_1 + \lambda_3}{\lambda_1} P_3(i, j),$$

or,

$$P_2(i, j) = [\pm 1 - (\lambda_1 + \lambda_3) P_3(i, j)] \frac{1}{\lambda_1 + \lambda_2}. \tag{7}$$

Using (6), (7), and the fact that the off-diagonal entries of P_3 are $\pm \frac{1}{m}$

$$G^{\dagger}(i,j) = \pm\frac{\lambda_1}{(\lambda_1+\lambda_2)^2} + \left(\pm\frac{1}{m}\right)\left[\frac{\lambda_1}{\lambda_1+\lambda_3} - \frac{\lambda_1(\lambda_1+\lambda_3)}{(\lambda_1+\lambda_2)^2}\right], \quad i \neq j. \tag{8}$$

From (8), the absolute values of the off-diagonal entries of $G^{\dagger}$ can take only two values, and this can be justified as follows. The expression inside the bracket in the second term on the right side of (8) cannot be zero due to Lemma 3. Since Q has three distinct eigenvalues, Q cannot have all its off-diagonal entries equal to 1 or all equal to -1 due to Lemma 4. Thus the off-diagonal entries of G take both values $\pm\frac{1}{\lambda_1}$. If P_3 and G have the exact same or the exact opposite sign distribution in their off-diagonal entries, then, from (7) and the given assumption, the off-diagonal entries of P_2 will equal $\pm c$ for some constant c and have the same sign distribution as that of P_3 or G. In that case, $P_2P_3 \neq 0$, which contradicts the Spectral Theorem. Thus P_3 and G cannot have the exact same or the exact opposite sign distribution. All this suggests that the *absolute* values of the off-diagonal entries of $G^{\dagger}$ take only two values, and the canonical dual is a 2-angle frame.

(b) Suppose that $-\lambda_1$ is a simple eigenvalue. Since the multiplicity of the minimum eigenvalue is 1, the frame is in $\mathbb{R}^{m-1}$.

If $\mathbf{1}$ is an eigenvector for $-\lambda_1$, then $Q = -\lambda_1\frac{J}{m} + \lambda_2P_2 + \lambda_3P_3$. Using the fact that $\frac{J}{m} + P_2 + P_3 = I$ in (5) gives

$$G = \frac{\lambda_1+\lambda_2}{\lambda_1}I - \frac{\lambda_1+\lambda_2}{\lambda_1}\frac{J}{m} + \frac{\lambda_3-\lambda_2}{\lambda_1}P_3.$$

The Gram matrix of the dual then becomes

$$G^{\dagger} = \frac{\lambda_1}{\lambda_1+\lambda_2}P_2 + \frac{\lambda_1}{\lambda_1+\lambda_3}P_3 = \frac{\lambda_1}{\lambda_1+\lambda_2}(I-\frac{J}{m}) + \lambda_1(\frac{1}{\lambda_1+\lambda_3} - \frac{1}{\lambda_1+\lambda_2})P_3.$$

Equating the diagonal and off-diagonal entries of G and $G^{\dagger}$, one can conclude that in this case the dual is 2-angle and equal norm.

Next suppose that $-\lambda_1$ is a simple eigenvalue with a regular eigenvector v that is not $\mathbf{1}$. This time

$$G = \frac{\lambda_1+\lambda_2}{\lambda_1}I - \frac{\lambda_1+\lambda_2}{\lambda_1}P_1 + \frac{\lambda_3-\lambda_2}{\lambda_1}P_3 \tag{9}$$

and

$$G^{\dagger} = \frac{\lambda_1}{\lambda_1+\lambda_2}I - \frac{\lambda_1}{\lambda_1+\lambda_2}P_1 + \frac{\lambda_1(\lambda_2-\lambda_3)}{(\lambda_1+\lambda_3)(\lambda_1+\lambda_2)}P_3. \tag{10}$$

Note that $P_1 = \frac{1}{m}vv^{\mathrm{T}}$ with diagonal entries all equal to $\frac{1}{m}$. Thus, in (9), the matrices G, I, and P_1 all have constant diagonal. This implies that P_3 also has constant diagonal. Using this in (10) shows that $G^\dagger$ also has constant diagonal, i.e., the canonical dual frame is equal norm.

Solving for P_3 in (9), and using the fact that the off-diagonal entries of P_1 are $\pm\frac{1}{m}$, gives

$$P_3(i,j) = \frac{1}{\lambda_3 - \lambda_2}\left[\pm 1 \pm \frac{\lambda_1 + \lambda_2}{m}\right], \quad \text{for } i \neq j. \tag{11}$$

Substituting (11) in (10), gives for $i \neq j$

$$G^\dagger(i,j) = \mp\frac{1}{m}\lambda_1 \frac{2\lambda_1 + \lambda_2 + \lambda_3}{(\lambda_1 + \lambda_2)(\lambda_1 + \lambda_3)} \mp \frac{\lambda_1}{(\lambda_1 + \lambda_2)(\lambda_1 + \lambda_3)}. \tag{12}$$

Due to Lemma 3, the first term on the right of (12) cannot be zero. Thus

$$G^\dagger(i,j) = \mp\frac{1}{m}C_2 \mp C_3,$$

where C_2 and C_3 are nonzero constants. This implies that the canonical dual is a 2-angle frame. □

Suppose that a signature matrix Q with three distinct eigenvalues has an irrational eigenvalue $\lambda + \sqrt{\mu}$. Then the other two eigenvalues of Q are $\lambda - \sqrt{\mu}$ and some $k \in \mathbb{Z}$ [13, 25]. The following result proved in [12] will be used.

Lemma 5 (Corollary 5.6 [12]) *Let Q be an $m \times m$ signature matrix with three distinct eigenvalues, at least one of which is irrational. If m is odd, then the eigenvalues of Q are*

$$[-\sqrt{m}]^{(m-1)/2},\ [0]^1,\ [\sqrt{m}]^{(m-1)/2}.$$

Theorem 3.3 *Let Q be an $m \times m$ signature matrix with three distinct eigenvalues, at least one of which is irrational. Let Φ denote any corresponding equiangular frame of m vectors in $\mathbb{R}^d$.*

(i) If m is odd, then the canonical dual is an equal norm frame with at most $m-1$ angles, and $d = \frac{m+1}{2}$.
(ii) Let Q have eigenvalues

$$[-k]^{m-2n}, [a - \sqrt{b}]^n, [a + \sqrt{b}]^n$$

with minimum eigenvalue $-k$, $k \in \mathbb{Z}^+$, $a \in \mathbb{Q}$, $b \in \mathbb{Q}^+$, and $m - 2n > 1$. Let the number of distinct moduli in the irrational part of the projection matrix of the eigenspace of either $a + \sqrt{b}$ or $a - \sqrt{b}$ be p. Then the canonical dual has at most $2p$ angles, and $d = 2n$.

Proof

(i) Due to Lemma 5, the eigenvalues of Q in this case are

$$[-\sqrt{m}]^{(m-1)/2},\ [0]^1,\ [\sqrt{m}]^{(m-1)/2}.$$

Since the multiplicity of the minimum eigenvalue is $\frac{m-1}{2}$, the value of d is given by

$$d = m - \frac{m-1}{2} = \frac{m+1}{2}.$$

Denote the projection matrices of $-\sqrt{m}$ and $\sqrt{m}$ by P_1 and $P_{\hat{1}}$, respectively. By the Spectral Theorem, $Q = -\sqrt{m}P_1 + \sqrt{m}P_{\hat{1}}$. Note that due to properties of eigenvectors corresponding to irrational eigenvalues, P_1 and $P_{\hat{1}}$ are irrational conjugates of each other, and can be written as $P_1 = P_a + P_b,\ P_{\hat{1}} = P_a - P_b$ where the (i, j)th entries of P_a and P_b are given by

$$P_a(i, j) = a_{ij} \in \mathbb{Q}$$

$$P_b(i, j) = 0 \quad \text{or} \quad \pm\sqrt{b_{ij}},\ b_{ij} \in \mathbb{Q},\ b_{ij}\ \text{notaperfectsquare}.$$

It follows that $Q = -2\sqrt{m}P_b$. The Gram matrix of Φ and the Gram matrix of the canonical dual are given by

$$G = I + \frac{1}{\sqrt{m}}Q = I - P_1 + P_{\hat{1}} = I - 2P_b, \tag{13}$$

$$G^{\dagger} = I - P_1 - \frac{1}{2}P_{\hat{1}} = I - \frac{3}{2}P_a - \frac{1}{2}P_b, \tag{14}$$

respectively. Using the relations $P_1^2 = P_1,\ P_{\hat{1}}^2 = P_{\hat{1}},\ P_{\hat{1}}P_1 = 0,\ P_1P_{\hat{1}} = 0$, one gets $P_a = 2P_b^2$ and thus

$$G^{\dagger} = I - 3P_b^2 - \frac{1}{2}P_b. \tag{15}$$

The off-diagonal entries of G are $\pm\frac{1}{\sqrt{m}}$, and its diagonal entries are all equal to 1. Equating the (i, j)th entries of the matrices in (13) then gives

$$P_b(i, j) = \begin{cases} 0 & \text{if } i = j \\ \pm\frac{1}{2\sqrt{m}} & \text{if } i \neq j. \end{cases}$$

Let $\beta := \frac{1}{2\sqrt{m}}$. The diagonal entries of P_b^2 are then all equal to $(m-1)\beta^2$. Since the diagonal entries of P_b are all equal to zero, this means from (15) that $G^\dagger$ has constant diagonal, and that the canonical dual is equal norm.
The absolute values of the off-diagonal entries of P_b^2 can take *at most* $\frac{m-1}{2}$ distinct values given by

$$\{(m-2)\beta^2, (m-4)\beta^2, \ldots, \beta^2\}.$$

This combined with the fact that $I - \frac{1}{2}P_b$ can take the values $1 \pm \frac{\beta}{2}$ means that the absolute values of the off-diagonal entries of $G^\dagger$ can take at most $m-1$ distinct values.

(ii) Now the multiplicity of the minimum eigenvalue is $m-2n$, and so d equals $2n$. Let P_2 and $P_{\hat{2}}$ denote the projection matrices of $a+\sqrt{b}$ and $a-\sqrt{b}$, respectively. As in part(i), these can be written as $P_2 = P_a + P_b$, $P_{\hat{2}} = P_a - P_b$, Let P_1 denote the projection matrix of $-k$. Since $m-2n > 1$, $\mathbf{1}$ is not a basis for the eigenspace of $-k$. Thus $P_1 \neq \frac{1}{m}J$, and the number of angles in the canonical dual cannot be determined by Theorem 3.2. By the Spectral Theorem

$$Q = -kP_1 + (a+\sqrt{b})(P_a + P_b) + (a-\sqrt{b})(P_a - P_b).$$

Using $P_1 + P_2 + P_{\hat{2}} = I$ gives

$$G = I + \frac{1}{k}Q = \frac{2}{k}\left((k+a)P_a + \sqrt{b}P_b\right).$$

The Gram matrix of the canonical dual is the pseudo inverse

$$G^\dagger = \frac{2k}{(k+a)^2 - b}\left[\frac{kG}{2} - 2\sqrt{b}P_b\right].$$

The result then follows from the fact that since Φ is equiangular, the off-diagonal entries of G are either $\frac{1}{k}$ or $-\frac{1}{k}$.

□

Then existence of signature matrices satisfying the conditions of Theorems 3.2 and 3.3 has been discussed in [12] and [25].

Due to the algebraic properties of signature matrices [25], one cannot expect to generalize the above results to any arbitrary number of distinct eigenvalues of Q. In the context of regular graphs, signature matrices with four eigenvalues are discussed in [29], and for this case, Theorem 3.3 can be extended as follows.

Theorem 3.4 *Let Q be an $m \times m$ signature matrix with four distinct eigenvalues*

$$[a-\sqrt{b}]^n, [a+\sqrt{b}]^n, [-k_1]^{m-2n-1}, [k_2]^1$$

with minimum eigenvalue $-k_1$, *where* $k_1, k_2 \in \mathbb{Z}^+$, $a \in \mathbb{Q}$, $b \in \mathbb{Q}^+$. *Suppose that* $\mathbf{1}$ *is an eigenvector of* Q *corresponding to* k_2. *Let the number of distinct moduli in the purely irrational part of the projection matrix of the eigenspace of either* $a + \sqrt{b}$ *or* $a - \sqrt{b}$ *be* p. *If* Φ *is any equiangular frame corresponding to* Q, *then the canonical dual is a frame in* $\mathbb{R}^{2n+1}$ *having at most* $2p$ *angles.*

Proof The projection matrix of the eigenspace of k_2 is $\frac{J}{m}$. The spectral decomposition of Q is

$$Q = -k_1 P_1 + k_2 \frac{J}{m} + (a + \sqrt{b}) P_2 + (a - \sqrt{b}) P_{\tilde{2}}.$$

The proof then follows in an identical manner as Theorem 3.3 by noting that the off-diagonal entries of J are all one. □

Acknowledgments The author is grateful to Prof. Ole Christensen for many useful discussions, and to the Department of Applied Mathematics and Computer Science, Technical University of Denmark for hospitality during a visit when some of this work was done. The author would like to thank Dr. Rae Young Kim for providing many helpful comments on this work. Finally, the author wishes to thank the anonymous reviewer for his/her thorough efforts in reading and commenting on the manuscript.

References

1. Bodmann, B.G., Elwood, H.J.: Complex equiangular Parseval frames and Seidel matrices containing pth roots of unity. Proc. Amer. Math. Soc. **138**(12), 4387–4404 (2010)
2. Bodmann, B.G., Paulsen, V.I., Tomforde, M.: Equiangular tight frames from complex Seidel matrices containing cube roots of unity. Linear Algebra Appl., **430** (1), 396–417 (2009)
3. Christensen, O.: Frames and pseudo-inverses. J. Math. Anal. Appl., **195** (2), 401–414 (1995)
4. Christensen, O.: An Introduction to Frames and Riesz Bases. Second expanded edition, Birkhäuser, Boston, Basel, Berlin (2016)
5. Christensen, O., Datta, S., Kim, R.Y.: Equiangular frames and generalizations of the Welch bound to dual pairs of frames. Linear Multilinear Algebra. **68**(12), 2495–2505 (2020)
6. Christensen, O., Datta, S., Kim, R.Y.: Duals and approximate duals of equiangular frames. In preparation.
7. Christensen, O., Powell, A., Xiao, X. C.: A note on finite dual frame pairs. Proc. Amer. Math. Soc. **140** (11), 3921–3930 (2012)
8. Datta, S., Oldroyd, J.: Construction of k-angle tight frames. Numer. Funct. Anal. Optim. **37** (8), 975–989 (2016)
9. Delsarte, P., Goethals, J.M., Seidel, J.J.: Bounds for systems of lines, and Jacobi polynomials. Philips Res. Rep. **30** (3) Issue in honour of C.J. Bouwkamp, 91–105 (1975)
10. Duffin, R.J., Schaeffer, A.C.: A class of nonharmonic Fourier series. Trans. Amer. Math. Soc. **72** (2), 341–366 (1952)
11. Fickus, M., Mixon, D.G., Tremain, J.C.: Steiner equiangular tight frames. Linear Algebra Appl. **436**(5), 1014–1027 (2012)
12. Greaves, G.: Equiangular lines systems and switching classes containing regular graphs. Linear Algebra Appl. **536**, 31–51 (2018)
13. Greaves, G., Koolen, J.H., Munemasa, A., Szöllősi, F.: Equiangular lines in Euclidean spaces. J. Combin. Theory Ser. A. **138**, 208–235 (2016)

14. Hoggar, S.G.: t-designs in projective spaces. European J. Combin. **3** (3), 233–254 (1982)
15. Holmes, R.B., Paulsen, V.I.: Optimal frames for erasures. Linear Algebra Appl. **377**, 31-51 (2004)
16. Klappenecker, A., Rötteler, M.: Mutually unbiased bases are complex projective 2-designs. In: Proceedings of the International Symposium on Information Theory, pp. 1740–1744, (2005)
17. Lemmens,P.W.H, Seidel, J.J.: Equiangular lines. J. Algebra. **24** (3), 494–512 (1973)
18. Massey, J.L., Mittelholzer, T.: Welch's bound and sequence sets for code-division multiple-access systems. In: Capocelli, R., De Santis, A., Vaccaro, U. (eds) Sequences II: Methods in Communication, Security and Computer Science, pp. 63–78. Springer-Verlag, New York (1991)
19. Renes, J.M., Blume-Kohout, R., Scott, A.J., Caves, C.M.: Symmetric informationally complete quantum measurements. J. Math. Phys. **45**(6), 2171–2180 (2004)
20. Roy, A., Scott, A.J.: Weighted complex projective 2-designs from bases: Optimal state determination by orthogonal measurements. J. Math. Phys. **48** (072110), (2007)
21. Sarwate, D.V.: Meeting the Welch bound with equality. In: Helleseth, T., Niederreiter, H.(eds) Sequences and Their Applications, DMTCS Series, chapter11. Springer-Verlag (1999)
22. Scott, A.J: Tight informationally complete quantum measurements. J. Phys. A **39**, 13507–13530 (2006)
23. Strohmer, T., Heath Jr., R.W.: Grassmannian frames with applications to coding and communication. Appl. Comput. Harmon. Anal. **14**(3), 257–275 (2003)
24. Sustik, M.A., Tropp, J.A., Dhillon, I.S., Heath Jr., R.W.: On the existence of equiangular tight frames. Linear Algebra Appl. **426**(2-3), 619–635 (2007)
25. Szöllősi, F., Östergård, R. J.: Enumeration of Seidel matrices. European J. Combin. **69**, 169–184 (2018)
26. Tóth, L.F.: Distribution of points in the elliptic plane. Acta Math. Hungar. **16**, 437–440 (1965)
27. Tropp, J.A.: Complex equiangular tight frames. In: Proceedings of SPIE 2005 (Wavelets XI), pp. 590412.01–590412.11. (2005)
28. Vale, R., Waldron, S.: Tight frames and their symmetries. Constr. Approx. **21** (1), 83–112 (2004)
29. van Dam, E.R.: Regular graphs with four eigenvalues. Linear Algebra Appl. **226/228**, 139–162 (1995)
30. Waldron, S.: On the construction of equiangular frames from graphs. Linear Algebra Appl. **431** (11), 2228–2242 (2009)
31. Welch, L.R.: Lower bounds on the maximum cross correlation of signals. IEEE Trans. Inform. Theory **20** (3), 397–399 (May 1974)
32. Xia, P., Zhou, S., Giannakis, G.B.: Achieving the Welch bound with difference sets. IEEE Trans. Inform. Theory **51** (5), 1900–1907 (2005)

Wavelet Sets for Crystallographic Groups

Kathy D. Merrill

Abstract Single wavelet sets, and thus single wavelets, are shown to exist for the actions of all crystallographic groups on $\mathbb{R}^2$ under all integer dilations. Examples of such sets satisfying the additional requirement that they are finite unions of convex sets are presented for each of the groups under dilation by two.

1 Introduction

Classical wavelets are finite sets of functions in $L^2(\mathbb{R}^n)$ whose integer translations and expansive matrix dilations give an orthonormal basis for the space. A generalization developed in [1] replaces $L^2(\mathbb{R}^n)$ with an abstract Hilbert space, and uses a group action and powers of a compatible operator in place of the integer translations and matrix dilations. In this paper, we consider such wavelets using the action of a 2-dimensional crystallographic group $\mathcal{G}$ on $L^2(\mathbb{R}^2)$ together with an integer dilation δ. Examples of crystallographic wavelets were first given in [15, 16], and [17] under the rubric of composite dilations. The theory of these wavelets was later worked out in general for all 2-dimensional crystallographic groups in [20] and [21]. Recent related work by Barbieri et al. [4, 5] has successfully used subspaces invariant under crystallographic groups to approximate large data sets.

Many of the examples of wavelets for crystallographic groups given in the literature are Haar-type wavelets, that is, wavelets that are normalized characteristic functions of sets. (See e.g. [8, 9, 14, 19–21].) In this paper, we focus instead on wavelet set wavelets, whose Fourier transforms are characteristic functions. Specifically, write $\mathcal{F}$ for the 2-dimensional Fourier transform,

$$\widehat{f}(\xi) = (\mathcal{F} f)(\xi) = \int_{\mathbb{R}^2} f(z) e^{-2\pi i \langle z, \xi \rangle} dz,$$

K. D. Merrill (✉)
Department of Mathematics, Colorado College, Colorado Springs, CO, USA
e-mail: kmerrill@coloradocollege.edu

M. Hirn et al. (eds.), *Excursions in Harmonic Analysis, Volume 6*, Applied and Numerical Harmonic Analysis, https://doi.org/10.1007/978-3-030-69637-5_10

and $\mathcal{F}^{-1} f = \check{f}$ for the inverse Fourier transform.

Definition 1 A finite collection of sets $\{W_1, \cdots, W_L\}$ is called an L-wavelet set if $\{\mathcal{F}^{-1}(\mathbf{1}_{W_1}), \cdots, \mathcal{F}^{-1}(\mathbf{1}_{W_L})\}$ is a wavelet, so that $\{\delta^j(\gamma(\mathcal{F}^{-1}(\mathbf{1}_{W_l})))\}$ forms an orthonormal basis for $L^2(\mathbb{R}^2)$, with $-\infty < j < \infty$, $\gamma \in \mathcal{G}$, and $1 \leq l \leq L$.

For us, $\mathcal{G}$ will be a two dimensional crystallographic group, and δ an integer scalar dilation. A set W is called a single wavelet set, or simply a *wavelet set* if $L = 1$; it is called a *multiwavelet set* if $L > 1$. Examples of (multi)wavelet sets for some of the semidirect product crystallographic groups were given in [7, 15] and [16]. An example of a multiwavelet set for a crystallographic group that is not a semidirect product appears in [23].

Most of these known examples of wavelet sets for the two dimensional crystallographic groups with integer dilations have been multiwavelets. (See [7] for an exception.) In fact, most of the integer dilation crystallographic wavelets of any type that have appeared in the literature are multiwavelets. This is similar to the early history of classical wavelets in $L^2(\mathbb{R}^2)$, before Dai et al. [11] showed in the late 1990s that classical single wavelet sets, and thus single wavelets, exist for all dilations. In this paper we prove an analogous result for the crystallographic groups. That is, we show that single wavelet sets, and thus single wavelets, exist for all the two dimensional crystallographic groups for all integer dilations. Our result does not require that the integer dilation δ be compatible with the crystallographic group $\mathcal{G}$ in the sense developed in [1, 15], and [21]. For dilations that do satisfy the compatibility condition, it is shown in [3] that single wavelet sets can be used to decompose the wavelet representations of the crystallographic groups into direct integrals of irreducibles. (For more about the compatibility condition, see Sect. 2.)

The general process for building single wavelet sets for the crystallographic groups depends on the same iterative methods used for classical single wavelet sets. Thus, as in the classical case, the crystallographic examples built using the existence theorem have a complicated geometric structure requiring an infinite number of convex pieces. However, just as in the classical case [22], simple single wavelet sets, i.e. sets that consist of a finite number of convex pieces, can be found in many cases. In this paper, we find simple single wavelet sets for all crystallographic groups under dilation by two. Wavelets based on these sets are fairly simple to encode for image compression. Keith Taylor [27] reports supervising an honors thesis that performed such computations by breaking these wavelet sets into triangles. Of course the slow decay of the associated wavelets prevents rapid convergence, but such wavelets are of theoretical interest, and may also be easier to smooth because of their simple nature. (See e.g. [24].)

The paper is organized as follows. In Sect. 2, we review needed facts about the crystallographic groups. Section 3 develops necessary and sufficient conditions for a set to be a crystallographic wavelet set, and contains the proof that such sets always exist for integer dilations. Section 4 has examples of simple single wavelet sets under dilation by two for all 17 crystallographic groups.

2 Preliminaries

A two dimensional crystallographic group $\mathcal{G}$, also called a wallpaper group, is a discrete subgroup of $\mathbb{R}^2 \rtimes O_2$, where O_2 is the 2-dimensional orthogonal group. An element $[x, L] \in \mathcal{G}$, with $x \in \mathbb{R}^2$ and $L \in O_2$ acts on $\mathbb{R}^2$ by $[x, L] \cdot z = L(x + z)$, and on $L^2(\mathbb{R}^2)$ by $[x, L] \cdot f(z) = f(L^{-1}z - x)$. Multiplication in the group is given by

$$[x, L] \cdot [y, M] = [M^{-1}x + y, LM].$$

Such a group represents a possible symmetry pattern in the plane that is repeated in two independent directions. Accordingly, each crystallographic group $\mathcal{G}$ has a normal subgroup $\mathcal{N}$ consisting of a lattice of translations isomorphic to $\mathbb{Z}^2$. The quotient $\mathcal{G}/\mathcal{N}$ is isomorphic to $\mathcal{B} = \{S \in O_2 : [x, S] \in \mathcal{G} \text{ for some } x \in \mathbb{R}^2\}$, which is a finite group called the point group.

There are 17 two dimensional crystallographic groups [13]. This can be shown by first proving that the only possible elements of the point groups are rotations by $\frac{2\pi}{n}$, for $n = 1, 2, 3, 4, 6$, and reflections. Thirteen of the groups are semidirect products of their point group and lattice. Such crystallographic groups are called symmorphic. The four non-symmorphic groups have essential glide reflections, which are simultaneous reflections and non-lattice translations. These glides prevent the separation of the action of the lattice and that of the point group. The traditional names of the wallpaper groups, which were first introduced by crystallographers, are sequences of four symbols specifying cell type, highest order rotation, and essential reflections and glides. In this paper, we will use a common shortened form of these names. (See e.g. [26] for a description of all 17 wallpaper groups and an explanation of their names.)

Table 1 gives explicit descriptions of all the wallpaper groups and their point groups in terms of one possible set of generators for each. These descriptions depend on a specific choice of lattice. We choose $\mathbb{Z}^2$ whenever it is possible to do so, thus for all groups not containing rotation by 3 or 6. In the latter case, we choose the lattice $\mathcal{L}$ generated by $(1, 0)$ and $(\frac{1}{2}, \frac{\sqrt{3}}{2})$. To make the lattice apparent, we include two independent translations in the list of generators in the table, even if this causes redundancy among the generators. We write ρ_n for rotation ccw by $\frac{2\pi}{n}$, τ_v for translation by v, σ_v for reflection in the line determined by v, and I for the 2×2 identity matrix. The non-symmorphic groups in Table 1 are pg, pmg, pgg, and p4g; these groups include glides as generators. The information presented in the table, as well as more detail about the groups, can be found, for example, in [12] or [25].

Since we will be discussing wavelets for the two dimensional crystallographic groups, we will be interested in the interaction between the groups and a dilation. We will restrict our attention to integer dilations. For $d \in \mathbb{Z}$, $d \geq 2$, write δ_d for the dilation operator acting on $L^2(\mathbb{R}^2)$ by

$$\delta_d f(z) = d f(dz).$$

Table 1 The two dimensional crystallographic groups

Name	Group description	Point group
p1	$\langle \tau_{(1,0)},\ \tau_{(0,1)} \rangle$	$\langle I \rangle$
p2	$\langle \tau_{(1,0)},\ \tau_{(0,1)},\ \rho_2 \rangle$	$\langle \rho_2 \rangle$
pm	$\langle \tau_{(1,0)}\ ,\ \tau_{(0,1)},\ \sigma_{(0,1)} \rangle$	$\langle \sigma_{(0,1)} \rangle$
pg	$\langle \tau_{(1,0)},\ \tau_{(0,1)},\ [(0, \frac{1}{2}), \sigma_{(0,1)}] \rangle$	$\langle \sigma_{(0,1)} \rangle$
pmm	$\langle \tau_{(1,0)},\ \tau_{(0,1)},\ \rho_2,\ \sigma_{(0,1)} \rangle$	$\langle \rho_2,\ \sigma_{(0,1)} \rangle$
pmg	$\langle \tau_{(1,0)},\ \tau_{(0,1)},\ \rho_2,\ [(0, \frac{1}{2}), \sigma_{(0,1)}] \rangle$	$\langle \rho_2,\ \sigma_{(0,1)} \rangle$
pgg	$\langle \tau_{(1,0)},\ \tau_{(0,1)},\ \rho_2,\ [(\frac{1}{2}, \frac{1}{2}), \sigma_{(0,1)}] \rangle$	$\langle \rho_2,\ \sigma_{(0,1)} \rangle$
p4	$\langle \tau_{(1,0)},\ \tau_{(0,1)},\ \rho_4 \rangle$	$\langle \rho_4 \rangle$
p4m	$\langle \tau_{(1,0)},\ \tau_{(0,1)},\ \rho_4,\ \sigma_{(1,1)} \rangle$	$\langle \rho_4,\ \sigma_{(1,1)} \rangle$
p4g	$\langle \tau_{(1,0)},\ \tau_{(0,1)},\ \rho_4,\ [(\frac{1}{2}, \frac{1}{2}), \sigma_{(1,1)}] \rangle$	$\langle \rho_4,\ \sigma_{(1,1)} \rangle$
cm	$\langle \tau_{(1,0)},\ \tau_{(0,1)},\ \sigma_{(1,1)} \rangle$	$\langle \sigma_{(1,1)} \rangle$
cmm	$\langle \tau_{(1,0)},\ \tau_{(0,1)},\ \rho_2,\ \sigma_{(1,1)} \rangle$	$\langle \rho_2,\ \sigma_{(1,1)} \rangle$
p3	$\langle \tau_{(1,0)},\ \tau_{(\frac{1}{2}, \frac{\sqrt{3}}{2})},\ \rho_3 \rangle$	$\langle \rho_3 \rangle$
p31m	$\langle \tau_{(1,0)},\ \tau_{(\frac{1}{2}, \frac{\sqrt{3}}{2})},\ \rho_3,\ \sigma_{(1,0)} \rangle$	$\langle \rho_3,\ \sigma_{(1,0)} \rangle$
p3m1	$\langle \tau_{(1,0)},\ \tau_{(\frac{1}{2}, \frac{\sqrt{3}}{2})},\ \rho_3,\ \sigma_{(0,1)} \rangle$	$\langle \rho_3,\ \sigma_{(0,1)} \rangle$
p6	$\langle \tau_{(1,0)},\ \tau_{(\frac{1}{2}, \frac{\sqrt{3}}{2})},\ \rho_6 \rangle$	$\langle \rho_6 \rangle$
p6m	$\langle \tau_{(1,0)},\ \tau_{(\frac{1}{2}, \frac{\sqrt{3}}{2})},\ \rho_6,\ \sigma_{(\frac{\sqrt{3}}{2}, \frac{1}{2})} \rangle$	$\langle \rho_6,\ \sigma_{(\frac{\sqrt{3}}{2}, \frac{1}{2})} \rangle$

Other papers in the literature about crystallographic wavelets have raised the issue of compatibility between dilations and the group action. For δ_d to be *compatible* with a crystallographic group $\mathcal{G}$ requires that $\delta_d^{-1}\mathcal{G}\delta_d \subseteq \mathcal{G}$. This condition ensures that the group acts in a coherent way on the layers of the associated multiresolution structure. (See e.g. [1, 23].) It turns out that all integer dilations are compatible with symmorphic groups, but only odd dilations are compatible with the non-symmorphic groups [3, 20, 21]. However, as this paper is concerned only with creating wavelet sets, the issue of compatibility does not arise. Thus, we will consider all integer dilations.

3 Crystallographic Wavelet Sets

A well known result for conventional wavelet sets states that W is a wavelet set for dilation by d if and only if W tiles the plane almost everywhere under both translation by the integers and dilation by d. A corresponding result for crystallographic groups is given by the following theorem.

Theorem 3.1 *Let $\mathcal{G}$ be a crystallographic group with lattice $\mathcal{N}$ and point group $\mathcal{B}$. Then W is a wavelet set for $\mathcal{G}$ and dilation by d if and only if all of the following conditions hold*

(i) W tiles $\mathbb{R}^2$ *a.e. under translation by* $\mathcal{N}$*; that is,*

$$\sum_{k \in \mathcal{N}} \mathbf{1}_W(\xi + k) = 1 \quad a.e.\ \xi \in \mathbb{R}^2.$$

(ii) $S(W) \cap S'(W)$ *has measure* 0 *for* $S \neq S' \in \mathcal{B}$

(iii) $\cup_{S \in \mathcal{B}} S(W)$ *tiles* $\mathbb{R}^2$ *a.e. under dilation by d; that is,*

$$\sum_{j \in \mathbb{Z}} \mathbf{1}_{\cup_{S \in \mathcal{B}} S(W)}(d^j \xi) = 1 \quad a.e.\ \xi \in \mathbb{R}^2.$$

Proof Writing $\widehat{\psi} = \mathbf{1}_W$, we will show that $\{\widehat{\delta^j}\widehat{\gamma}\widehat{\psi} : j \in \mathbb{Z}, \gamma \in \mathcal{G}\}$ is an orthonormal basis for $L^2(\mathbb{R}^2)$ if and only if the three conditions hold. A straightforward calculation yields

$$([x, S]^\wedge f)(\xi) = e^{-2\pi i \langle x, S^*(\xi) \rangle} f(S^*(\xi)). \tag{1}$$

Thus we see that $\{[\ell, I]^\wedge \widehat{\psi} : \ell \in \mathcal{N}\} = \{e^{2\pi i \langle \ell, \cdot \rangle} \mathbf{1}_W : \ell \in \mathcal{N}\}$ gives an orthonormal basis for $L^2(W)$ if and only if (i) holds.

We write $\mathcal{G} = \{[\ell + c_S, S] : \ell \in \mathcal{N}, S \in \mathcal{B}\}$, where $c_S \in \mathbb{R}^2 \setminus \mathcal{N}$ is a constant for each S, and $c_S = 0$ if $S = I$. Applying Equation (1) for a fixed $S \in \mathcal{B}$, we have

$$[\ell + c_S, S]^\wedge \widehat{\psi}(\xi) = e^{-2\pi i \langle S(c_S), \xi \rangle} e^{-2\pi i \langle \ell, S^*(\xi) \rangle} \mathbf{1}_W(S^*(\xi)).$$

This shows that condition (i) is equivalent to $\{[\ell + c_S, S]^\wedge \widehat{\psi} : \ell \in \mathcal{N}\}$ forming an orthonormal basis for $L^2(S(W))$.

Using this, we see that conditions (i) and (ii) together imply that $\{\widehat{\gamma}\widehat{\psi} : \gamma \in \mathcal{G}\}$ forms an orthonormal basis for $L^2(\cup_{S \in \mathcal{B}} S(W))$. Since

$$(\widehat{\delta} f)(\xi) = \delta^{-1} f(\xi).$$

we see that including condition (iii) implies that $\{\widehat{\delta^j}\widehat{\gamma}\widehat{\psi} : j \in \mathbb{Z}, \gamma \in \mathcal{G}\}$ is an orthonormal basis for $L^2(\mathbb{R}^2)$.

Conversely, if W is a wavelet set for $\mathcal{G}$, the orthonormality of $\{\widehat{\gamma}\widehat{\psi} : \gamma \in \mathcal{G}\}$ implies condition (ii), and the orthonormality and completeness of $\{\widehat{\delta^j}\widehat{\gamma}\widehat{\psi} : j \in \mathbb{Z}, \gamma \in \mathcal{G}\}$ imply both (iii) and the fact that $\{[\ell + c_S, S]^\wedge \widehat{\psi} : \ell \in \mathcal{N}\}$ forms an orthonormal basis for $L^2(S(W))$. We have shown the latter to be equivalent to condition (i). □

Remark 1 A similar proof shows that $\mathbf{1}_{W_1}, \mathbf{1}_{W_2}, \ldots \mathbf{1}_{W_L}$ are Fourier transforms of an $L-$wavelet for $\mathcal{G}$ and dilation by d if and only if all of the following hold:

(i) Each W_l tiles $\mathbb{R}^2$ a.e. under translation by $\mathcal{N}$
(ii) $S(W_l) \cap S'(W_{l'})$ has measure 0 unless $S = S'$ and $l = l'$
(iii) $\cup_{l=1}^{L} \cup_{S \in \mathcal{B}} S(W)$ tiles $\mathbb{R}^2$ a.e. under dilation by d.

Remark 2 In the case of symmorphic (semidirect product) crystallographic groups, the theory of composite dilations applies. For this case, Gu, Labate, Lim, Weiss, and Wilson [17] give necessary and sufficient conditions for Parseval frame multiwavelet sets that are similar to conditions (i) and (iii) of Theorem 3.1. Condition (ii) is not needed because their theorem describes Parseval wavelets rather than orthonormal wavelets. The authors also give a strategy for building orthonormal multiwavelet sets. They show this strategy cannot create single wavelet sets for reasons similar to those that prevent classical single MRA wavelets in $L^2(\mathbb{R}^2)$.

Corollary 1 *If two crystallographic groups $\mathcal{G}$ and $\mathcal{G}'$ differ in replacing reflections by glide reflections that mirror over the same line, then $\mathcal{G}$ and $\mathcal{G}'$ have the same wavelet sets.*

Using Theorem 3.1, we can build single wavelet sets for crystallographic groups from conventional subspace wavelet sets. Let $M \subseteq \mathbb{R}^2$ be a finite union of polar sectors of the form $\{r(\cos\phi, \sin\phi) : \theta_1 \leq \phi \leq \theta_2, r \geq 0\}$, and let $\mathcal{H}$ be the Hilbert space defined by $\mathcal{F}(\mathcal{H}) = L^2(\mathbb{R}^2)\mathbf{1}_M$. A set $W \subseteq M$ is called a *subspace wavelet set* for $\mathcal{H}$ if $\mathbf{1}_W$ is the Fourier transform of a wavelet for $\mathcal{H}$. A more general form of such sets appears throughout the literature, and their properties are thoroughly described in [10]. A subspace wavelet set W for a set M of the form above has the property that its integer translates tile $\mathbb{R}^2$, while its dilates by d tile M. When combined with information about the point group given in Table 1, such a set is perfectly situated to satisfy Theorem 3.1. Thus, the result from [10] that subspace wavelet sets always exist can be used to prove the following theorem.

Theorem 3.2 *Single wavelet sets exist for each of the 2-dimensional crystallographic groups for any integer dilation d.*

Proof By the general theorem in [11], as well as numerous constructions in, e.g. [2, 6, 11], we know that conventional single wavelet sets exist for all dilations. These sets in turn are crystallographic wavelet sets for the group p1. For the other crystallographic groups, we will use conventional subspace wavelet sets, which are shown to exist for all dilations and dilation invariant subspaces in [10].

First, let $\mathcal{G}$ be a 2-dimensional crystallographic group with lattice $\mathbb{Z}^2$ and point group $\mathcal{B}$. If M is a polar sector that tiles the plane under the action of $\mathcal{B}$, and if W is a subspace wavelet set for M, then by Theorem 3.1, W is a crystallographic wavelet set for $\mathcal{G}$. Specifically, for the groups p2, pm, and pg, we can take M to be the right half plane; for the group cm, we take M to be one of the half planes determined by the line at angle $\frac{\pi}{4}$ above the horizontal axis; for pmm, pmg, pgg, and p4, M is the first quadrant; for cmm, the quadrant between $\theta_1 = -\frac{\pi}{4}$ and $\theta_2 = \frac{\pi}{4}$; and for p4m and p4g, the first octant.

The last 5 groups in Table 1, which involve three-fold or six-fold rotation, contain translations by the hexagonal lattice $\mathcal{L}$ determined by integer combinations of $(1, 0)$ and $(\frac{1}{2}, \frac{\sqrt{3}}{2})$, rather than by $\mathbb{Z}^2$. Define $L = \begin{pmatrix} 1 & \frac{1}{2} \\ 0 & \frac{\sqrt{3}}{2} \end{pmatrix}$ and $L' = \begin{pmatrix} 1 & -\frac{1}{2} \\ 0 & \frac{\sqrt{3}}{2} \end{pmatrix}$. Then if the set W tiles under translation by the integers, both LW and $L'W$ tile under

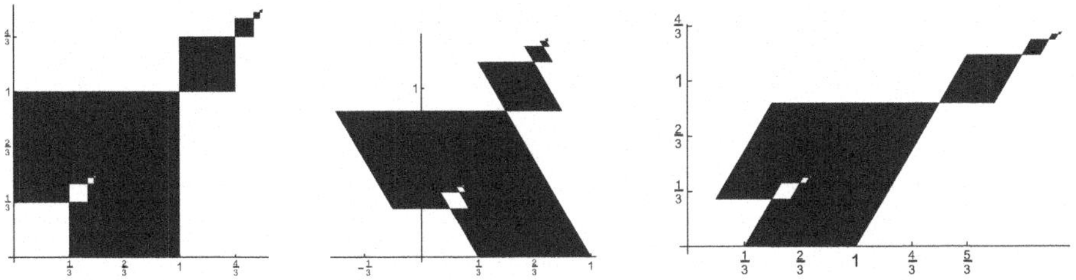

Fig. 1 Dilation by 3 wavelet set for p4, pmm, pmg, and pgg on the left, for p3 in the center, and for p6 on the right

translation by the hexagonal lattice. If W tiles the polar sector $\theta \in [0, \theta_2]$ under dilation, LW tiles $\theta \in [0, \frac{2\theta_2}{3}]$, and $L'W$ tiles $\theta \in [0, \frac{4\theta_2}{3}]$. Thus, if W is a subspace wavelet set for the first quadrant, LW and $L'W$ are crystallographic wavelet sets for p6 and p3 respectively. If W is a subspace wavelet set for the first octant, LW and $L'W$ are crystallographic wavelet sets for p6m and p31m respectively. Neither LW nor $L'W$ gives a wavelet set for p3m1 since in the first case, the union of the point group images do not tile $\mathbb{R}^2$ by dilation, and in the second, the point group images are not disjoint. However, if we instead let W be a subspace wavelet for the union of the first and fifth octants, then LW does satisfy all three conditions of Theorem 3.1 for p3m1. □

Example 1 We use the iterative method from [2], Theorem 3, to create a subspace wavelet set for the first quadrant under dilation by 3. As in [2], we first build a scaling set E such that $W = 3E \setminus E$ is a subspace wavelet set for the first quadrant. We start the iterative algorithm with $E_0 = [0, \frac{1}{3})^2$. Each successive step scales the previous piece by $\frac{1}{3}$, and then translates it by $(\frac{1}{3}, \frac{1}{3})$. The resulting scaling set is $E = \cup_{n=1}^{\infty} \left([0, \frac{1}{3^n})^2 + (\sum_{j=1}^{n-1} \frac{1}{3^j}, \sum_{j=1}^{n-1} \frac{1}{3^j}) \right)$. The conventional subspace wavelet set, $W = 3E \setminus E$, is shown on the left in Fig. 1. As mentioned in the proof of Theorem 3.2, this is a crystallographic wavelet set for pmm, pmg, pgg, and p4. We create wavelet sets for p3 and p6, by multiplying W by the lattice matrices $L' = \begin{pmatrix} 1 & -\frac{1}{2} \\ 0 & \frac{\sqrt{3}}{2} \end{pmatrix}$ and $L = \begin{pmatrix} 1 & \frac{1}{2} \\ 0 & \frac{\sqrt{3}}{2} \end{pmatrix}$ respectively. The results are shown in the center and right of Fig. 1.

In the next section, we will find simple wavelet sets, that is, wavelet sets that are a finite union of convex sets, for all 17 crystallographic groups under dilation by 2. Many of the sets are easily modified to other dilations. In particular, the constructions below can be modified to give simple wavelet sets for dilation by 3, the smallest dilation that is compatible with all 17 crystallographic groups, for all but the groups p3 and p3m1.

4 Simple Crystallographic Wavelet Sets

We first remark that crystallographic wavelet sets for the group p1 are the same as conventional wavelet sets. Simple conventional wavelet sets, and thus simple crystallographic wavelet sets for p1, are given in [22] for any integer dilation d. We will now use the ideas from the previous section to find simple wavelet sets under dilation by 2 for the other 16 groups. Simple conventional subspace wavelet sets are sometimes difficult to find for the polar regions described in Theorem 3.2. Thus, in some cases, we will instead need to break up conventional simple multiwavelet sets.

For example, for crystallographic groups whose point groups have order 2, we find simple single wavelet sets by partitioning conventional 2-wavelet sets. Such sets tile under dilation and give two-fold tilings under translation. The sets in Fig. 2 were shown to be 2-wavelet sets for translation by $\mathbb{Z}^2$ and dilation by 2 in [23].

For pg and pm, which we know have the same wavelet sets by Corollary 1, we divide the diamond annulus 2-wavelet set on the left in Fig. 2 into two disjoint pieces, each of which tiles under translation, such that one is the vertical reflection of the other. Specifically, let $W^{\mathrm{pm}} = \mathrm{conv}\{(0, \frac{1}{3}), (0, \frac{2}{3}), (2, 0), (1, 0)\} \cup \mathrm{conv}\{(1, 0), (2, 0), (\frac{1}{2}, -\frac{1}{2}), (0, -\frac{1}{3})\} \cup \mathrm{conv}\{(0, -\frac{1}{3}), (0, -\frac{2}{3}), (-\frac{1}{2}, -\frac{1}{2})\}$, as shown on the left in Fig. 3. For p2, we start with the same 2-wavelet set, and this time divide it into two pieces such that each tiles under translation and one is rotation by π of the other. The resulting wavelet set W^{p2}, which is formed by translating the lower left triangle of W^{pm} by (0, 1), is shown in the center of Fig. 3. Tiling under translation can be established for each by piecewise integer translating them into $[0, 1] \times [-\frac{1}{3}, \frac{2}{3}]$. The other conditions of Theorem 3.1 follow from the fact that the left hand 2-wavelet set in Fig. 2 is the disjoint union of each of these sets acted on by their point groups.

Our ability to break this conventional 2-wavelet set into two eligible pieces for the groups pm, pg, and p2 required it to be symmetric with respect to the reflection and rotation in those point groups. For cm, we use instead the conventional 2-wavelet set shown on the right in Fig. 2, which is symmetric with respect to $\sigma_{(1,1)}$. We divide

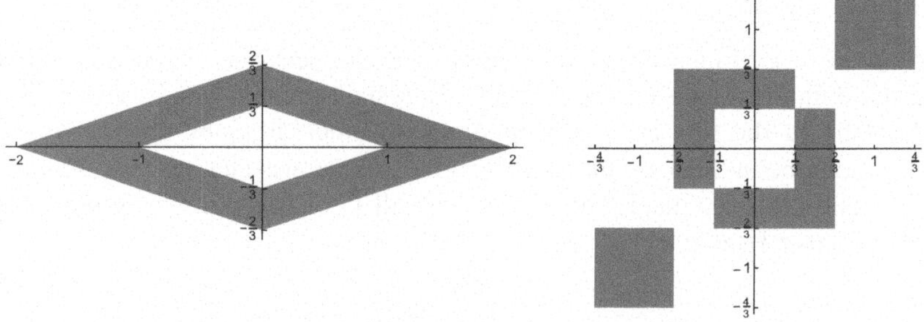

Fig. 2 Conventional 2-wavelet sets for dilation by 2

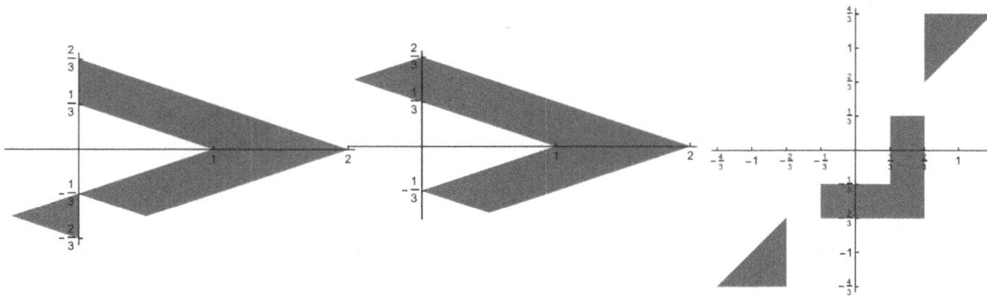

W^{pm} W^{p2} W^{cm}

Fig. 3 Simple wavelet sets for crystallographic groups with order 2 point groups

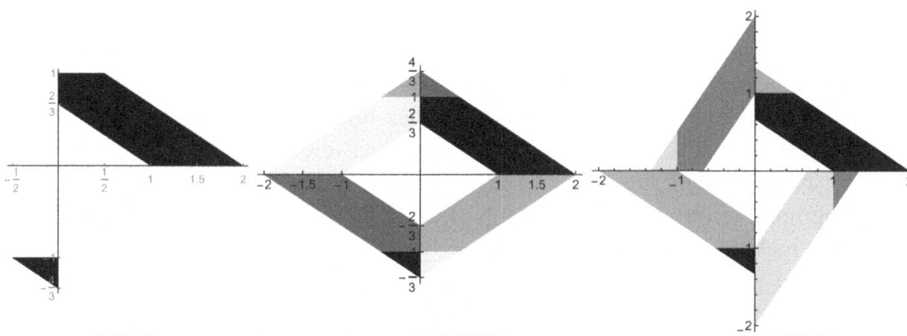

W^{pmm} $\cup_{S\in\mathcal{B}^{\mathrm{pmm}}} S(W^{\mathrm{pmm}})$ $\cup_{S\in\mathcal{B}^{\mathrm{p4}}} S(W^{\mathrm{pmm}})$

Fig. 4 Simple wavelet set for pmm, pmg, pgg, and p4

this 2-wavelet set into two disjoint pieces such that each is the reflection of the other, choosing so that each piece tiles under integer translation. Specifically, let $W^{\mathrm{cm}} = \mathrm{conv}\{(\frac{2}{3}, \frac{2}{3}), (\frac{2}{3}, \frac{4}{3}), (\frac{4}{3}, \frac{4}{3})\} \cup \mathrm{conv}\{(-\frac{2}{3}, -\frac{2}{3}), (-\frac{2}{3}, -\frac{4}{3}), (-\frac{4}{3}, -\frac{4}{3})\} \cup \mathrm{conv}\{(-\frac{1}{3}, -\frac{1}{3}), (-\frac{1}{3}, -\frac{2}{3}), (\frac{2}{3}, -\frac{2}{3}), (\frac{2}{3}, \frac{1}{3}), (\frac{1}{3}, \frac{1}{3}), (\frac{1}{3}, -\frac{1}{3}), (-\frac{1}{3}, -\frac{1}{3})\}$. By Theorem 3.1, W^{cm}, shown on the right in Fig. 3, is a wavelet set for cm.

We now alter the conventional 2-wavelet set of Fig. 2 to create a conventional 4-wavelet set that can be used for four of the five crystallographic groups whose point groups have order 4. To do this, we apply the matrix $\left(\begin{smallmatrix} 1 & 0 \\ 0 & 2 \end{smallmatrix}\right)$ to the 2-wavelet set. This converts the set of Fig. 2 into a rhombic annulus that tiles under translation by the lattice spanned by $(1, 0)$ and $(0, 2)$. As this new set, shown in the center image of Fig. 4, still tiles under dilation by 2, and 4-tiles under translation by the integers, it is a conventional 4-wavelet for this dilation.

Break this rhombic annulus into four translation tiling pieces that are reflections of one another in the x and y axes, as shown in the center image of Fig. 4, to give a single wavelet set for pmm, pmg, and pgg. The wavelet set W^{pmm}, which is shown on the left of Fig. 4, is of the form $\mathrm{conv}\{(0, \frac{2}{3}), (0, 1), (\frac{1}{2}, 1), (2, 0), (1, 0)\} \cup \mathrm{conv}\{(-\frac{1}{2}, -1), (0, -1), (0 - \frac{4}{3})\}$. To see that W^{pmm} tiles under integer translation,

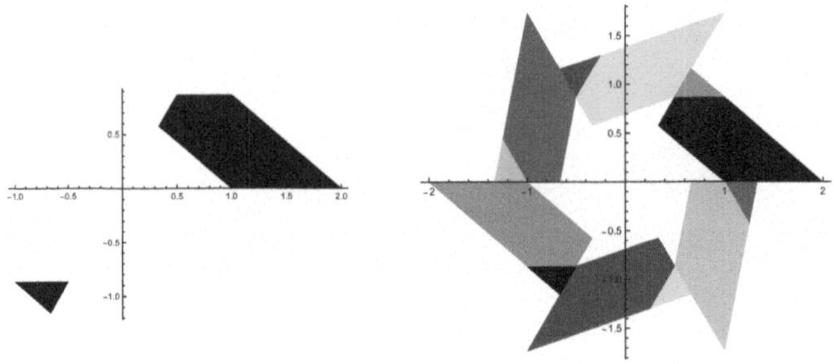

Fig. 5 Simple wavelet set $W^{\text{p6}} = LW^{\text{pmm}}$ for p6, and its six-fold rotations

piecewise integer translate it into $[0, 1]^2$. The set W^{pmm} is thus a wavelet set for pmm, pmg, and pgg since the rhombic 4-wavelet set is the disjoint union of their point group acting on W^{pmm}. The right image in Fig. 4 shows $\cup_{S\in\mathcal{B}^{\text{p4}}} S(W^{\text{pmm}})$ for p4, which is also a disjoint union that tiles under dilation by 2. Thus W^{pmm} is also a wavelet set for p4. The conventional 4-wavelet set in the center of Fig. 4 does not have enough symmetry to give a simple wavelet set for cmm, the final crystallographic group with point group of order 4. We will present one derived by a different method later in this section.

As in the previous section, the simple wavelet set for the group p4 can be multiplied by the change of lattice $L = \begin{pmatrix} 1 & \frac{1}{2} \\ 0 & \frac{\sqrt{3}}{2} \end{pmatrix}$ to give a wavelet set for p6. The set $W^{\text{p6}} = LW^{\text{pmm}}$ is shown on the left in Fig. 5, along with its rotations under powers of ρ_6 on the right. The fact that W^{p6} tiles under translation by $\mathcal{L}$ again follows immediately from the fact that W^{pmm} tiles under translation by $\mathbb{Z}^2$. The other two requirements of Theorem 3.1 follow by noting that multiplication by L sends the polar sector $\theta \in [0, \frac{\pi}{2}]$ to the polar sector $\theta \in [0, \frac{\pi}{3}]$, and preserves rotation by π and dilation by 3. Unlike in Sect. 3, multiplying W^{pmm} by L' does not produce a wavelet set for p3. This is because W^{pmm} has a piece that has been rotated by π from its first quadrant position, and p3 is not compatible with rotation by π.

To find a simple wavelet set for p3 and also for rest of the remaining seven crystallographic groups, we turn to the conventional first quadrant subspace wavelet set provided by the following lemma.

Lemma 1 *Let* $A = \cup_{j=0}^{2}[\frac{2j}{3}, \frac{2(j+1)}{3})]^2$. *Then the set* $A \setminus \frac{1}{2}A$ *gives a conventional subspace wavelet set for the first quadrant under dilation by* 2.

Proof Since $\frac{1}{2}A \subseteq A$, the set $A \setminus \frac{1}{2}A$ tiles under dilation by 2. Note that $A \setminus \frac{1}{2}A$ has area equal to 1. Thus, using a classical result from harmonic analysis (see e.g.[18]), we can establish tiling under integer translation by showing that $\widehat{\mathbf{1}}_{A\setminus\frac{1}{2}A}$ vanishes on $\mathbb{Z}^2 \setminus 0$. Assume k, ℓ are both nonzero. (The argument for one nonzero is similar.) We have

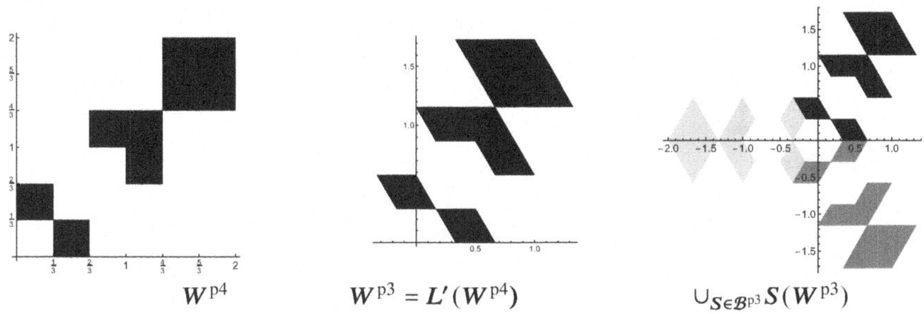

Fig. 6 Alternative simple wavelet set W^{p4} for pmm, pmg, pgg, and p4; and simple wavelet set for p3 together with its rotations

$$\widehat{\mathbf{1}}_A(k,\ell) = \frac{e^{-\frac{2}{3}i\pi(k+\ell)}\sin(\frac{2k\pi}{3})\sin(\frac{2\ell\pi}{3})}{k\ell\pi^2}\sum_{j=0}^{2} e^{\frac{-2\pi i 2j(k+\ell)}{3}}.$$

The sum at the end of this expression is 0 except when $k+\ell \in 3\mathbb{Z}$. The formula for $\widehat{\mathbf{1}}_{\frac{1}{2}A}(k,\ell)$ is similar (with $\frac{2}{3}$ replaced by $\frac{1}{3}$), and thus is also 0 except when $k+\ell \in 3\mathbb{Z}$. Thus, it remains to address the case $k+\ell \in 3\mathbb{Z}$. In this case, write $\ell = 3n-k$ and note that $\widehat{\mathbf{1}}_A(k,\ell)-\widehat{\mathbf{1}}_{\frac{1}{2}A}(k,\ell)$ has a factor of $\sin(\frac{2k\pi}{3})\sin(\frac{2(3n-k)\pi}{3})-e^{-i\pi n}\sin(\frac{k\pi}{3})\sin(\frac{(3n-k)\pi}{3})$, which is 0. □

This conventional first quadrant subspace wavelet set W^{p4}, shown on the left in Fig. 6, gives an alternative simple crystallographic wavelet set for p4, pmm, pmg, and pgg. (To differentiate it from the previous simple wavelet set for these groups, we name it W^{p4}.) Unlike W^{pmm}, it can be multiplied by the matrix L', as in the previous section, to give a simple wavelet set for p3. W^{p3} is shown in the center image of Fig. 6, and together with its rotates on the right. We will see that variations of the set W^{p4} can also be used to create simple dilation 2 wavelet sets for all the remaining crystallographic groups.

First, to find a wavelet set for the group cmm, we divide W^{p4} into two sets: the portion W_a^{p4}, which has integer translates that fill the upper and lower quarter triangles of the unit square $[0,1]^2$ (shown in black on the left in Fig. 7), and the portion W_b^{p4}, which has integer translates that fill the left and right quarter triangles (shown in gray). We then reflect W_b^{p4} over the x-axis into the fourth quadrant. Since the $\sigma_{(1,0)}W_b^{\mathrm{p4}}$ also has integer translates filling the left and right quarter triangles, the set $W^{\mathrm{cmm}} = W_a^{\mathrm{p4}} \cup \sigma_{(1,0)}W_b^{\mathrm{p4}}$, shown in the center image of Fig. 7, tiles under translation by the integers. The right image of Fig. 7 shows the disjoint union of W^{cmm} and its reflections in the diagonals, which tiles the plane under dilation by 2. Thus W^{cmm} is a simple wavelet set for cmm.

Next, we will build wavelet sets for crystallographic groups whose point groups have order 8 out of a conventional subspace 2-wavelet set for the first quadrant.

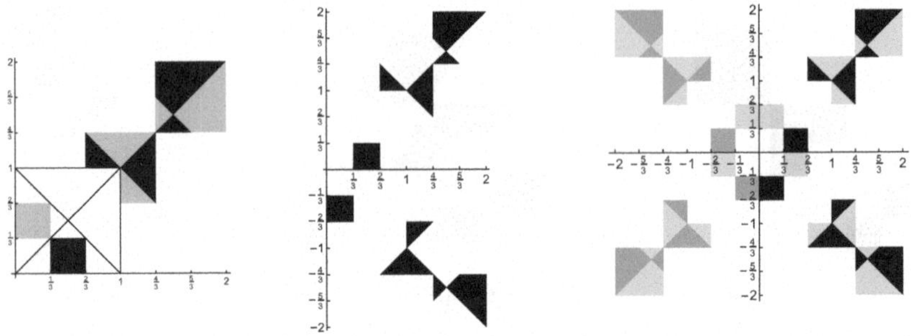

$W^{\mathrm{p4}} = W_a^{\mathrm{p4}} \cup W_b^{\mathrm{p4}}$ $W^{\mathrm{cmm}} = W_a^{\mathrm{p4}} \cup \sigma_{(1,0)} W_b^{\mathrm{p4}}$ $\cup_{S \in \mathcal{B}^{\mathrm{cmm}}} S W^{\mathrm{cmm}}$

Fig. 7 Building a simple wavelet set for cmm.

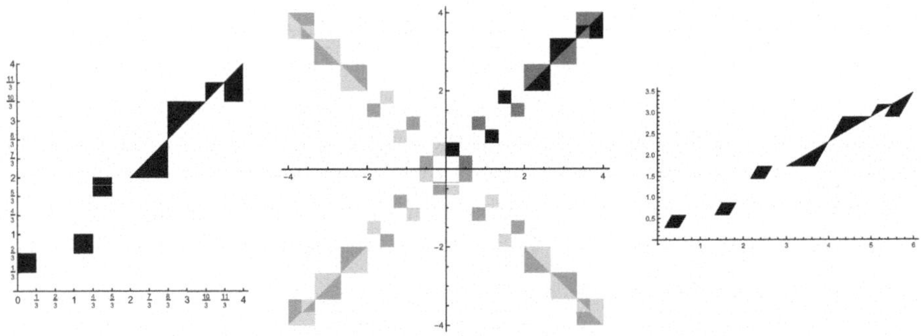

W^{p4m} $\cup_{S \in \mathcal{B}^{\mathrm{p4m}}} S W^{\mathrm{p4m}}$ $W^{\mathrm{p6m}} = L W^{\mathrm{p4m}}$

Fig. 8 Simple wavelet set for p4m and p4g, acted on by point group, and simple wavelet set for p6m

For this set, we use two copies of W^{p4} along the main diagonal, slightly altered by an integer translation of three of the smaller squares. Specifically, consider the set $C = \left(A \cup \tau_{(2,2)} A\right) \setminus \frac{1}{2}\left(A \cup \tau_{(2,2)} A\right)$, where A is as in Lemma 1 . The set C gives a two-fold tiling of the first quadrant under integer translation since it can be formed from integer translations of pieces of two copies of W^{p4}. Since $\frac{1}{2}\left(A \cup \tau_{(2,2)} A\right) \subseteq \left(A \cup \tau_{(2,2)} A\right)$, it also tiles under dilation by 2. This set appears in the first quadrant of the center image in Fig. 8.

The simple wavelet set for p4m and p4g is formed by taking W^{p4m} to be a set that tiles the first quadrant under integer translation and has the disjoint union $W^{\mathrm{p4m}} \cup \sigma_{(1,1)} W^{\mathrm{p4m}} = C$. Specifically, let $W^{\mathrm{p4m}} = \mathrm{conv}\{(0, \frac{1}{3}), (0, \frac{2}{3}), (\frac{1}{3}, \frac{2}{3}), (\frac{1}{3}, \frac{1}{3})\}$ $\cup$ $\mathrm{conv}\{(1, \frac{2}{3}), (1, 1), (\frac{4}{3}, 1), (\frac{4}{3}, \frac{2}{3})\}$ $\cup$ $\mathrm{conv}\{(\frac{4}{3}, 2), (\frac{5}{3}, 2), (\frac{5}{3}, \frac{5}{3}), (\frac{4}{3}, \frac{5}{3})\}$ $\cup$ $\mathrm{conv}\{(2, 2), (\frac{8}{3}, 2), (\frac{8}{3}, \frac{8}{3})\}$ $\cup$ $\mathrm{conv}\{(\frac{8}{3}, \frac{8}{3}), (\frac{8}{3}, \frac{10}{3}), (\frac{10}{3}, \frac{10}{3})\}$ $\cup$ $\mathrm{conv}\{(\frac{10}{3}, \frac{10}{3}), (\frac{10}{3}, \frac{11}{3}), (\frac{11}{3}, \frac{11}{3})\}$ $\cup$ $\mathrm{conv}\{(\frac{11}{3}, \frac{11}{3}), (4, 4), (4, \frac{10}{3}), (\frac{11}{3}, \frac{10}{3})\}$. (See set on the left in Fig. 8).

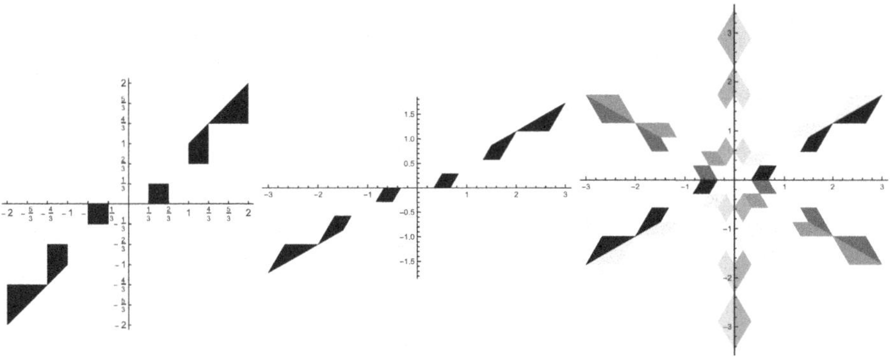

D $\quad W^{\mathrm{p3m1}} = LD$ $\quad \cup_{S \in \mathcal{B}^{\mathrm{p3m1}}} SW^{\mathrm{p3m1}}$

Fig. 9 Building a simple wavelet set for p3m1 and p31m

Tiling under integer translation can be established by piecewise integer translating W^{p4m} into $[0, 1]^2$. The effect of $\sigma_{(1,1)}$ on W^{p4m} is shown in dark gray in the center graph of Fig. 8, and then the combined effect of this reflection and the four-fold rotations in the lighter shades. A wavelet set for p6m can be found by mapping W^{p4m} by the lattice matrix L. The result, shown on the right in Fig. 8, meets all the requirements of Theorem 3.1 by a similar argument as used for p6.

The wavelet set for p4m can also be multiplied by the lattice matrix L' to give one for p31m. Instead, we use a subspace wavelet set for the first and fifth octants to produce a wavelet set for both p31m and p3m1. To get the subspace wavelet set, we modify the first quadrant conventional subspace wavelet set W^{p4} by deleting the portion in the second octant, and including instead a copy of the first octant rotated by π. The resulting set D, shown on the left in Fig. 9, is easily seen to tile the plane under integer translation and the first and fifth octants by dilation by 2, and so is a conventional subspace wavelet set for those two octants. The set LD, shown in the center image of Fig. 9 is then a wavelet set for p3m1 and also for p31m. The graph on the right in Fig. 9 shows the union of the wavelet set with its reflection and rotations, which dilates to a hexagonal star annulus.

References

1. Baggett, L., Carey, A., Moran, W., Ohring, P.: General existence theorems for orthonormal wavelets, an abstract approach. Publ. Res. Inst. Math. Sci. **31**, 95–111 (1995)
2. Baggett, L., Medina, H., Merrill, K.: Generalized multiresolution analyses and a construction procedure for all wavelet sets in $\mathbb{R}^n$. J. Fourier Anal. Appl. **5**, 563–573 (1999)
3. Baggett, L., Merrill, K., Packer, J., Taylor, K.: A direct integral decomposition of the wavelet representation for crystallographic groups, preprint.
4. Barbieri, D., Cabrelli, C., Hernández, E., Molter, U.: Approximation by group invariant subspaces, preprint
https://arxiv.org/abs/1907.08300 (2019)

5. Barbieri, D., Cabrelli, C., Hernández, E., Molter, U.: Optimal translational-rotational invariant dictionaries for images, preprint https://arxiv.org/abs/1909.01887 (2019)
6. Benedetto, J.J., Leon, M.: The construction of single wavelets in d-dimensions. J. Geom. Anal. **11**, 1–15 (2001)
7. Blanchard, J.: Minimally supported frequency composite dilation wavelets. J. Fourier Anal. Appl. **15**, 796–815 (2009)
8. Blanchard, J., Krishtal, I.: Matricial filters and crystallographic composite dilation wavelets. Math. Comput. **81**, 905–922 (2012)
9. Blanchard, J., Steffen, K.: Crystallographic Haar-type composite dilation wavelets. In: Cohen, J., Zyed, A.I. (eds.) Wavelets and Multiscale Analysis: Theory and Applications, pp 83–108. Birkhäuser, Boston (2011)
10. Dai, X., Diao, Y., Gu, Q., Han, D.: The existence of subspace wavelet sets. J. Comput. Appl. Math. **155**, 83–90 (2003)
11. Dai, X., Larson, D., Speegle, D.: Wavelet sets in $\mathbb{R}^n$. J. Fourier Anal. Appl. **3**, 451–456 (1997)
12. Farris, F.: Creating Symmetry. Princeton University Press (2015)
13. Federov, E.: Symmetry in the plane. Zapiski Rus. Mineralog. Obščestva, Ser. 2 **28**, 345–390 (1891)
14. González, A. L., Moure, M. C.: Crystallographic Haar wavelets. J. Fourier Anal. Appl. **17**, 1119–1137 (2011)
15. Guo, K., Labate, D., Lim, W.-Q, Weiss, G., Wilson, E.: Wavelets with composite dilations. Electron. Res. Announc. Am. Math. Soc. **10**, 78–87 (2004)
16. Guo, K., Labate, D., Lim, W., Weiss, G., Wilson, E.: The theory of wavelets with composite dilations. In: Heil, C. (ed.) Harmonic Analysis and Applications, pp. 231–250. Birkhäuser, Boston (2006)
17. Guo, K., Labate, D., Lim, W.-Q, Weiss, G., Wilson, E.: Wavelets with composite dilations and their MRA properties, Appl. Comput. Harmon. Anal., **20**, 202-236 (2006)
18. Kolountzakis, M.: Lattice tilings by cubes: whole, notched and extended. Electron. J. Combinatorics **5**, R14 (1998)
19. Krishtal, I., Robinson, B., Weiss, G., Wilson, E.: Some simple Haar-type wavelets in higher dimensions. J. Geom. Anal. **17**, 87–96 (2007)
20. MacArthur, J.: Compatible dilations and wavelets for the wallpaper groups. Ph.D. thesis, Dalhousie University, in preparation.
21. MacArthur, J., Taylor, K.: Wavelets with crystal symmetry shifts. J. Fourier Anal. Appl. **17**, 1109–1118 (2011)
22. Merrill, K.: Simple wavelet sets for scalar dilations in $L^2(\mathbb{R}^2)$. In: Jorgensen, P., Merrill, K., Packer, J. (eds.) Wavelets and Frames: a Celebration of the Mathematical Work of Lawrence Baggett, pp. 177–192. Birkhäuser, Boston (2008)
23. Merrill, K.: Generalized Multi-Resolution Analyses. Lecture Notes in Applied and Numerical Harmonic Analysis, Birkhäuser, Basel (2018)
24. Merrill, K.: Smooth well-localized Parseval wavelets based on wavelet sets in $\mathbb{R}^2$. Contemp. Math. **464**, 161–175 (2008)
25. Morandi, P.J.: The Classification of Wallpaper Patterns: From Cohomology to Escher's Tesselations. New Mexico State University, Las Cruces (2007)
26. Schattschneider, D.: The plane symmetry groups, their recognition and notation. Amer. Math. Monthly **85**, 439–450 (1978)
27. Taylor, K.: private communication.

Discrete Translates in Function Spaces

Alexander Olevskii and Alexander Ulanovskii

Abstract This paper is a more complete version of the lecture presented by the first author at the Fourier Analysis and Applications Conference celebrating John Benedetto's 80th Birthday in the University of Maryland, September 19–21, 2019. We discuss different aspects of the completeness property of translates in $L^p(\mathbb{R})$ and in more general Banach function spaces. In particular, we describe a wide class of Banach spaces that can be generated by uniformly discrete translates of a single function.

1 Completeness of Translates

In what follows, we will use the standard form of Fourier transform on $\mathbb{R}$:

$$\widehat{\varphi}(x) := \int_{\mathbb{R}} e^{-2\pi i t x} \varphi(t)\, dt.$$

Given a function φ, consider the set of its translates

$$\{\varphi(t - s) : s \in \mathbb{R}\}. \tag{1}$$

The classical Wiener's theorems provide necessary and sufficient conditions on a function φ whose translates (1) span the space $L^1(\mathbb{R})$ or $L^2(\mathbb{R})$.

A. Olevskii
Tel Aviv University, Tel Aviv, Israel

A. Ulanovskii (✉)
Stavanger University, Stavanger, Norway
e-mail: alexander.ulanovskii@uis.no

M. Hirn et al. (eds.), *Excursions in Harmonic Analysis, Volume 6*, Applied and Numerical Harmonic Analysis, https://doi.org/10.1007/978-3-030-69637-5_11

Theorem 1.1 ([21])

(i) *Let $\varphi \in L^1(\mathbb{R})$. The set (1) spans $L^1(\mathbb{R})$ if and only if $\widehat{\varphi}$ has no zeros.*
(ii) *Let $\varphi \in L^2(\mathbb{R})$. The set (1) spans $L^2(\mathbb{R})$ if and only if $\widehat{\varphi}$ does not vanish almost everywhere on $\mathbb{R}$.*

In both cases, this result can be stated as follows: The translates of φ span the space if and only if the zero set

$$Z(\widehat{\varphi}) := \{x \in \mathbb{R} : \widehat{\varphi}(x) = 0\}$$

of its Fourier transform is small in a certain sense.

Wiener conjectured (see [21, p. 93]) that a similar property would hold for the L^p-spaces, at least when $1 < p < 2$. This problem has been studied by Segal [20], Beurling [2], Pollard [17], Herz [7], Newman [11], and other authors (see [6, 9, 19]).

Beurling [2] proved that the translates of a function $\varphi \in L^p(\mathbb{R}) \cap L^1(\mathbb{R})$ span $L^p(\mathbb{R})$ provided the Hausdorff dimension of $Z(\widehat{\varphi})$ is less than $2(p-1)/p$. This condition is sharp, but not necessary.

Moreover, it was proved by Lev and Olevskii that the spanning property of the translates of $\varphi \in L^p(\mathbb{R})$ cannot be characterized in terms of the zero set of $\widehat{\varphi}$.

Theorem 1.2 ([10]) *For every p, $1 < p < 2$, there are two functions $\varphi_1, \varphi_2 \in (L^1 \cap L^p)(\mathbb{R})$ such that*

(i) $Z(\widehat{\varphi_1}) = Z(\widehat{\varphi_2})$*;*
(ii) *the set of translates of φ_1 spans $L^p(\mathbb{R})$, while the set of translates of φ_2 does not.*

2 Completeness of Discrete Translates in $L^2(\mathbb{R})$

Sometimes, even discrete translates of a function may span a space.

Let Λ be a discrete subset of $\mathbb{R}$. Given a function φ, consider the set of its Λ-translates

$$\{\varphi(t - \lambda) : \lambda \in \Lambda\}. \tag{2}$$

Definition 1 We say that Λ is a completeness spectrum (for translates) if there exists $\varphi \in L^2(\mathbb{R})$ such that the set (2) spans $L^2(\mathbb{R})$. The function φ is called a Λ-generator.

One may ask: Which discrete sets Λ *are completeness spectra*?

Observe that the Fourier transform of $\varphi(t - \lambda)$ is the function $e^{2\pi i \lambda x}\widehat{\varphi}(x)$. Hence, using the Fourier transform, the above question can be re-formulated as follows: *For which Λ there exists a function $G \in L^2(\mathbb{R})$ such that the system*

$$\{G(t)e^{2\pi i\lambda t} : \lambda \in \Lambda\} \tag{3}$$

spans the whole space $L^2(\mathbb{R})$?

Consider two examples:

$$\Lambda_1 := \{\sqrt{n} : n \in \mathbb{Z}_+\}, \quad \Lambda_2 := \mathbb{Z}.$$

Due to the periodicity of the exponential functions $\exp(2\pi inx)$, $n \in \mathbb{Z}$, one may check that no system (3) with $\Lambda = \mathbb{Z}$ may span $L^2(\mathbb{R})$. This means that Λ_2 is not a completeness spectrum. On the other hand, Λ_1 is so "large" that it is a completeness spectrum (see a simple proof in [14, p. 112]).

In connection with these examples, one may wonder what is "most important" for a set Λ to be a completeness spectrum, its arithmetic structure, or its density? In particular, can a "small perturbation" of the set of integers be a completeness spectrum?

2.1 Uniformly Discrete Spectra

Recall that a set $\Lambda \subseteq \mathbb{R}$ is called uniformly discrete (u.d.) if

$$\inf_{\lambda,\lambda'\in\Lambda,\lambda\neq\lambda'} |\lambda - \lambda'| > 0.$$

For a period of time, it was apparently believed that u.d. sets Λ are "not sufficiently dense" to be completeness spectra, see the discussion in [18]. It turned out that this is not the case.

Let us call Λ an almost integer set if

$$\Lambda := \{n + \gamma(n) : n \in \mathbb{Z}, 0 < |\gamma(n)| = o(1), |n| \to \infty\}. \tag{4}$$

Almost integer sets can be viewed as "small perturbations" of the set of integers.

Theorem 2.1 ([12]) *Every almost integer set Λ is a completeness spectrum.*

The proof is based on a "small denominators" argument, see also [14], lec. 11.

2.2 Schwartz Generators

One may also consider the "quality" of Λ-generators. Is it possible, for a given Λ in Theorem 2.1, to find a generator φ that is a Schwartz function, $\varphi \in \mathcal{S}(\mathbb{R})$? In general, the answer is negative. However, it is positive provided the perturbations $\gamma(n)$ in Theorem 2.1 decrease exponentially fast.

Theorem 2.2 ([13]) *Let Λ be an almost integer spectrum in (4) where γ_n are exponentially small*

$$0 < |\gamma_n| < r^n, \quad n \in \mathbb{Z}, \quad \textit{for some} \quad 0 < r < 1. \tag{5}$$

Then, Λ admits a generator $\varphi \in \mathcal{S}(\mathbb{R})$.

This result may seem surprising, since if the perturbations $\gamma(n)$ are exponentially small, we are "closer" to the limit case $\Lambda = \mathbb{Z}$, when no generator exists. The precise threshold in this problem is the quasi-analytic decay. For this and some other relative results, see [14, lec. 11].

3 Completeness of Discrete Translates in $L^p(\mathbb{R})$

Let us consider the completeness of translates in the Banach spaces $L^p(\mathbb{R})$, $p \geq 1$. Let us extend Definition 1 to this case.

Definition 2 We say that Λ is an L^p-completeness spectrum, if there exists $\varphi \in L^p(\mathbb{R})$ (a Λ-generator) such that the set (2) spans $L^p(\mathbb{R})$.

Which sets Λ are completeness spectra for $L^p(\mathbb{R})$? The Fourier transform in $L^p(\mathbb{R})$ for $p \neq 2$ does not have such nice properties as in the case $p = 2$. That is why a reformulation in terms of the completeness of exponentials in weighted spaces is not always effective.

Observe that it is "easier" for a set to be an L^p-completeness spectrum for large values of p. Namely, the following is true (see [4]): Assume Λ is an L^p-completeness spectrum. Then, it is an L^q-completeness spectrum, for every $q > p$. We will see that there are indeed sharp contrasts between the results for different values of p. Below we illustrate this on the following three cases $p = 1$, $1 < p \leq 2$, and $p > 2$.

3.1 The Case $p > 2$

Recall that $\mathbb{Z}$ is not a completeness spectrum for $L^2(\mathbb{R})$. However, rather surprisingly, $\mathbb{Z}$ is an L^p-completeness spectrum, for every $p > 2$.

Theorem 3.1 ([1]) *For every $p > 2$, there is a smooth function $\varphi \in L^p(\mathbb{R}) \cap L^2(\mathbb{R})$ such that family*

$$\{\varphi(x - n) : n \in \mathbb{Z}\}$$

is complete and minimal in $L^p(\mathbb{R})$.

Recall that a system is complete and minimal if it ceases to be complete after a removal of any element.

Hence, $\Lambda = \mathbb{Z}$ admits an L^p-generator for every $p > 2$ (while it does not for $p \leq 2$).

For the proof, see [1] or [14, lec. 12].

3.2 *The Case* $p = 1$

This is the only value of p, for which a characterization of the completeness spectra is known.

Theorem 3.2 ([5]) *A set* Λ *is an* L^1*-completeness spectrum if and only if it has infinite Beurling–Malliavin density.*

Recall that the Beurling–Malliavin (BM) density was introduced by Beurling and Malliavin [3] in connection with the completeness problem for exponential systems. Given a discrete set $\Lambda \subseteq \mathbb{R}$, denote by

$$E(\Lambda) := \{e^{2\pi i\lambda t} : \lambda \in \Lambda\}$$

the corresponding exponential system. The completeness radius of Λ is defined as the supremum over $r \geq 0$ such that $E(\Lambda)$ is complete in $L^2(-r, r)$. The classical Beurling and Malliavin theorem [3] states that the BM density of Λ is equal to its completeness radius.

Hence, Theorem 3.2 is equivalent to the statement that Λ is an L^1-completeness spectrum for translates if and only if $E(\Lambda)$ is complete in $L^2(-r, r)$, for every $r > 0$. See the proof in [5] or in [14, lec. 12].

It follows from Theorem 3.2 and the definition of Beurling–Malliavin density that no u.d. set Λ is an L^1-completeness spectrum.

3.3 *The Case* $1 < p < 2$

This case is much less investigated. In particular, it was an open problem whether an L^p-completeness spectrum can be uniformly discrete. It was recently proved to be the case, see [15]. A more general result is discussed in the next section.

4 Discrete Translates in Function Spaces

Above we focused on the completeness property for translates in $L^p(\mathbb{R})$, $p \geq 1$. We now consider this problem for general Banach function spaces X on $\mathbb{R}$.

Definition 3 A set $\Lambda \subseteq \mathbb{R}$ is called an X-completeness spectrum, if a function $\varphi \in X$ exists (a Λ-generator) such that every Λ-translate $\varphi(t - \lambda), \quad \lambda \in \Lambda$, is an element of X, and the set of all finite linear combinations of these translates forms a dense subset of X.

More precisely, we will consider the following:

Problem Which Banach function spaces X admit u.d. completeness spectra?

In other words, we ask which X can be generated by u.d. translates of a single function. We will consider this problem for the class of Banach function spaces X on $\mathbb{R}$ satisfying two assumptions:

(I) The Schwartz space $\mathcal{S}(\mathbb{R})$ is embedded in X continuously and densely.

It follows that the elements of the dual space X^* are tempered distributions in the sense of Loran Schwartz. It is well known that the Fourier transform of any tempered distribution is again a tempered distribution. If g is a tempered distribution, its spectrum, spec g, is the closed support of its Fourier transform.

Our second assumption forbids elements of X^* to have spectrum in $\mathbb{Z}$:

(II) No non-trivial element $g \in X^*$ has spectrum lying in $\mathbb{Z}$.

One may check that if every function in X is integrable, then condition (II) is not fulfilled. In fact, for "reasonable" Banach spaces X, the converse is also true, see examples below. In particular, one may check that every space $L^p(\mathbb{R}), p \geq 1$, satisfies assumption (I). A space $L^p(\mathbb{R})$ satisfies assumption (II) if and only if $p > 1$.

Theorem 4.1 ([16]) *Every Banach space X satisfying (I) and (II) can be spanned by uniformly discrete translates of a single function.*

Moreover, we will show that the generating function can be chosen "universally," see Theorem 4.2 below.

Observe that we do not require the spaces X in Theorem 4.1 to be translation invariant.

Below we present an explicit construction of generators in this result. This construction is based on a certain uniqueness theorem for the Fourier transform of distributions with "deep zeros" on $\mathbb{Z}$.

4.1 Examples

Below we present several examples of Banach function spaces for which Theorem 4.1 is applicable.

1. Symmetric spaces. A symmetric space (see [8]) X on $\mathbb{R}$ is a Banach space of measurable functions such that $L^1(\mathbb{R}) \cap L^\infty(\mathbb{R}) \subseteq X \subseteq L^1(\mathbb{R}) + L^\infty(\mathbb{R})$ and the following is true: If $f \in X$ and $|\varphi(x)| \leq |f(x)|, x \in \mathbb{R}$, then $\varphi \in X$. Also, every "rearrangement" of $|f|$ (obtained by an invertible measure preserving transformation) belongs to X and has the same norm.

Proposition 1 ([16]) *A separable symmetric space X can be spanned by uniformly discrete translates of a single function if and only if it contains a function f, $\int_{\mathbb{R}} |f| = \infty$.*

2. Sobolev spaces. Let, as usual, $W_p^l(\mathbb{R})$, $p \geq 1, l \in \mathbb{N}$, denote the Sobolev space. We have the following proposition.

Proposition 2 ([16]) *The space $W_p^l(\mathbb{R})$ can be spanned by uniformly discrete translates of a single function if and only if $p > 1$.*

3. Weighted $\mathbf{L^1}$-spaces. One may check that Theorem 4.1 holds for every space

$$X = L^1(w, \mathbb{R}) := \{f : \|f\|_w = \int_{\mathbb{R}} |f(x)|w(x)\,dx < \infty\},$$

where $w > 0$ is a bounded weight vanishing at infinity. Observe that such spaces X are not (in general) translation invariant.

4.2 *The Construction*

Definition 4 Denote by K the class of functions $\Phi \in S(\mathbb{R})$ that are positive outside $\mathbb{Z}$ and satisfy

$$|\Phi^{(k)}(t)| \leq C_{1,k} e^{-C_{2,k}(|t| + \frac{1}{\min_{n\in\mathbb{Z}} |t-n|})}, \quad t \in \mathbb{R}, k = 0, 1, \ldots, \tag{6}$$

where $C_{j,k}$ are the positive constants depending on Φ.

Condition (6) means that Φ and all its derivatives have "deep zeros" on the set of integers and at infinity.

Next, let $\widehat{K}$ denote the class of the Fourier transforms $\varphi = \widehat{\Phi}$ of functions $\Phi \in K$. Since every Φ has an exponential decay at $\pm\infty$, every element of $\widehat{K}$ is a function analytic and bounded in some horizontal strip $\{z \in \mathbb{C} : |\Im z| < C\}$.

Theorem 4.1 is an immediate consequence of the following.

Theorem 4.2 ([16]) *Let $\varphi \in \widehat{K}$, $\varphi \neq 0$. Then, the set of translates $\{\varphi(t - \lambda) : \lambda \in \Lambda\}$ is complete in X, for every Λ satisfying (4) and (5) and every X satisfying (I) and (II).*

Recall that conditions (4) and (5) mean that Λ is an "exponentially small perturbation" of $\mathbb{Z}$.

Fix any non-trivial function $\varphi \in \widehat{K}$ and denote by $\varphi * S'(\mathbb{R})$ the class of all tempered distributions $f = \varphi * g$, $g \in S'(\mathbb{R})$, where $S'(\mathbb{R})$ denotes the space of tempered distributions. Every element of this class is a function analytic and of polynomial growth in some horizontal strip. Let Φ and G be the Fourier transforms of φ and g, respectively. We see that the class $\varphi * S'(\mathbb{R})$ consists of the inverse Fourier transforms of the distributions $\Phi \cdot G$, where $G \in S'(\mathbb{R})$. By (6), the latter class can be viewed as "distributions with deep zeros" at the integers and infinity.

4.3 Uniqueness Theorem for Distributions with Deep Zeros

The crucial step in our proof of Theorem 4.2 is the following uniqueness result that shows that every set Λ satisfying (4) and (5) is a uniqueness set for the class $\varphi * S'(\mathbb{R})$.

Theorem 4.3 ([16]) *Assume that $\varphi \in \widehat{K}$, $g \in S'(\mathbb{R})$ and that Λ satisfies (4) and (5). If $(\varphi * g)|_\Lambda = 0$, then $\varphi * g = 0$.*

The proof is quite involved. Let us illustrate this theorem on the simple case $g = \delta_0$. Then, the theorem states that condition $\varphi|_\Lambda = 0$ implies $\varphi = 0$.

First, we prove that condition $\varphi|_\Lambda = 0$ implies $\varphi|_\mathbb{Z} = 0$.

Recall that φ is analytic in some strip. Use this and conditions (4) and (5) to deduce that the assumption $\varphi|_\Lambda = 0$ implies that the sequence $\varphi(n)$, $n \in \mathbb{Z}$, has an exponential decay.

It follows that the periodic function

$$H(t) := \sum_{k\in\mathbb{Z}} \varphi(n)e^{2\pi i n t}$$

admits an analytic extension to some horizontal strip. By the Poisson summation formula,

$$H(t) = \sum_{n\in\mathbb{Z}} \Phi(n-t).$$

From (6) (with $k = 0$), we see that H has an exponential zero at the origin, so that $H = 0$. This yields $\varphi|_\mathbb{Z} = 0$.

One can check that the above argument can be applied to $\varphi', \varphi'', \ldots,$ to get $\varphi^{(k)}|_\mathbb{Z} = 0, k = 1, 2, \ldots$. Hence, $\varphi = 0$.

4.4 Proof of Theorem 4.2

Fix a non-trivial function $\varphi \in \widehat{K}$. For simplicity, assume that $\varphi(-t) = \varphi(t)$. Assume the translates $\{\varphi(t-\lambda), \lambda \in \Lambda\}$ are not complete in X. Then, there is a functional $g \in X^*$ "orthogonal" to them, which means

$$(\varphi * g)|_\Lambda = 0.$$

By Theorem 4.3, $\varphi * g = 0$. This means that $\Phi \cdot G = 0$. Since Φ does not vanish outside of $\mathbb{Z}$, we see that $G = 0$ on $\mathbb{R} \setminus \mathbb{Z}$. Hence, $\operatorname{spec} g \subseteq \mathbb{Z}$. Applying property (II), we conclude that $g = 0$.

Open Problem Does there exist a set of translates of a single function that is complete and minimal in $L^2(\mathbb{R})$?

4.5 Two Generators

We have seen that not every space X satisfying (I) can be spanned by u.d. translates of a single function. In particular, $L^1(\mathbb{R})$ is such a space.

However, two functions "working together" may do "much more" then a single one:

Theorem 4.4 ([16]) *There is a pair of functions $\varphi_1, \varphi_2 \in S(\mathbb{R})$ whose Λ-translates span every space X satisfying assumption (I), for every Λ satisfying (4) and (5).*

This shows an advantage of collective work!

References

1. Atzmon, A., Olevskii, A.: Completeness of integer translates in function spaces on $\mathbb{R}$. *J. Approx. Theory* 87, no. 3, 291–327 (1996)
2. Beurling, A.: On a closure problem. Ark. Mat. 1, 301–303 (1951). See also: *The Collected Works of Arne Beurling*, in: Harmonic Analysis, vol. 2, Harmonic Analysis. Birkhäuser, Boston (1989)
3. Beurling, A., Malliavin, P.: On the closure of characters and the zeros of entire functions. Acta Math. 118, 79–93 (1967)
4. Blank, N.: Generating sets for Beurling algebras. J. Approx. Theory 140, no. 1, 61–70 (2006)
5. Bruna, J., Olevskii, A., Ulanovskii, A.: Completeness in $L^1(\mathbb{R})$ of discrete translates. Rev. Mat. Iberoam. 22, no. 1, 1–16 (2006)
6. Edwards, R. E.: Spans of translates in $L^p(G)$. J. Austral. Math. Soc. 5 216–233 (1965)
7. Herz, C. S.: A note on the span of translations in L^p. Proc. Amer. Math. Soc. 8, 724–727 (1957)
8. Kreĭn, S. G., Petunin, Ju. I., Semenov, E. M.: Interpolation of linear operators. Translations of Mathematical Monographs, vol. 54. AMS, Providence, R.I., (1982)
9. Kinukawa, M.: A note on the closure of translations in L^p. Tôhoku Math. J. 18, 225–231 (1966)
10. Lev, N., Olevskii, A.: Wiener's "closure of translates" problem and Piatetski–Shapiro's uniqueness phenomenon. Ann. of Math. (2) 174, no. 1, 519–541 (2011)
11. Newman, D. J.: The closure of translates in l^p. Amer. J. Math. 86 651–667 (1964)
12. Olevskii, A.: Completeness in $L^2(\mathbb{R})$ of almost integer translates. C. R. Acad. Sci. Paris Sér. I Math. 324, no. 9, 987–991 (1997)
13. Olevskii, A., Ulanovskii, A.: Almost integer translates. Do nice generators exist?. J. Fourier Anal. Appl. 10, no. 1, 93–104 (2004)
14. Olevskii, A., Ulanovskii, A.: Functions with Disconnected Spectrum: Sampling, Interpolation, Translates. AMS, University Lecture Series, 65 (2016)
15. Olevskii, A., Ulanovskii, A.: Discrete translates in $L^p(\mathbb{R})$. Bull. Lond. Math. Soc. 50, no. 4, 561–568 (2018)
16. Olevskii, A., Ulanovskii, A.: Discrete translates in function spaces. Anal. Math. 44, no. 2, 251–261 (2018)
17. Pollard, H.: The closure of translations in L^p. Proc. Amer. Math. Soc. 2, 100–104 (1951)

18. Ramanathan, J., Steger, T.: Incompleteness of sparse coherent states. Appl. Comput. Harmon. Anal. 2, 148–153 (1995)
19. Rosenblatt, J. M., Shuman, K. L.: Cyclic functions in $L^p(R)$, $1 < p < \infty$. J. Fourier Anal. Appl. 9, 289–300 (2003)
20. Segal, I. E.: The span of the translations of a function in a Lebesgue space. Proc. Nat. Acad. Sci. U. S. A. 30, 165–169 (1944)
21. Wiener, N.: Tauberian Theorems. Annals of Math. 33, (1), 1–100 (1932)

Part IV
Sampling and Signal Processing

This part is dedicated to the rich interplay between harmonic analysis, signal processing, and sampling theory. While these fields have been intertwined for many decades, the synergy between the fields persists to the present day and continues to result in novel research developments. In what follows we present a "sampling" of these results, drawn from some of the foremost experts in the world.

In the first chapter of this part (Chap. "Local-to-Global Frames and Applications to the Dynamical Sampling Problem"), Aldroubi, Cabrelli, Molter, and Petrosyan consider conditions on frames so that local reconstruction properties of the frame imply stable recovery properties, globally. They apply their results to the dynamical sampling problem, which has spurred an active research program due to its relevance in numerous applications.

Next, Cordero, de Gosson, Dörfler, and Nicola approach problems in signal processing using Born–Jordan-type distributions. By generalizing the Wigner distribution they are able to reduce artifacts in certain images, thus motivating their study and opening up new research questions.

In Chap. "Sampling by Averages and Average Splines on Dirichlet Spaces and on Combinatorial Graphs" Isaac Pesenson introduces Paley–Wiener spaces of functions on strictly local regular Dirichlet spaces and obtains provable recovery guarantees for sampling such functions from their local averages, in contrast to standard pointwise sample recovery guarantees. These results are then extended combinatorial graphs. In both cases the results rely on Poincaré-type inequalities.

Subsequently, Díaz Martin, Medri, and Molter take a look at the fields of Dynamical Sampling and Control Theory. In their article they describe how the two fields are in fact closely related, showing that many of the topics in one field are studied in the other field, albeit using different nomenclatures. They develop a dictionary between the two fields, thus giving the reader a means by which to move between them.

This part concludes with Chap. "Linear Multiscale Transforms Based on Even-Reversible Subdivision Operators", in which Dyn and Zhuang extend the construction of linear multiscale transforms from interpolatory subdivision operators to so-called even-reversible operators. They show that many common subdivision operators are even-reversible, including non-interpolatory subdivision operators with spline or pseudo-spline masks.

Local-to-Global Frames and Applications to the Dynamical Sampling Problem

Akram Aldroubi, Carlos Cabrelli, Ursula Molter, and Armenak Petrosyan

Abstract In this paper, we consider systems of vectors in a Hilbert space $\mathcal{H}$ of the form $\{g_{jk} : j \in J, k \in K\} \subseteq \mathcal{H}$, where J and K are countable sets of indices. We find conditions under which the local reconstruction properties of such a system extend to global stable recovery properties on the whole space. As a particular case, we obtain new local-to-global results for systems of type $\{A^n g\}_{g \in \mathcal{G}, 0 \leq n \leq L}$ arising in the dynamical sampling problem.

1 Introduction

The notion of frame was originally introduced by Duffin and Schaffer [23] in the context of non-harmonic analysis, in particular exponential systems of the form $\{e^{2\pi i \lambda_i x}\}_{i \in I}$, where $\{\lambda_i\}_{i \in I} \subseteq \mathbb{R}$ were considered. However, the notion found wide applications in signal processing research, and it has attracted a great deal of interest from the mathematics as well as the engineering communities during the last several decades.

Definition 1 A system of vectors $\mathcal{F} = \{f_i\}_{i \in I} \subseteq \mathcal{H}$ is said to be a frame for the Hilbert space $\mathcal{H}$ if there exist two constants $\alpha, \beta > 0$ such that

A. Aldroubi
Department of Mathematics, Vanderbilt University, Nashville, TN, USA
e-mail: akram.aldroubi@vanderbilt.edu

C. Cabrelli · U. Molter
Departamento de Matemática, FCEyN, UBA and IMAS CONICET, Buenos Aires, Argentina
e-mail: cabrelli@dm.uba.ar; umolter@dm.uba.ar

A. Petrosyan (✉)
Computational and Applied Mathematics Group, Oak Ridge National Laboratory, Oak Ridge, TN, USA
e-mail: petrosyana@ornl.gov; arm.petros@gmail.com

M. Hirn et al. (eds.), *Excursions in Harmonic Analysis, Volume 6*, Applied and Numerical Harmonic Analysis, https://doi.org/10.1007/978-3-030-69637-5_12

$$\alpha\|f\|^2 \leq \sum_{i\in I} |\langle f, f_i\rangle|^2 \leq \beta\|f\|^2,$$

for all $f \in \mathcal{H}$. If only the upper bound holds, then the system is called a Bessel system.

Frames are a generalization of bases. In particular, if $\{f_i\}_{i\in I}$ is a frame for $\mathcal{H}$, then any vector $f \in \mathcal{H}$ has the representation $f = \sum_{i\in I} \langle f, f_i\rangle \widetilde{f_i}$, where $\{\widetilde{f_i}\}_{i\in I}$ is another frame called the canonical dual frame. However, unlike bases, the vectors in a frame $\{f_i\}_{i\in I}$ may have linear dependencies, thus resulting in a redundant representation (see [17] for more details). If the frame system is not a basis, then a frame $\{f_i\}_{i\in I}$ has many dual frames other than the canonical dual frame.

In signal processing models, often the system $\mathcal{F}$ consists of functions, defined on the domain $\Gamma = \mathbb{Z}$ or $\mathbb{R}$. The functions f_i are usually localized . By that, we mean that every point in the domain is contained in the supports of only finitely many functions from the system. This may happen, for example, when $\mathcal{F}$ is generated by the shifts of a finite number of compactly supported functions. In this setting, we may be able to divide the domain into smaller patches such that every function f can be recovered on each patch, from inner products with those functions from $\mathcal{F}$ whose support overlap with that patch, i.e., f can be recovered from local information. For that reason, we consider systems $\mathcal{F}$ re-indexed to have the form $\{g_{jk} : j \in J, k \in K\} \subseteq \mathcal{H}$, where the index $j \in J$ corresponds to the patch j and the index in $k \in K$ corresponds to indexing within each patch. The index sets I, J, K will always be at most countable. The main interest in this note is to find conditions that guarantee stable global reconstruction of any function on the whole domain Γ, given that the system has local recovery properties. In that regard, among other results, we prove the following theorem.

Theorem *Let $\mathcal{H}$ be a Hilbert space and $\{P_j\}_{j\in\mathbb{N}}$ be projectors acting on $\mathcal{H}$ satisfying $P_r P_j = 0$ for $r \neq j$, and also $\sum_j P_j = \mathbf{1}$, where convergence is in the strong operator topology. Assume that $\{P_j g_{jk} : k \in K\}$ form a frame for $P_j\mathcal{H}$ with uniform frame bounds α, β, and $\|P_r g_{jk}\| \leq |c_k(j-r)|$ with $c_k \in \ell^1(\mathbb{Z})$ for all $k \in K$. If $\sum_{k\in K} \|c_k\|_1 < \alpha$, then $\{g_{jk}\}_{j\in J, k\in K}$ is a frame for $\mathcal{H}$ with frame bounds $\alpha - \sum_{k\in K} \|c_k\|_1$ and $\beta + \sum_{k\in K} \|c_k\|_1$.*

As a main application of our results, we study the dynamical sampling (DS) problem where one wants to recover an unknown signal f from spatial–temporal samples $\{\langle f, A^n g\rangle\}_{g\in\mathcal{G}, 0\leq n\leq L}$, where $L < \infty$, and $\mathcal{G}$ is a collection of functions defined in Γ. Here, Γ is either $\mathbb{R}$ (continuous case) or $\mathbb{Z}$ (discrete case), and A is an operator acting on $L^2(\Gamma)$. This leads to the study of frame properties of the iterative system

$$\{A^n g\}_{g\in\mathcal{G}, 0\leq n\leq L}. \tag{1}$$

The dynamical sampling problem was introduced in [5], and it is an active area of research in the applied harmonic analysis community, with a range of potential fields of applications [7]. Early results in DS considered convolution operators on the spaces $l^2(\mathbb{Z}_d)$, $l^2(\mathbb{Z})$ and $L^2(\mathbb{R})$ with the samples taken on a sparse uniform grid [5, 6]. In [9, 27], authors allowed the operator A to be unknown too. For an operator theoretic approach to the dynamical sampling problem, see [2, 4, 8, 11, 25]. The case when $f \in l^2(\mathbb{N})$, $A = B^{-1}DB$, and $D = \sum_j \lambda_j P_j$ is an infinite diagonal matrix can be found in [4]. For general bounded normal operators, see [11–13]. The authors in [18–22] consider general frames and investigate conditions under which they can be represented as iterated systems arising in dynamical sampling. In [10], authors study the problem of phaseless recovery from dynamical samples. The scalability properties of the system arising in DS is investigated in [1].

Our main results in this paper are proved for general class of systems, and the results are applied to dynamical sampling. In particular, we consider dynamical systems where the system $\mathcal{G}$ is localized in space (e.g., when it is generated by the shift of a single compactly supported function), and A is a compactly supported convolution operator. More specifically, we obtain the following theorem as corollary of the results in Sect. 2.

Theorem *Let a be a finite sequence, and consider the convolution operator $Af = a * f$ for $f \in \ell^2(\mathbb{Z})$. Let $I = \{-N, \dots, N\}$ and $I_j = I + N_j \subseteq \mathbb{Z}$ be such that $\cup_j I_j = \mathbb{Z}$, and $I_j \cap I_k = \emptyset$ for $|j - k| > M$. Assume that $G_0 = \{\chi_{I_j} a^n(\cdot - l) : l \in \Omega \subseteq I\}$ is a frame for $\ell^2(I)$ (χ_{I_j} is the characteristic function of I_j), and let S_k be the shift operator by the integer k. Then, $\cup_j S_{N_j} G_0$ is a frame for $\ell^2(\mathbb{N})$.*

The paper is organized as follows. In Sect. 2, we prove our main results concerning general systems with local-to-global indexing. In Sect. 3, we apply the results from Sect. 2 to iterative system of vectors arising from dynamical sampling problem.

2 Main Results for General Systems

In this section, we prove our main theorems concerning local-to-global frame properties of the systems $\{g_{jk} : j \in J,\ k \in K\} \subseteq \mathcal{H}$, where J and K are countable (finite or infinite). Here, $g_{j,k}$ are some vectors, not necessarily functions on $\mathbb{R}$ or $\mathbb{Z}$. The local patches are replaced with projection operators $\{P_j\}_{j\in J}$ or more generally with operators $\{T_j\}_{j\in J} \subseteq B(\mathcal{H})$. Our first proposition gives a way to construct a global frame from local frames . This is a generalization of Theorem 5.3 from [3] written in terms of g-frames as introduced in [26] (see also [26, Theorem 3.1]). A sequence $\{T_j\}_{j\in J} \subseteq B(\mathcal{H})$ is a g-frame if there exist positive constants A, B such that $A\|f\|^2 \leq \sum_j \|T_j f\|^2 \leq B\|f\|^2$ for all $f \in \mathcal{H}$.

Proposition 1 *If the sequence $\{T_j\}_{j\in J} \subseteq B(\mathcal{H})$ is a g-frame, T_j has closed range for each j, and the set $\{T_j g_{jk} : k \in K\}$ forms a frame for $T_j(\mathcal{H})$ with uniform frame* *bounds α, β, then $\{T_j^* T_j g_{jk} : j \in J, k \in K\}$ is a frame for $\mathcal{H}$ with frame bounds $\alpha A, \beta B$.*

Conversely, if $\{T_j g_{jk} : k \in K\}$ forms a frame for $T_j(\mathcal{H})$ with uniform frame bounds α, β, and $\{T_j^* T_j g_{jk} : j \in J, k \in K\}$ is a frame for $\mathcal{H}$ with frame bounds $\alpha A, \beta B$, then

$$\frac{\alpha}{\beta} A\|f\|^2 \leq \sum_j \|T_j f\|^2 \leq \frac{\beta}{\alpha} B\|f\|^2 \quad \textit{for all } f \in \mathcal{H}.$$

Proof Let $f \in \mathcal{H}$. Then,

$$\alpha\|T_j f\|^2 \leq \sum_{k\in K} |\langle T_j f, T_j g_{jk}\rangle|^2 = \sum_{k\in K} |\langle f, T_j^* T_j g_{jk}\rangle|^2.$$

Summing over $j \in J$, we get

$$\alpha A\|f\|^2 \leq \alpha \sum_{j\in J} \|T_j f\|^2 \leq \sum_{j\in J}\sum_{k\in K} |\langle f, T_j^* T_j g_{jk}\rangle|^2,$$

which establishes the lower frame bound for the set $\{T_j^* T_j g_{jk} : j \in J, k \in K\}$. A similar calculation yields the upper bound.

To prove the second part, let $f \in \mathcal{H}$. Then,

$$\sum_{k\in K} |\langle f, T_j^* T_j g_{jk}\rangle|^2 = \sum_{k\in K} |\langle T_j f, T_j g_{jk}\rangle|^2 \leq \beta\|T_j f\|^2.$$

Summing over $j \in J$, we get

$$\alpha A\|f\|^2 \leq \sum_{j\in J}\sum_{k\in K} |\langle f, T_j^* T_j g_{jk}\rangle|^2 \leq \beta \sum_{j\in J} \|T_j f\|^2.$$

Thus, $\frac{\alpha}{\beta} A\|f\|^2 \leq \sum_{j\in J} \|T_j f\|^2$. Similar calculations yield the upper bound. □

For the next proposition, we recall the definition of fusion frames (see [14–16, 24] and the references therein) and completeness for a system of orthogonal projectors.

Definition 2 A system of orthogonal projections $\{P_j\}_{j\in J}$ is said to be a fusion frame if there exist positive constants $m, M > 0$ such that

$$m\|f\|^2 \leq \sum_j \|P_j f\|^2 \leq M\|f\|^2, \quad f \in \mathcal{H}. \tag{2}$$

Definition 3 A system of projections $\{P_j\}_{j\in \mathbb{N}}$ is said to be complete if $P_j f = 0$ for all j implies that $f = 0$.

Using the fact that for orthogonal projectors P_j, $P_j = P_j^2$, and $P_j = P_j^*$, Proposition 1 immediately yields the following well-known fact.

Proposition 2 *If $\{P_j\}_{j\in J}$ is a fusion frame with bounds A, B, and the set $\{P_j g_{jk} : k \in K\}$ forms a frame for $P_j(\mathcal{H})$ with uniform frame bounds α, β, then $\{P_j g_{jk} : j \in J, k \in K\}$ is a frame for $\mathcal{H}$ with frame bounds αA, βB.*

Conversely, if $\{P_j g_{jk} : k \in K\}$ forms a frame for $P_j(\mathcal{H})$ with uniform frame bounds α, β, and $\{P_j g_{jk} : j \in J, k \in K\}$ is a frame for $\mathcal{H}$ with frame bounds αA, βB, then $\{P_j\}_{j\in J}$ is a fusion frame with bounds $\frac{\alpha}{\beta}A$, and $\frac{\beta}{\alpha}B$.

Theorem 2.1 *Assume that $\{P_j\}_{j\in\mathbb{N}}$ is such that $P_jP_k = P_kP_j$ for all $j, k \in \mathbb{N}$, and $P_jP_k = 0$ for $|j - k| \geq \mathcal{I}$ for some positive integer $\mathcal{I}$. If $\{P_j\}_{j\in\mathbb{N}}$ is complete, then $\{P_j\}_{j\in\mathbb{N}}$ is a fusion frame.*

Proof The set $\{P_j\}_{j\in\mathbb{N}}$ is the union of $\mathcal{I}$ sets $\{P_j\}_{j\in\mathbb{N}} = \bigcup\limits_{k=1}^{\mathcal{I}} \{P_{k+\mathcal{I}j}\}_{j\in\mathbb{N}}$. Using the fact that for each k, $\sum\limits_{j\in\mathbb{N}} P_{k+\mathcal{I}j}$ is an orthogonal projection, we get that for $f \in \mathcal{H}$

$$\begin{aligned}
\|\sum_{j\in\mathbb{N}} P_j f\|^2 &= \|\sum_{k=1}^{\mathcal{I}} \sum_{j\in\mathbb{N}} P_{k+\mathcal{I}j} f\|^2 \\
&\leq \sum_{k=1}^{\mathcal{I}} \|\sum_{j\in\mathbb{N}} P_{k+\mathcal{I}j} f\|^2 \\
&\leq \mathcal{I}\|f\|^2.
\end{aligned}$$

To prove the lower bound, we proceed by induction. We first note that, because P_1, P_2 commute, P_1P_2 is an orthogonal projector on $P_1(\mathcal{H}) \cap P_2(\mathcal{H})$. Let $Q_1 = P_1$ and $Q_2 = P_2 - P_2Q_1$. Then, $Q_2^* = Q_2$, $Q_2^2 = Q_2$, and $Q_1Q_2 = 0$. Moreover, since $P_1Q_2 = 0$, we get

$$\begin{aligned}
\|Q_1 f\|^2 + \|Q_2 f\|^2 = \|P_1 f\|^2 + \|(P_2 - P_2P_1) f\|^2 &= \|P_1 f\|^2 + \|(\mathbf{1} - P_1)P_2 f\|^2 \\
&\leq \|P_1 f\|^2 + \|P_2 f\|^2.
\end{aligned}$$

For $n > 2$, let $Q_n = P_n - P_n \sum\limits_{k=1}^{n-1} Q_k$. Then, using the fact that $R = \sum\limits_{k=1}^{n-1} Q_k$ commutes with P_n, it is not difficult to verify that $Q_n^* = Q_n$, and $Q_n^2 = Q_n$ and that $Q_nR = 0$. Assume that $Q_jQ_k = 0$ for $j, k < n$. Then, applying Q_l to both sides of the equation $Q_nR = 0$, we get that $Q_nQ_l = 0$ for all $l < n$. Thus, $Q_jQ_k = 0$ for $j \neq k$. Using induction again, we have

$$\|Rf\|^2 + \|Q_n f\|^2 = \sum_{j=1}^{n-1} \|Q_j f\|^2 + \|(\mathbf{1} - R)P_n f\|^2 \leq \sum_{j=1}^{n-1} \|P_j f\|^2 + \|P_n f\|^2.$$

Hence,

$$\sum_{j\in\mathbb{N}} \|Q_j f\|^2 \leq \sum_{j\in\mathbb{N}} \|P_j f\|^2.$$

Since $Q_1 = P_1$, $Q_n = P_n - P_n R$ for $n \geq 2$, we get that $\{Q_j\}$ is also complete whenever P_j is complete. Hence,

$$\|f\|^2 = \sum_{j\in\mathbb{N}} \|Q_j f\|^2 \leq \sum_{j\in\mathbb{N}} \|P_j f\|^2.$$

□

Combining Proposition 2 with Theorem 2.1, we obtain the following corollary.

Corollary 1 *Assume that* $\{P_j\}_{j\mathbb{N}}$ *is complete and is such that* $P_j P_k = P_k P_j$ *for all* $j, k \in \mathbb{N}$*, and* $P_j P_k = 0$ *for* $|j - k| \geq M$ *for some positive integer* M*. Furthermore, assume that* $\{P_j g_{jk} : k \in K\}$ *from a frame for* $P_j\mathcal{H}$ *with uniform frame bounds* α, β *and that* $P_j g_{jk} = g_{jk}$ *for all* j, k*. Then, the set* $\{g_{jk}\}_{j\in J, k\in K}$ *is a frame for* $\mathcal{H}$.

In the previous corollary, the assumption that $P_j g_{jk} = g_{jk}$ is rather strong and does not always happen in practice. The next theorem offers a way to overcome this assumption.

Theorem 2.2 *Let* $\{P_j\}_{j\in\mathbb{N}}$ *be the orthogonal projectors satisfying* $P_r P_j = 0$ *for* $r \neq j$*, and* $\sum_j P_j = \mathbf{1}$ *where convergence is in the strong operator topology. Assume that* $\{P_j g_{jk} : k \in K\}$ *forms a frame for* $P_j\mathcal{H}$ *with uniform frame bounds* α, β*, and* $\|P_r g_{jk}\| \leq |c_k(j - r)|$ *with* $c_k \in \ell^1(\mathbb{Z})$*. If* $\sum_{k=1}^{K} \|c_k\|_1 < \alpha$*, then* $\{g_{jk}\}_{j\in J, k\in K}$ *is a frame for* $\mathcal{H}$ *with frame bounds* $\alpha - \sum_{k=1}^{K} \|c_k\|_1$ *and* $\beta + \sum_{k=1}^{K} \|c_k\|_1$.

Proof For the lower bound, we have

$$\begin{aligned}
\Big(\sum_{j,k} |\lambda f, g_{jk}\rangle|^2\Big)^{1/2} &= \Big(\sum_{j,k} |\lambda \sum_{l} P_l f, g_{jk}\rangle|^2\Big)^{1/2} \\
&= \Big(\sum_{j,k} |\lambda P_j f, g_{jk}\rangle + \lambda \sum_{l\neq j} P_l f, g_{jk}\rangle|^2\Big)^{1/2} \\
&\geq \Big(\sum_{j,k} |\lambda P_j f, g_{jk}\rangle|^2\Big)^{1/2} - \Big(\sum_{j,k} |\lambda \sum_{l\neq j} P_l f, g_{jk}\rangle|^2\Big)^{1/2}
\end{aligned}$$

$$\geq \Big(\alpha \sum_j \|P_j f\|^2\Big)^{1/2} - \Big(\sum_{j,k} \big|\sum_{l \neq j} \|P_l f\| |c_k(l-j)|\big|^2\Big)^{1/2}$$

$$= \alpha^{1/2}\|f\| - \Big(\sum_{j,k} \big|\sum_{l \neq j} \|P_l f\| |c_k(l-j)|\big|^2\Big)^{1/2}$$

$$\geq \alpha^{1/2}\|f\| - \Big(\sum_k \sum_l \|P_l f\|^2 \|c_k\|_1^2\Big)^{1/2}$$

$$= \alpha^{1/2}\|f\| - \Big(\sum_k \|c_k\|_1^2\Big)^{1/2} \|f\|.$$

A similar calculation gives the upper bound. □

3 Application to DS

In this section, we apply our results from Sect. 2 to the dynamical sampling setting. As a direct corollary of Theorem 2.2, we get the following theorem.

Theorem 3.3 *Let $a \in \ell_1(\mathbb{Z})$ be such that for a set $\Omega \subseteq I = \{0, \ldots, N\}$, the system $\{a^k * \delta_i|_I : i \in \Omega, k = 0, \ldots, K\}$ is a frame in the space $\ell^2(I)$ with frame bounds α and β, where a^k denotes the k times convolution of a and $\delta_i \in \ell^2(I)$ is given by $\delta_i(i) = 1$ and $\delta_i(j) = 0$ for $j \neq i$. Let $I_j = I - jN$ and $\Omega_j = \Omega - jN$. If*

$$\gamma = \sum_{i \in \Omega} \sum_{k=0,\ldots,K} \sum_{j \in \mathbb{Z}\setminus\{0\}} \|a^k * \delta_i|_{I_j}\|_{\ell^2(I_j)} < \alpha,$$

then the system $\{a^k * \delta_i : i \in \cup_{j\in\mathbb{Z}}\Omega_j, k = 0, \ldots, K\}$ is a frame for $\ell^2(\mathbb{Z})$ with frame bounds $\alpha - \gamma$ and $\beta + \gamma$.

To insure the assumptions in the above theorem are realistic, below we construct an example where they actually hold.

Example 1 Let $I = \{0, 1, 2\}$, $\Omega = \{0\}$, and

$$a(i) = \begin{cases} 1, & i = 0 \\ \tau, & i = 1 \\ \tau^2, & i = 2 \\ 0, & \text{otherwise} \end{cases},$$

where $\tau > 0$. Then, notice that

$$a * a(i) = \begin{cases} 1, & i = 0 \\ 2\tau, & i = 1 \\ 3\tau^2, & i = 2 \\ 2\tau^3, & i = 3 \\ \tau^4, & i = 4 \\ 0, & \text{otherwise} \end{cases}.$$

Hence,

$$\{a^k * \delta_0|_I : k = 0, 1, 2\} = \{(1, 0, 0), (1, \tau, \tau^2), (1, 2\tau, 3\tau^2)\},$$

which are linearly independent for $\tau > 0$ since for the matrix

$$\Phi = \begin{pmatrix} 1 & 0 & 0 \\ 1 & \tau & \tau^2 \\ 1 & 2\tau & 3\tau^2 \end{pmatrix}$$

$\det(\Phi) = 5\tau^3$ and hence are a basis in $\ell^2(I)$.

The quantity γ in Theorem 3.3 is equal to $2\tau^3 + \tau^4$. For sufficiently small values of τ, it can be checked that the lower frame bound of the system, which is the smallest singular value of Φ, is larger than γ and thus the conditions of Theorem 3.3 are satisfied. For the value of $\tau = 0.1$, the lower frame bound is approximately 0.0040, whereas $\gamma = 0.0021$.

The next two propositions concern frames from dynamical sampling from diagonal and convolutional operators, and their proofs follow from Corollary 1.

Proposition 3 *Let D be an infinite diagonal matrix acting on $\ell^2(\mathbb{N})$. Assume that $D \in B(\ell^2)$. Let $\{I_j\}_{j\in\mathbb{N}} \subseteq \mathbb{N}$ be such that, $|I_j| \leq L$, $\cup_j I_j = \mathbb{N}$, and $I_j \cap I_k = \emptyset$ for $|j - k| > M$. Let D_j be the submatrix of D indexed by I_j, and let $G_j \subseteq \ell^2(\mathbb{N})$ be a set of vectors such that, for any $g \in G_j$, supp $g \subseteq I_j$ and $\{D_j^n g\}_{g\in G_j, 0\leq n<L}$ is a frame for $\ell^2(I_j)$ with frame bounds α, β independent of j. Then, the set $\{D^n g : 0 \leq n < L, g \in \cup_{j\in\mathbb{N}} G_j\}$ is a frame for $\ell^2(\mathbb{N})$.*

As a specific instance, let us consider the next example.

Example 2 Assume that the entries of the diagonal matrix D are periodic on the diagonal, i.e., $D_{ii} = f(i)$, where $f(i + p) = f(i)$ for $i \in \mathbb{N}$. This case reduces each submatrix D_j of D in the proposition to be one of finitely many possible matrices. For each of these matrices, we construct the corresponding G_j. Since the construction uses finitely many matrices and finitely many associated vectors, we can find frame bounds α, β that are independent of j, and the conditions in Proposition 3 will be satisfied.

The next proposition addresses the case of convolutional operators that cover a large class of operators of practical interest including the diffusion operator on $\ell^2(\mathbb{Z})$.

Proposition 4 *Let a be a finite sequence, and consider the convolution operator $Af = a * f$ for $f \in \ell^2(\mathbb{Z})$. Let $I = \{-N, \dots, N\}$, $\Omega \subseteq I$, and $I_j = I + N_j \subseteq \mathbb{Z}$ be such that $\cup_j I_j = \mathbb{Z}$, and $I_j \cap I_k = \emptyset$ for $|j - k| > M$. Assume that $G_0 = \{\chi_{I_j} a^n(\cdot - l) : l \in \Omega\}$ is a frame for $\ell^2(I)$ (χ_{I_j} is the characteristic function of I_j), and let S_k be the shift operator by the integer k. Then, $\cup_j S_{N_j} G_0$ is a frame for $\ell^2(\mathbb{N})$.*

Acknowledgments The research of A. Aldroubi is supported in part by NSF grant DMS-1322099. C. Cabrelli and U. Molter were partially supported by grants: UBACyT 20020170100430BA, PICT 2014-1480 (ANPCyT), and CONICET PIP 11220110101018. A. Petrosyan acknowledges support by the Oak Ridge National Laboratory, which is operated by UT-Battelle, LLC., for the U.S. Department of Energy under Contract DE-AC05-00OR22725.

References

1. Aceska, R. and Kim, Y.: Scalability of frames generated by dynamical operators. *Frontiers in Applied Mathematics and Statistics*, 3(22), 2017.
2. Aldroubi, A., Cabrelli, C., Cakmak, A. F., Molter, U., and Petrosyan, A.: Iterative actions of normal operators. *Journal of Functional Analysis*, 272(3):1121–1146, 2017.
3. Aldroubi, A., Cabrelli, C., and Molter, U.: Wavelets on irregular grids with arbitrary dilations matrices and frame atoms for $L^2(\mathbb{R}^d)$. *Applied and Computational Harmonic Analysis*, 17,119–140, 2004.
4. Aldroubi, A., Cabrelli, C., and Molter, U., and Tang, S.: Dynamical Sampling. *Applied and Computational Harmonic Analysis*, 42(3): 378-401, 2017
5. Aldroubi, A., Davis, J., and Krishtal, I.: Dynamical sampling: time-space trade-off. *Applied and Computational Harmonic Analysis*, 34(3):495–503, 2013.
6. Aldroubi, A., Davis, J., and Krishtal, I.: Exact reconstruction of signals in evolutionary systems via spatiotemporal trade-off. *Journal of Fourier Analysis and Applications*, 21(1),11–31, 2015.
7. Aldroubi, A., Huang, L., Krishtal, I., Lederman, R.R., Ledeczi, A., and Volgyesi, P.: Dynamical sampling with additive random noise. *Sampling Theory in Signal and Image Processing*, Vol. 17, pp. 153–182, 2018.
8. Aldroubi, A., Huang, L., and Petrosyan, A.: Frames induced by the action of continuous powers of an operator. *Journal of Mathematical Analysis and Applications* 478 (2), 1059–1084, 2018.
9. Aldroubi, A. and Krishtal, I.: Krylov subspace methods in dynamical sampling. *STSIP, Shannon Centennial Volume*, Vol. 15, pp. 9–20, 2016.
10. Aldroubi , A., Krishtal, I., and Tang, S.: Phaseless reconstruction from space–time samples. *Applied and Computational Harmonic Analysis*, 2018.
11. Aldroubi, A. and Petrosyan, A.: Dynamical sampling and systems from iterative actions of operators. In *Frames and Other Bases in Abstract and Function Spaces*, pages 15–26. Springer, 2017.
12. Cabrelli, C., Molter, U., Paternostro, V., and Philipp, F.: Finite sensor dynamical sampling. In *Sampling Theory and Applications (SampTA), 2017 International Conference on*, pages 50–54. IEEE, 2017.
13. Cabrelli, C., Molter, U., Paternostro, V., and Philipp, F.: Dynamical sampling on finite index sets. *Journal D'Analyse Mathématique, (JAMA)*, 140:637–667, 2020.

14. Cahill, J., Casazza, P.G., Ehler, M., and Li., Sh.: Tight and random nonorthogonal fusion frames. In *Trends in harmonic analysis and its applications*, volume 650 of *Contemporary Mathematics*, pages 23–36. American Mathematical Society, Providence, RI, 2015.
15. Cahill, J., Casazza, P.G., and Li., S.: Non-orthogonal fusion frames and the sparsity of fusion frame operators. *Journal of Fourier Analysis and Applications*, 18(2):287–308, 2012.
16. Casazza, P.G., Kutyniok, G., and Li., Sh.: Fusion frames and distributed processing. *Applied and Computational Harmonic Analysis*, 25(1):114–132, 2008.
17. Christensen, O.: *An Introduction to Frames and Riesz Bases (Applied and Numerical Harmonic Analysis)*. Birkhäuser, 2016.
18. Christensen, O. and Hasannasab, M.: Operator representations of frames: boundedness, duality, and stability. *Integral Equations and Operator Theory*, 88(4), 483–499, 2017.
19. Christensen, O. and Hasannasab, M.: Frame properties of systems arising via iterated actions of operators. *Applied and Computational Harmonic Analysis*, 46(3): 664–673, 2018.
20. Christensen, O., Hasannasab, M., and Philipp, F.: Frame properties of operator orbits. *Mathematische Nachrichten*, 293(1):52–66, 2020.
21. Christensen, O., Hasannasab, M., and Rashidi, E.: Dynamical sampling and frame representations with bounded operators. *Journal of Mathematical Analysis and Applications*, 463(2): 634–644, 2018.
22. Christensen, O., Hasannasab, M., and Stoeva, D.T.: Operator representations of sequences and dynamical sampling. *Sampling. Theory in Signal and Image Process*, 17(1): 29–42, 2018.
23. Duffin, R.J. and Schaeffer, S. C.: A class of nonharmonic Fourier series. *Transactions of the American Mathematical Society*, 72:341–366, 1952.
24. Kaushik, Sh.K. and Kumar, V.: On fusion frames in Banach spaces. *Georgian Mathematical Journal*, 18(1):121–130, 2011.
25. Philipp, F.: Bessel orbits of normal operators. *Journal of Mathematical Analysis and Applications*, 448(2):767–785, 2017.
26. Sun, W.: G-frames and g-Riesz bases. *Journal of Mathematical Analysis and Applications*, 322(1):437–452, 2006.
27. Tang, S.: System identification in dynamical sampling. *Advances in Computational Mathematics*, 43(3):555–580, 2017.

Signal Analysis Using Born–Jordan-Type Distributions

Elena Cordero, Maurice de Gosson, Monika Dörfler, and Fabio Nicola

Abstract In this chapter, we exhibit recent advances in signal analysis via time–frequency distributions. New members of the Cohen class, generalizing the Wigner distribution, reveal to be effective in damping artefacts of some signals. We will survey their main properties and drawbacks and present open problems related to such phenomena.

MSC 42B10, 42B37

1 Introduction

The Wigner distribution (Wigner transform, Wigner function or Wigner–Ville distribution) has a long tradition that started as a probability quasi-distribution in 1932 with Eugene Wigner's ground-breaking paper [29]. In 1948, it was reinvented by Jean Ville in [19] as a quadratic representation of the local time–frequency energy of a signal. This was the starting point for its numerous applications in signal analysis: from electrical engineering and communication theory to any field involving the problem of treating signals: seismology, biology, medicine, etc.

Given two functions $f, g \in L^2(\mathbb{R}^d)$, their (cross-)Wigner distribution is defined to be

E. Cordero (✉)
Università di Torino, Dipartimento di Matematica, Torino, Italy
e-mail: elena.cordero@unito.it

M. de Gosson · M. Dörfler
University of Vienna, Faculty of Mathematics, Wien, Austria
e-mail: maurice.de.gosson@univie.ac.at; monika.doerfler@univie.ac.at

F. Nicola
Politecnico di Torino, Dipartimento di Scienze Matematiche, Torino, Italy
e-mail: fabio.nicola@polito.it

M. Hirn et al. (eds.), *Excursions in Harmonic Analysis, Volume 6*, Applied and Numerical Harmonic Analysis, https://doi.org/10.1007/978-3-030-69637-5_13

$$W(f,g)(x,\omega)=\int_{\mathbb{R}^d} e^{-2\pi i\omega y} f(x+\tfrac{1}{2}y)\overline{g(x-\tfrac{1}{2}y)}dy \tag{1.1}$$

($\omega y=\omega\cdot y$ denoting the scalar product in $\mathbb{R}^d$). If $f=g$, we set $Wf:=W(f,f)$, named the Wigner distribution of f. The quadratic nature of the Wigner distribution Wf causes the appearance of interferences between the distinct components of the signal. These interferences are artefacts and do not reflect true signal components; compare the idealized TF representation in Fig. 1 below. This representation is ideal because it combines the beneficial properties of linear representations, namely no unwanted interferences, with those of the quadratic Wigner distribution, namely no smearing in time and frequency. Roughly speaking, if a signal f is the sum of two components f_1, f_2, then its Wigner distribution becomes

$$W(f_1+f_2)=Wf_1+Wf_2+2\mathcal{R}eW(f_1,f_2).$$

The cross-term $W(f_1,f_2)$ appearing above produces unwanted interferences, which are thus also called ghost terms: they do not correspond to actual signal components. This will be made clear in the following examples.

Recall the translation and modulation operators:

$$T_xf(y)=f(y-x),\quad M_\omega f(y)=e^{2\pi iy\omega}f(y),\quad x,\omega\in\mathbb{R}^d,$$

which combined are called time–frequency shifts:

$$\pi(z)f(y)=M_\omega T_xf(y)=e^{2\pi iy\omega}f(y-x),\quad z=(x,\omega).$$

We can show such unwanted phenomenon for a signal f that is the sum of four Gaussian atoms, that is, time–frequency shifts of the Gaussian function. For example, in dimension $d=1$, consider the Gaussian function $\varphi(t)=e^{-\pi t^2}$ and the signal f that is the sum of the following 4 time–frequency shifts of φ:

$$f=\pi(20,0.25)\varphi+\pi(40,0.15)\varphi+\pi(40,0.35)\varphi+\pi(60,0.25)\varphi=\sum_{i=1}^{4}\varphi_i. \tag{1.2}$$

Figure 1 is an *idealized TF representation* but does not correspond to anything computable from the signal. The Wigner distribution Wf is represented in Fig. 2, and we observe the following. Since

$$W(f,f)=\sum_{i=1}^{4}W\varphi_i+\sum_{i=1}^{4}\sum_{i\neq j}\mathcal{R}eW(\varphi_i,\varphi_j), \tag{1.3}$$

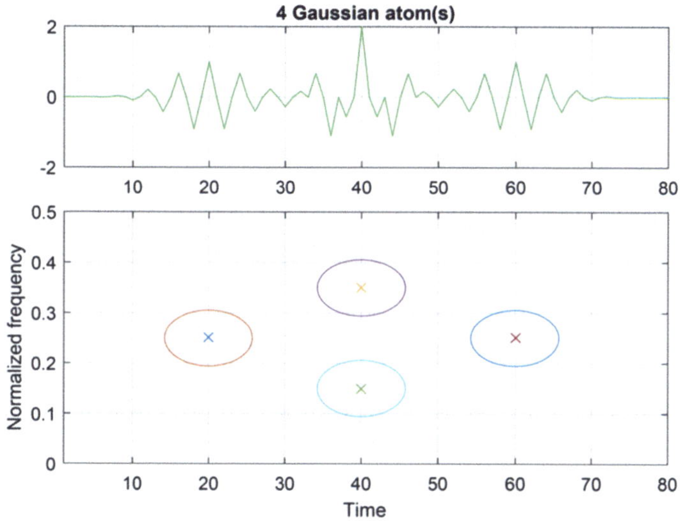

Fig. 1 An idealized TF representation of the sum of four Gaussian windows

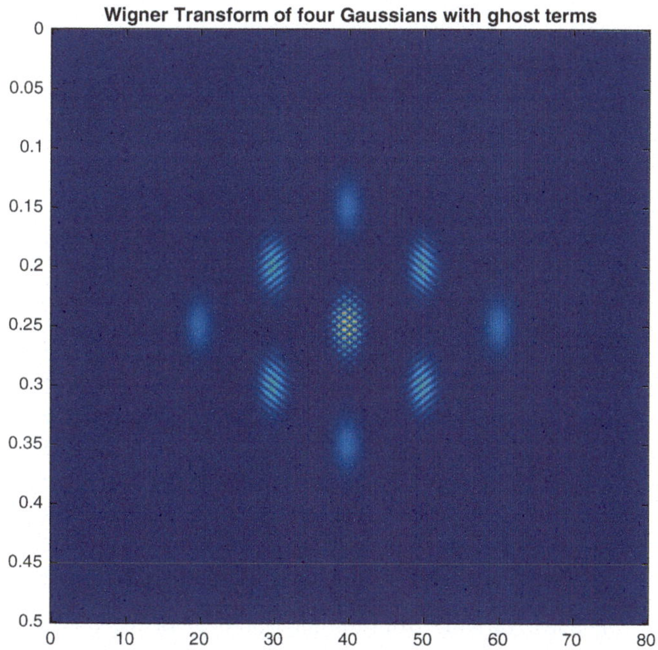

Fig. 2 Wigner distribution of the sum of four Gaussian windows

and since with $\varphi_i(t) = \pi(x_i, \omega_i)\varphi$, the cross-term $W(\varphi_i, \varphi_j)$ is given by

$$W(\varphi_i, \varphi_j)(x, \omega) = 2e^{[-\frac{\pi}{2}(x-\frac{x_i+x_j}{2})^2 - 2\pi(\omega-\frac{\omega_i+\omega_j}{2})^2]} e^{2\pi i[(\omega_i+\omega_j)x-(x_i-x_j)(\frac{\omega_i+\omega_j}{2})]}. \tag{1.4}$$

As depicted in Fig. 2, we thus see six cross-terms, of which two overlap in the centre of the four Gaussian windows.

The four spots placed at the vertices of the rhombus are the real time–frequency content of the signal, whereas the other five spots represent the "ghost frequencies", as defined in [3], arising from the interferences of the four components of the signal.

To overcome this undesirable phenomenon, *reduced interference distributions* were proposed by Leon Cohen in [6], see also the textbooks [7, 18].

The idea underneath the introduction of the so-called Cohen class is to reduce the interferences by a smoothing operation obtained using a convolution product. Precisely, we define the Cohen class a follows.

Definition 1 A member of the Cohen class Qf is a quadratic time–frequency representation obtained by convolving the Wigner function Wf with a distribution $\theta \in \mathcal{S}'(\mathbb{R}^{2d})$ (called Cohen kernel), that is,

$$Qf = Wf * \theta, \quad f \in \mathcal{S}(\mathbb{R}^d). \tag{1.5}$$

For $f \in \mathcal{S}(\mathbb{R}^d)$, it is easy to check that $Wf \in \mathcal{S}(\mathbb{R}^{2d})$ (see, e.g., [17]) so that Qf is a well-defined tempered distribution for every $f \in \mathcal{S}(\mathbb{R}^d)$.

A possible choice for the Cohen kernel is $\theta = \mathcal{F}_\sigma\Theta^1$, with $\mathcal{F}_\sigma\Theta^1$ being the symplectic Fourier transform of the function

$$\Theta^1(x, \omega) = \operatorname{sinc}(x\omega) = \begin{cases} \dfrac{\sin(\pi x\omega)}{\pi x\omega} & \text{for } x\omega \neq 0 \\ 1 & \text{for } x\omega = 0. \end{cases} \tag{1.6}$$

In this way, we obtain the Born–Jordan (BJ) distribution:

$$Q^1 f = Wf * \mathcal{F}_\sigma(\Theta^1), \quad f \in L^2(\mathbb{R}^d), \tag{1.7}$$

see [3, 6–8, 10, 18, 24] and the references therein.

Such distribution was first introduced in 1925 as a quantization rule by the physicists M. Born and P. Jordan [5] and later widely employed by engineers for its *smoothing effects*, cf. the textbook [18].

Examples for tests and real-world signals show in the BJ distribution:

(a) "ghost frequencies" (arising from the interferences of the several components that do not share the same time or frequency localization) are damped very well, see [3, 4];
(b) the interferences arranged along the horizontal and vertical directions are substantially kept;
(c) the noise is, on the whole, reduced.

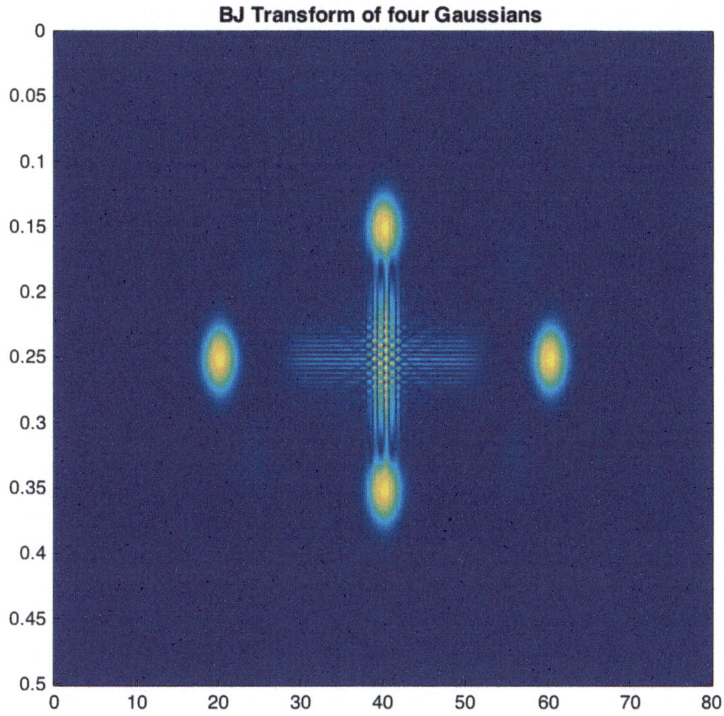

Fig. 3 The BJ distribution of the signal f in (1.2)

Figure 3 represents the BJ distribution of the signal f in (1.2). The smoothing effect of the BJ distribution can be seen quite clearly. But still there are artefacts that one would like to damp. This is the main purpose of the contribution [13], where new distributions from the Cohen class were proposed and studied. Here, we present the most relevant issues of the contribution above, highlighting the mathematical explanation of the smoothing effects of the distributions.

Following the suggestions of Jean-Pierre Gazeau, in the manuscript [13], new interesting Cohen kernels were proposed using the B-spline functions B_n.

The sequence of B-splines $\{B_n\}_{n\in\mathbb{N}_+}$ is defined inductively as follows. The first B-spline is

$$B_1(t) = \chi_{\left[-\frac{1}{2},\frac{1}{2}\right]}(t),$$

and the spline B_{n+1} is

$$B_{n+1}(t) = (B_n * B_1)(t) = \int_{\mathbb{R}} B_n(t-y)B_1(y)dy = \int_{-\frac{1}{2}}^{\frac{1}{2}} B_n(t-y)dy. \tag{1.8}$$

The spline B_n is a piecewise polynomial of degree at most $n-1$, $n \in \mathbb{N}_+$, and satisfying $B_n \in C^{n-2}(\mathbb{R})$, $n \geq 2$. For the main properties, we refer, e.g., to [16].

The sinc function is recaptured as $\mathrm{sinc}(\xi) = \mathcal{F}B_1(\xi)$, and by induction we infer

$$\mathrm{sinc}^n(\xi) = \mathcal{F}B_n(\xi), \quad n \in \mathbb{N}_+. \tag{1.9}$$

This suggests the following definition.

Definition 2 For $n \in \mathbb{N}$, the n-th Born–Jordan kernel on $\mathbb{R}^{2d}$ is defined by

$$\Theta^n(x,\omega) = \mathrm{sinc}^n(x\omega), \quad (x,\omega) \in \mathbb{R}^{2d}. \tag{1.10}$$

The related Born–Jordan distribution of order n (BJDn) is

$$Q^n f = Wf * \mathcal{F}_\sigma(\Theta^n), \quad f \in L^2(\mathbb{R}^d). \tag{1.11}$$

The cross-BJDn is given by

$$Q^n(f,g) = W(f,g) * \mathcal{F}_\sigma(\Theta^n), \quad f,g \in L^2(\mathbb{R}^d). \tag{1.12}$$

We write $Q^n(f,f) = Q^n f$, for every $f \in L^2(\mathbb{R}^d)$.

For $n = 0$, $\Theta^0 \equiv 1$ and $\mathcal{F}_\sigma(1) = \delta$, so that $Q^0 f = Wf$, the Wigner distribution of the signal f.

In Fig. 4, we computed the Born–Jordan distribution of order 3 of the signal in (1.2).

Notice that this new class is a subset of the Cohen classes, containing the Wigner and classical BJ distributions. This subclass can be applied to signal processing since the mathematical explanations of their smoothing properties testify the numerical evidences of dumping artefacts in many examples, and such reduction increases with n. We survey the different facets of this phenomenon, referring mainly to the theoretical results in [13], highlighted by new pictures in this chapter.

First, in Figs. 5, 6, and 7, we show the Wigner, BJ, and BJ3 distributions of a rotation of the original signal in (1.2). Figure 8 shows a comparison of the Wigner transform, the Born–Jordan transform and the fifth Born–Jordan transform of another sum of four time–frequency shifts of Gaussian functions. It is clearly visible that the amount of cross-term suppression increases by applying higher-order smoothing.

This suggests that such distributions could be successfully applied in signal processing.

In the following sections, we provide a rigorous mathematical explanation of the regularity and smoothness properties of Q^n; the notion of Fourier Lebesgue wave-front set will play the central role in showing the damping of interferences of Q^n in comparison with the Wigner distribution. The use of function spaces from time–frequency analysis, combined with microlocal analysis techniques, is the key tool of our proofs.

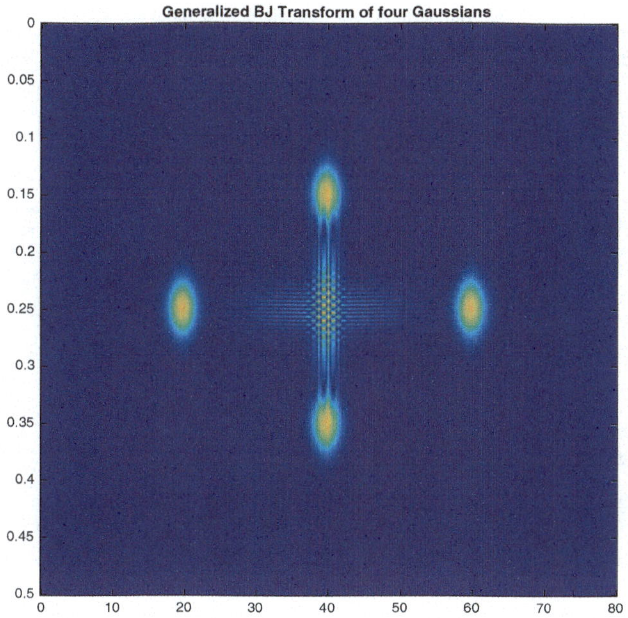

Fig. 4 The BJ distribution of order 3 of the signal in (1.2)

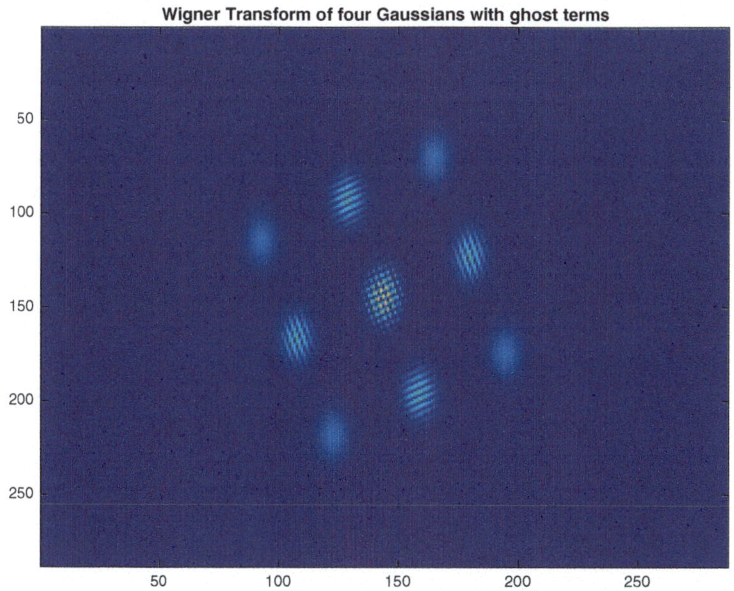

Fig. 5 The Wigner transform shows 9 spots, only 4 spots represent the real signal, and the others are ghost terms

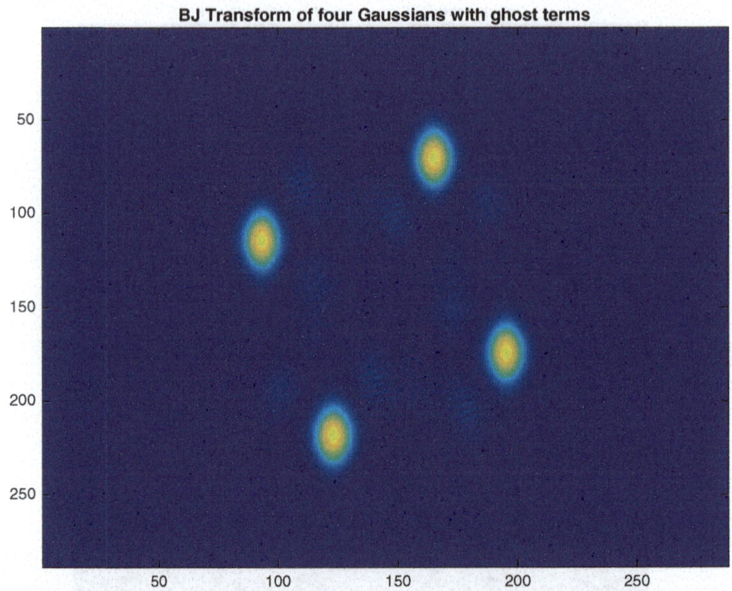

Fig. 6 The BJ transform of the previous signal gives a better information than the Wigner one. Here, one sees 4 spots that represent the time–frequency content of the signal though few little shades appear as interferences

1.1 Notation

We use $x\omega = x \cdot \omega = x_1\omega_1 + \ldots + x_d\omega_d$ for the scalar product in $\mathbb{R}^d$. The brackets $\langle \cdot, \cdot \rangle$ denote the inner product in $L^2(\mathbb{R}^d)$ or the duality pairing between Schwartz functions and temperate distributions (antilinear in the second argument). For functions f, g, we write $f \lesssim g$ if $f(x) \leq Cg(x)$ for every x and some constant $C > 0$, and similarly for $\gtrsim$. The notation $f \asymp g$ means $f \lesssim g$ and $f \gtrsim g$. We write $C_c^\infty(\mathbb{R}^d)$ for the class of smooth functions on $\mathbb{R}^d$ with compact support. The notation σ stands for the standard symplectic form on the phase space $\mathbb{R}^{2d} \equiv \mathbb{R}^d \times \mathbb{R}^d$; the phase space variable is denoted by $z = (x, \omega)$ and the dual variable by $\zeta = (\zeta_1, \zeta_2)$. By definition, $\sigma(z, \zeta) = Jz \cdot \zeta = \omega \cdot \zeta_1 - x \cdot \zeta_2$, where

$$J = \begin{pmatrix} 0_{d\times d} & I_{d\times d} \\ -I_{d\times d} & 0_{d\times d} \end{pmatrix}.$$

The Fourier transform of a function f in $\mathbb{R}^d$ is

$$\mathcal{F}f(\omega) = \widehat{f}(\omega) = \int_{\mathbb{R}^d} e^{-2\pi i x\omega} f(x)\, dx,$$

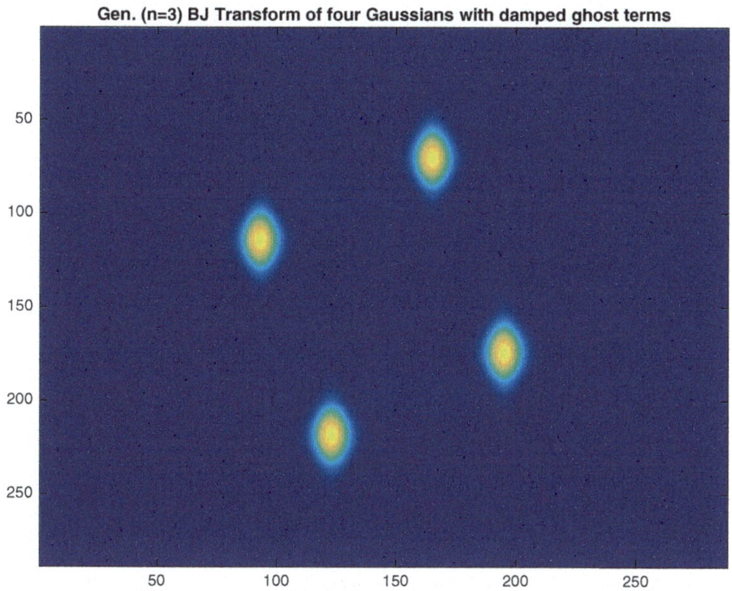

Fig. 7 The generalized BJ transform (order $n = 3$) of the same signal shows the best time–frequency content of it. The interferences are almost disappeared

and the symplectic Fourier transform of a function F in the phase space $\mathbb{R}^{2d}$ is

$$\mathcal{F}_\sigma F(\zeta) = \int_{\mathbb{R}^{2d}} e^{-2\pi i\sigma(\zeta,z)} F(z)\, dz.$$

The symplectic Fourier transform is an involution, i.e., $\mathcal{F}_\sigma(\mathcal{F}_\sigma F) = F$. Moreover, $\mathcal{F}_\sigma F(\zeta) = \mathcal{F} F(J\zeta)$.

Observe that $\Theta^n(J(\zeta_1, \zeta_2)) = \Theta^n(\zeta_1, \zeta_2)$ so that

$$\mathcal{F}_\sigma(\Theta^n) = \mathcal{F}(\Theta^n), \quad \forall n \in \mathbb{N}_+. \tag{1.13}$$

2 Time–Frequency and Microlocal Analysis Methods

In this section, we recall the function spaces from time–frequency analysis and the wave-front set from microlocal analysis that play the key role in this study.

Modulation Spaces Let us first recall another very popular time–frequency representation: *the short-time Fourier transform (STFT)*. Fix a Schwartz function

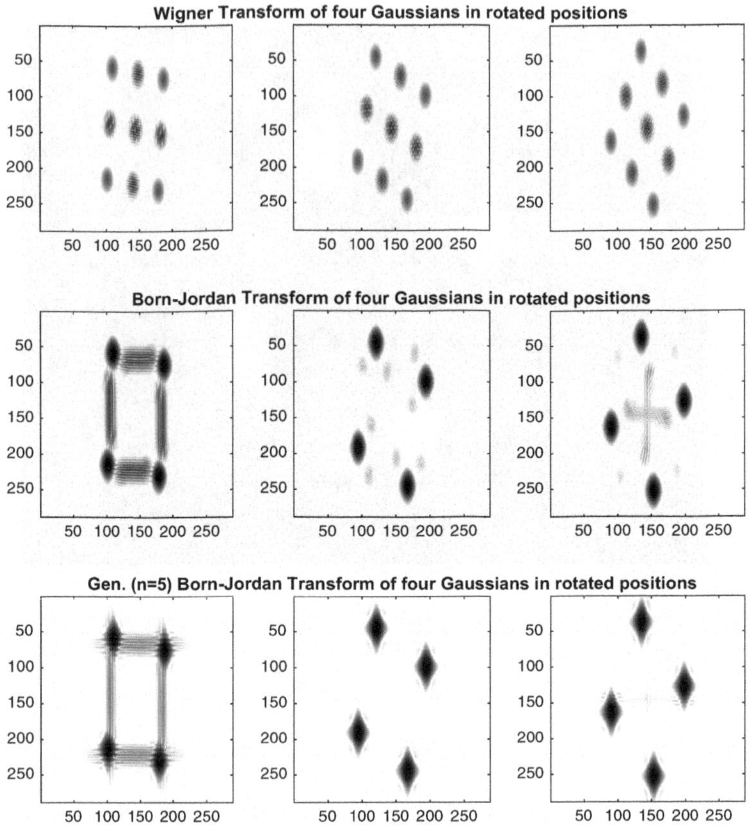

Fig. 8 Sum of four Gaussian functions in rotated positions: Comparison of Wigner distribution, Born–Jordan and generalized Born–Jordan distributions of order $n = 5$

$g \in \mathcal{S}(\mathbb{R}^d) \setminus \{0\}$ (the so-called window), and then the short-time Fourier transform of $f \in c\mathcal{S}'(\mathbb{R}^d)$ is

$$V_g f(x, \omega) = \int_{\mathbb{R}^d} f(y)\, \overline{g(y - x)}\, e^{-2\pi i y\omega}\, dy, \quad (x, \omega) \in \mathbb{R}^{2d}. \tag{2.1}$$

Consider $1 \leq p, q \leq \infty$. The *modulation space* $M^{p,q}(\mathbb{R}^d)$ consists of all tempered distributions $f \in \mathcal{S}'(\mathbb{R}^d)$ such that

$$\|f\|_{M^{p,q}} := \left(\int_{\mathbb{R}^d} \left(\int_{\mathbb{R}^d} |V_g f(x, \omega)|^p \, dx \right)^{q/p} d\omega \right)^{1/q} < \infty \tag{2.2}$$

(with obvious modifications for $p = \infty$ or $q = \infty$). We write $M^p(\mathbb{R}^d)$ instead of $M^{p,p}(\mathbb{R}^d)$. The modulation spaces are Banach spaces for any $1 \le p, q \le \infty$, and every non-zero $g \in \mathcal{S}(\mathbb{R}^d)$ yields an equivalent norm in (2.2).

Modulation spaces were introduced in [21] and are now available in textbooks, see, e.g., [23]. They include as special cases several function spaces arising in harmonic analysis. In particular for $p = q = 2$, we have

$$M^2(\mathbb{R}^d) = L^2(\mathbb{R}^d),$$

whereas $M^1(\mathbb{R}^d)$ is the Feichtinger algebra $S_0(\mathbb{R}^d)$, cf. [22, 23].

In the notation $M^{p,q}$, the exponent p is a measure of decay at infinity (on average) in the scale of spaces ℓ^p, whereas the exponent q is a measure of smoothness in the scale $\mathcal{F}L^q$.

Other instances of modulation spaces, also known as *Wiener amalgam spaces*, are obtained by exchanging the order of integration in (2.2). Namely, for $p, q \in [1, \infty)$, the modulation space $W(\mathcal{F}L^p, L^q)(\mathbb{R}^d)$ is the subspace of tempered distributions $f \in \mathcal{S}'(\mathbb{R}^d)$ such that

$$\|f\|_{W(\mathcal{F}L^p,L^q)(\mathbb{R}^d)} := \left(\int_{\mathbb{R}^d}\left(\int_{\mathbb{R}^d}|V_g f(x,\omega)|^p\,d\omega\right)^{q/p}dx\right)^{1/q} < \infty$$

(with obvious changes for $p = \infty$ or $q = \infty$). Using Parseval's identity in (2.1), we infer the fundamental identity of time–frequency analysis

$$V_g f(x, \omega) = e^{-2\pi i x\omega} V_{\widehat{g}} \widehat{f}(\omega, -x),$$

hence

$$|V_g f(x, \omega)| = |V_{\widehat{g}} \widehat{f}(\omega, -x)| = |\mathcal{F}(\widehat{f}\, T_\omega \overline{\widehat{g}})(-x)|,$$

so that

$$\|f\|_{M^{p,q}} = \left(\int_{\mathbb{R}^d} \|\widehat{f}\, T_\omega \overline{\widehat{g}}\|_{\mathcal{F}L^p}^q\, d\omega\right)^{1/q} = \|\widehat{f}\|_{W(\mathcal{F}L^p,L^q)}.$$

Hence, Wiener amalgam spaces can be viewed as the image under Fourier transform of modulation spaces: $\mathcal{F}(M^{p,q}) = W(\mathcal{F}L^p, L^q)$.

We will frequently use the following product property of Wiener amalgam spaces ([20, Theorem 1 (v)]): For $1 \le p, q \le \infty$,

$$\textit{if}\, f \in W(\mathcal{F}L^1, L^\infty)\ \textit{and}\, g \in W(\mathcal{F}L^p, L^q)\ \textit{then}\, fg \in W(\mathcal{F}L^p, L^q). \tag{2.3}$$

Taking $p = 1, q = \infty$, we obtain that $W(\mathcal{F}L^1, L^\infty)(\mathbb{R}^{2d})$ is an algebra under pointwise multiplication.

Proposition 1 *Let* $1 \le p, q \le \infty$ *and* $A \in GL(d, \mathbb{R})$*. Then, for every* $f \in W(\mathcal{F}L^p, L^q)(\mathbb{R}^d)$*,*

$$\|f(A\cdot)\|_{W(\mathcal{F}L^p,L^q)} \le C|\det A|^{(1/p-1/q-1)}(\det(I + A^*A))^{1/2}\|f\|_{W(\mathcal{F}L^p,L^q)}. \tag{2.4}$$

In particular, for $A = \lambda I$, $\lambda > 0$*,*

$$\|f(A\cdot)\|_{W(\mathcal{F}L^p,L^q)} \le C\lambda^{d\left(\frac{1}{p}-\frac{1}{q}-1\right)}(\lambda^2 + 1)^{d/2}\|f\|_{W(\mathcal{F}L^p,L^q)}. \tag{2.5}$$

In the sequel, we shall use the following fact [12, Lemma 5.1].

Lemma 1 *Let* $\chi \in C_c^\infty(\mathbb{R})$*. Then, the function* $\chi(\zeta_1\zeta_2)$ *belongs to* $W(\mathcal{F}L^1, L^\infty)(\mathbb{R}^{2d})$*.*

Wave-Front Set for Fourier–Lebesgue Spaces For $s \in \mathbb{R}$, the Sobolev space $H^s(\mathbb{R}^d)$ is constituted by the distributions $f \in \mathcal{S}'(\mathbb{R}^d)$ such that

$$\|f\|_{H^s} := \|\widehat{f}(\omega)\langle\omega\rangle^s\|_{L^2} < \infty, \tag{2.6}$$

where $\langle\omega\rangle = (1 + |\omega|^2)^{1/2}$. The H^s wave-front set allows to quantify the regularity of a function/distribution in the Sobolev scale, at any given point and direction. This is achieved by microlocalizing the definition of the H^s norm in (2.6) as follows (cf. [25, Chapter XIII]).

Definition 3 Given a tempered distribution $f \in \mathcal{S}'(\mathbb{R}^d)$, its wave-front set $WF_{H^s}(f) \subseteq \mathbb{R}^d \times (\mathbb{R}^d \setminus \{0\})$ is the set of points $(x_0, \omega_0) \in \mathbb{R}^d \times \mathbb{R}^d$, $\omega_0 \ne 0$, where the following condition is *not* satisfied: for some cut-off function $\varphi \in C_c^\infty(\mathbb{R}^d)$ with $\varphi(x_0) \ne 0$ and some open conic neighbourhood of $\Gamma \subseteq \mathbb{R}^d \setminus \{0\}$ of ω_0, we have

$$\|\mathcal{F}[\varphi f](\omega)\langle\omega\rangle^s\|_{L^2(\Gamma)} < \infty.$$

More generally, one can start from the Fourier–Lebesgue spaces $\mathcal{F}L_s^q(\mathbb{R}^d)$, $s \in \mathbb{R}$, $1 \le q \le \infty$, which is the space of distributions $f \in \mathcal{S}'(\mathbb{R}^d)$ such that the norm

$$\|f\|_{\mathcal{F}L_s^q(\mathbb{R}^d)} = \|\widehat{f}(\omega)\langle\omega\rangle^s\|_{L^q(\mathbb{R}^d)} \tag{2.7}$$

is finite.

Arguing exactly as in Definition 3 with the space L^2 replaced by L^q, one can introduce the corresponding notion of wave-front set $WF_{\mathcal{F}L_s^q}(f)$.

Definition 4 Given $f \in \mathcal{S}'(\mathbb{R}^d)$, its wave-front set $WF_{\mathcal{F}L_s^q}(f) \subseteq \mathbb{R}^d \times (\mathbb{R}^d \setminus \{0\})$ is the set of points $(x_0, \omega_0) \in \mathbb{R}^d \times \mathbb{R}^d$, $\omega_0 \ne 0$, where the following condition *is not satisfied*: for some cut-off function φ (i.e., φ is smooth and compactly supported on $\mathbb{R}^d$), with $\varphi(x_0) \ne 0$, and some open conic neighbourhood $\Gamma \subseteq \mathbb{R}^d \setminus \{0\}$ of ω_0, it holds

$$\|\mathcal{F}[\varphi f](\omega)\langle\omega\rangle^s\|_{L^q(\Gamma)} < \infty. \tag{2.8}$$

Observe that $WF_{\mathcal{F}L^2_s}(f) = WF_{H^s}(f)$ is the standard H^s wave-front set in Definition 3.

For our purposes, we recall some basic results about the action of constant coefficient linear partial differential operators on such wave-front set (cf. [26]). Given the operator

$$P = \sum_{|\alpha|\leq m} c_\alpha \partial^\alpha, \quad c_\alpha \in \mathbb{C},$$

for $1 \leq q \leq \infty$, $s \in \mathbb{R}$, $f \in \mathcal{S}'(\mathbb{R}^d)$,

$$WF_{\mathcal{F}L^q_s}(Pf) \subseteq WF_{\mathcal{F}L^q_{s+m}}(f).$$

Consider now the inverse inclusion. We say that $\zeta \in \mathbb{R}^d$, $\zeta \neq 0$, is noncharacteristic for the operator P if

$$\sum_{|\alpha|=m} c_\alpha \zeta^\alpha \neq 0.$$

The following result is a microlocal version of the classical regularity result of elliptic operators (see [26, Corollary 1 (2)]).

Proposition 2 *Let $1 \leq q \leq \infty$, $s \in \mathbb{R}$ and $f \in \mathcal{S}'(\mathbb{R}^d)$. Let $z \in \mathbb{R}^d$, and assume that $\zeta \in \mathbb{R}^d \setminus \{0\}$ is noncharacteristic for P. Then, if $(z, \zeta) \notin WF_{\mathcal{F}L^q_s}(Pf)$, we have $(z, \zeta) \notin WF_{\mathcal{F}L^q_{s+m}}(f)$.*

3 Time–Frequency Analysis of the n-th Born–Jordan Kernel

In this section, we summarize the main results of the topic, obtained in the papers [11–13]. The Born–Jordan kernel Θ^1 in (1.6) satisfies

$$\Theta^1 \in W(\mathcal{F}L^1, L^\infty)(\mathbb{R}^{2d}),$$

cf. [12].

The previous property is true for any Θ^n, $n \in \mathbb{N}_+$, as shown below (see [13]).

Proposition 3 *For $n \in \mathbb{N}_+$, the function Θ^n in* (1.10) *belongs to the Wiener algebra $W(\mathcal{F}L^1, L^\infty)(\mathbb{R}^{2d})$.*

The properties of the kernels above yield the following result.

Theorem 3.1 *Let $f \in \mathcal{S}'(\mathbb{R}^d)$ be a signal, with $Wf \in M^{p,q}(\mathbb{R}^{2d})$ for some $1 \le p, q \le \infty$. Then, $Q^n f \in M^{p,q}(\mathbb{R}^{2d})$, for every $n \in \mathbb{N}_+$.*

Proof We need to show that $Q^n f$ is in $M^{p,q}(\mathbb{R}^{2d})$. Taking the symplectic Fourier transform in (1.7), this is equivalent to

$$\Theta^n \mathcal{F}_\sigma(Wf) = \Theta^n Af \in W(\mathcal{F}L^p, L^q),$$

where $\mathcal{F}_\sigma(Wf) = Af$ is called the ambiguity function of f, see, e.g., [23]. The claim is attained using the product property (2.3): by Proposition 3, the function $\Theta^n \in W(\mathcal{F}L^1, L^\infty)$ and by assumption $Wf \in M^{p,q}(\mathbb{R}^{2d})$ so that we infer $\mathcal{F}(Wf) \in W(\mathcal{F}L^p, L^q)$. Finally, $\mathcal{F}_\sigma(Wf)(\zeta) = \mathcal{F}(Wf)(J\zeta) \in W(\mathcal{F}L^p, L^q)$ by Proposition 1. □

The previous statement holds in greater generality and can be rephrased for members in the Cohen class as follows.

Theorem 3.2 *Let $f \in \mathcal{S}'(\mathbb{R}^d)$ be a signal, with $Wf \in M^{p,q}(\mathbb{R}^{2d})$ for some $1 \le p, q \le \infty$ and the Cohen kernel (cf. (1.5)) $\theta \in M^{1,\infty}(\mathbb{R}^{2d})$. Then, the corresponding distribution Qf is in $M^{p,q}(\mathbb{R}^{2d})$.*

Proof It is the consequence of the following convolution relation for modulation spaces in [9]:

$$M^{p,q}(\mathbb{R}^{2d}) * M^{1,\infty}(\mathbb{R}^{2d}) \hookrightarrow M^{p,q}(\mathbb{R}^{2d}),$$

for any $1 \le p, q \le \infty$. □

The chirp function $F(\zeta_1, \zeta_2) = e^{2\pi i \zeta_1 \zeta_2}$ enjoys the following property, see [12, 22].

Proposition 4 *The function $F(\zeta_1, \zeta_2) = e^{2\pi i \zeta_1 \zeta_2}$ belongs to $W(\mathcal{F}L^1, L^\infty)(\mathbb{R}^{2d})$.*

By Proposition 4 and by the dilation properties for Wiener amalgam spaces in (2.4), we obtain

Corollary 1 *For $\zeta = (\zeta_1, \zeta_2)$, consider the function $F_J(\zeta) = F(J\zeta) = e^{-2\pi i \zeta_1 \zeta_2}$. Then, $F_J \in W(\mathcal{F}L^1, L^\infty)(\mathbb{R}^{2d})$.*

4 Smoothness of the Born–Jordan Distribution of Order *n*

In this section, we survey the results in [13, Sec. 5], comparing the smoothness of the Born–Jordan distribution of order n with the Wigner distribution. To measure the smoothness of a signal f, we use modulation spaces. They detect the concentration of functions and distributions on the time–frequency plane. Roughly speaking, a distribution $f \in \mathcal{S}'(\mathbb{R}^d)$ belongs to $M^{p,q}(\mathbb{R}^d)$ if it decays at infinity like a function in $L^p(\mathbb{R}^d)$, whereas it displays a $\mathcal{F}L^q(\mathbb{R}^d)$-local regularity.

For the following global result, we use the notation

$$\nabla_x \cdot \nabla_\omega := \sum_{j=1}^{d} \frac{\partial^2}{\partial x_j \partial \omega_j}. \tag{4.1}$$

Theorem 4.1 *Let $f \in \mathcal{S}'(\mathbb{R}^d)$ be a signal, with $Wf \in M^{p,q}(\mathbb{R}^{2d})$ for some $1 \leq p, q \leq \infty$. Then, for any $n \in \mathbb{N}_+$,*

$$Q^n f \in M^{p,q}(\mathbb{R}^{2d})$$

and

$$(\nabla_x \cdot \nabla_\omega)^n Q^n f \in M^{p,q}(\mathbb{R}^{2d}). \tag{4.2}$$

Proof The claim $Q^n f \in M^{p,q}(\mathbb{R}^{2d})$ is proven in Theorem 3.1.

We now prove (4.2). Taking the symplectic Fourier transform, we see that it is sufficient to prove that

$$(\zeta_1\zeta_2)^n \operatorname{sinc}^n(\zeta_1\zeta_2)\mathcal{F}_\sigma Wf = \frac{1}{\pi^n} \sin^n(\pi\zeta_1\zeta_2)\mathcal{F}_\sigma Wf \in W(\mathcal{F}L^p, L^q).$$

Using

$$\sin(\pi\zeta_1\zeta_2) = \frac{e^{\pi i\zeta_1\zeta_2} - e^{-\pi i\zeta_1\zeta_2}}{2i} \tag{4.3}$$

and applying Proposition 4, Corollary 1 and Proposition 1, with the scaling $\lambda = 1/\sqrt{2}$, we get $\sin(\pi\zeta_1\zeta_2) \in W(\mathcal{F}L^1, L^\infty)$.

Hence, for $n = 1$,

$$\frac{1}{\pi} \sin(\pi\zeta_1\zeta_2)\mathcal{F}_\sigma Wf \in W(\mathcal{F}L^p, L^q)$$

by the product property (2.3). Assume now that, for a certain $n \in \mathbb{N}_+$,

$$\frac{1}{\pi^n} \sin^n(\pi\zeta_1\zeta_2)\mathcal{F}_\sigma Wf \in W(\mathcal{F}L^p, L^q).$$

Then,

$$\frac{1}{\pi^{n+1}} \sin^{n+1}(\pi\zeta_1\zeta_2)\mathcal{F}_\sigma Wf = \underbrace{\frac{1}{\pi} \sin(\pi\zeta_1\zeta_2)}_{\in W(\mathcal{F}L^1, L^\infty)} \cdot \underbrace{\frac{1}{\pi^n} \sin^n(\pi\zeta_1\zeta_2)\mathcal{F}_\sigma Wf}_{\in W(\mathcal{F}L^p, L^q)} \in W(\mathcal{F}L^p, L^q),$$

by (4.3) and the product property (2.3) again. By induction, we attain the result. □

We can now provide the mathematical explanation of the Q^n's smoothing effects.

Theorem 4.2 *Let $f \in \mathcal{S}'(\mathbb{R}^d)$ be a signal, with $Wf \in M^{\infty,q}(\mathbb{R}^{2d})$ for some $1 \leq q \leq \infty$. Let $(z, \zeta) \in \mathbb{R}^{2d} \times \mathbb{R}^{2d}$, with $\zeta = (\zeta_1, \zeta_2)$ satisfying $\zeta_1 \cdot \zeta_2 \neq 0$. Then,*

$$(z, \zeta) \notin WF_{\mathcal{F}L^q_{2n}}(Q^n f).$$

Proof Consider $n \in \mathbb{N}_+$. We will apply Proposition 2 to the $2n$-th order operator P^n, where P is defined in (4.1). The noncharacteristic directions for P^n are given by the vectors $\zeta = (\zeta_1, \zeta_2) \in \mathbb{R}^d \times \mathbb{R}^d$, satisfying $\zeta_1 \cdot \zeta_2 \neq 0$. By (4.2) (with $p = \infty$), we have

$$WF_{\mathcal{F}L^q}(P^n Q^n f) = \emptyset$$

because $\varphi F \in \mathcal{F}L^q$ if $\varphi \in C_c^\infty(\mathbb{R}^{2d})$ and $F \in M^{\infty,q}(\mathbb{R}^{2d})$ (with $F = P^n Q^n f$). This implies

$$(z, \zeta) \notin WF_{\mathcal{F}L^q}(P^n Q^n f), \quad \forall (z, \zeta) \text{ such that } \zeta = (\zeta_1, \zeta_2), \ \zeta_1 \cdot \zeta_2 \neq 0.$$

Since ζ is noncharacteristic for the operator P^n, by Proposition 2, we infer

$$(z, \zeta) \notin WF_{\mathcal{F}L^q_{2n}}(Q^n f)$$

for every $z \in \mathbb{R}^{2d}$. □

Roughly speaking, if the Wigner distribution Wf has local regularity $\mathcal{F}L^q$ and some control at infinity, then $Q^n f$ is smoother, possessing $s = 2n$ **additional derivatives**, at least in the directions $\zeta = (\zeta_1, \zeta_2)$ satisfying $\zeta_1 \cdot \zeta_2 \neq 0$. In dimension $d = 1$, this condition reduces to $\zeta_1 \neq 0$ and $\zeta_2 \neq 0$. Hence, this result explains the smoothing phenomenon of such distributions, which involves all the directions except those of the coordinates axes.

This is the reason why the interferences of two components that do not share the same time or frequency localization come out substantially reduced.

Observe that for $n = 1$, we recapture the damping phenomenon of the classical Born–Jordan distribution (cf. [12, Theorem 1.2]).

For signals in $L^2(\mathbb{R}^d)$, the previous result can be rephrased in terms of the classical Hörmander's wave-front set.

Corollary 2 *Let $f \in L^2(\mathbb{R}^d)$, so that $Wf \in L^2(\mathbb{R}^{2d})$. Let (z, ζ) be as in the statement of Theorem 4.2. Then, $(z, \zeta) \notin WF_{H^{2n}}(Q^n f)$, that is the distribution $Q^n f$ has regularity H^{2n} at z and in the direction ζ.*

Proof We apply Theorem 4.2 with $q = 2$. In fact, for $f \in L^2(\mathbb{R}^d)$, Moyal's formula gives $Wf \in L^2(\mathbb{R}^{2d}) = M^{2,2}(\mathbb{R}^d) \subseteq M^{\infty,2}(\mathbb{R}^{2d})$, by inclusion relations for modulation spaces. Observe that the $\mathcal{F}L^2_{2n}$ wave-front set coincides with the H^{2n} wave-front set. □

What about smoothing effects in the directions $\zeta = (\zeta_1, \zeta_2)$: $\zeta_1 \cdot \zeta_2 = 0$?

It seems that the smoothing effects do not occur in these directions, as Fig. 1 shows. The mathematical explanation in terms of modulation spaces is below. Roughly speaking, if we assume that the distribution $Q^n f$ of a signal f has a local regularity $\mathcal{F}L^{q_1}$ better than the $\mathcal{F}L^{q_2}$-Wigner one, then necessarily it must hold $q_1 \geq q_2$. Hence, the best we could expect is $q_1 = q_2$, that is, the same regularity.

Theorem 4.3 *Suppose that for some* $1 \leq p, q_1, q_2 \leq \infty$, $n \in \mathbb{N}_+$ *and* $C > 0$, *it occurs*

$$\|Q^n f\|_{M^{p,q_1}} \leq C\|Wf\|_{M^{p,q_2}}, \tag{4.4}$$

for every $f \in \mathcal{S}(\mathbb{R}^d)$. *Then,* $q_1 \geq q_2$.

Proof The main steps are as follows. We will test the estimate (4.4) using rescaled Gaussian functions $f(x) = \varphi(\lambda x)$, with $\lambda > 0$ large parameter. Restricting to a neighbourhood of $\zeta_1 \cdot \zeta_2 = 0$, the constrain $q_1 \geq q_2$ must be satisfied.

An easy computation yields

$$W(\varphi(\lambda\,\cdot))(x,\omega) = 2^{d/2}\lambda^{-d}\varphi(\sqrt{2}\lambda\, x)\varphi(\sqrt{2}\lambda^{-1}\,\omega). \tag{4.5}$$

For every $1 \leq p, q \leq \infty$, the above formula gives

$$\|W(\varphi(\lambda\,\cdot))\|_{M^{p,q}} = 2^{d/2}\lambda^{-d}\|\varphi(\sqrt{2}\lambda\,\cdot)\|_{M^{p,q}}\|\varphi(\sqrt{2}\lambda^{-1}\,\cdot)\|_{M^{p,q}}.$$

By the dilation properties of Gaussians (first proved in [28, Lemma 1.8], see also [14, Lemma 3.2]),

$$\|W(\varphi(\lambda\,\cdot))\|_{M^{p,q}} \asymp \lambda^{-2d+d/q+d/p} \quad \text{as } \lambda \to +\infty. \tag{4.6}$$

We now study the $M^{p,q}$-norm of the BJDn $Q^n(\varphi(\lambda\,\cdot))$. It will be estimated from below obtaining the same expansion as in (4.6). In detail,

$$\|Q^n(\varphi(\lambda\,\cdot))\|_{M^{p,q}} = \|\mathcal{F}_\sigma(\Theta^n) * W(\varphi(\lambda\,\cdot))\|_{M^{p,q}}.$$

By taking the symplectic Fourier transform and using Lemma 1 and the product property (2.3), we have

$$\begin{aligned}\|\mathcal{F}_\sigma(\Theta^n) * W(\varphi(\lambda\,\cdot))\|_{M^{p,q}} &\asymp \|\Theta^n\mathcal{F}_\sigma[W(\varphi(\lambda\,\cdot))]\|_{W(\mathcal{F}L^p,L^q)}\\ &\gtrsim \|\Theta^n(\zeta_1,\zeta_2)\chi(\zeta_1\zeta_2)\mathcal{F}_\sigma[W(\varphi(\lambda\,\cdot))]\|_{W(\mathcal{F}L^p,L^q)}\end{aligned}$$

for any $\chi \in C_c^\infty(\mathbb{R})$ and $n \in \mathbb{N}_+$. Choosing the function χ compactly supported in the interval $[-1/4, 1/4]$ and $\chi \equiv 1$ in the interval $[-1/8, 1/8]$ (the latter condition will be used later), we write

$$\chi(\zeta_1\zeta_2) = \chi(\zeta_1\zeta_2)\Theta^n(\zeta_1, \zeta_2)\Theta^{-n}(\zeta_1, \zeta_2)\widetilde{\chi}(\zeta_1\zeta_2),$$

with $\widetilde{\chi} \in C_c^\infty(\mathbb{R})$ supported in $[-1/2, 1/2]$ and $\widetilde{\chi} = 1$ on $[-1/4, 1/4]$, therefore on the support of χ. Since by Lemma 1 the function $\Theta^{-n}(\zeta_1, \zeta_2)\widetilde{\chi}(\zeta_1\zeta_2)$ belongs to $W(\mathcal{F}L^1, L^\infty)$, by the product property, the last expression can be estimated from below as

$$\|\chi(\zeta_1\zeta_2)\|_{W(\mathcal{F}L^p, L^q)} \gtrsim \|\chi(\zeta_1\zeta_2)\mathcal{F}_\sigma[W(\varphi(\lambda\,\cdot))]\|_{W(\mathcal{F}L^p, L^q)}.$$

Finally in the proof of [12, Theorem 1.4], it was shown

$$\|\chi(\zeta_1\zeta_2)\mathcal{F}_\sigma[W(\varphi(\lambda\,\cdot))]\|_{W(\mathcal{F}L^p, L^q)} \gtrsim \lambda^{-2d+d/p+d/q} \quad \text{as } \lambda \to +\infty. \tag{4.7}$$

Comparing (4.7) with (4.6), we obtain the desired conclusion. □

5 Conclusion and Perspectives

The generalized Born–Jordan distributions presented in this chapter produce an improvement in the damping of unwanted artefacts of some signals as the one represented in Fig. 1. For other pictures of signals where the smoothing effects are visible, we refer to the original paper [13].

Let us underline that the emergence of interferences is a well-known drawback of any quadratic representation, the distributions BJDn are not immune to this phenomenon, as it can be seen in Figures 9 and 10, which are short extracts from real music signals. For comparison, we also used the spectrogram, which is another member of the Cohen class (see [3] and references therein), given by

$$|V_f f(x, \omega)|^2, \quad f \in L^2(\mathbb{R}^d).$$

Many other alternative time–frequency distributions have been proposed for different practical purposes, and we refer the interested reader to the textbooks [7, 18].

If we consider linear perturbations of the Wigner distribution, introduced and studied in [1, 2], then it turns out they do not provide effective damping of artefacts. In fact, a negative answer concerning reduction of interferences is shown in [15].

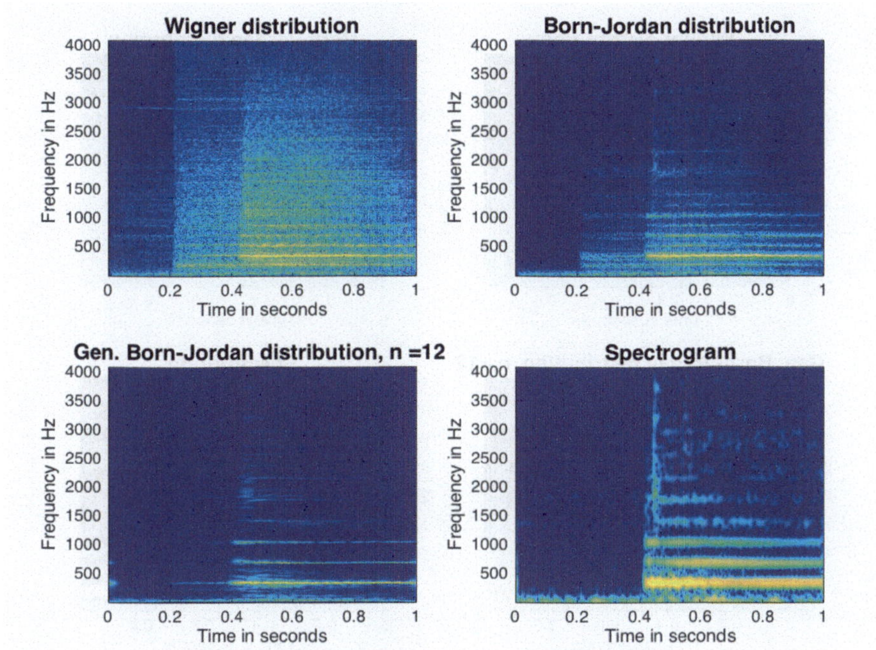

Fig. 9 A short extract of a musical signal

Although it is clear there is no time–frequency distribution that is the best time–frequency representation to analyse any kind of signal, it is an open problem to find the right distribution for a certain class of signals. For signals that are sums of time–frequency shifts of Gaussians, the representations BJDs work quite well, as shown in the present chapter. This suggests the following question:

Concerning the members of the Cohen class, what are the best possible kernels for damping artefacts?

Technical Notes The figures in this chapter were produced using LTFAT (Large Time–Frequency Analysis Toolbox), cf. [27].

Acknowledgement Monika Dörfler has been supported by the Vienna Science and Technology Fund (WWTF) through project MA14-018.

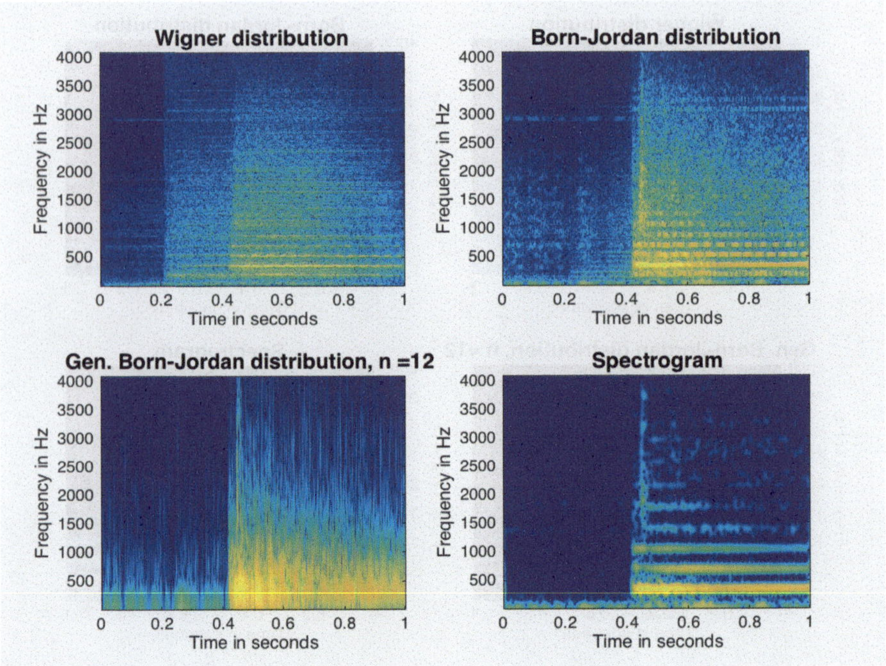

Fig. 10 An extract of another musical signal

References

1. Bayer, D.: Bilinear Time-Frequency Distributions and Pseudodifferential Operators. PhD Thesis, University of Vienna (2010)
2. Bayer, D., Cordero, E., Gröchenig, K., Trapasso, S. I.: Linear perturbations of the Wigner transform and the Weyl quantization in *Advances in Microlocal and Time-Frequency Analysis*, Applied and Numerical Harmonic Analysis, Birkhäuser/Springer, (2020) ISBN 978-3-030-36138-9
3. Boggiatto, P., De Donno, G., Oliaro, A.: Time-frequency representations of Wigner type and pseudo-differential operators. Trans. Amer. Math. Soc., **362**(9), 4955–4981 (2010)
4. Boggiatto, P., Carypis, E., Oliaro, A.: Windowed-Wigner representations in the Cohen class and uncertainty principles. J. Geom. Anal., **23**(4), 1753–1779 (2013)
5. Born, M., Jordan, P.: Zur Quantenmechanik, Zeits. Physik, **34**, 858–888 (1925)
6. Cohen, L.: Generalized phase-space distribution functions, J. Math. Phys., **7**, 781–786 (1966)
7. Cohen, L.: *Time Frequency Analysis: Theory and Applications*. Prentice Hall (1995)
8. Cohen, L.: *The Weyl operator and its generalization*. Springer Science & Business Media, (2012)
9. Cordero E., Gröchenig, K.: Time-frequency analysis of Localization operators, J. Funct. Anal., **205**(1), 107–131 (2003)
10. Cordero, E., de Gosson, M., Nicola, F.: On the invertibility of Born–Jordan quantization, J. Math. Pures Appl., **105**, 537–557 (2016)
11. Cordero, E., de Gosson, M., Nicola, F.: Time-frequency Analysis of Born–Jordan Pseudodifferential Operators, J. Funct. Anal., **272**(2), 577–598 (2017)

12. Cordero, E., de Gosson, M., Nicola, F.: On the reduction of the interferences in the Born–Jordan distribution, Appl. Comput. Harmon. Anal., **44**(2), 230–245 (2018)
13. Cordero, E., de Gosson, M., Döfler, M., Nicola, F.: Generalized Born–Jordan Distributions and Applications. Adv. Comput. Math., **46**(51), (2020). https://doi.org/10.1007/s10444-020-09788-w
14. Cordero, E., Nicola, F.: Metaplectic representation on Wiener amalgam spaces and applications to the Schrödinger equation, J. Funct. Anal., **254**, 506–534 (2008)
15. Cordero, E., Trapasso, S. I.: Linear perturbations of the Wigner distribution and the Cohen class, Anal. Appl., **18**(3), 385–422, (2020)
16. Christensen, O.: *An Introduction to Frames and Riesz Bases*. Birkhäuser, Basel, (2016)
17. de Gosson, M.: *The Wigner transform*. Advanced Textbooks in Mathematics, World Scientific Publishing Co. Pte. Ltd. Hackensack, NJ (2017)
18. Hlawatsch, F., Auger, F.: *Time-Frequency Analysis*. Wiley (2008).
19. Ville, J.: Théorie et Applications de la Notion de Signal Analytique. Câbles et Transmission, **2**, 61–74 (1948)
20. Feichtinger, H. G.: Banach convolution algebras of Wiener's type, In Proc. Conf. *Function, Series, Operators*, Budapest August 198, Colloq. Math. Soc. János Bolyai, **35** 509–524, North-Holland, Amsterdam (1983)
21. Feichtinger, H. G.: *Modulation spaces on locally compact abelian groups*, Technical Report, University Vienna (1983), and also in "Wavelets and Their Applications", M. Krishna, R. Radha, S. Thangavelu, editors, Allied Publishers, 99–140 (2003)
22. Feichtinger, H. G.: On a new Segal algebra, Monatshefte Math., **92**(4), 269–289 (1981).
23. Gröchenig, K.: *Foundation of Time-Frequency Analysis*, Birkhäuser, Boston MA (2001)
24. M. de Gosson, *Born–Jordan Quantization: Theory and Applications.*, Springer (2016)
25. Hörmander, L.: *Lectures on nonlinear hyperbolic differential equations*, Springer (1997)
26. Pilipović, S., Teofanov, N., Toft, J.: Wave front sets in Fourier Lebesgue spaces, Rend. Sem. Mat. Univ. Pol. Torino **66**(4), 299–319 (2008)
27. Prusa, Z., Sondergaard, P. L., Holighaus, N., Wiesmeyr, C., Balazs, P.: The Large Time-Frequency Analysis Toolbox 2.0. Sound, Music, and Motion, Lecture Notes in Computer Science (2014), pp 419-442
28. Toft, J.: Continuity properties for modulation spaces, with applications to pseudo-differential calculus. I, J. Funct. Anal., **207**(2), 399–429 (2004)
29. Wigner, E.: On the quantum correction for thermodynamic equilibrium. Phys. Rev. **40**, 799–755 (1932)

Sampling by Averages and Average Splines on Dirichlet Spaces and on Combinatorial Graphs

Isaac Z. Pesenson

Dedicated to the 80th Birthday of John Benedetto.

Abstract In the framework of a strictly local regular Dirichlet space $\mathbf{X}$ we introduce the subspaces PW_{ω}, $\omega > 0$, of Paley–Wiener functions of bandwidth ω. It is shown that every function in PW_{ω}, $\omega > 0$, is uniquely determined by its average values over a family of balls $B(x_j, \rho)$, $x_j \in \mathbf{X}$, which form an admissible cover of $\mathbf{X}$ and whose radii are comparable to $\omega^{-1/2}$. The entire development heavily depends on some Poincaré-type inequalities. In the second part of the paper we realize the same idea in the setting of a weighted combinatorial finite or infinite graph G. We have to treat the case of graphs separately since the Poincaré inequalities we are using on them are somewhat different from the Poincaré inequalities in the first part.

1 Introduction

We consider a metric measure space $\mathbf{X}$ with doubling property and a self-adjoint operator $\mathcal{L}$ in $L_2(\mathbf{X})$ which governs local geometry of $\mathbf{X}$ through a Poincare-type inequality. In fact, we are working in the environment of the so-called strictly local regular Dirichlet spaces [1, 3, 6, 23–25]. Following [16–18] we introduce the subspaces $PW_{\omega}(\mathcal{L})$, $\omega > 0$, of Paley–Wiener functions in $L_2(\mathbf{X})$ and then show that every function f in $PW_{\omega}(\mathcal{L})$, $\omega > 0$, is uniquely determined by its average values over a family of balls $B(x_j, \rho)$, $x_j \in \mathbf{X}$, which form an admissible cover of $\mathbf{X}$ and whose radii are comparable to $\omega^{-1/2}$. Reconstruction methods for reconstruction of an $f \in PW_{\omega}(\mathcal{L})$ from appropriate set of its averages are introduced. One method is using language of Hilbert frames. Another one is using average variational interpolating splines which are constructed in the setting of

I. Z. Pesenson (✉)
Department of Mathematics, Temple University, Philadelphia, PA, USA
e-mail: pesenson@temple.edu

M. Hirn et al. (eds.), *Excursions in Harmonic Analysis, Volume 6*, Applied and Numerical Harmonic Analysis, https://doi.org/10.1007/978-3-030-69637-5_14

metric measure spaces. It is shown that every Paley–Wiener function is a limit of specific variational splines which have the same averages over balls $B(x_j, \rho)$ as f.

In the second part of the paper we realize the same idea in the setting of a weighted combinatorial finite or infinite countable graph G. Although the second part is very close to the first one ideologically, it is somewhat different technically and can be read independently of the first part. Note, that a point wise sampling on combinatorial graphs was developed in [20–22].

In both parts of the paper we strongly rely on the local Poincaré (2.5) and a global Poincaré (3.7) and a local Poincaré-type (8.2) and a global Poincaré-type (8.3) inequalities. It is worth to notice that these inequalities in the case of graphs are not exactly the same as in the case of Dirichlet spaces. It explains why we treat graphs separately. A detailed comparison of Poincaré inequalities we are using on Dirichlet spaces and on graphs is given in Remark 4.

It should be mentioned that the idea to use certain local information (other than values at sampling point) for reconstruction of bandlimited functions on graphs was explored in [14, 15, 26, 27]. However, their results and methods are very different from ours. We also want to mention that methods of the present paper are similar to methods of our paper [19] in which sampling by average values and average splines were developed on Riemannian manifolds.

2 Metric Measure Spaces and Poincare Inequality

In this section we describe a class of metric measure spaces we are going to work with. Actually, we assume many properties but the most principal of them are: the doubling property (2.1), existence of a self-adjoint non-negative operator in the corresponding $L_2(\mathbf{X})$, and the Poincaré inequality (2.5).

Main Assumptions About Metric Measure Space

A homogeneous space in the sense of Coifman and Weiss [4, 5] is a triple $(\mathbf{X}, dist, \mu)$ where $\mathbf{X}$ is a set which is equipped with a metric $dist$ and a positive Radon measure μ such that the so-called doubling condition holds. Namely, there exists a $D > 0$ such that for every open ball $B(x, r)$ with center $x \in \mathbf{X}$ and radius $r > 0$

$$0 < |B(x, 2r)| \leq 2^D |B(x, r)|, \quad \text{where} \quad |B(x, s)| = \mu(B(x, s)). \tag{2.1}$$

It is very common to assume that $dist$ defines a locally compact separable topology on $\mathbf{X}$.

In the first part of our paper we will work with the so-called *strictly local regular Dirichlet spaces.* We just outline the main points of this framework and refer to [1, 3, 6, 23–25] where all the missing details can be found.

The major assumption is that the real space $L_2(\mathbf{X}, \mu)$ is equipped with a non-negative self-adjoint operator $\mathcal{L}$ whose domain $\mathcal{D}(\mathcal{L})$ is dense in $L_2(\mathbf{X}, \mu)$. We consider associated symmetric bilinear form $\mathcal{E}$ with domain $\mathcal{D}(\mathcal{E}) \supseteq \mathcal{D}(\mathcal{L})$

which is defined by the formula

$$\mathcal{E}(f,g) = \langle \mathcal{L}f, g\rangle = \langle f, \mathcal{L}g\rangle = \mathcal{E}(g,f),\ f,g \in \mathcal{D}(\mathcal{E}).$$

Assuming that $\mathcal{E}$ is a strongly local regular Dirichlet form one can show existence of a signed measure $d\Gamma$ which is a bilinear map defined on $\mathcal{D}(\mathcal{E}) \times \mathcal{D}(\mathcal{E})$ such that

$$\mathcal{E}(f,g) = \int_{\mathbf{X}} d\Gamma(f,g),\ f,g \in \mathcal{D}(\mathcal{E}).$$

In particular, for the non-negative square root $\mathcal{L}^{1/2}$ one has

$$\|\mathcal{L}^{1/2}f\|^2 = \mathcal{E}(f,f) = \int_{\mathbf{X}} d\Gamma(f,f),\ f \in \mathcal{D}(\mathcal{E}). \tag{2.2}$$

Remark 1 In one of the most real life situations when $\mathbf{X}$ is a Riemannian manifold and $\mathcal{L}$ is the corresponding Laplace–Beltrami operator the measure $d\Gamma(f,g)$ is given by a formula

$$d\Gamma(f,g)(x) = \Gamma(f,g)(x)d\mu, \tag{2.3}$$

where the function $\Gamma(f,g)$ for smooth compactly supported f and g is defined by the formula

$$\Gamma(f,g) = \frac{1}{2}\left(\mathcal{L}(fg) - f\mathcal{L}g - g\mathcal{L}f\right). \tag{2.4}$$

In fact, the last formula holds even in general case if one imposes some extra conditions on the domain of $\mathcal{L}$.

Our next important assumption is about **a local Poincaré inequality**.

Assumption 1 *There exists constant $C > 0$ such that for any $f \in \mathcal{D}(\mathcal{L})$ and any ball $B(x,r)$, $x \in \mathbf{X}$, of a sufficiently small radius r the following Poincare inequality holds:*

$$\int_{B(x,r)} |f(y) - f_B|^2 d\mu(y) \le Cr^2 \int_{B(x,r)} \Gamma(f,f)d\mu, \tag{2.5}$$

where

$$f_B = \frac{1}{|B(x,r)|}\int_{B(x,r)} f(x)d\mu(x).$$

The above assumptions constitute our framework for the next section. In Sect. 5 we will add another assumption to this list.

3 A Global Poincaré Inequality

3.1 A Covering Lemma

Lemma 1 *There exists a constant* $N = N(\mathbf{X})$ *and for every sufficiently small* $\rho > 0$ *there exists a set of points* $\mathbf{X}_\rho = \{x_j\},\ x_j \in \mathbf{X}$, *such that*

1. *balls* $B(x_j, \rho/2)$ *are disjoint* $B(x_j, \rho/2) \cap B(x_i, \rho/2) = \emptyset,\ j \neq i$,
2. *multiplicity of the cover* $\left\{B(x_j, \rho)\right\}$ *is not greater than* $N(\mathbf{X})$.

Proof Let us choose a family of disjoint balls $B(x_i, \rho/2)$ such that there is no ball $B(x, \rho/2), x \in M$, which has empty intersections with all balls from our family. Then the family $B(x_i, \rho)$ is a cover of M. Every ball from the family $\{B(x_i, \rho)\}$ that has non-empty intersection with a particular ball $B(x_j, \rho)$ is contained in the ball $B(x_j, 2\rho)$. Since any two balls from the family $\{B(x_i, \rho/2)\}$ are disjoint, it gives the following estimate for the index of multiplicity N of the cover $\{B(x_i, \rho)\}$:

$$N \leq \sup_{y\in\mathbf{X}} \frac{|B(y, 2\rho)|}{\inf_{x\in B(y,2\rho)} |B(x, \rho/2)|}. \tag{3.1}$$

Let us notice that for any $m \in \mathbf{N}$ the doubling property (2.1) implies

$$\left|B(x, 2^m r)\right| \leq 2^{mD}|B(x, r)|,\ x \in \mathbf{X},\ r > 0,\ \rho > 0,$$

and if for a $\rho > 0$ one has $2^m \leq \rho < 2^{m+1}$, then

$$|B(x, \rho r)| < |B(x, 2^{m+1} r)| \leq 2^{(m+1)D}|B(x, r)| \leq 2^D \rho^D |B(x, r)|. \tag{3.2}$$

Now, the last inequality along with the obvious inclusion $B(x, r) \subseteq B(y, r + dist(x, y))$ implies

$$|B(x, r)| \leq 2^D \left(1 + dist(x, y)/r\right)^D |B(y, r)|,\ x, y \in \mathbf{X},\ r > 0. \tag{3.3}$$

Let us return to the inequality (3.1). Since $x \in B(y, 2\rho)$ we have $dist(x, y) \leq 2\rho$ and according to (3.3)

$$|B(y, \rho/2)| \leq 2^D \left(1 + \frac{dist(x, y)}{\rho/2}\right)^D |B(x, \rho/2)| \leq 10^D |B(x, \rho/2)|.$$

In other words, for any $y \in \mathbf{X}$ and every $x \in B(y, 2\rho)$ one has

$$|B(x, \rho/2)| \geq 10^{-D}|B(y, \rho/2)|. \tag{3.4}$$

This inequality along with (3.2) allows to continue estimation of $N(\mathbf{X})$:

$$N(\mathbf{X}) \leq 10^D \sup_{y\in\mathbf{X}} \frac{|B(y,2\rho)|}{|B(x,\rho/2)|} \leq 10^D \sup_{y\in\mathbf{X}} \frac{8^D|B(y,\rho/2)|}{|B(y,\rho/2)|} = 80^D.$$

Lemma is proved.

□

Definition 1 Every set $X_\rho = \{x_j\}$, $x_j \in \mathbf{X}$, $\rho > 0$, which satisfies properties of the previous lemma will be called a **metric ρ-lattice**.

3.2 A Global Poincaré Inequality and Its Implications

We will need the inequality (3.5) below. One has for all $\alpha > 0$

$$|A|^2 \leq (|A-B|+|B|)^2 \leq |A-B|^2 + 2|A-B||B| + |B|^2,$$

and $2|A-B||B| \leq \alpha^{-1}|A-B|^2 + \alpha|B|^2$, $\alpha > 0$, which imply the inequality

$$(1+\alpha)^{-1}|A|^2 \leq \alpha^{-1}|A-B|^2 + |B|^2,\ \alpha > 0. \tag{3.5}$$

Let $X_\rho = \{x_j\}$ be a ρ-lattice and $\zeta_j(x) = |B(x_j,\rho)|^{-1}\chi_j(x)$, where χ_j is characteristic function of $B(x_j,\rho)$. We have

$$|f|^2 \leq (1+\alpha)\left|f - \langle f,\zeta_j\rangle\right|^2 + \frac{1+\alpha}{\alpha}\left|\langle f,\zeta_j\rangle\right|^2,\ \alpha > 0,$$

and then

$$\|f\|^2 \leq \sum_j \int_{B(x_j,\rho)} |f|^2 \leq (1+\alpha)\sum_j \int_{B(x_j,\rho)} |f - \langle f,\zeta_j\rangle|^2 + \frac{1+\alpha}{\alpha}\sum_j |\langle f,\xi_j\rangle|^2,$$

$$\xi_j = |B(x_j,\rho)|^{-1/2}\chi_j. \tag{3.6}$$

Since by the very definition of a ρ-lattice the corresponding cover by balls $B(x_j,\rho)$ has finite multiplicity $\leq N(\mathbf{X})$ we obtain according to (2.5) and (2.2)

$$\sum_j \int_{B(x_j,\rho)} |f - \langle f,\zeta_j\rangle|^2 \leq C\rho^2 \sum_j \int_{B(x_j,\rho)} d\Gamma(f,f) \leq$$

$$CN\rho^2 \int_{\mathbf{X}} \left|\mathcal{L}^{1/2}f\right|^2 \leq CN(\mathbf{X})\rho^2\|\mathcal{L}^{1/2}f\|^2.$$

Thus we can formulate the following result.

Theorem 3.1 *For every ρ-lattice $\{x_j\}$ and the corresponding set of functions $\{\xi_j\}$ defined in (3.6) the following inequality holds for any $f \in \mathcal{D}(\mathcal{L}^{1/2})$*

$$\|f\|^2 \le (1+\alpha)CN(\mathbf{X})\rho^2\|\mathcal{L}^{1/2}f\|^2 + \frac{1+\alpha}{\alpha}\sum_j |\langle f, \xi_j\rangle|^2, \; \alpha > 0, \tag{3.7}$$

where C is the same as in (2.5). We call this inequality ***a global Poincaré inequality***.

As a consequence we obtain the following statement of many zeros.

Theorem 3.2 *For every ρ-lattice $\{x_j\}$ and the corresponding set of functions $\{\xi_j\}$ defined in (3.6) the following inequality holds for any $f \in \mathcal{D}(\mathcal{L}^{1/2})$ such that $\langle f, \xi_j\rangle = 0$ for all j:*

$$\|f\|^2 \le CN(\mathbf{X})\rho^2\|\mathcal{L}^{1/2}f\|^2, \tag{3.8}$$

where C is the same as in (2.5).

To obtain another important consequence of Theorem 3.1 we need the following lemma which was proved in [7].

Lemma 2 *If T is a self-adjoint operator in a Hilbert space and for some φ in the domain of T*

$$\|\varphi\| \le A + a\|T\varphi\|, \; a > 0,$$

then for all $k = 2^l$, $l = 0, 1, 2, \ldots$

$$\|\varphi\| \le kA + 8^{k-1}a^k\|T^k\varphi\|, \tag{3.9}$$

as long as φ belongs to the domain of T^k.

Theorem 3.3 *For every ρ-lattice X_ρ and every $m = 2^l, l = 0, 1, 2, \ldots$ there exist constants $c_1 = c_1(\mathbf{X}, \rho, m)$, $C_1 = C_1(\mathbf{X})$ such that*

$$c_1\left(\|f\|^2 + \|\mathcal{L}^{m/2}f\|^2\right)^{1/2} \le \left(\sum_j |\langle f, \xi_j\rangle|^2 + \|\mathcal{L}^{m/2}f\|^2\right)^{1/2} \le$$

$$C_1\left(\|f\|^2 + \|\mathcal{L}^{m/2}f\|^2\right)^{1/2}. \tag{3.10}$$

Proof Consider (3.7) with $\alpha = 1$ and apply Lemma 2. It will show existence of a $\widetilde{C} = \widetilde{C}(\mathbf{X}, \rho, m)$ such that

$$\|f\|^2 \leq \widetilde{C}\left(\sum_j |\langle f, \xi_j\rangle|^2 + \|\mathcal{L}^{m/2}f\|^2\right),$$

which implies left side of (3.10). By using the Hölder inequality we obtain

$$\sum_j |\langle f, \xi_j\rangle|^2 = \sum_j \left| |B(x_j, \rho)|^{-1/2} \int_{B(x_j,\rho)} f \right|^2 \leq$$

$$\sum_j \int_{B(x_j,\rho)} |f|^2 \leq N(\mathbf{X})\|f\|^2, \tag{3.11}$$

where $N(\mathbf{X})$ is from Lemma 1. This inequality implies the right-hand side of (3.10). Theorem is proved.

□

4 Frames of Averages in Paley–Wiener Spaces

4.1 Paley–Wiener Functions in $L_2(\mathbf{X})$

Since $\mathcal{L}$ is a self-adjoint non-negative definite operator in the Hilbert space $L_2(\mathbf{X})$ it has a uniquely defined non-negative self- adjoint square root $\sqrt{\mathcal{L}}$. By using Spectral Theorem for $\sqrt{\mathcal{L}}$ and associated operational calculus one can define the projector $\mathbf{1}_{[0,\,\omega]}(\sqrt{\mathcal{L}})$, where $\mathbf{1}_{[0,\,\omega]}$ is the characteristic function of the interval $[0, \omega]$.

Definition 2 We will say that a function $f \in L_2(\mathbf{X})$ belongs to the *Paley–Wiener space* $PW_\omega(\mathcal{L})$ if it belongs to the range of the projector $\mathbf{1}_{[0,\,\omega]}(\sqrt{\mathcal{L}})$.

Next we denote by $\mathcal{H}^k$ the domain of $\mathcal{L}^{k/2}$. It is a Banach space, equipped with the graph norm $\|f\|_k = \|f\| + \|\mathcal{L}^{k/2}f\|$. The next theorem contains generalizations of several results from classical harmonic analysis (in particular the Paley–Wiener theorem). It follows from our results in [7] and [8].

Theorem 4.1 *The spaces $PW_\omega(\mathcal{L})$ have the following properties:*

1. *the space $PW_\omega(\mathcal{L})$ is a linear closed subspace in $L_2(\mathbf{X})$.*
2. *the space $\bigcup_{\omega>0} PW_\omega(\mathcal{L})$ is dense in $L_2(\mathbf{X})$;*
3. *(Bernstein inequality) $f \in PW_\omega(\mathcal{L})$ if and only if $f \in \mathcal{H}^\infty = \bigcap_{k=1}^\infty \mathcal{H}^k$, and the following Bernstein inequalities hold true:*

$$\|\mathcal{L}^{s/2}f\| \leq \omega^s\|f\| \quad \textit{for all } s \in \mathbf{R}_+. \tag{4.1}$$

4.2 Sampling and Hilbert Frames

A set of vectors $\{\theta_v\}$ in a Hilbert space H is called a Hilbert frame if there exist constants $a, b > 0$ such that for all $f \in H$

$$a\|f\|_2^2 \leq \sum_v |\langle f, \theta_v \rangle|^2 \leq b\|f\|_2^2. \tag{4.2}$$

The largest a and smallest b are called respectively the lower and the upper frame bounds and the ratio b/a is known as the tightness of the frame. If $a = b$ then $\{\theta_v\}$ is a *tight* frame, and if $a = b = 1$ it is called a *Parseval* frame. Parseval frames are similar in many respects to orthonormal bases. For example, if all members of a Parseval frame are unit vectors, then it is an orthonormal basis.

According to the general theory of Hilbert frames [9, 11] the frame inequality (4.2) implies that there exists a dual frame $\{\Theta_v\}$ (which is not unique in general) for which the following reconstruction formula holds:

$$f = \sum_v \langle f, \theta_v \rangle \Theta_v. \tag{4.3}$$

In general it is not easy to find a dual frame. For this reason one can resort to the following frame algorithm (see [11], Ch. 5) which performs reconstruction by iterations.

Given a relaxation parameter $0 < \mu < \frac{2}{b}$, set $\eta = \max\{|1 - \mu a|, |1 - \mu b|\} < 1$. Let $f_0 = 0$ and define recursively

$$f_n = f_{n-1} + \mu \Phi(f - f_{n-1}), \tag{4.4}$$

where Φ is the frame operator which is defined on H by the formula $\Phi f = \sum_\nu \langle f, \xi_\nu \rangle \xi_\nu$. In particular, $f_1 = \mu \Phi f = \mu \sum_j \langle f, \xi_\nu \rangle \xi_\nu$. Then $\lim_{n \to \infty} f_n = f$ with a geometric rate of convergence, that is,

$$\|f - f_n\| \leq \eta^n \|f\|. \tag{4.5}$$

In particular, for the choice $\mu = \frac{2}{a+b}$ the convergence factor is

$$\eta = \frac{b - a}{a + b}.$$

4.3 Frames of Averages in Paley–Wiener Spaces

Theorem 4.2 *For a given $\alpha > 0$ there exists a constant $0 < \nu = \nu(\mathbf{X}, \alpha) < 1$ such that for any $\omega > 0$, every metric ρ-lattice $\mathbf{X}_\rho = \{x_j\}$ with $\rho = \frac{\nu}{\sqrt{\omega}}$ the corresponding set of functionals*

$$f \to \langle f, \xi_j \rangle = |B(x_j, \rho)|^{-1/2} \int_{B(x_j,\rho)} f \tag{4.6}$$

is a frame in $PW_\omega(\mathcal{L})$, i. e.

$$\frac{\alpha}{2+2\alpha}\|f\|^2 \leq \sum_j |\langle f, \xi_j \rangle|^2 \leq N(\mathbf{X})\|f\|^2,\ \alpha > 0.v \tag{4.7}$$

Proof Using Bernstein inequality (4.1) and global Poincaré inequality (3.7) one obtains for $f \in PW_\omega(\mathcal{L})$

$$\|f\|^2 \leq (1+\alpha)CN(\mathbf{X})(\rho\omega^{1/2})^2\|f\|^2 + \frac{1+\alpha}{\alpha}\sum_j |\langle f, \xi_j \rangle|^2. \tag{4.8}$$

If for a fixed $\alpha > 0$ one picks

$$\nu = \frac{1}{\sqrt{2(1+\alpha)CN(\mathbf{X})}},$$

then by choosing ρ which satisfies $\rho = \nu\omega^{-1/2}$ we can move the first term on the right in (4.8) to the left side to obtain

$$\|f\|^2 \leq \left(\frac{2}{\alpha}+2\right)\sum_j |\langle f, \xi_j \rangle|^2,\ \alpha > 0. \tag{4.9}$$

Opposite inequality follows from (3.11). Theorem is proven.

□

Corollary 1 *For a given $\alpha > 0$ there exists a constant $0 < \nu = \nu(\mathbf{X}, \alpha) < 1$ such that for any $\omega > 0$ and every metric ρ-lattice $\mathbf{X}_\rho = \{x_j\}$ with $\rho = \nu\omega^{-1/2}$ every function f in $PW_\omega(\mathbf{X})$ is completely determined by its averages (4.6) and can be reconstructed from them in a stable way by using formula (4.3). One can also use frame algorithm described by (4.4).*

5 Average Variational Splines on Metric Measure Spaces

5.1 Construction of Average Variational Splines

We consider a sufficiently small $\rho > 0$, a fixed ρ-lattice $\{x_j\}$ and the corresponding cover $\{B(x_j, \rho)\}$. Let

$$\mathcal{A}_j(f) = \mathcal{A}_{B(x_j,\rho)}(f) = \langle f, \xi_j \rangle = |B(x_j, \rho)|^{-1/2} \int_{B(x_j,\rho)} f,$$

where $\xi_j = |B(x_j, \rho)|^{-1/2}\chi_j(x)$, χ_j being the characteristic function of $B(x_j, \rho)$. Since all lattices have a uniform multiplicity $N(\mathbf{X})$ one has that if $f \in L_2(\mathbf{X})$ then $\{\mathcal{A}_j(f)\} \in l_2$:

$$\sum_j \left|\mathcal{A}_j(f)\right|^2 \leq \sum_j \int_{B(x_j,\rho)} |f|^2 \leq N(\mathbf{X})\|f\|^2.$$

Definition 3 For a sufficiently large $k \in \mathbf{N}$ and a sequence $\mathbf{s} = \{s_j\}_{j\in J} \in l_2$ the set of all functions in $\mathcal{D}(\mathcal{L}^{k/2}) \subseteq L_2(\mathbf{X})$ such that $\mathcal{A}_j(f) = s_j$ will be denoted by $Z_{\mathbf{s}}^{k/2}$. In particular, the subspace $Z_{\mathbf{0}}^{k/2}$ corresponds to the trivial sequence $\mathbf{0} = \{0, 0, \ldots, \}$.

We introduce $\mathcal{P}_{k/2} : \mathcal{D}(\mathcal{L}^{k/2}) \to l_2$ defined as

$$\mathcal{P}_{k/2} f = \left\{\mathcal{A}_j(f)\right\} \in l_2,\ f \in \mathcal{D}(\mathcal{L}^{k/2}).$$

In general, every map $\mathcal{P}_{k/2}$ depends on k and on a lattice. To simplify our framework we adopt the following assumption which holds true in many important cases.

Assumption 2 *There exists a $k_0 \in \mathbf{N}$ such that for every $k > k_0$ and every lattice with sufficiently small $\rho > 0$ the image of $\mathcal{D}(\mathcal{L}^{k/2}) \subseteq L_2(\mathbf{X})$ under $\mathcal{P}_{k/2}$ is the entire l_2.*

We consider the following optimization problem:

For a given lattice $\{x_j\}$ and a sequence $\mathbf{s} = \{s_j\}_{j\in J} \in l_2$ find a function f in $Z_{\mathbf{s}}^{k/2}$, $k > k_0$, which minimizes the functional

$$u \to \|\mathcal{L}^{k/2} u\|. \tag{5.1}$$

Theorem 5.1 *If the **Assumptions 1, 2** are satisfied, then the optimization problem has a unique solution for every sequence $\mathbf{s} = \{s_j\}_{j\in J} \in l_2$ and every $k > k_0$*

Proof The same problem for functional

$$u \to \left(\|u\|^2 + \|\mathcal{L}^{k/2} u\|^2\right)^{1/2},\ k > k_0,\ u \in \mathcal{D}(\mathcal{L}^{k/2}), \tag{5.2}$$

can be solved easily. For a given sequence $\{s_j\} \in l_2$ consider a function f in $\mathcal{D}(\mathcal{L}^{k/2})$ such that $\mathcal{A}_j(f) = s_j$. Such function exists due to the last assumption. Let Pf denote the orthogonal projection of this function f in the Hilbert space $\mathcal{D}(\mathcal{L}^{k/2})$ with the natural inner product

$$\langle f, g\rangle + \left\langle \mathcal{L}^{k/2} f,\ \mathcal{L}^{k/2} g\right\rangle \tag{5.3}$$

on the subspace $Z_{\mathbf{0}}^{k/2}$ with the norm $\left(\|u\|^2 + \|\mathcal{L}^{k/2} u\|^2\right)^{1/2}$. Then the function $g = f - Pf$ will be the unique solution of the above minimization problem for the functional (5.2). Both existence and uniqueness follow from the fact that any two functions in $Z_{\mathbf{s}}$ differ by an $h \in Z_{\mathbf{0}}^{k/2}$. Indeed,

1. to show uniqueness notice that for any f_1 in $Z_{\mathbf{s}}$ one has $f_1 = f + h$, where $h \in Z_{\mathbf{0}}$ and since $h = Ph$

$$f_1 - Pf_1 = f + h - P(f + h) = f - Pf;$$

2. minimality follows from

$$\|g + h\|^2 + \|\mathcal{L}^{k/2}(g + h)\|^2 = \left(\|g\|^2 + \|\mathcal{L}^{k/2}g\|^2\right) + \left(\|h\|^2 + \|\mathcal{L}^{k/2}h\|^2\right),$$

where orthogonality of g to $Z_{\mathbf{0}}^{k/2}$ with respect to (5.3) was used.

The problem with functional $u \to \|\mathcal{L}^{k/2}u\|$ is that it is not a norm. But we already proved in Theorem 3.3 that the inner product (5.3) is equivalent to the inner product

$$\sum_j |\mathcal{A}_j(f)\mathcal{A}_j(g)|^2 + \left\langle \mathcal{L}^{k/2}f, \right\rangle \mathcal{L}^{k/2}g \Big\rangle. \tag{5.4}$$

Thus the above procedure can be applied to the Hilbert space with the inner product (5.4) and it clearly proves existence and uniqueness of the solution of our minimization problem for the functional $u \to \|\mathcal{L}^{k/2}u\|$. Theorem is proved.

□

Definition 4 For a given $\mathbf{s} = \{s_j\}$ the corresponding unique solution will be called the variational spline and denoted as $S_k(\mathbf{s})$.

We have the following characterization of the above optimization problem.

Theorem 5.2 *For a sequence $s = \{s_j\}_{j \in J} \in l_2$ a function $f \in Z_{\mathbf{s}}^{k/2}$, $k > k_0$, minimizes the functional (5.1) if and only if $\mathcal{L}^{k/2}f$ is orthogonal to $\mathcal{L}^{k/2}Z_0$ in $L_2(\mathbf{X})$.*

Proof If $f \in Z_{\mathbf{s}}^{k/2}$, $k > k_0$, then any other function in $Z_{\mathbf{s}}^{k/2}$ has the form $f + h$ for some $h \in Z_{\mathbf{0}}$. If $\mathcal{L}^{k/2}f$ is orthogonal to $\mathcal{L}^{k/2}Z_0$, then

$$\|\mathcal{L}^{k/2}(f + h)\|^2 = \|\mathcal{L}^{k/2}f\|^2 + \|\mathcal{L}^{k/2}h\|^2,$$

which shows that f is the minimizer. Conversely, let $f \in Z_{\mathbf{s}}^{k/2}$, $k > k_0$ and f minimizes (5.1). For any $h \in Z_0$, $\lambda \in \mathbb{R}$ one has

$$\|\mathcal{L}^{k/2}(f + \lambda h)\|^2 = \|\mathcal{L}^{k/2}f\|^2 + 2\lambda\langle\mathcal{L}^{k/2}f, \mathcal{L}^{k/2}h\rangle + \lambda^2\|\mathcal{L}^{k/2}h\|^2.$$

It shows that the vector f can be a minimizer only if $\langle \mathcal{L}^{k/2} f, \mathcal{L}^{k/2} h\rangle = 0$ since otherwise the minimizer would be given by the formula

$$f - \frac{\langle \mathcal{L}^{k/2} f, \mathcal{L}^{k/2} h\rangle}{\|\mathcal{L}^{k/2} f\|^2} h.$$

□

5.2 Interpolation and Approximation by Average Splines

Definition 5 Consider a sufficiently small $\rho > 0$, a fixed ρ-lattice $\{x_j\}$, and the corresponding cover $\left\{B(x_j, \rangle\rho)\right\}$. For $f \in E_k$, $k > k_0$, we denote by $S_k(f)$ the solution of the minimization problem such that $S_k(f) - f \in Z_0$. We say that $S_k(f)$ is a variational spline which interpolates f by its average values on balls $\left\{B(x_j, \rangle\rho)\right\}$.

Note, that this terminology is justified since $S_k(f) - f \in Z_0$ if and only if for every j

$$\int_{B(x_j,\rho)} S_k(f) = \int_{B(x_j,\rho)} f.$$

The following Lemma was proved in [17, 18].

Lemma 3 *If T is a self-adjoint non-negative operator in a Hilbert space H and for an $\varphi \in H$ and a positive $a > 0$ the following inequality holds true*

$$\|\varphi\| \leq a\|T\varphi\|,$$

then for the same $\varphi \in H$, and all $k = 2^l, l = 0, 1, 2, \ldots$ the following inequality holds

$$\|\varphi\| \leq a^k \|T^k \varphi\|, \tag{5.5}$$

as long as φ in the domain of T^k.

Proof By the spectral theory [2] there exists a direct integral of Hilbert spaces

$$X = \int_0^\infty X(\tau) dm(\tau)$$

and a unitary operator F from H onto X, which transforms domain of $T^t, t \geq 0$, onto $X_t = \{x \in X | \tau^t x \in X\}$ with norm

$$\|T^t f\|_H = \left(\int_0^\infty \tau^{2t} \|Ff(\tau)\|^2_{X(\tau)} dm(\tau)\right)^{1/2}$$

and $F(T^t f) = \tau^t(Ff)$. According to our assumption we have for a particular $\varphi \in H$

$$\int_0^\infty |F\varphi(\tau)|^2 dm(\tau) \leq a^2 \int_0^\infty \tau^2 |F\varphi(\tau)|^2 dm(\tau)$$

and then for the interval $\mathcal{I} = \mathcal{I}(0, a^{-1})$ we have

$$\int_{\mathcal{I}} |F\varphi(\tau)|^2 dm(\tau) + \int_{[0,\infty]\setminus\mathcal{I}} |F\varphi|^2 dm(\tau) \leq$$

$$a^2 \left(\int_{\mathcal{I}} \tau^2 |F\varphi|^2 dm(\tau) + \int_{[0,\infty]\setminus\mathcal{I}} \tau^2 |F\varphi|^2 dm(\tau) \right).$$

Since $a^2\tau^2 < 1$ on $\mathcal{I}(0, a^{-1})$

$$0 \leq \int_{\mathcal{I}} \left(|F\varphi|^2 - a^2\tau^2|F\varphi|^2 \right) dm(\tau) \leq \int_{[0,\infty]\setminus\mathcal{I}} \left(a^2\tau^2|F\varphi|^2 - |F\varphi|^2 \right) dm(\tau).$$

This inequality implies the inequality

$$0 \leq \int_{\mathcal{I}} \left(a^2\tau^2|F\varphi|^2 - a^4\tau^4|F\varphi|^2 \right) dm(\tau) \leq \int_{[0,\infty]\setminus\mathcal{I}} \left(a^4\tau^4|F\varphi|^2 - a^2\tau^2|F\varphi|^2 \right) dm(\tau)$$

or

$$a^2 \int_0^\infty \tau^2 |F\varphi|^2 dm(\tau) \leq a^4 \int_0^\infty \tau^4 |F\varphi|^2 dm(\tau),$$

which means

$$\|\varphi\| \leq a\|T\varphi\| \leq a^2\|T^2\varphi\|.$$

Now, by using induction one can finish the proof of the Lemma. The Lemma is proved. □

Remark 2 By using Lemma 2 with $A = 0$ one could have (3.9) with $A = 0$ however, the inequality (5.5) is stronger.

Theorem 5.3 *For every $k = 2^l$, $l \in \mathbb{N}$, and every $f \in \mathcal{D}(\mathcal{L}^{k/2})$ the following inequality takes place:*

$$\|f - S_k(f)\|^2 \leq 4C_0^k \rho^{2k} \|\mathcal{L}^{k/2} f\|^2. \tag{5.6}$$

In particular, if $f \in PW_\omega(\mathcal{L})$ then

$$\|f - S_k(f)\|^2 \leq (C_1\rho\omega)^{2k} \|f\|^2, \tag{5.7}$$

where $C_1 = 2^{1/k}C_0^{1/2} < 2C_0^{1/2}$.

Proof If $k = 2^l$, then since $f - S_k(f)$ belongs to Z_0 by (3.8) we have

$$\|f - S_k(f)\|^2 \leq CN(\mathbf{X})\rho^2\|\mathcal{L}^{1/2}(f - S_k(f))\|^2.$$

Using Lemma 3 with $T = \mathcal{L}^{1/2}$ we obtain

$$\|f - S_k(f)\|^2 \leq C_0^k\rho^{2k}\|\mathcal{L}^{k/2}(f - S_k(f))\|^2,$$

with $C_0 = CN(\mathbf{X})$ and the minimality of $S_k(f)$ gives

$$\|f - S_k(f)\|^2 \leq 4C_0^k\rho^{2k}\|\mathcal{L}^{k/2}f\|^2.$$

Combining this inequality with the Bernstein inequality (4.1) we are coming to (5.7). Theorem is proven.

□

As a summary we have the following statement.

Theorem 5.4 *Assume that the **Assumptions 1, 2** are satisfied. Consider* $PW_\omega(\mathcal{L})$, $\omega > 0$, *and* $\rho < (C_1\omega)^{-1}$, *where* C_1 *is from (5.7). For every* ρ-*lattice* $\{x_j\}$ *and the corresponding set of functions* $\{\xi_j\}$ *defined in (3.6) every* $f \in PW_\omega(\mathcal{L})$, $\omega > 0$, *is uniquely determined by the set of values* $\{\langle f, \xi_j\rangle\}$ *and can be reconstructed by the formula*

$$\lim_{k\to\infty} S_k(f) = f, \tag{5.8}$$

where the rate of convergence is

$$\|f - S_k(f)\| \leq \gamma^k\|f\|, \tag{5.9}$$

with $\gamma = C_1\rho\omega < 1$.

6 Average Sampling and Average Splines on Combinatorial Graphs

6.1 Analysis on Graphs

Let G denote an undirected weighted graph, with a finite or countable number of vertices $V(G)$ and weight function $w : V(G) \times V(G) \to \mathbb{R}_0^+$. w is symmetric, i.e., $w(u, v) = w(v, u)$, and $w(u, u) = 0$ for all $u, v \in V(G)$. The edges of the graph are the pairs (u, v) with $w(u, v) \neq 0$.

Our assumption is that for every $v \in V(G)$ the following finiteness condition holds:

$$w(v) = \sum_{u \in V(G)} w(u, v) < \infty. \tag{6.1}$$

Let $L_2(G)$ denote the space of all complex-valued functions with the inner product

$$\langle f, g \rangle = \sum_{v \in V(G)} f(v)\overline{g(v)}$$

and the norm

$$\|f\|_2 = \left(\sum_{v \in V(G)} |f(v)|^2 \right)^{1/2}.$$

Definition 6 The weighted gradient norm of a function f on $V(G)$ is defined by

$$\|\nabla f\| = \left(\sum_{u,v \in V(G)} \frac{1}{2} |f(u) - f(v)|^2 w(u, v) \right)^{1/2}.$$

The set of all $f : G \to \mathbb{C}$ for which the weighted gradient norm is finite will be denoted as $\mathcal{D}^2(\nabla)$.

Remark 3 The factor $\frac{1}{2}$ makes up for the fact that every edge (i.e., every *unordered* pair (u, v)) enters twice in the summation. Note also that loops, i.e. edges of the type (u, u), in fact do not contribute.

We intend to prove Poincaré-type estimates involving weighted gradient norm. In the case of a *finite* graph and $L_2(G)$-space the weighted Laplace operator $\mathcal{L}$: $L_2(G) \to L_2(G)$ is introduced via

$$(\mathcal{L}f)(v) = \sum_{u \in V(G)} (f(v) - f(u))w(v, u) . \tag{6.2}$$

This graph Laplacian is a well-studied object; it is known to be a positive-semidefinite self-adjoint operator (matrix).

According to Theorem 8.1 and Corollary 8.2 in [12] if for an *infinite* graph there exists a $C > 0$ such that the degrees are uniformly bounded

$$w(u) = \sum_{u \in V(G)} w(u, v) \leq C \tag{6.3}$$

then operator which is defined by (6.2) on functions with compact supports has a unique positive-semidefinite self-adjoint *bounded* extension $\mathcal{L}$ which is acting according to (6.2). It means in particular that its domain is the entire $L_2(G)$.

What is really important for us is that in both of these cases for the non-negative square root $\mathcal{L}^{1/2}$ one has the equality

$$\|\mathcal{L}^{1/2} f\| = \|\nabla f\| \tag{6.4}$$

for all $f \in \mathcal{D}^2(\nabla)$. This fact is not difficult to show directly (see [13] for finite graphs and [10] for infinite which satisfy (6.3)).

Lemma 4 *For all $f \in L_2(G)$ contained in the domain of $\mathcal{L}$, we have*

$$\|\mathcal{L}^{1/2} f\|^2 = \|\nabla f\|^2 . \tag{6.5}$$

For $f \in PW_\omega(\mathcal{L})$, this implies

$$\|\nabla f\| = \|\mathcal{L}^{1/2} f\| \leq \sqrt{\omega}\|f\|. \tag{6.6}$$

7 A Poincaré-Type Inequality for Finite Graphs

It is known that for every finite connected graph $\lambda_0 = 0$ as a simple eigenvalue of the Laplace operator L and the corresponding eigenfunction is a constant on the entire graph. Given a connected and finite graph G and a function $f \in L_2(G)$ we consider its average

$$f_G = \frac{1}{|G|} \sum_{v \in V(G)} f(v).$$

The notation $a\mathbf{1}$ is used for a constant function $f(v) = a$ for all $v \in G$.

Theorem 7.1 *For every connected and finite graph G (which contains more than one vertex) the following Poincaré-type inequality holds:*

$$\sum_{u \in V(G)} |f(u) - f_G \mathbf{1}|^2 \leq \frac{1}{\lambda_1} \|\nabla f\|^2 = \frac{1}{\lambda_1} \|\mathcal{L}^{1/2} f\|^2, \; f \in L_2(G), \tag{7.1}$$

where λ_1 is the first nonzero eigenvalue of $\mathcal{L}$.

Proof Note, that the average of the function $f - f_G\mathbf{1}$ is zero:

$$\sum_{u \in V(G)} \left(f(u) - \left(\frac{1}{|G|} \sum_{v \in V(G)} f(v) \right) \mathbf{1} \right) = \sum_{u \in V(G)} f(u) - \sum_{v \in V(G)} f(v) = 0.$$

If λ_1 is the first nonzero eigenvalue of $\mathcal{L}$, then $\sqrt{\lambda_1}$ is the first nonzero eigenvalue of the non-negative square root $\mathcal{L}^{1/2}$. Since function $f - f_G\mathbf{1}$ is orthogonal to constants it implies

$$\|f - f_G\mathbf{1}\| \le \frac{1}{\sqrt{\lambda_1}}\|\mathcal{L}^{1/2}(f - f_G\mathbf{1})\| = \frac{1}{\sqrt{\lambda_1}}\|\mathcal{L}^{1/2} f\|. \tag{7.2}$$

But according to Lemma 4 it gives $\|f - f_G\mathbf{1}\| \le \frac{1}{\sqrt{\lambda_1}}\|\nabla f\|$. Theorem is proven. □

8 A Local and Global Poincare-Type Inequalities for Finite and Infinite Graphs

Let G be a finite or infinite and countable connected graph and $\Omega \subseteq V(G)$ is a finite and connected subset of vertices which we will treat as an **induced** graph and will denote by the same letter Ω. We remind that this means that the set of vertices of such graph, which will be denoted as $V(\Omega)$, is exactly the set of vertices in Ω and the set of edges are all edges in G whose both ends belong to Ω. Let Δ_Ω be the Laplace operator constructed according to (6.2) for such induced graph Ω. The first nonzero eigenvalue of the operator Δ_Ω will be denoted as $\lambda_{1,\Omega}$. Let $w_\Omega(u, v),\ u, v \in V(\Omega)$, and

$$w_\Omega(v) = \sum_{u \in V(\Omega)} w(u, v),\ v \in V(\Omega),$$

be the corresponding weight functions. Obviously, $w(u) \ge w_\Omega(u)$.

Suppose that $\Xi = \{\Omega_j\}$ is a disjoint cover of $V(G)$ by connected and finite subgraphs Ω_j. We define functions ξ_j by the formula

$$\xi_j = \frac{1}{\sqrt{|\Omega_j|}}\chi_j, \tag{8.1}$$

where χ_j is the characteristic function of Ω_j, and $|\Omega_j|$ is the number of vertices in Ω_j. We will be interested in functionals on $L_2(G)$ defined by these functions

$$f \mapsto \langle f, \xi_j\rangle = \frac{1}{\sqrt{|\Omega_j|}} \sum_{v \in V(\Omega_j)} f(v),\ f \in L_2(G).$$

We will also need functions

$$\zeta_j = \frac{1}{|\Omega_j|}\chi_j,$$

and corresponding functionals

$$f \mapsto \langle f, \zeta_j \rangle = \frac{1}{|\Omega_j|} \sum_{v \in V(\Omega_j)} f(v), \ f \in L_2(G).$$

By using these notations we formulate the next two theorems. Our **local Poincaré-type inequality** is the following.

Theorem 8.1 *Let G be a finite or infinite and countable graph. Let $\Omega \subseteq V(G)$ be a finite* ***induced subgraph****. Let Δ_Ω be the Laplace operator of the induced graph Ω whose first nonzero eigenvalue is $\lambda_{1,\Omega}$. Then for every $f \in L_2(G)$ the following inequality holds:*

$$\sum_{u \in V(\Omega)} |f(u) - f_\Omega \chi_\Omega(u)|^2 \leq \frac{1}{\lambda_{1,\Omega}} \|\nabla_\Omega f\|_\Omega^2, \tag{8.2}$$

where

$$\|\nabla_\Omega f\|_\Omega^2 = \sum_{u,v \in V(\Omega)} \frac{1}{2} |f(u) - f(v)|^2 \, w(u, v).$$

Clearly, it is a direct consequence of Theorem 7.1. The next Theorem contains what we call **a global Poincaré-type inequality**.

Theorem 8.2 *Let G be a connected finite or infinite and countable graph. Suppose that $\Xi = \{\Omega_j\}$ is a disjoint cover of $V(G)$ by connected and finite subgraphs Ω_j. Let Δ_j be the Laplace operator of the* ***induced*** *graph Ω_j whose first nonzero eigenvalue is $\lambda_{1,j}$. We assume that there exists a nonzero lower boundary of all $\lambda_{1,j}$:*

$$\Lambda = \Lambda_\Xi = \inf_j \lambda_{1,j} > 0.$$

In these notations the following inequality holds for every $f \in L_2(G)$ and every $\alpha > 0$:

$$\|f\|^2 \leq \frac{1+\alpha}{\alpha} \frac{1}{\Lambda_\Xi} \|\mathcal{L}^{1/2} f\|^2 + (1+\alpha) \sum_j |\langle f, \xi_j \rangle|^2 . \tag{8.3}$$

Proof One has

$$\|f\|^2 = \sum_{v \in V(G)} |f(v)|^2 = \sum_j \left(\sum_{v \in V(\Omega_j)} |f(v)|^2 \right). \tag{8.4}$$

For every $u \in V(\Omega_j)$ we apply (3.5) to obtain the next inequality in which $f_{\Omega_j} = \langle f, \zeta_j \rangle$

$$|f(u)|^2 \leq \frac{1+\alpha}{\alpha}\left|f(u) - f_{\Omega_j}\chi_j(u)\right|^2 + (1+\alpha)\left|f_{\Omega_j}\chi_j(u)\right|^2, \tag{8.5}$$

which holds for every positive $\alpha > 0$. By using (8.2) we obtain

$$\sum_{u\in V(\Omega_j)} |f(u)|^2 \leq \frac{1+\alpha}{\alpha}\sum_{u\in V(\Omega_j)} |f(u)-f_{\Omega_j}\chi_j(u)|^2+(1+\alpha)\sum_{u\in V(\Omega_j)}\left|f_{\Omega_j}\chi_j(u)\right|^2 \leq$$

$$\frac{1+\alpha}{\alpha}\frac{1}{\lambda_{1,j}}\|\nabla_j f\|^2_{\Omega_j} + (1+\alpha)|\Omega_j|\left|f_{\Omega_j}\right|^2. \tag{8.6}$$

Summation over j gives the following inequality:

$$\|f\|^2 \leq \frac{1+\alpha}{\alpha}\frac{1}{\Lambda_\Xi}\sum_j \|\nabla_j f\|^2_{\Omega_j} + (1+\alpha)\sum_j |\Omega_j|\left|\langle f, \zeta_j\rangle\right|^2 =$$

$$\frac{1+\alpha}{\alpha}\frac{1}{\Lambda_\Xi}\sum_j\left(\sum_{u,v\in V(\Omega_j)}\frac{1}{2}|f(u)-f(v)|^2 w_j(u,v)\right) + (1+\alpha)\sum_j\left|\langle f, \xi_j\rangle\right|^2, \ \alpha > 0. \tag{8.7}$$

Since for all j one has that $w(u, v) = w_j(u, v)$, $u, v \in V(\Omega_j)$, and since sets Ω_j are disjoint, it is obvious that the first term in the last line is not greater than

$$\frac{1+\alpha}{\alpha}\frac{1}{\Lambda_\Xi}\|\nabla f\|^2.$$

It gives

$$\|f\|^2 \leq \frac{1+\alpha}{\alpha}\frac{1}{\Lambda_\Xi}\|\nabla f\|^2 + (1+\alpha)\sum_j\left|\langle f, \xi_j\rangle\right|^2, \ \alpha > 0, \tag{8.8}$$

and by applying Theorem 6.2 we obtain (8.3). Theorem is proved. □

Remark 4 What follows is a list of differences and similarities between Poincaré inequalities we used on Dirichlet space and on combinatorial graphs.

1. The Poincaré inequality on Dirichlet spaces (2.5) is formulated for balls, while a local Poincaré-type inequality for graphs (8.2) is for **any finite induced subgraph**.
2. There is a radius squared on the right-hand side in (2.5) and reciprocal of the first eigenvalue of the Laplacian on a corresponding induced subgraph on the right-hand side in (8.2). Note, that (8.2) is sharp. It obviously shows that if a graph has

the property that for every ball $B(x,r)$ the following inequality holds:

$$\frac{1}{\lambda_{1,B(x,r)}} \le Cr^2,\ C>0,$$

then our (8.2) implies a "regularly" looking inequality.

3. In fact, one can implement in case of graphs the formulas (2.3) and (2.4) to see that

$$\Gamma(f,f)(v)=\frac{1}{2}\sum_{u\in V(G)}|f(v)-f(u)|^2 w(u,v),$$

and

$$\sum_{v\in V(G)}\Gamma(f,f)(v)=\sum_{v\in V(G)}\left(\frac{1}{2}\sum_{u\in V(G)}|f(v)-f(u)|^2 w(u,v)\right)=\|\nabla f\|^2.$$

In this case the integral $\int_{B(x,r)}\Gamma(f,f)d\mu$ on the right-hand side of (2.5) corresponds to the quantity

$$\sum_{v\in\Omega}\left(\frac{1}{2}\sum_{u\in V(G)}|f(v)-f(u)|^2 w(u,v)\right),$$

which is clearly not smaller than $\|\nabla_\Omega f\|_\Omega^2$ which appears on the right-hand side of (8.2).

9 A Sampling Theorem and a Reconstruction Methods Using Frames

Theorem 9.1 *If the assumptions of the previous Theorem hold, then the set of functionals $\{\zeta_j\}$ is a frame in any space $PW_\omega(\mathcal{L})$ as long as $\Lambda_\Xi > \frac{1+\alpha}{\alpha}\omega$. In other words, if*

$$\gamma=\frac{1+\alpha}{\alpha}\frac{\omega}{\Lambda_\Xi}<1,\ \alpha>0, \tag{9.1}$$

then

$$\frac{(1-\gamma)}{(1+\alpha)}\|f\|^2\le\sum_j|\langle f,\xi_j\rangle|^2\le\|f\|^2. \tag{9.2}$$

Proof Indeed, if $f \in PW_\omega(\mathcal{L})$, then by the Bernstein inequality (6.6) the (8.3) can be rewritten as

$$\|f\|^2 \leq \frac{1+\alpha}{\alpha}\frac{\omega}{\Lambda_\Xi}\|f\|^2 + (1+\alpha)\sum_j \left|\left\langle f, \xi_j\right\rangle\right|^2.$$

If (9.1) holds, then one has

$$0 < (1-\gamma)\|f\|^2 \leq (1+\alpha)\sum_j \left|\left\langle f, \xi_j\right\rangle\right|^2.$$

On the other hand, since

$$\left|\sum_{v\in V(\Omega_j)} f(v)\right|^2 \leq \left|\Omega_j\right|\left(\sum_{v\in V(\Omega_j)} |f(v)|^2\right),$$

one has

$$\sum_j \left|\left\langle f, \xi_j\right\rangle\right|^2 = \sum_j \frac{1}{|\Omega_j|}\left|\sum_{v\in V(\Omega_j)} f(v)\right|^2 \leq \sum_j \left(\sum_{v\in V(\Omega_j)} |f(v)|^2\right) \leq \|f\|^2.$$

Theorem is proven. □

Note, that for the classical Paley–Wiener spaces on the real line the inequalities similar to (9.2) in the case when $\{\xi_j\}$ are delta functions were proved by Plancherel and Polya. Today they are better known as the frame inequalities. Now we can formulate sampling theorem based on average values.

Theorem 9.2 *Under the same conditions and notations as above every function $f \in PW_\omega(\mathcal{L})$ is uniquely determined by its averages $\left\langle f, \xi_j\right\rangle$ and can be reconstructed from this set of values in a stable way.*

9.1 Reconstruction Algorithms in Terms of Frames

What we just proved in the previous section is that under the same assumptions as above the set of functionals $f \to \left\langle f, \xi_j\right\rangle$ is a frame in the subspace $PW_\omega(\mathcal{L})$. This fact allows to apply formula (4.3) which describes a stable method of reconstruction of a function $f \in PW_\omega(\mathcal{L})$ from a set of samples $\{\left\langle f, \xi_j\right\rangle\}$. Another possibility for reconstruction is to use the frame algorithm given by (4.4).

10 Average Variational Splines and a Reconstruction Algorithm

10.1 Variational Interpolating Splines

As in the previous sections we assume that G is a connected finite or infinite and countable graph and $\Xi = \{\Omega_j\}$ is a disjoint cover of $V(G)$ by connected and finite subgraphs Ω_j.

For a given sequence $\mathbf{v} = \{v_j\} \in l_2$ and functionals (8.1) the set of all functions in $L_2(G)$ such that $\langle f, \xi_j \rangle = v_j$ will be denoted by $Z_{\mathbf{v}}^{k/2}$. In particular, $Z_{\mathbf{0}}^{k/2}$ corresponds to the sequence of zeros. We consider the following optimization problem:

For a given sequence $\mathbf{v} = \{v_j\} \in l_2$ *find a function* f *in the set* $Z_{\mathbf{v}}^{k/2} \subseteq L_2(G)$ *which minimizes the functional* $u \to \|\mathcal{L}^{k/2}u\|$, $u \in Z_{\mathbf{v}}^{k/2}$. Similarly to Theorem 5.1 one can prove the following.

Theorem 10.1 *Under the above assumptions the optimization problem has a unique solution for every k.*

Proof Using Theorem 8.2 one can justify the following algorithm:

1. Pick any function $f \in Z_{\mathbf{v}}^{k/2}$.
2. Construct $P_0 f$ where P_0 is the orthogonal projection of f onto $Z_{\mathbf{0}}^{k/2}$ with respect to the inner product

$$\langle f, g \rangle_k = \sum_j \langle f, \xi_j \rangle \langle g, \xi_j \rangle + \langle \mathcal{L}^{k/2} f, \mathcal{L}^{k/2} g \rangle.$$

3. The function $f - P_0 f$ is the unique solution to the given optimization problem.

□

Definition 7 For $f \in L_2(G)$ the interpolating variational spline is denoted by $S_k(f)$ and it is the solution of the minimization problem such that $S_k(f) - f \in Z_{\mathbf{0}}^{k/2}$.

One can easily prove the following characterization of variational splines:

Theorem 10.2 *A function* $u \in L_2(G)$ *is a variational spline if and only if* $\mathcal{L}^k u$ *is orthogonal to* $\mathcal{L}^k Z_{\mathbf{0}}^{k/2}$.

10.2 Reconstruction Using Splines

By applying this Lemma and using the same reasoning as in the first part of the paper one can prove the following reconstruction theorem. Below we are keeping notations of Theorem 9.1.

Theorem 10.3 *If the assumptions of Theorem 9.1 are satisfied and in particular*

$$\gamma = \frac{1+\alpha}{\alpha}\frac{\omega}{\Lambda_\Xi} < 1,\ \alpha > 0,$$

then any function f in $PW_\omega(\mathcal{L})$, $\omega > 0$, can be reconstructed from a set of its averages $\{\langle f, \xi_j\rangle\}$ using the formula

$$f = \lim_{k\to\infty} S_k(f),\ k = 2^l,\ l = 0, 1, \ldots,$$

and the error estimate is

$$\|f - S_k(f)\| \le 2\gamma^k \|f\|,\ k = 2^l,\ l = 0, 1, \ldots. \tag{10.1}$$

Proof Pick an $k = 2^l$, $l = 0, 1, 2, \ldots$. and apply (8.3) to the function $f - S_k(f)$. It gives for every $\alpha > 0$:

$$\|f - S_k(f)\| \le \frac{1+\alpha}{\alpha}\frac{1}{\Lambda_\Xi}\|\mathcal{L}^{1/2}(f - S_k(f))\|^2.$$

By Lemma 3 it implies the following inequality:

$$\|f - S_k(f)\| \le \left(\frac{1+\alpha}{\alpha}\frac{1}{\Lambda_\Xi}\right)^k \|\mathcal{L}^{k/2}(f - S_k(f))\|^2.$$

Using minimization property of $S_k(f)$ and the Bernstein inequality (4.1) for $f \in PW_\omega(\mathcal{L})$ one obtains (10.1) with

$$\gamma = \frac{1+\alpha}{\alpha}\frac{\omega}{\Lambda_\Xi}.$$

The assertion follows from the assumption that $\gamma < 1$. □

11 Example: Lattice $\mathbb{Z}$

Let us consider a one-dimensional infinite lattice $\mathbb{Z} = \{\ldots, -1, 0, 1, \ldots\}$ as an unweighted graph. The dual group of the commutative additive group $\mathbb{Z}$ is the one-dimensional torus. The corresponding Fourier transform $\mathcal{F}$ on the space $L_2(\mathbb{Z})$ is defined by the formula

$$\mathcal{F}(f)(\xi) = \sum_{k\in\mathbb{Z}} f(k)e^{ik\xi},\ f \in L_2(\mathbb{Z}), \xi \in [-\pi, \pi).$$

It gives a unitary operator from $L_2(\mathbb{Z})$ on the space $L_2(\mathbb{T}) = L_2(\mathbb{T}, d\xi/2\pi)$, where $\mathbb{T}$ is the one-dimensional torus and $d\xi/2\pi$ is the normalized measure. One can verify the following formula:

$$\mathcal{F}(\mathcal{L}f)(\xi) = 4\sin^2\frac{\xi}{2}\mathcal{F}(f)(\xi).$$

The next result is obvious.

Theorem 11.1 *The spectrum of the Laplace operator $\mathcal{L}$ on the one-dimensional lattice $\mathbb{Z}$ is the interval* $[0, 4]$. *A function f belongs to the space $PW_\omega(\mathbb{Z})$, $0 \leq \omega \leq 4$, if and only if the support of $\mathcal{F}f$ is a subset of $[-\pi, \pi)$ on which $4\sin^2\frac{\xi}{2} \leq \omega$.*

We consider the cover $\Xi = \{\Omega_j\}$ of $\mathbb{Z}$ by disjoint sets $\Omega_j = \{j, \ j+1\}$ where j runs over all even integers: $\{\ldots, -2, 0, 2, \ldots\} = 2\mathbf{Z}$. We treat every Ω_j as an induced graph whose set of vertices is $V(\Omega_j) = \{j, \ j+1\}$, $j \in 2\mathbf{Z}$, and which has only one edge $(j, \ j+1)$. Functional ξ_j takes form

$$\langle f, \xi_j \rangle = \frac{1}{\sqrt{2}}\left(f(j) + f(j+1)\right), \ j \in 2\mathbf{Z}, \ f \in \ell^2(\mathbb{Z}). \tag{11.1}$$

One can check that spectrum of the Laplace operator Δ_j on Ω_j defined by (6.2) contains just two values $\{0, 2\}$. Thus $\Lambda_\Xi = 2$. For an $0 < \omega < 4$ and $\alpha > 0$ condition (9.1) takes form

$$\gamma = \frac{1+\alpha}{\alpha}\frac{\omega}{2} < 1. \tag{11.2}$$

Note, that since for a large α the fraction $(1+\alpha)/\alpha$ is close to 1 the condition implies that one can have any $0 < \omega < 2$. As an application of Theorem 9.2 we obtain the following result.

Theorem 11.2 *For every $0 < \omega < 2$ every function $f \in PW_\omega(\mathcal{L})$ is uniquely determined by its average values (11.1) and can be reconstructed from them in a stable way.*

In particular, if instead of infinite graph $\mathbb{Z}$ one would consider a path graph $\mathbb{Z}_N$ whose eigenvalues are given by formulas $2 - 2\cos\frac{k\pi}{N}$, $k = 0, 1, \ldots, N-1$, the last Theorem would mean that any eigenfunction with eigenvalue from a lower half of the spectrum is uniquely determined and can be reconstructed from averages (11.1).

Remark 5 In connection with this example it is worth to notice the following. Let us introduce the restriction operator

$$\mathcal{R}^S_\omega : PW_\omega(\mathcal{L}) \to L_2(S), \ S \subseteq V(\mathbb{Z}),$$

where

$$\mathcal{R}_{\omega}^{S}(f) = f|_{S}, \ f \in PW_{\omega}(\mathbb{Z}).$$

Theorem 11.3 *For any finite set of vertices $S \subseteq \mathbb{Z}$ and for* ***every*** *$\omega > 0$ the restriction operator $\mathcal{R}_{\omega}^{S}$ is surjective.*

References

1. Albeverio, S.: Theory of Dirichlet forms and applications. (English summary) Lectures on probability theory and statistics (Saint-Flour, 2000), 1106, Lecture Notes in Math., 1816, Springer, Berlin, 2003.
2. Birman, B., Solomyak, M.: *Spectral theory of selfadjoint operators in Hilbert space*, D. Reidel Publishing Co., Dordrecht, 1987.
3. Bouleau, N., Hirsch, F.: Dirichlet forms and analysis on Wiener space. Walter de Gruyter, Berlin, New York (1991)
4. Coifman, R., Weiss, G.: *Analyse harmonique non-commutative sur certains espaces homog'enes*, Lecture Notes in Math. 242, Springer-Verlag, 1971.
5. Coifman, R., Weiss, G.: *Extensions of Hardy spaces and their use in analysis*, Bull. Amer. Math. Soc. 83 (1977), 569–645.
6. Fukushima, M. Oshima, Y., Takeda, M.: *Dirichlet forms and symmetric Markov processes*, De Gruyter Studies in Mathematics vol 19, Walter De Gruyter and Co., Berlin (1994)
7. Pesenson, I.: *A sampling theorem on homogeneous manifolds*, Trans. Amer. Math. Soc. **352** (2000), no. 9, 4257–4269.
8. Pesenson, I.: *Poincare-type inequalities and reconstruction of Paley–Wiener functions on manifolds,* J. of Geometric Analysis, (4), 1, (2004), 101–121.
9. Duffin, R., Schaeffer, A.: *A class of nonharmonic Fourier series*, Trans. AMS, 72, (1952), 341–366.
10. Führ, H., Pesenson, I.: *Poincaré and Plancherel-Polya inequalities in harmonic analysis on weighted combinatorial graphs*, SIAM J. Discrete Math. 27 (2013), no. 4, 2007–2028.
11. Gröchenig, K.: *Foundations of Time-Frequency Analysis*, Birkhäuser, 2001.
12. Haeseler, S., Keller, M., Lenz, D., Wojciechowski, R.: *Laplacians on infinite graphs: Dirichlet and Neumann boundary conditions*, J. Spectr. Theory 2 (2012), no. 4, 397–432.
13. Mohar, B.: *Some applications of Laplace eigenvalues of graphs*, in G. Hahn and G. Sabidussi, editors, *Graph Symmetry: Algebraic Methods and Applications (Proc. Montrŕeal 1996)*, volume 497 of Adv. Sci. Inst. Ser. C. Math. Phys. Sci., pp. 225–275, Dordrecht (1997), Kluwer.
14. Parada-Mayorga, A.: *Blue Noise and Optimal Sampling on Graphs*. Thesis (Ph.D.) University of Delaware. 2019. 239 pp. ISBN: 978-1687-97469-3.
15. Parada-Mayorga, A., Lau, D. L., Giraldo, J. H., Arce, G. R.: *Blue-noise sampling on graph*, IEEE Trans. Signal Inform. Process. Netw. 5 (2019), no. 3, 554–569.
16. Pesenson, I.: *The best approximation in a representation space of a Lie group*, Dokl. Acad. Nauk USSR, v. 302/5 (1988), 1055–1059; (Engl. Transl. in Soviet Math. Dokl. 38/2 (1989), 384–388)
17. Pesenson, I.: *Sampling of Paley–Wiener functions on stratified groups*, J. Four. Anal. Appl. 4 (1998), 269–280.
18. Pesenson, I.: *Sampling of band limited vectors*, J. Fourier Anal. Appl. 7/1 (2001), 93–100.
19. Pesenson, I.: *Poincaré-type inequalities and reconstruction of Paley–Wiener functions on manifolds,* J. Geometric Anal. 4/1 (2004), 101–121.
20. Pesenson, I.: *Sampling in Paley–Wiener spaces on combinatorial graphs*, Trans. Amer. Math. Soc. 360 (2008), no. 10, 5603–5627.

21. Pesenson, I. Z.: *Variational splines and Paley–Wiener spaces on combinatorial graphs*, Constr. Approx. 29 (2009), no. 1, 1–21.
22. Pesenson, I. Z., Pesenson, M. Z.: *Sampling, filtering and sparse approximations on combinatorial graphs*, J. Fourier Anal. Appl. 16 (2010), no. 6, 921–942.
23. Sturm, K. T.: *Analysis on local Dirichlet spaces I. Recurrence, conservativeness and Lp-Liouville properties*, J. Reine Angew. Math. 456 (1994), 173–196.
24. Sturm, K. T.: *Analysis on local Dirichlet spaces II. Upper Gaussian estimates for the fundamental solutions of parabolic equations*, Osaka J. Math. 32 (1995), 275–312.
25. Sturm, K. T.: *Analysis on local Dirichlet spaces III. The parabolic Harnack inequality*, J. Math. Pures Appl. 75 (1998), 273–297.
26. Tremblay, N., Borgnat, P.: *Subgraph-based filterbanks for graph signals*, IEEE Trans. Signal Process. 64 (2016), no. 15, 3827–3840.
27. Wang, X., Liu, P., Gu, Y.: *Local-Set-Based Graph Signal Reconstruction*, IEEE Transactions on Signal Processing 63 (2015), no. 9, 2432–2444.

Dynamical Sampling: A View from Control Theory

Rocío Díaz Martín, Ivan Medri, and Ursula Molter

Abstract In this contribution we establish a dictionary between terms in two different areas in mathematics: Dynamical Sampling and Control Theory. The purpose is to show that many of the topics studied in either field are also addressed in the other one, just with a different terminology. We further analyze the relations between discrete-time and continuous-time versions of the problem, using results from both fields.

In addition, we will also differentiate between a discretization of the continuous-time dynamical system and a discrete-time dynamical system itself.

1 Introduction

In the so-called Dynamical Sampling problem a signal that is evolving in time through the action of an operator A is spatially sub-sampled at multiple times, and one seeks to reconstruct it from these spatial–temporal samples, thereby exploiting time evolution (see, e.g., [1–4, 6, 18, 22]). In some of the literature a possible connection between *Dynamical Sampling* and *Control Theory* is mentioned. However, so far there is no explicit formulation of this fact.

R. Díaz Martín
IAM - CONICET, CABA, Argentina

Universidad Nacional de Córdoba, Córdoba, Argentina
e-mail: rpd0109@famaf.unc.edu.ar

I. Medri
Department of Mathematics, Vanderbilt University, Nashville, TN, USA

U. Molter (✉)
Dto. de Matemática, FCEyN, IMAS – CONICET and Universidad de Buenos Aires, Ciudad Universitaria, CABA, Argentina
e-mail: umolter@conicet.gov.ar

M. Hirn et al. (eds.), *Excursions in Harmonic Analysis, Volume 6*, Applied and Numerical Harmonic Analysis, https://doi.org/10.1007/978-3-030-69637-5_15

In this chapter we aim to construct bridges between these two areas, since they are two different ways of looking at the exact same phenomena. What one community calls *dynamical sampling* the other one calls *observability*. Both fields have been thoroughly explored, and it is our intention to compare what has been proven in each of them, in order to provide a kind of *dictionary* that allows to translate the results from one field to the other. We hope that establishing these equivalences will allow both communities to benefit from each other. For example, whereas Dynamical Sampling has mostly focused on discrete-time systems, Observability has mostly focused on continuous-time systems.

This brings us to the second purpose of this chapter. We analyze relations between discrete-time and continuous-time dynamical systems using results from both fields and our dictionary. It is in fact true that the relation is completely understood when dealing with finite-dimensional spaces [12, 16]. For infinite dimensions, the topic is much more subtle. Detailed proofs and further results will also be exhibited in the paper [14].

We organize this chapter as follows: In Sect. 2 we recall the definitions from Control Theory and Dynamical Sampling, highlighting all the instances in which they coincide and provide the link between them. Specifically, in Sect. 2.1 we include the background on Control Theory where the notion of Observability comes from, in Sect. 2.2 we do the same but for the case of Dynamical Sampling, and in Sect. 2.3 we will use the previous definitions to relate these theories.

In Sect. 3 we analyze some results from both, Control Theory and Dynamical Sampling. We will consider mostly infinite-dimensional spaces. Some general theorems can be proved independently of the time being finite/infinite or the dynamics being continuous/discrete. For more specialized results, we then separate the theory into finite and infinite sampling times: Sects. 3.1 and 3.2, respectively.

2 Control Theory and Dynamical Sampling: Overview and Relations

We first summarize without proofs the basic facts about controllability and observability. We will be concerned mainly about the latter. However, since they are dual notions (as will be clear from Theorem 2.1) it is useful to know how to translate from one to the other when studying the observability problem. We then establish the classical problem in dynamical sampling and its generalizations. As a conclusion, one can see that in certain settings the Dynamical Sampling problem can be considered as a particular case of the Observability problem.

2.1 Control Theory Notions: Discrete and Continuous Systems

Let us begin with some terminology of control theory of linear dynamical systems (see, e.g., [13, 19, 23]). For the purpose of this exposition, we will not work under the most general hypothesis. In particular, we will restrict the definitions to Hilbert spaces and continuous operators. For more general settings, i.e., Banach spaces instead of Hilbert spaces and densely defined close operators instead of bounded operators, we refer the interested reader to the above cited references. We will always consider both, the discrete- and the continuous-time systems.

Let X, Y, and $\mathcal{U}$ be complex separable Hilbert spaces. We call X the *state space*, Y the *observation space*, and $\mathcal{U}$ the *control space*. Let $A : X \to X$, $B : X \to Y$, and $C : \mathcal{U} \to X$ be linear and bounded mappings. They are called *dynamic operator*, *observation operator*, and *control operator*, respectively.

The operator A is the infinitesimal generator of the *uniformly continuous* one-parameter group $\{e^{tA}\}_{t\in\mathbb{R}}$, where $e^{tA} := \sum_{n=0}^{\infty} \frac{(tA)^n}{n!}$ and the convergence is in operator norm (cf. [21]). We say that A *generates an exponentially stable dynamic* if, for all $t \geq 0$, $\|e^{tA}\| \leq e^{t\omega}$ for some $\omega < 0$. This is equivalent to $\sup\{\mathrm{Re}(\lambda) : \lambda \in \sigma(A)\} < 0$, where $\sigma(A)$ denotes the spectrum of A (cf. [15]). When the parameter t is discrete, we say that A *generates a strong asymptotically stable dynamic* if $\|A\| < 1$ (this implies that $A^k x \to 0$ when $k \to \infty$ for each $x \in X$).

We will work with *continuous linear systems* of the form:

$$\begin{cases} \dot{x}(t) = Ax(t) + Cu(t), & x(0) = x_0, \qquad t \in [0, \tau), \\ y(t) = Bx(t), \end{cases} \tag{2.1}$$

where τ could be finite or $\tau = \infty$. Consider the Hilbert space of square-integrable vector-valued functions

$$L^2([0,\tau),\mathcal{U}) := \left\{ u : [0,\tau) \to \mathcal{U} : \int_0^\tau \|u(t)\|_{\mathcal{U}}^2 \, dt < \infty \right\}.$$

For any *initial state* $x_0 \in X$ and $u \in L^2([0,\tau),\mathcal{U})$, the *general mild solution* (cf. [21]) to (2.1) is given by

$$x(t) = e^{tA}x_0 + \int_0^t e^{(t-s)A}Cu(s)ds. \tag{2.2}$$

For each finite time $0 \leq t \leq \tau$, the operators $\Theta(t) : L^2([0,\tau),\mathcal{U}) \to X$, given by $\Theta(t)u := \int_0^t e^{(t-s)A}Cu(s)ds$, are called *controllability maps* (cf. [23]).

Analogously, we will consider *discrete linear systems* of the form:

$$\begin{cases} x(k+1) = Ax(k) + Cu(k), & x(0) = x_0, \\ y(k) = Bx(k), \end{cases} \tag{2.3}$$

where $k \in \mathbb{Z}_+ := \mathbb{N} \cup \{0\}$. We denote the Hilbert space of square-summable vector-valued functions by

$$\ell^2(\Gamma, \mathcal{U}) := \left\{ u : \Gamma \to \mathcal{U} : \sum_{k=0}^{\infty} \|u(k)\|_{\mathcal{U}}^2 < \infty \right\},$$

where $\Gamma = \{0, 1, \ldots, \gamma\}$ for some $\gamma \in \mathbb{Z}_+$ or $\gamma = \infty$. For $x_0 \in X$ and $u \in \ell^2(\Gamma, \mathcal{U})$, the general solution of Eq. (2.3) is given by

$$x(k+1) = A^{k+1}x_0 + \sum_{j=0}^{k} A^{k-j}Cu(j).$$

In this case we define the controllability maps $\Theta(k) : \ell^2(\Gamma, \mathcal{U}) \to X$ (for each $k \in \Gamma$) as $\Theta(k)u := \sum_{j=0}^{k} A^{k-j}Cu(j)$.

One may say that *to observe* the system (2.1) (resp., (2.3)) means to recover the initial data x_0 by knowing only the output function $t \mapsto y(t)$ (resp., $k \mapsto y(k)$). On another side, *to control* system (2.1) or (2.3) means to act on the system with a suitable function u in such a way that starting from a given initial state x_0 the systems attain at a time τ the given final state $x(\tau) = x_1$. Since there are several notions of observability and controllability, in what follows we define some of them and put special emphasis on the duality between observability and controllability.

2.1.1 Controllability

For infinite-dimensional systems (that is when the state space X is of infinite dimension), there are two important controllability notions: *exact* and *approximate*. For simplicity, we consider only finite time $\tau < \infty$ ($\gamma < \infty$). At first, we restrict our attention to continuous systems of the form (2.1).

The *set of reachable states from* $x_0 \in X$ *at finite time* τ is defined as

$$\mathcal{A}(\tau)x_0 := \{x(\tau) : \text{ there exist } u \in L^2([0,\tau), \mathcal{U}) \text{ such that system (2.1) admits the solution } x\}.$$

Definition 2.1 The pair (A, C) is called:

1. *exactly controllable at finite time* τ if for every $x_0 \in X$, $\mathcal{A}(\tau)x_0 = X$;
2. *approximately controllable at finite time* τ if for every $x_0 \in X$, $\mathcal{A}(\tau)x_0$ is dense in X.

We denote them briefly by ECO and ACO, respectively. Note that for these definitions the operator B in system (2.1) is irrelevant.

Remark 2.1 [19, Corollary 1.2.3] Using system (2.2), we have that the set of reachable states is the affine space

$$\mathcal{A}(\tau)x_0 = e^{\tau A}x_0 + \left\{\Theta(\tau)u : u \in L^2([0,\tau),\mathcal{U})\right\}.$$

Also, for arbitrary $x_0, x_0' \in X$, we have $\mathcal{A}(\tau)x_0 = X$ if and only if $\mathcal{A}(\tau)x_0' = X$. Respectively, $\mathcal{A}(\tau)x_0$ is dense in X if and only if $\mathcal{A}(\tau)x_0'$ is dense in X for all $x \in X$. In particular, setting $x_0 = 0$, we have that the pair (A, C) is exactly controllable (respectively, approximately controllable) at moment τ if and only if $\Theta(\tau)$ is surjective (respectively, $\Theta(\tau)$ has dense image in X).

For the discrete case (2.3), we have exactly the same definitions of controllability after changing $\tau \in [0,\infty)$ to $\gamma \in \mathbb{Z}_+$ and replacing $L^2([0,\tau),\mathcal{U})$ by $\ell^2(\{0, 1, \ldots, \gamma\}, \mathcal{U})$.

2.1.2 Observability

As in the case of controllability, for infinite-dimensional systems, there are two important observability notions, each depending on time. These are called *exact* and *approximate* observability. In this section we introduce these notions and explore how they are related to each other for linear systems of the form (2.1) or (2.3).

For the case of observability, we will look at the system (2.1) with the control term $C \equiv 0$. We will state the definitions for the case of continuous systems. The necessary changes for the discrete case are given at the end of the section.

Since we are interested in the output function of the system (2.1), we introduce the *output operator until time* τ or *observability map* $\Psi_\tau : X \to L^2([0,\tau), Y)$ given by

$$(\Psi_\tau x_0)(t) := Be^{tA}x_0 \qquad \forall x_0 \in X.$$

Before the main definitions of the section, an *admissibility* condition for the operator B must be given (see, e.g., [23]).

Definition 2.2 (Admissible observation operator) An operator $B \in \mathcal{B}(X, Y)$ is called an *admissible observation operator at time* τ for $\{e^{tA}\}$ if Ψ_τ is continuous.

Remark 2.2

1. B is admissible for $\{e^{tA}\}$ at time $\tau > 0$ if and only if there exists a constant $K_\tau \geq 0$ such that

$$\int_0^\tau \|Be^{tA}x_0\|_Y^2 dt \leq K_\tau^2 \|x_0\|_X^2 \qquad \forall x_0 \in X.$$

2. Since $B : X \to Y$ is a bounded linear operator, Ψ_τ is continuous for every finite time $0 < \tau < \infty$. Indeed,

$$\int_0^\tau \|Be^{tA}x\|^2 dt \leq \frac{e^{2\|A\|\tau} - 1}{2\|A\|}\|B\|^2\|x\|^2.$$

3. If A generates an exponentially stable dynamic, then $\|\Psi_\tau\| \leq K$ for all $0 \leq \tau < \infty$. Therefore, in this case, Ψ_τ is continuous for $\tau = \infty$ and we have

$$\int_0^\infty \|Be^{tA}x\|^2 dt \leq -\frac{1}{2\omega}\|B\|^2\|x\|^2,$$

where $\omega < 0$ is such that $\|e^{tA}\| \leq e^{\omega t}$ for all $t \geq 0$.

Definition 2.3 Let $\tau \in (0, \infty)$ or $\tau = \infty$. Let B be an admissible observation operator at time τ.

1. The pair (A, B) is called *exactly observable at time* τ if Ψ_τ is bounded from below.
2. The pair (A, B) is called *approximately observable at time* τ if $\operatorname{Ker} \Psi_\tau = \{0\}$.

To shorten notation, we write EOB and AOB for the exactly and approximately observable conditions, respectively.

Remark 2.3 [23, Remark 6.1.2] Exact observability implies approximate observability.

These definitions have some equivalent formulations (see, e.g., [23, Chapter 6]). First, note that B is admissible and (A, B) is exactly observable if and only if Ψ_τ is bounded above and below. Hence, by the Open Mapping Theorem this is equivalent to say that Ψ_τ is invertible on its image. In this case one can recover the initial state x_0 from the observations $y = \Psi_\tau x_0$ exactly as $x_0 = \Psi_\tau^\dagger y$, where $\Psi_\tau^\dagger := (\Psi_\tau^* \Psi_\tau)^{-1}\Psi_\tau^*$ is the Moore–Penrose pseudoinverse of Ψ_τ. These observations yield the following proposition.

Proposition 2.1 *Equivalent definitions of exact observability at time* τ*:*

1. *The pair* (A, B) *is EOB at time* τ.
2. *Any initial state* $x_0 \in X$ *can be expressed from the corresponding output function* $y = \Psi_\tau x_0$ *via a bounded operator. That is, there exists a bounded operator* $E_\tau : L^2([0, \tau), Y) \to X$ *such that* $E_\tau(\Psi_\tau x_0) = x_0$ *for all* x_0 *in* X.
3. *The observability Grammian for time* τ, $Q_\tau := \Psi_\tau^* \Psi_\tau$, *is strictly positive. That is, there exists* $c > 0$ *such that* $\langle Q_\tau x, x \rangle > c\|x\|^2$ *for every* $x \neq 0$.

For approximate observable systems, we have that if the observation function $y = \Psi_\tau x_0$ is zero, then $x_0 = 0$. This is sometimes called a *unique continuation property*. Even though this does not imply that Ψ_τ is bounded below (and therefore with bounded inverse), it does guaranty that we have only one possible initial state for every output.

Proposition 2.2 *Equivalent definitions of approximate observability at time* τ*:*

1. *The pair* (A, B) *is AOB at time* τ.
2. *The observability Grammian for time* τ *satisfies* $\operatorname{Ker} Q_\tau = \{0\}$.

Finally, for discrete-time dynamical systems we have the same notions of observability replacing e^{tA} by A^k. Specifically, the observability map is given by

$$\Psi_\gamma : X \longrightarrow \ell^2(\{0, 1, \ldots, \gamma\}, Y)$$
$$(\Psi_\gamma x_0)(k) := BA^k x_0,$$

where we can take $\gamma = \infty$ every time τ is allowed to be infinite.

2.1.3 Connections Between Observability and Control

Let us now compare observability and control. We will consider two dynamical systems: one determined by A and the other by A^*. Since in our setting X and Y are Hilbert spaces, their duals can be identified with themselves. We summarize both dynamics in the following table.

Discrete time	Continuous time
Observability problem	
$\begin{cases} x(k+1) = Ax(k), & x(0) = x_0 \\ y(k) = Bx(k) \end{cases}$ (2.4)	$\begin{cases} \dot{x}(t) = Ax(t), & x(0) = x_0 \\ y(t) = Bx(t) \end{cases}$ (2.5)
Observability map	
$\Psi_\gamma : X \longrightarrow \ell^2(\{0, 1, \ldots, \gamma\}, Y)$ $(\Psi_\gamma x)(k) = BA^k x$	$\Psi_\tau : X \longrightarrow L^2([0, \tau), Y)$ $(\Psi_\tau x)(t) = Be^{tA} x$
Observation	
$y = \Psi_\gamma x_0$	$y = \Psi_\tau x_0$
Control problem	
$\begin{cases} x(k+1) = A^* x(k) + B^* u(k) \\ x(0) = x_0 \end{cases}$ (2.6)	$\begin{cases} \dot{x}(t) = A^* x(t) + B^* u(t) \\ x(0) = x_0 \end{cases}$ (2.7)
Controllability maps	
$\Theta(k) : \ell^2(\{0, 1, \ldots, \gamma\}, Y) \longrightarrow X$ $\Theta(k)u = \sum_{j=0}^{k} (A^*)^{k-j} B^* u(j)$	$\Theta(t) : L^2([0, \tau), Y) \longrightarrow X$ $\Theta(t)u = \int_0^t e^{(t-s)A^*} B^* u(s) ds$
Solution	
$x(k+1) = (A^*)^{k+1} x_0 + \Theta(k)u$	$x(t) = e^{tA^*} x_0 + \Theta(t)u$

In systems (2.4) and (2.5) the state space is X, the space of observations is Y, and B is the observation operator. In systems (2.6) and (2.7) the state space is again X, $\mathcal{U} := Y$ is the space of controls, and $C := B^*$ is the control operator. One can check that $\Psi_\tau^* = \Theta(\tau)$, and $\Psi_\gamma^* = \Theta(\gamma)$. Using this, Propositions 2.1 and 2.2, and Remark 2.1, one can conclude the following theorem.

Theorem 2.1 *Let $A \in \mathcal{B}(X)$, $B \in \mathcal{B}(X, Y)$ and $\tau < \infty$ (resp., $\gamma < \infty$ for the discrete case).*

1. *The pair (A, B) is exactly **observable** at time τ if and only if the dual pair (A^*, B^*) is exactly **controllable** at time τ (resp., γ).*
2. *The pair (A, B) is approximately **observable** at time τ (resp., γ) if and only if the dual pair (A^*, B^*) is approximately **controllable** at time τ (resp., γ).*

For a detailed proof of 1, see [19, Theorem 1.3.4], and for 2, see [19, Theorem 1.3.3]. In view of this theorem, we say that the systems (2.4) and (2.6) (resp., (2.5) and (2.7)) are dual to each other.

This duality will be useful later when trying to relate dynamical systems and control theory. It is worth noticing though that for this duality we use strongly that τ (resp., γ) is finite. For infinite time systems, the duality is not as simple. Hence, in that case we do not define the control problem because working directly with the observability formulation is much more convenient for our task.

2.2 The Dynamical Sampling Problem: Discrete and Continuous

The classical sampling and reconstruction problem consists in recovering a scalar-valued function from the knowledge of its values at certain points of the spatial domain X. In the dynamical sampling problem, the set of *space samples* is replaced by a set of *space-time samples*.

The dynamical sampling problem, originally introduced in [4], can be stated as follows (see, e.g., [3]): Let $x : X \to \mathbb{C}$ be a function in a complex separable Hilbert space $(X, \langle\cdot,\cdot\rangle)$, and assume that x evolves through an evolution linear bounded operator $A : X \to X$ so that the function at time $k \in \mathbb{Z}_+$ has evolved to become $x^{(k)} := A^k x$. The space X can be identified with $\ell^2(I)$ where $I = \{1, \ldots, d\}$ in the finite-dimensional case and $I = \mathbb{N}$ in the infinite-dimensional case. Let $\{e_i\}_{i \in I}$ denote the standard basis of $\ell^2(I)$. The *time-space sample* at time $k \in \mathbb{Z}_+$ and location $i \in I$ is the value

$$x^{(k)}(i) := \langle A^k x, e_i \rangle.$$

In this way to each pair $(i, k) \in I \times \mathbb{Z}_+$ a sample value $x^{(k)}(i)$ is associated. Using this notation, the dynamical sampling problem can then be described as: Under which conditions on the operator A, and a set $S \subseteq I \times \mathbb{Z}_+$, can every $f \in \ell^2(I)$ be recovered in a stable way from the samples in S, that is from $\{x^{(k)}(i) : (i, k) \in S\}$.

Without using the identification with $\ell^2(I)$, sampling a function $x \in X$ can be associated to computing its components along a fixed family of vectors $\mathcal{G} \subseteq X$. In that case, the dynamical sampling problem is to recover x (in a stable way) from the samples

$$\{\langle A^k x, g\rangle\}_{k\in\{0,1,\ldots,\gamma\}, g\in\mathcal{G}}, \tag{2.8}$$

where $\gamma \in \mathbb{Z}_+$ or $\gamma = \infty$. Note that this is not the most general formulation of the problem since the way it is stated implies that the sampling set $\mathcal{G}$ remains the same for every time k. In the case where X is finite dimensional, the article [3] also considered slightly more general sampling schemes.

From now on, let us assume that $\mathcal{G}$ is a countable set (either finite or infinite). For each $g \in \mathcal{G}$, let S_g be the operator from X to $X_g := \ell^2(\{0, 1, \ldots, \gamma\})$, defined by $(S_g x)(k) := \langle A^k x, g\rangle$ and define $\mathcal{S}$ to be the operator $\mathcal{S} := \bigoplus_{g\in\mathcal{G}} S_g$. Then x can be recovered from the samples (2.8) in a *stable way* if and only if there exist constants $c_1, c_2 > 0$ such that

$$c\|x\|_X^2 \leq \|\mathcal{S}x\|^2 = \sum_{g\in\mathcal{G}} \|S_g x\|_{X_g}^2 \leq c_2\|x\|_X^2.$$

Or equivalently,

$$c_1\|x\|^2 \leq \sum_{g\in\mathcal{G}}\sum_{k=0}^{\gamma} |\langle x, (A^k)^* g\rangle|^2 \leq c_2\|x\|^2. \tag{2.9}$$

That is, if the set of vectors

$$\{(A^*)^k g\}_{t\in\{0,1,\ldots,\gamma\}, g\in\mathcal{G}} \tag{2.10}$$

is a *frame* for X (cf. [3, Lemma 1.2]). Therefore, the dynamical sampling problem can also be stated as follows.

Problem 1 Let X be a complex separable Hilbert space and $\mathcal{G} \subseteq X$ a countable set of vectors. Let A be a bounded linear operator and $\gamma \in \mathbb{Z}_+ \cup \{\infty\}$. Under which conditions is the set $\{(A^*)^k g\}_{t\in\{0,1,\ldots,\gamma\}, g\in\mathcal{G}}$ a frame for X.

From this starting point, in later developments, the dynamical sampling community focused, amid other things, on generalizing the discrete iterations of the operator A to continuous scenarios. In particular, in [5] the authors, working with the notion of *continuous frames*, dealt with the problem of determining when the set

$$\{(A^*)^t g\}_{t\in[0,\tau), g\in\mathcal{G}} \tag{2.11}$$

is a *semi-continuous frame* or at least a *complete* or *Bessel* system. The interval $[0, \tau)$ is allowed to be finite or infinite. In the previously mentioned article $(A^*)^t$ was defined using functional calculus for A being a normal operator. Under special conditions on the operator A, the dynamic given by A^t is equivalent to the one given by e^{tA}, which resembles the continuous dynamical systems studied previously.

Let us recall the definitions of continuous frames and Bessel systems.

Definition 2.4 (*Continuous frame-Bessel*) Let X be a complex Hilbert space and let (Ω, μ) be a measure space with positive measure μ. A mapping $F : \Omega \to X$ is called a *continuous frame* with respect to (Ω, μ), if

1. F is *weakly measurable*, i.e., $\omega \mapsto \langle x, F(\omega)\rangle$ is a measurable function on Ω for all $x \in X$;
2. there exist positive constants c_1 and c_2 such that

$$c_1\|x\|^2 \leq \int_\Omega |\langle x, F(\omega)\rangle|^2 \, d\mu(\omega) \leq c_2\|x\|^2, \text{ for all } x \in X. \tag{2.12}$$

Here the largest (smallest) constant c_1 (c_2) is called lower (upper) continuous frame bounds. Further, F is called a *tight* continuous frame if $c_1 = c_2$.

The mapping F is called *Bessel* if the second equality (2.12) holds. In this case, c_2 is called a Bessel constant.

Notice that if F from Definition 2.4 satisfies 1. and the Bessel condition, then the map

$$B : X \longrightarrow L^2(\Omega)$$
$$Bx(\omega) := \langle x, F(\omega)\rangle$$

is well-defined, linear, and bounded. It is called the *analysis operator* of the family $\{F(\omega)\}_{\omega\in\Omega}$. The adjoint operator B^* is called the *synthesis operator* of the family $\{F(\omega)\}_{\omega\in\Omega}$. If F is a continuous frame for X, then the so-called frame operator B^*B is invertible.

Since we work with at most countable sets $\mathcal{G}$, we can take $\Omega = \mathcal{G} \times [0, \tau)$ (resp., $\Omega = \mathcal{G} \times \{0, 1, \ldots, \gamma\}$), and μ as the product of the counting measure on $\mathcal{G}$ and the Lebesgue measure on $[0, \tau)$ (resp., the counting measure on $\{0, 1, \ldots, \gamma\}$). In this case, F is called a *semi-continuous frame* (resp., frame) and Eq. (2.12) applied to (2.11) becomes

$$c_1\|x\|^2 \leq \sum_{g\in\mathcal{G}} \int_0^\tau |\langle x, (A^*)^t g\rangle|^2 dt \leq c_2\|x\|^2, \text{ for all } x \in X \tag{2.13}$$

(resp., Eq. (2.12) applied to Eq. (2.10) becomes Eq. (2.9)).

2.3 Relations Between the Dynamical Sampling and Observability Problem

We will establish relations between Dynamical Sampling and Control/Observability. To highlight the statements, we set them in box form.

Let X be an arbitrary complex separable Hilbert space and $A \in \mathcal{B}(X)$. We first show the connection between the set of vectors $\mathcal{G}$ from Problem 1 in the Dynamical Sampling setting and the observation operator B from Eq. (2.4) or (2.5). Given a countable set of vectors $\mathcal{G} \subseteq X$, we consider the observation space $Y = \ell^2(|\mathcal{G}|)$ and $B : X \to Y$ the analysis operator of $\mathcal{G}$, i.e., $Bx := (\langle x, g \rangle)_{g \in \mathcal{G}}$. If $\mathcal{G}$ is a Bessel set of vectors in X, then $B \in \mathcal{B}(X, Y)$. Conversely, let Y be any complex separable Hilbert space, and let $B \in \mathcal{B}(X, Y)$. If we consider $\{e_i\}_{i \in \mathbb{N}}$ an orthonormal basis for Y, then we can write

$$Bx = \sum_{i \in \mathbb{N}} \langle Bx, e_i \rangle_Y e_i = \sum_{i \in \mathbb{N}} \langle x, B^* e_i \rangle_X e_i, \qquad \forall x \in X. \tag{2.14}$$

Taking $\mathcal{G} := \{B^* e_i\}_{i \in \mathbb{N}}$, one may think of B as an operator from X to $\ell^2(\mathbb{N})$ of the form $x \mapsto (\langle x, B^* e_i \rangle)_{i \in \mathbb{N}}$. Since B is bounded, $\mathcal{G}$ is a Bessel set of vectors in X.

From now on, we think the observation operator B as the analysis operator of a set of sampling vectors $\mathcal{G}$.

Let $\gamma \in \mathbb{Z}_+ \cup \{\infty\}$. The observability map $\Psi_\gamma : X \to \ell^2(\{0, 1, \ldots, \gamma\}, Y)$ of the system (2.4) takes the form:

$$\begin{aligned}(\Psi_\gamma x_0)(k) &= BA^k x_0 \\ &= \left(\langle A^k x_0, g \rangle \right)_{g \in \mathcal{G}} \\ &= \left(\langle x_0, (A^*)^k g \rangle \right)_{g \in \mathcal{G}}.\end{aligned}$$

Hence, for discrete time, the **observability map** Ψ_γ associated to the dynamical system (2.4) is the **analysis operator** of $\{(A^*)^k g\}_{t \in \{0,1,\ldots,\gamma\}, g \in \mathcal{G}}$.

We have, for all $x \in X$,

$$\|\Psi_\gamma x\|^2_{\ell^2(\{0,1,\ldots,\gamma\}, Y)} = \sum_{k=0}^{\gamma} \|BA^k x\|^2 = \sum_{k=0}^{\gamma} \sum_{g \in \mathcal{G}} |\langle x, (A^*)^k g \rangle|^2. \tag{2.15}$$

Therefore, Problem 1 is the same as deciding whether the pair (A, B) is exactly observable. This is a consequence of Eq. (2.15): In Definition 2.3 the bounded below and above conditions for the observation map Ψ_γ are exactly formula (2.9).

The pair (A, B) associated to the discrete system (2.4) is **exactly observable** at time γ if and only if the family of vectors $\{(A^*)^k g\}_{t\in\{0,1,\dots,\gamma\},g\in\mathcal{G}}$ is a **frame** for X.

The **observability Grammian** $Q_\gamma = \Psi_\gamma^*\Psi_\gamma$ is the **frame operator** associated to $\{(A^*)^k g\}_{t\in\{0,1,\dots,\gamma\},g\in\mathcal{G}}$.

Moreover, with the exact same argument, (A, B) will be AOB if and only if the set (2.10) is complete in X.

The pair (A, B) associated to the discrete system (2.4) is **approximately observable** at time γ if and only if the family of vectors $\{(A^*)^k g\}_{t\in\{0,1,\dots,\gamma\},g\in\mathcal{G}}$ is a **complete system** in X.

It is worth noticing that even though it is not explicitly stated in the Dynamical Sampling problem that the set $\mathcal{G}$ needs to be Bessel, it is a necessary condition as can be deduced from the upper bound of Eq. (2.9). Also, this necessary condition implies that the operator B is admissible (see Remark 2.2).

The operator B is **admissible** for $\{A^k\}_{k\in\mathbb{Z}_+}$ at time γ if and only if $\{(A^*)^k g\}_{t\in\{0,1,\dots,\gamma\},g\in\mathcal{G}}$ is a **Bessel system** in X.

Analogously, let $\tau \in [0,\infty)\cup\{\infty\}$. Under the same specifications on the space Y and the operator B, the observability map $\Psi_\tau : X \to L^2([0,\tau), Y)$ takes the form:

$$(\Psi_\tau x_0)(t) = \left(\langle x_0, e^{tA^*} g\rangle\right)_{g\in\mathcal{G}}$$

and its norm is given by $\|\Psi_\tau x\|^2 = \int_0^\tau \sum_{g\in\mathcal{G}} |\langle x, e^{tA^*} g\rangle|^2 dt$. Thus, the exact (resp., approximate) observability conditions on the pair (A, B) in the continuous case (2.5) turn to be the problem of deciding whether the set $\{e^{tA^*} g\}_{t\in[0,\tau),g\in\mathcal{G}}$ is a continuous frame (resp., complete system).

The operator B is **admissible** for $\{e^{tA}\}_{t\in[0,\infty)}$ at time τ if and only if $\{e^{tA^*}g\}_{t\in[0,\tau),g\in\mathcal{G}}$ is a **Bessel system** in X.
The observability map Ψ_τ associated to the dynamical system (2.5) is the **analysis operator** of $\{(e^{tA^*}g\}_{t\in\{0,1,\ldots,\gamma\},g\in\mathcal{G}}$.

The pair (A, B) associated to the continuous system (2.5) is **approximately observable** at time τ if and only if the family of vectors $\{e^{tA^*}g\}_{t\in[0,\tau),g\in\mathcal{G}}$ is a **complete system** in X.

The pair (A, B) associated to the continuous system (2.5) is **exactly observable** at time τ if and only if the family of vectors $\{e^{tA^*}g\}_{t\in[0,\tau),g\in\mathcal{G}}$ is a **semi-continuous frame** for X.

The **observability Grammian** $Q_\tau = \Psi_\tau^*\Psi_\tau$ is the **frame operator** associated to $\{e^{tA^*}g\}_{t\in[0,\tau),g\in\mathcal{G}}$.

Remark 2.4 The problem of Dynamical Sampling for continuous time was originally proposed for $\{A^t\}$. When A is a self-adjoint and strictly positive operator, the operator A^t coincide with $e^{t\widetilde{A}}$ for an appropriate choice of $\widetilde{A} \in \mathcal{B}(X)$, for all $t \geq 0$. When A is not self-adjoint, it is not clear how to find (or decide if there exists) an appropriate $\widetilde{A}$. However, most results from dynamical sampling for A^t can be adapted to e^{tA}. Since our main purpose in this chapter is to explore the connections between Control Theory and Dynamical Sampling, we chose to work with the family of vectors

$$\{e^{tA^*}g\}_{t\in[0,\tau),g\in\mathcal{G}} \qquad \text{instead of} \qquad \{(A^*)^t g\}_{t\in[0,\tau),g\in\mathcal{G}},$$

adjusting the results from [5].

3 A Comparison Between Results from Control Theory and Dynamical Sampling

In finite dimension spaces X, conditions for exact and approximate observability and the frame condition, for the discrete and continuous problems, have been found in both areas and are completely understood. We could mention [11] from the

Observability perspective and [3] from the Dynamical Sampling one, among others. Surprisingly, the discrete and continuous cases, as well as the exact and approximate observability conditions, are all the same (cf. [12]). This case is also simple in another way: For finite-dimensional spaces, letting the system evolve infinite time is never necessary and adds no value, so only finite time sampling schemes shall be considered.

In infinite-dimensional cases, the relation between the different conditions is much more involved. The exact and approximate observability notions are no longer equivalent, the discrete and continuous case have different necessary and sufficient conditions, and sampling infinitely in time can prove to be an advantage in certain cases.

From the point of view of Control Theory, the analysis of all these aspects is quite complete when the dynamic operator A has an unconditional basis of eigenvectors. In general the results are proved for continuous scenarios and the analogous results are folklore for the discrete case. We aim to make them explicit every time it is possible. From the point of view of Dynamical Sampling, more things can be added since the only assumption on the operator is to be normal. Moreover, conditions on the discrete case are always explicitly stated. Using results from both fields, it can be proven that some of the relations between the continuous setting and discrete setting that are true in the finite-dimensional case do not hold in the infinite-dimensional case.

We recall next the definition of unconditional basis and Riesz sequences, since we will often use operator A whose eigenvectors are an unconditional basis for the space.

Definition 3.1 A sequence of vectors $\{\varphi_n\}$ in a Hilbert space X is said to be

1. an *unconditional sequence* if there exist positive constants c_1, c_2 such that for every finite sequence (a_n) it holds that

$$c_1 \sum |a_n|^2 \|\varphi_n\|^2 \leq \| \sum a_n \varphi_n \|^2 \leq c_2 \sum |a_n|^2 \|\varphi_n\|^2; \tag{3.1}$$

2. a *Riesz sequence* if there exist positive constants c_1, c_2 such that for every finite sequence (a_n) it holds that

$$c_1 \sum |a_n|^2 \leq \| \sum a_n \varphi_n \|^2 \leq c_2 \sum |a_n|^2. \tag{3.2}$$

If $\{\varphi_n\}$ is a complete system in X, it is called

1. an *unconditional basis* if it satisfies (3.1);
2. a *Riesz basis* if it satisfies (3.2).

In [19] a fundamental theorem [Theorem 3.3.2] was proved in terms of controllability. We restate it here using the duality between control and observability that has been discussed in 2.1.3. It is written in a general way so that it includes discrete and continuous dynamics as well as finite and infinite time. For that reason, we will denote by Ψ the observability map, and it will be understood that it takes the form of the discrete Ψ_γ or continuous Ψ_τ depending on the context. In this formulation it provides general conditions on exact observability that can then be specialized for each case. Later, in Proposition 3.1 we will make these specializations explicit.

Theorem 3.1 *Let X be a complex separable Hilbert space and $A \in \mathcal{B}(X)$. Let $\{\varphi_n\}_{n\geq 1}$ be the eigenvectors of A*

$$A\varphi_n = -\lambda_n \varphi_n, \qquad n \geq 1,$$

and suppose that they constitute an unconditional basis in X. Consider

$$\mathcal{E}_n := \Psi\varphi_n \qquad n \geq 1. \tag{3.3}$$

The pair (A, B) is exactly observable if and only if

1. *$\{\mathcal{E}_n\}$ is an unconditional sequence and*
2. *$\|\mathcal{E}_n\| \asymp \|\varphi_n\|$, that is, there exist positive constants c_1 and c_2 such that*

$$c_1\|\varphi_n\| \leq \|\mathcal{E}_n\| \leq c_2\|\varphi_n\|.$$

Proof First, we assume that $\{\mathcal{E}_n\}$ is an unconditional sequence satisfying $\|\mathcal{E}_n\| \asymp \|\varphi_n\|$. Since $\{\varphi_n\}_{n\geq 1}$ is an unconditional basis, for each $x \in X$ we have

$$x = \sum_{n\geq 1} a_n\varphi_n, \quad \text{with } \|x\|^2 \asymp \sum_{n\geq 1} |a_n|^2\|\varphi_n\|^2 < \infty.$$

Then, by the continuity and linearity of Ψ

$$(\Psi x)(t) = \sum_{n\geq 1} a_n\mathcal{E}_n(t).$$

By hypothesis, we get

$$\|\Psi x\|^2 \asymp \sum_{n\geq 1} |a_n|^2\|\mathcal{E}_n\|^2 \asymp \sum_{n\geq 1} |a_n|^2\|\varphi_n\|^2 \asymp \|x\|^2.$$

That is, the observability map is bounded from above and below, so (A, B) is EOB.

For the converse, the first statement is a consequence of two facts. One is that exact observability is equivalent to saying that the observation operator Ψ is an isomorphism on its image. The second is that topological isomorphisms map unconditional bases into unconditional bases. Therefore, Ψ will map the set $\{\varphi_n\}_{n\geq 1}$ into an unconditional basis of the image of Ψ, $\Psi(X)$, and in particular into an unconditional sequence. Thus, $\{\mathcal{E}_n\}$ is an unconditional sequence. Moreover, since exact observability also means that Ψ is bounded above and below, there exist constants $c_1, c_2 > 0$ such that

$$c_1\|\varphi_n\|^2 \leq \|\Psi\varphi_n\|^2 = \|\mathcal{E}_n\|^2 \leq c_2\|\varphi_n\|^2.$$

□

Corollary 3.1 *Under the notation of Theorem 3.1, suppose in addition that* $\{\varphi_n\}_{n\geq 1}$ *is an Riesz basis in* X*. The pair* (A, B) *is exactly observable if and only if* $\{\mathcal{E}_n\}$ *is a Riesz sequence.*

Proposition 3.1 *Under the hypothesis of Theorem 3.1,* $\|\mathcal{E}_n\| \asymp \|\varphi_n\|$ *if and only if*

- *for continuous and finite time* $\mathcal{T} = [0, \tau)$*:*

$$\sup_{n\geq 1} |\operatorname{Re}(\lambda_n)| < \infty, \tag{3.4}$$

$$\inf_{n\geq 1} \frac{\|B\varphi_n\|}{\|\varphi_n\|} > 0, \tag{3.5}$$

- *for continuous and infinite time* $\mathcal{T} = [0, \infty)$*: there exist constants* $c_1, c_2 > 0$ *such that for all* $n \geq 1$

$$\operatorname{Re}(\lambda_n) > 0, \tag{3.6}$$

$$c_1\sqrt{2\operatorname{Re}(\lambda_n)} \leq \frac{\|B\varphi_n\|}{\|\varphi_n\|} \leq c_2\sqrt{2\operatorname{Re}(\lambda_n)}, \tag{3.7}$$

- *for discrete and infinite time* $\mathcal{T} = \mathbb{Z}_+$*: there exist constants* $c_1, c_2 > 0$ *such that for all* $n \geq 1$

$$|\lambda_n| < 1, \tag{3.8}$$

$$c_1\sqrt{1 - |\lambda_n|^2} \leq \frac{\|B\varphi_n\|}{\|\varphi_n\|} \leq c_2\sqrt{1 - |\lambda_n|^2}. \tag{3.9}$$

Condition (3.4) is trivially satisfied if A is a bounded operator. In general this proposition and the last theorem still hold for closed operators with almost the same proof.

Proof We start the proof for the continuous finite case. Let us begin assuming $\|\mathcal{E}_n\| \asymp \|\varphi_n\|$. That is, there exist $c_1, c_2 > 0$, independent of n, such that

$$c_1\|\varphi_n\|^2 \leq \|\mathcal{E}_n\|^2_{L^2([0,\tau),Y)} = \frac{e^{-2\operatorname{Re}(\lambda_n)\tau} - 1}{-2\operatorname{Re}(\lambda_n)}\|B\varphi_n\|^2 \leq c_2\|\varphi_n\|^2. \tag{3.10}$$

Using that B is bounded and the left inequality, we conclude

$$c_1 \leq \frac{e^{-2\operatorname{Re}(\lambda_n)\tau} - 1}{-2\operatorname{Re}(\lambda_n)}\|B\|^2.$$

Therefore, $\sup_{n\geq 1}\{\operatorname{Re}(\lambda_n)\} < \infty$. Since $\|e^{tA}\| \leq e^{t\|A\|}$, we also know that $\sup_{n\geq 1}\{\operatorname{Re}(-\lambda_n)\} < \infty$. This proves that $\sup_{n\geq 1}|\operatorname{Re}(\lambda_n)| < \infty$. Using this and the left bound again, we obtain

$$c_1\|\varphi_n\|^2 \leq \frac{e^{-2\operatorname{Re}(\lambda_n)\tau}-1}{-2\operatorname{Re}(\lambda_n)}\|B\varphi_n\|^2 \leq K\|B\varphi_n\|^2,$$

where K is a positive constant. Therefore,

$$\inf_{n\geq 1}\frac{\|B\varphi_n\|}{\|\varphi_n\|} > 0.$$

The reverse implication is straightforward by noticing that there exist nonzero positive constants c_1, c_2 and d_1, d_2 such that

$$c_1\|\varphi_n\| \leq \|B\varphi_n\| \leq c_2\|\varphi_n\|, \qquad \text{and}$$

$$d_1 < \frac{e^{-2\operatorname{Re}(\lambda_n)\tau}-1}{-2\operatorname{Re}(\lambda_n)} < d_2, \qquad \text{for all } n.$$

Consequently, (3.10) holds.

The second and the third items are easy to prove. Just consider that the analogs of (3.10) are

$$c_1\|\varphi_n\|^2 \leq \|\mathcal{E}_n\|^2_{L^2([0,\infty),Y)} = \left(\int_0^\infty e^{-2\operatorname{Re}(\lambda_n)t}dt\right)\|B\varphi_n\|^2 \leq c_2\|\varphi_n\|^2,$$

and

$$c_1\|\varphi_n\|^2 \leq \|\mathcal{E}_n\|^2_{\ell^2(\mathbb{Z}_+,Y)} = \left(\sum_{k=0}^\infty |\lambda_n|^{2k}\right)\|B\varphi_n\|^2 \leq c_2\|\varphi_n\|^2.$$

Then, for the first case, $e^{-2\operatorname{Re}(\lambda_n)t} \in L^2([0,\infty))$ for all $n \geq 1$ so $\operatorname{Re}(-\lambda_n) < 0$ and $\int_0^\infty e^{-2\operatorname{Re}(\lambda_n)t}dt = \frac{1}{2\operatorname{Re}(\lambda_n)}$. This implies (3.7). For the second case, the geometric series converges if and only if $|\lambda_n| < 1$. Therefore (3.9) must hold. The reverse implications are trivial. □

Remark 3.1 Suppose that we are under the hypothesis of Theorem 3.1. The exact observability implies the following necessary conditions on the point spectrum $\sigma_p(A)$ of the dynamic operator A:

- For continuous and finite time $\mathcal{T} = [0,\tau)$, there exists $\alpha > 0$ such that

$$\sigma_p(A) \subseteq \{\lambda \in \mathbb{C} : -\alpha < \operatorname{Re}(\lambda) < \alpha\}.$$

- For continuous and infinite time $\mathcal{T} = [0, \infty)$, there exists $\alpha > 0$ such that

$$\sigma_p(A) \subseteq \{\lambda \in \mathbb{C} : -\alpha < \mathrm{Re}(\lambda) < 0\}.$$

- For discrete and infinite time $\mathcal{T} = \mathbb{Z}_+$,

$$\sigma_p(A) \subseteq \{\lambda \in \mathbb{C} : |\lambda| < 1\} =: \mathbb{D}.$$

In the remainder of this subsection we will separate the finite time case from the infinite time.

3.1 Finite Time

We will consider, unless specifically stated differently, pairs of the form (A, B) associated to dynamical systems as (2.5) for $0 < \tau < \infty$. The results from Dynamical Sampling that we will cite are proved in the literature for evolution operators of the form $\{A^t\}_{t\in[0,\tau)}$ instead of $\{e^{tA}\}_{t\in[0,\tau)}$. We will state them using the dynamics given by $\{e^{tA}\}_{t\in[0,\tau)}$ to show the relation with Control Theory. We recall that, in general, it is not true that $e^{tA} = (e^A)^t$, so the adaptations are not immediate. As a rule of thumb, most results will follow just by assuming that it is true, and the technical details and full proofs of these facts should be considered in each case and are presented in [14].

Some considerations have to be mentioned. Since e^{tA} is always defined for bounded operators (even for non-normal ones), the normality hypothesis that is often required in Dynamical Sampling is not further necessary. In a similar fashion, since e^{tA} is always invertible, the invertibility of A will not be required.

Taking all this into account, we will present most results written as similar to the original ones as possible.

From Theorem 3.1, we know that exact observability results can be derived from checking whether $\{\mathcal{E}_n\}$ is an unconditional sequence of the corresponding space. In general, the conditions for that to happen for finite time require results that exceed the aim of this presentation. The interested reader can find results in this direction in [7, 8, 17, 19, 20]. However, without appealing to these conditions, still several things can be said for finite time observability.

In particular, one desires the amount of sampling points (both spatial and temporal) to be as *small* as possible. We compile some theorems that claim something in this regard.

The first result establishes that every time one can recover the initial state of the system by sampling continuously during a finite time, one can choose to sample intermittently in time without losing information or stability. We could call this process as a discretization in the sampling scheme or observability problem. Notice that this gives no relation between the discrete and continuous problems.

It is just a middle ground where one still has a system evolving continuously but samples discretely. The following result is almost verbatim of [5, Theorem 5.4] when changing A^t to e^{tA}.

Theorem 3.2 *Let X be a complex separable Hilbert space, $A \in \mathcal{B}(X)$, and $\mathcal{G}$ a Bessel system of vectors in X. The following are equivalent:*

1. $\{e^{tA^*}g\}_{g\in\mathcal{G},t\in[0,\tau)}$ *is a semi-continuous frame for X.*
2. *There exists $\delta > 0$ such that for any finite set $T = \{t_i : i = 1,\dots,n\}$ with $0 = t_1 < t_2 < \dots < t_n \le t_{n+1} = \tau$ and $|t_{i+1} - t_i| < \delta$ for all $i \in \{1,\dots,n\}$, the system $\{e^{tA^*}g\}_{g\in\mathcal{G},t\in T}$ is a frame for X.*
3. *There exists a finite partition $T = \{t_i : i = 1,\dots,n\}$ of $[0,\tau)$ with $0 = t_1 < t_2 < \dots < t_n \le \tau$ such that $\{e^{tA^*}g\}_{g\in\mathcal{G},t\in T}$ is a frame for X.*

The last theorem and its corollary can be further generalized for non-Hilbert spaces in the context of the observability problem. Its proof follows the arguments given in [5] but without requiring the inner product of the space X. In particular, one can work with a B that is not the analysis operator of any set of vectors.

As a conclusion, when in the presence of systems that are exactly observable in finite time, there is no necessity to sample continuously. As a side product, one can also conclude that when we have observable systems in finite time, it is necessary to sample at an infinite quantity of spatial points. (Otherwise, the discretized version will be a finite set of vectors and it can never be a frame.) That is obviously a disadvantage that we will try to solve recurring to a trade-off between infinite space and infinite time sampling schemes. The following corollary is an adaptation of [5, Corollary 5.3].

Corollary 3.2 *Let X be an infinite-dimensional Hilbert space, and let $A \in \mathcal{B}(X)$. If $\{e^{tA^*}g\}_{g\in\mathcal{G},t\in[0,\tau)}$ is a frame for X, then $|\mathcal{G}| = +\infty$.*

Here is another point where both communities have reached similar results. Without explicitly recurring to the discretization of the sampling process, a similar result was also established in [19, Theorem 1.3.19] for the controllability problem. We rewrite it to be directly readable for the observability problem. Its original version should be translated using the duality relationship of Theorem 2.1.

Theorem 3.3 *Let X be an infinite-dimensional Hilbert space, and let $A \in \mathcal{B}(X)$. If the observation operator $B \in \mathcal{B}(X, Y)$ is compact, then the pair (A, B) cannot be exactly observable at any moment $\tau < \infty$. In particular, if B is of finite rank, then the pair (A, B) cannot be exactly observable at any moment $\tau < \infty$.*

Another important question beyond the discretization problem is how much time one has to sample to have observability. In this direction one can analyze the dependence on time of three different properties of systems (A, B). The first one is the admissibility condition on B, and the second and third are the approximate and exact observability of the pair.

It can be proved that the admissibility condition on B is independent of the sampling time. We have the following theorem that is inspired by [5, Theorem 4.3] but without requiring that A is normal.

Theorem 3.4 *Let X, Y be a complex separable Hilbert spaces and $A \in \mathcal{B}(X)$.*

1. *A countable set of vectors $\mathcal{G} \subseteq X$ is a Bessel system in X, if and only if $\{e^{tA^*}g\}_{g\in\mathcal{G},t\in[0,\tau)}$ is a Bessel system in X, for any $\tau < \infty$.*
2. *A linear map $B : X \to Y$ is continuous if and only if it is admissible for $\{e^{tA}\}$ at some finite time τ.*

Moreover, statements 1. and 2. are equivalent.

Proof We will first prove the equivalence of the statements using the relations given in Sect. 2.3. Given $\mathcal{G} \subseteq X$ a Bessel system, we have that its analysis operator $Bx := (\langle x, g\rangle)_{g\in\mathcal{G}}$ is linear and bounded from X to $\ell^2(|\mathcal{G}|)$. Reciprocally, given $B \in \mathcal{B}(X, Y)$, as in (2.14), $Bx = \sum_{i\in\mathbb{N}}\langle Bx, e_i\rangle_Y e_i = \sum_{i\in\mathbb{N}}\langle x, B^*e_i\rangle_X e_i$, then B is isometrically isomorphic to the analysis operator of the set $\mathcal{G} := \{B^*e_i\}_{i\in\mathbb{N}}$ for an orthonormal basis $\{e_i\}_{i\in\mathbb{N}}$ of Y.

Further, as the analysis operator B of $\mathcal{G} \subseteq X$ is admissible for $\{e^{tA}\}_{t\in[0,\infty)}$ at time τ if and only if $\{e^{tA^*}g\}_{t\in[0,\tau),g\in\mathcal{G}}$ is a Bessel system in X we conclude that 1. and 2. are equivalent.

Let us now prove 2. By Remark 2.2, every $B \in \mathcal{B}(X, Y)$ is admissible at any finite time.

Conversely, let $\tau_0 < \infty$ be such that B is admissible for $\{e^{tA}\}$ at τ_0. If $\tau_0 = 0$, there is nothing to prove. For $\tau_0 > 0$, it is immediate that B is admissible for $\{e^{tA}\}$ at any time $0 < \tau \leq \tau_0$. We consider the auxiliary operator $T_\tau : X \to X$ defined by

$$T_\tau x := \int_0^\tau e^{tA}x\, dt,$$

for some appropriate choice of $0 < \tau \leq \tau_0$. Since $\tau \leq \tau_0 < \infty$, we can show by using elementary properties of the Bochner integral that $T \in \mathcal{B}(X)$. Indeed,

$$\|T_\tau x\| = \left\| \int_0^\tau e^{tA}x\, dt \right\| \leq \int_0^\tau \|e^{tA}x\|\, dt \leq \tau_0 e^{\tau_0\|A\|}\|x\| \qquad \forall x \in X.$$

To finish the proof, notice that if there exists $0 < \tau \leq \tau_0$ such that T_τ is invertible, then B is bounded if and only if BT_τ is.

To prove that T_τ is invertible for some $0 < \tau \leq \tau_0$, we will show that for an appropriate choice of τ, 0 does not belong to the spectrum of T_τ, $\sigma(T_\tau)$. If $h(t, z) := \frac{e^{tz}-1}{z} = \sum_{k=1}^{\infty}\frac{t^k z^{k-1}}{k!}$, for $t > 0$ and $z \in \mathbb{C}$, by the fundamental theorem of calculus for Bochner integrals, we have

$$h(\tau, A)x = h(\tau, A)x - h(0, A)x = \int_0^\tau \frac{d}{dt}h(t, A)x\, dt = \int_0^\tau e^{tA}x\, dt = T_\tau x.$$

In addition, since $e^{tA}x$ is Bochner integrable as a function of t, it is easy to see that for each $g \in X$ $\langle T_\tau x, g\rangle = \int_0^\tau \langle e^{tA}x, g\rangle\, dt$.

Now, choosing $0 < \tau' \leq \tau_0$ sufficiently small such that $h(\tau', \cdot)$ is non-vanishing for all $z \in \sigma(A)$, we have $\sigma(T_{\tau'}) = h(\tau', \sigma(A))$, and hence $0 \notin \sigma(T_{\tau'})$.

To show that for this choice of τ' the operator $BT_{\tau'}$ is bounded, we again look at B as the analysis operator of a set of vectors $\mathcal{G}$, $Bx := (\langle x, g\rangle)_{g\in\mathcal{G}}$. Using Hölder's inequality, the admissibility of B, and properties of Bochner integrals, we obtain that for every $x \in X$

$$\begin{aligned}\|BT_{\tau'}x\|^2 &= \sum_{g\in\mathcal{G}} |\langle T_{\tau'}x, g\rangle|^2 = \sum_{g\in\mathcal{G}} \left| \int_0^{\tau'} \langle e^{tA}x, g\rangle\, dt \right|^2 \\ &\leq \sum_{g\in\mathcal{G}} \tau' \int_0^{\tau'} |\langle e^{tA}x, g\rangle|^2 dt = \tau' \int_0^{\tau'} \|Be^{tA}x\|^2 dt \leq \tau' K_{\tau'} \|x\|^2.\end{aligned}$$

□

Approximate observability is also independent of time (see [19, Theorem 1.3.1]). This result is known as Kalman's Theorem. As before, we state it here in the context of observability using Theorem 2.1.

Theorem 3.5 *[19, Theorem 1.3.1] Let X and Y be Hilbert spaces, $A \in \mathcal{B}(X)$, and $B \in \mathcal{B}(X, Y)$. The following are equivalent:*

1. *The pair (A, B) associated to the continuous system* (2.5) *is AOB at some finite time $0 < \tau_0 < \infty$.*
2. *The pair (A, B) associated to the continuous system* (2.5) *is AOB at any time $\tau \in (0, \infty)$.*
3. *The pair (A, B) associated to the continuous system* (2.5) *is AOB at infinite time.*
4. *The pair (A, B) associated to the discrete system* (2.4) *is AOB at infinite time.*

Moreover, the equivalence between the first three items holds for general dynamic operators A (not only for bounded ones).

Notice that Kalman's Theorem does not only prove the independence of time for finite time but for infinite time as well. Moreover, it also says that the property of being approximately observable is shared by both, the continuous and discrete problem, that is exactly the equivalence between 3. and 4. For approximate observability, all the possible schemes presented in this chapter are equivalent!

Under the Dynamical Sampling hypotheses, the previous points are exactly:

1. The family $\{e^{tA^*}g\}_{g\in\mathcal{G}, t\in[0,\tau_0)}$ is complete in X for some $0 < \tau_0 < \infty$.
2. The family $\{e^{tA^*}g\}_{g\in\mathcal{G}, t\in[0,\tau)}$ is complete in X for any $\tau \in (0, \infty)$.
3. The family $\{e^{tA^*}g\}_{g\in\mathcal{G}, t\in[0,\infty)}$ is complete in X.
4. The family $\{(A^*)^k g\}_{g\in\mathcal{G}, k\in\mathbb{Z}_+}$ is complete in X.

For exact observability, we only state a sufficient condition for time independence. It is done in the settings of Dynamical Sampling as an adaptation of [5, Theorem 5.5].

Theorem 3.6 *Let A be a self-adjoint operator and $\mathcal{G}$ be a countable set in a Hilbert space X. Then,* $\{e^{tA^*}g\}_{g\in\mathcal{G},t\in[0,1)}$ *is a semi-continuous frame in X if and only if* $\{e^{tA^*}g\}_{g\in\mathcal{G},t\in[0,\tau)}$ *is a semi-continuous frame in X for all finite positive τ. As a consequence, the self-adjoint Dynamical Sampling problem is independent of time.*

3.2 Infinite Time

When considering an infinite-dimensional state space, one of the disadvantages of finite time schemes is that we need to sample at an infinite set of spatial points (see Corollary 3.2). It is therefore of interest to look at infinite schemes but using only a finite number of spatial samples.

3.2.1 Exponentially Stable Dynamics

In this subsection we will show that for exponentially (strong asymptotically) stable dynamics it is not possible to have infinite time sampling schemes with only finitely many space samples.

Proposition 3.2 *Let X be a Hilbert space, and let $A \in \mathcal{B}(X)$ generate an exponentially (resp., strong asymptotically) stable dynamic. If the pair (A, B) associated to the continuous system* (2.5) *(discrete system* (2.4)*) is exactly observable at infinite time, then it is exactly observable at some finite time.*

Proof For the continuous system, the proof can be found in [23, Proposition 6.5.2]. Using the same arguments of that proof, one can show that if the discrete system (2.4) is exactly observable at infinite time, then it is observable at some finite time. Indeed, since there exist positive constants c_1 and c_2 such that

$$c_1\|x\|^2 \leq \sum_{k\in\mathbb{Z}_+} \|BA^k x\|^2 \leq c_2\|x\|^2,$$

for all $\gamma \in \mathbb{Z}_+$, the observability maps $\Psi_\gamma x = \sum_{k=0}^{\gamma} BA^k x$ are bounded operators. Therefore we only need to see that there exists $\gamma > 0$ such that the operator Ψ_γ is bounded from below. For this,

$$\sum_{k=0}^{\gamma} \|BA^k x\|^2 = \sum_{k\in\mathbb{Z}_+} \|BA^k x\|^2 - \sum_{k\geq\gamma+1} \|BA^k x\|^2$$

$$\geq c_1\|x\|^2 - \sum_{k\in\mathbb{Z}_+} \|BA^k A^{\gamma+1}x\|^2$$

$$\geq c_1\|x\|^2 - c_2\|A^{\gamma+1}x\|^2 \geq \left(c_1 - c_2\|A\|^{2(\gamma+1)}\right)\|x\|^2.$$

Since $\|A\| < 1$, we can take $\gamma \in \mathbb{Z}_+$ sufficiently large such that $(c_1 - c_2\|A\|^{2(\gamma+1)}) > 0$ and for such γ we obtain that Ψ_γ is bounded below. □

In the notation of Dynamical Sampling this proposition can be rewritten as follows: *Let $\mathcal{G}$ be a subset of vectors in X. If A generates an exponentially stable dynamic such that $\{e^{tA^*}g\}_{g\in\mathcal{G},t\in[0,\infty)}$ is a semi-continuous frame for X, then there exists $\tau < \infty$ for which $\{e^{tA^*}g\}_{g\in\mathcal{G},t\in[0,\tau)}$ is a semi-continuous frame for X. Accordingly, if A generates a strongly asymptotically stable dynamic such that $\{(A^*)^k g\}_{g\in\mathcal{G},k\in\mathbb{Z}_+}$ is a frame for X, then there exists $\gamma \in \mathbb{Z}_+$ such that $\{(A^*)^k g\}_{g\in\mathcal{G},n\in\{0,1,\ldots,\gamma\}}$ is a frame for X.*

As an immediate consequence of this proposition and Theorem 3.3, we obtain the following corollary.

Corollary 3.3 *Let A generate an exponentially (resp., strongly asymptotically) stable dynamic in an infinite-dimensional Hilbert space X. If the pair (A, B) associated to the system* (2.5) *(resp.,* (2.4)*) is EOB at infinite time, then B cannot be a compact operator. In particular, it cannot be a finite rank operator.*

Analogously, for the Dynamical Sampling context we recover Corollary 3.2: *If A generates an exponentially (resp., strong asymptotically) stable dynamic in an infinite-dimensional Hilbert space and $\{e^{tA^*}g\}_{g\in\mathcal{G},t\in[0,\infty)}$ is a semi-continuous frame (resp., $\{(A^*)^k g\}_{g\in\mathcal{G},k\in\mathbb{Z}_+}$ is a frame) for X, then $|\mathcal{G}| = \infty$.*

3.2.2 Non-exponentially Stable Dynamics

For non-exponentially (strong asymptotically) stable dynamic generator, the scenario is quite different. We can exhibit some results that allow to sample a signal at a finite amount of spatial points and recover it after an infinite time (continuous or discrete) evolution process.

We start with the case of discrete dynamics. For ease of the exposition, we show only the results for rank one operator B. Due to the Riesz Representation Theorem, they are equivalent to sampling at a single spatial point b. In this case, if the dynamic operator A has an orthonormal basis of eigenvectors the following theorem holds.

Theorem 3.7 *([3, Theorem 3.16]) Let A be a bounded linear operator in an infinite-dimensional Hilbert space X with an orthogonal basis of eigenvectors $\{\varphi_n\}_{n\geq 1}$ such that $A\varphi_n = -\lambda_n\varphi_n$. Let $b \in X$. Then the set $\{(A^*)^k b\}_{k\in\mathbb{Z}_+}$ is a frame for X if and only if*

1. $|\lambda_n| < 1$ *for all* $n \in \mathbb{N}$.
2. $|\lambda_n| \to 1$.

3. $\{\lambda_n\}$ *satisfies Carleson's condition in the disc:*

$$\inf_n \prod_{k\neq n} \left| \frac{\lambda_n - \lambda_k}{1 - \overline{\lambda}_n \lambda_k} \right| \geq \delta, \qquad \textit{for some } \delta > 0.$$

4. $0 < c_1 \leq \frac{|\langle \varphi_n, b\rangle|}{\|\varphi_n\|\sqrt{1-|\lambda_n|^2}} \leq c_2 < \infty$, *for some constants* c_1, c_2 *independent for all* $n \in \mathbb{N}$.

This theorem has its analogy in the observability question of discrete systems: *Let* X *and* A *be as in Theorem 3.7. Let* B *be a rank one operator represented by* $b \in X$*. Then the pair* (A, B) *associated to the discrete system* (2.4) *is EOB at infinite time if and only if conditions 1. to 4. in Theorem 3.7 hold.*

Note that this can be extended to the case of the existence of unconditional basis of eigenvectors, instead of an orthonormal one. This is based on the following facts: Given $A \in \mathcal{B}(X)$,

- X has an unconditional basis of eigenvectors of A if and only if it has a Riesz basis of eigenvectors of A. Indeed, if $\{\varphi_n\}$ is an unconditional basis of eigenvectors of A, then $\{\frac{\varphi_n}{\|\varphi_n\|}\}$ is a Riesz basis. The converse is trivial.
- X has a Riesz basis of eigenvectors of A if and only if A is conjugated by an isomorphism to an operator that has an orthonormal basis of eigenvectors. In effect, if $\{\varphi_n\}$ is a Riesz basis of eigenvectors of A, then there exists an isomorphism $T \in \mathcal{B}(X)$ such that $T\varphi_n = e_n$, where $\{e_n\}$ is an orthonormal basis of X and are the eigenvectors of the operator $\widetilde{A} := TAT^{-1} \in \mathcal{B}(X)$. The reciprocal follows similarly.

Moreover, for the cases when A is a normal operator, and under discrete dynamics, it is proved in [2] that if the set $\{(A^*)^k b\}_{k\in\mathbb{Z}_+}$ is a frame for X, then the operator A must be *diagonal.*

All this can be extended to the continuous case. If the dynamic operator A has an unconditional basis of eigenvectors, from Theorem 3.1, Proposition 3.1, and Corollary 3.3 we are lead to some necessary and sufficient conditions for observability of a continuous infinite time system.

Theorem 3.8 *Let* A *be a bounded linear operator in an infinite-dimensional Hilbert space* X *with an unconditional basis of eigenvectors* $\{\varphi_n\}_{n\geq 1}$ *such that* $A\varphi_n = -\lambda_n\varphi_n$*. Let* B *be a rank one operator represented by* $b \in X$*. Then the pair* (A, B) *associated to the continuous system* (2.5) *is EOB at infinite time if and only if*

1. $\mathrm{Re}(\lambda_n) > 0$ *for all* $n \in \mathbb{N}$.
2. $\mathrm{Re}(\lambda_n) \to 0$ *(A cannot generate an exponentially stable dynamic).*
3. $\{\mathcal{E}_n\}_{n\geq 1}$*, given by* $\mathcal{E}_n(t) = e^{-\lambda_n t}\langle \varphi_n, b\rangle$*, is an unconditional sequence of* $L^2([0,\infty), \mathbb{C})$.
4. $0 < c_1 \leq \frac{|\langle \varphi_n, b\rangle|}{\|\varphi_n\|\sqrt{2\,\mathrm{Re}(\lambda_n)}} \leq c_2 < \infty$, *for some constants* c_1, c_2 *independent of* n.

Translating this into the Dynamical Sampling notation, one can read: *Let X and A be as in Theorem 3.8. Let $b \in X$. Then, the set $\{e^{tA^*}b\}_{t\in[0,\infty)}$ is a semi-continuous frame for X if and only if conditions 1. to 4. in Theorem 3.8 hold.*

Proof Since $\text{rank}(B) = 1$, $Bx = \langle x, b\rangle$ for all $x \in X$. Thus, $\mathcal{E}_n = \Psi\varphi_n = e^{-\lambda_n t}\langle\varphi_n, b\rangle$. By Theorem 3.1, the system (2.5) is EOB if and only if condition 3. holds and $\mathcal{E}_n \asymp \varphi_n$. Moreover, using Proposition 3.1, $\mathcal{E}_n \asymp \varphi_n$ if and only if conditions 1. and 4. hold. This says that the the system (2.5) is EOB if and only if conditions 1., 3., and 4. hold. Further, by Corollary 3.3 2. is necessary. Moreover, 2. is a consequence of the other conditions: Since $\{\frac{\varphi_n}{\|\varphi_n\|}\}$ is a Riesz basis, using 4. we have

$$c_1^2\sum_n |2\,\text{Re}(\lambda_n)| \leq \sum_n |\langle\frac{\varphi_n}{\|\varphi_n\|}, b\rangle|^2 \leq \|b\|^2.$$

Therefore, $\text{Re}(\lambda_n) \to 0$. □

Notice that under the above hypothesis the functions $\mathcal{E}_n$ are scalar-valued. By [19, Theorem 4.2.2 and Remark 4.2.3] we have an equivalent formulation for condition 3: If $\{\lambda_n\}_{n\geq 1} \subseteq \mathbb{C}$ is such that $\text{Re}(\lambda_n) > 0$ for all $n \in \mathbb{N}$, then $\{e^{-\lambda_n t}\}_{n\geq 1}$ is an unconditional sequence of $L^2([0,\infty))$ if and only if the sequence $\{\lambda_n\}_{n\geq 1}$ satisfies the *half-plane Carleson's condition*

$$\prod_{k\neq n}\left|\frac{\lambda_n - \lambda_k}{\lambda_n - \overline{\lambda_k}}\right| \geq \delta > 0, \qquad \forall n \geq 1.$$

Moreover, the Carleson's condition in Theorem 3.7 is just an equivalent formulation for $\mathcal{E}_n(k) = \langle b, \varphi_n\rangle(\lambda_n)^k$ to be an unconditional sequence of $\ell^2([0,\infty))$ (cf. [3] and references therein).

It can also be shown that if A is normal, in order for the pair (A, B) associated to the continuous system (2.5) to be EOB at infinite time, the operator A must be diagonal [14].

It is worth noticing here that there is no equivalent formulation of Kalman's Theorem (Theorem 3.5) for exact observability instead of approximate observability. Therefore, $\{e^{tA^*}g\}_{g\in\mathcal{G},t\in[0,\infty)}$ is complete if and only if $\{(A^*)^k g\}_{g\in\mathcal{G},k\in\mathbb{Z}_+}$ is complete, but it is not true that $\{e^{tA^*}g\}_{g\in\mathcal{G},t\in[0,\infty)}$ is a semi-continuous frame for X if and only if $\{(A^*)^k g\}_{g\in\mathcal{G},k\in\mathbb{Z}_+}$ is a frame for X.

To construct a counterexample, it is enough to find an operator A that satisfies the conditions on Theorem 3.7 (in particular, condition 1.) but that fails to satisfy conditions of Theorem 3.8. Notice that one can start with a diagonal operator $\widetilde{A}$ for this purpose and then rotate the set of eigenvalues to have a new operator A that has at least one eigenvalue with $\text{Re}(\lambda_n) < 0$.

However, denoting by $\mathbb{C}_+ := \{z \in \mathbb{C} : \text{Re}(z) > 0\}$ and using the self-inverse bijection $M : \mathbb{D} \to \mathbb{C}_+$, given by

$$M(z) := \frac{1-z}{1+z},$$

it easy to notice that conditions 1. to 3. in Theorem 3.7 for $\{\lambda_n\}$ are in correspondence with conditions 1. to 3. in Theorem 3.8 for the sequence $\{M(\lambda_n)\}$.

Assuming that $|\lambda_n + 1| > \varepsilon > 0$ for all $n \in \mathbb{N}$, and defining $\widetilde{b} \in X$ satisfying $\langle \widetilde{b}, \varphi_n \rangle = \frac{\sqrt{2}}{|1+\lambda_n|} \langle b, \varphi_n \rangle$ for all $n \geq 1$, we obtain

$$\frac{|\langle b, \varphi_n \rangle|}{\|\varphi_n\| \sqrt{1 - |\lambda_n|^2}} = \frac{\sqrt{2} |\langle b, \varphi_n \rangle|}{\|\varphi_n\| \sqrt{2\,\mathrm{Re}(M(\lambda_n))}\, |1 + \lambda_n|}$$
$$= \frac{|\langle \widetilde{b}, \varphi_n \rangle|}{\|\varphi_n\| \sqrt{2 Re(M(\lambda_n))}}.$$

Therefore, condition 4. in Theorem 3.10 for b is equivalent to condition 4. in Theorem 3.9 for $\widetilde{b}$.

This yields: *The pair* (A, B) *associated to the discrete system* (2.4) *is EOB at infinite time if and only if the pair* $(M(A), \widetilde{B})$ *associated to the continuous system* (2.5) *is EOB at infinite time, where* $\widetilde{B}$ *is the analysis operator for* $\widetilde{b}$.

The condition $|\lambda_n + 1| > \varepsilon > 0$ is needed for A and $M(A)$ to be simultaneously bounded. It can be replaced without loss of generality by $|\lambda_n - z| > \varepsilon > 0$ for some $|z| = 1$ if we use the map M composed with a rotation of the circle.

As a final comment, it is worth noticing that, by Kalman's Theorem, and the fact that EOB implies AOB, conditions 1. to 4. in Theorems 3.7 and 3.8 are sufficient conditions for AOB under any type of dynamics. That is, for example, that conditions for the discrete system to be EOB automatically imply the continuous system to be AOB.

More general conditions for exact observability when not in the case of rank one operators can also be found in [9, 10, Theorem 2.3] and [19, Theorems 4.2.4 and 4.2.5].

The analogies between discrete and continuous cases will still follow the lines of the ones given in this chapter, although many properties turn out to be more difficult to handle than the rank one case. For details, we refer to [14].

Acknowledgments We thank the anonymous referee for his careful reading of the manuscript and many suggestions that improved the presentation of this chapter.

This project has mainly been carried out while R. Díaz Martín was holding a postdoctoral fellowship from CONICET, and I. Medri was holding postdoctoral fellowships from CONICET and Vanderbilt University. The authors thank those institutions for their support.

U. Molter was supported by grants UBACyT 20020170100430BA (University of Buenos Aires), PIP11220150100355 (CONICET), and PICT 2014-1480 (Ministerio de Ciencia, Tecnología e Innovación, Argentina).

References

1. Aldroubi, A., Aceska, R., Davis, J., Petrosyan, A.: Dynamical sampling in shift-invariant spaces, Contemp. Math. of the AMS **603**, 139–148 (2013).
2. Aldroubi, A., Cabrelli, C., Çakmak, A.F., Molter, U., Petrosyan, A.: Iterative actions of normal operators, J. Funct. Anal. **272**, 1121–1146 (2017).
3. Aldroubi, A., Cabrelli, C., Molter, U., Tang, S.: Dynamical Sampling, Appl. Comput. Harmon. Anal. **42**, 378–401 (2017).
4. Aldroubi, A., Davis, J., Krishtal, I.: Dynamical Sampling: Time Space Trade-off, Appl. Comput. Harmon. Anal. **34**, 495–503 (2013).
5. Aldroubi, A., Huang, L.X., Petrosyan, A.: Frames induced by the action of continuous power of an operator, J. Math. Anal. Appl. **478**, 1059–1084 (2019).
6. Aldroubi, A., Petrosyan, A.: Dynamical Sampling and Systems from Iterative Actions of Operators. In: Pesenson I., Le Gia Q., Mayeli A., Mhaskar H., Zhou DX. (eds) Frames and Other Bases in Abstract and Function Spaces. Applied and Numerical Harmonic Analysis. Birkhäuser, Cham (2017).
7. Avdonin, S.A., Ivanov, S.A.: Exponential Riesz bases of subspaces and divided differences, Algebra i Analiz **13**:3, 1–17 (2001); St. Petersburg Math. J. **13**:3, 339–351 (2002).
8. Avdonin, S.A., Ivanov, S.A.: Families of exponentials: the method of moments in controllability problems for distributed parameter systems. Cambridge University Press, (1995).
9. Cabrelli, C., Molter, U., Paternostro, V., Philipp, F.: Dynamical sampling on finite index sets, JAMA **140**, 637–667 (2020).
10. Cabrelli, C., Molter, U., Suárez, D.: Multi-orbital frames through model spaces. arXiv preprint arXiv:1908.11011v1 (2019).
11. Chen, C.T., Desoer, C.A.: Controllability and Observability of Composite Systems. IEEE Transactions on Automatic Control, **12**:4, 402–409 (1967).
12. Chen, C.T., Desoer, C.A.: A Proof of Controllability of Jordan Form State Equations. IEEE Transactions on Automatic Control (1968).
13. Coron, J.M.: Control and Nonlinearity. Mathematical Surveys and Monographs Volume 136, American Mathematical Society (2007).
14. Díaz Martín, R., Medri, I., Molter, U.: Continuous and discrete dynamical sampling. J. Math. Anal. Appl. **499**, 2 (2021).
15. Eisner, T.: Stability of operators and operator semigroups. Birkhäuser (2010).
16. Herman, R., Krener, A.: Nonlinear controllability and observability, IEEE Transactions on automatic control, **22**:5, 728–740 (1977).
17. Hruscev, S.V., Nikolskii, N.K., Pavlov, B.S.: Unconditional bases of exponentials and of reproducing kernels. Complex analysis and spectral theory. Springer, Berlin, Heidelberg, 214–335 (1981).
18. Lu, Y., Vetterli, M.: Spatial super-resolution of a diffusion field by temporal oversampling in sensor networks, In: Acoustics, Speech and Signal Processing, 2009. IEEE International Conference on ICASSP 2249–2252 (2009).
19. Nikolskii, N.K.: Operators, Functions, and Systems: An Easy Reading. Mathematical Surveys and Monographs Volume 93, American Mathematical Society (2002).
20. Nikolskii, N.K.: Treatise on the shift operator: spectral function theory. Springer Science and Business Media, Volume 273 (2012).
21. Pazy, A.: Semigroups of Linear Operators and Applications to Partial Differential Equations. Springer (1983).
22. Ranieri, J., Chebira, A., Lu, Y.M., Vetterli, M.: Sampling and reconstructing diffusion fields with localized sources, In: IEEE International Conference on Acoustics, Speech and Signal Processing (ICASSP), 4016–4019 (2011).
23. Tucsnak, M., Weiss, G. : Observation and control for operator semigroups. Springer Science and Business Media (2009).

Linear Multiscale Transforms Based on Even-Reversible Subdivision Operators

Nira Dyn and Xiaosheng Zhuang

Abstract Multiscale transforms for real-valued data, based on interpolatory subdivision operators, have been studied in recent years. They are easy to define and can be extended to other types of data, for example, to manifold-valued data. In this chapter, we define linear multiscale transforms, based on certain linear, non-interpolatory subdivision operators, termed "even-reversible." For such operators, we prove, using Wiener's lemma, the existence of an inverse to the linear operator defined by the even part of the subdivision mask and term it "even-inverse." We show that the non-interpolatory subdivision operators, with spline or pseudo-spline masks, are even-reversible and derive explicitly, for the quadratic and cubic spline subdivision operators, the symbols of the corresponding even-inverse operators. We also analyze properties of the multiscale transforms based on even-reversible subdivision operators, in particular, their stability and the rate of decay of the details.

1 Introduction

A multiscale transform is a ubiquitous way for representing data of the form $\{(t_k, f_k),\ k \in \mathbb{Z}\}$, with $t_k = kh$ for some positive h, and f_k a real value associated with the point t_k. A general approach to multiscale transforms was introduced in [21]. Such a linear representation can be obtained by discrete wavelet transforms (see e.g. [3]). Another common way for generating a multiscale transform is based on linear or non-linear interpolatory subdivision schemes (see e.g. [1, 6, 10, 19]). For a detailed analysis of several non-linear multiscale transforms, see [12] and references therein. The main applications of multiscale transforms are in data compression and denoising (see e.g. [1, 2, 6, 18–20, 28]). Multiscale transforms

N. Dyn
School of Mathematical Sciences, Tel-Aviv University, Tel Aviv, Israel
e-mail: niradyn@post.tau.ac.il

X. Zhuang (✉)
Department of Mathematics, City University of Hong Kong, Kowloon, Hong Kong
e-mail: xzhuang7@cityu.edu.hk

M. Hirn et al. (eds.), *Excursions in Harmonic Analysis, Volume 6*, Applied and Numerical Harmonic Analysis, https://doi.org/10.1007/978-3-030-69637-5_16

based on interpolatory subdivision operators are simple and rather intuitive. The linear ones were extended to manifold-valued data in [14, 15] and [24].

In this chapter, we extend the construction of multiscale transforms from interpolatory subdivision operators to a wider class of subdivision operators termed *even-reversible* operators and show that a large class of the subdivision operators studied in the literature are even-reversible. We derive properties of the multiscale transforms based on even-reversible subdivision operators, such as decay rates of the data generated by the transforms, and the stability of the transforms.

1.1 Multiscale Transform: From Interpolatory to Even-Reversible Subdivision

First, we present the multiscale transform based on an interpolatory subdivision operator $\mathcal{S}$. Let $\mathbf{f}$ denote a bi-infinite sequence with elements $\{f_k \in \mathbb{R},\ k \in \mathbb{Z}\}$. Since $\mathcal{S}$ is interpolatory,

$$(\mathcal{S}\mathbf{f})_{2k} = f_k, \quad k \in \mathbb{Z},$$

and it is straightforward to decimate data at level j, $\mathbf{f}^j$, to that at the coarser level $j-1$, by taking every second element. Then, the refinement of $\mathbf{f}^{j-1}$ by $\mathcal{S}$ is exact for the even elements of $\mathbf{f}^j$. Here is the pyramid of data generated by j *decomposition* steps of the multiscale transform based on $\mathcal{S}$,

$$\mathbf{f}^{\ell-1} = \{f_k^{(\ell-1)} = f_{2k}^{(\ell)},\ k \in \mathbb{Z}\},\ \mathbf{d}^\ell = \mathbf{f}^\ell - \mathcal{S}\mathbf{f}^{\ell-1},\ \ \ell = j, j-1, \ldots, 1. \tag{1.1}$$

The elements of $\mathbf{d}^\ell$ are termed *details* at level ℓ. The data $\mathbf{f}^j$ can be obtained exactly from the data of the pyramid by the *reconstruction* steps

$$\mathbf{f}^\ell = \mathcal{S}\mathbf{f}^{\ell-1} + \mathbf{d}^\ell, \quad \ell = 1, 2, \ldots, j. \tag{1.2}$$

Note that every second element of $\mathbf{d}^\ell$ vanishes, since $\mathcal{S}$ is interpolatory. Thus, this method yields details satisfying

$$\mathbf{d}_{2k}^\ell = 0, \quad k \in \mathbb{Z}. \tag{1.3}$$

Multiscale transforms based on non-interpolatory subdivision operators were studied before, with the decimation from level j to level $j-1$, defined by "reverse subdivision." In [26], $\mathbf{f}^{j-1} = \mathcal{D}\mathbf{f}^j$, where the decimation operator $\mathcal{D}$ is related to the subdivision operator $\mathcal{S}$ by a least squares fit of $\mathcal{S}\mathbf{f}^{j-1}$ to $\mathbf{f}^j$. This method is improved in [25], by minimizing a functional consisting of two terms; one is $\|\mathcal{S}(\mathcal{D}\mathbf{f}^j) - \mathbf{f}^j\|_2$, and the other is a roughness measure of $\mathbf{f}^{j-1} = (\mathcal{D}\mathbf{f}^j)$. It is clear that in this approach, the details do not satisfy (1.3). Specific ways to reverse Chaikin scheme and Catmull–Clark scheme are presented in [22] and [23], respectively.

Here, we propose to reverse only the even elements of $\mathbf{f}^j$, as in the case of interpolatory subdivision operators, namely for a given subdivision operator $\mathcal{S}$ to use a decimation operator $\mathcal{D}$ such that

$$(\mathcal{S}(\mathcal{D}\mathbf{f}^j))_{2k} = f_{2k}^{(j)}, \quad k \in \mathbb{Z}. \tag{1.4}$$

This is achieved by using Wiener's lemma [16], which guarantees the existence of such a decimation operator, under mild conditions on the mask of $\mathcal{S}$. We term such subdivision operators, for which $\mathcal{D}$ satisfying (1.4) exists, even-reversible and refer to $\mathcal{D}$ as the even-inverse of $\mathcal{S}$.

1.2 Outline

Section 2 consists of mathematical tools and results used in the chapter, such as Wiener's lemma. In Sect. 3, we first give a sufficient condition on the mask of a subdivision operator guaranteeing that the operator is even-reversible, then show that pseudo-spline subdivision operators are even-reversible, and derive explicit expression of the symbols of the even-inverse operators corresponding to the quadratic and cubic spline subdivision operators. In Sect. 4, we derive the decay rate of the details in the pyramid generated by the multiscale transform and analyze the stability of the transform. Some final remarks are given in Sect. 5, and some technical proofs are postponed to the Appendix.

2 Preliminaries

In this section, we introduce some necessary notation and known results related to subdivision operators and Wiener algebra, which are used in this chapter.

We denote by $l(\mathbb{Z})$ the space of real-valued sequences $\alpha : \mathbb{Z} \to \mathbb{R}$ and by $l_0(\mathbb{Z}) \subseteq l(\mathbb{Z})$ the space of sequences of finite support. For $p \in [1, \infty]$, $l_p(\mathbb{Z})$ denotes the usual space of l_p sequences. That is, $l_p(\mathbb{Z}) := \left\{\alpha \in l(\mathbb{Z}) : \|\alpha\|_p := \left(\sum_{k\in\mathbb{Z}} |\alpha_k|^p\right)^{1/p} < \infty\right\}$, $1 \leq p < \infty$, and $l_\infty(\mathbb{Z}) := \{\alpha \in l(\mathbb{Z}) : \|\alpha\|_\infty := \sup_{k\in\mathbb{Z}} |\alpha_k| < \infty\}$. Note that the inclusion $l_0(\mathbb{Z}) \subseteq l_1(\mathbb{Z}) \subseteq l_p(\mathbb{Z}) \subseteq l_q(\mathbb{Z}) \subseteq l_\infty(\mathbb{Z})$ holds for any $1 < p < q < \infty$.

We say that $\alpha \in l(\mathbb{Z})$ is a *mask* if $\alpha \in l_1(\mathbb{Z})$. A mask α is of finite support ($\alpha \in l_0(\mathbb{Z})$) if $\alpha_k = 0$ for all $|k| \geq N$ for some integer N. Given a mask α, we can define the (dyadic) *upscaling rule* or *subdivision operator* $\mathcal{S}_\alpha : l_\infty(\mathbb{Z}) \to l_\infty(\mathbb{Z})$ associated with α by

$$(\mathcal{S}_\alpha c)_k := \sum_{\ell\in\mathbb{Z}} \alpha_{k-2\ell} c_\ell, \quad k \in \mathbb{Z},\ c \in l_\infty(\mathbb{Z}). \tag{2.1}$$

It is easily seen that $\mathcal{S}_\alpha$ is a bounded linear operators on $l_\infty(\mathbb{Z})$ provided that $\alpha \in l_1(\mathbb{Z})$. The *subdivision scheme* based on $\mathcal{S}_\alpha$ is the repeated application of (2.1), generating a sequence of sequences $\mathcal{S}_\alpha^j c$, $j \in \mathbb{N}$, for $c \in l_\infty(\mathbb{Z})$. The subdivision scheme is said to be *convergent* if the sequence of piecewise linear interpolants to the data $\{(k2^{-j}, (\mathcal{S}_\alpha^j c)_k), k \in \mathbb{Z}\}$, $j \in \mathbb{N}$, is uniformly convergent for any $c \in l_\infty(\mathbb{Z})$. For more about subdivision schemes, see e.g. [9].

We next introduce convolution, downsampling, and upsampling operations. For two sequences $\alpha \in l_1(\mathbb{Z})$ and $c \in l_\infty(\mathbb{Z})$, we defined the *convolution* $\alpha * c \in l_\infty(\mathbb{Z})$ to be

$$(\alpha * c)_k := \sum_{\ell\in\mathbb{Z}} \alpha_\ell c_{k-\ell}, \quad k \in \mathbb{Z},$$

and the *downsampling operator* $\downarrow 2$ applied to a sequence c to be $(c \downarrow 2)_k := c_{2k}$, $k \in \mathbb{Z}$, as well as the *upsampling operator* $\uparrow 2$:

$$(c \uparrow 2)_k := \begin{cases} c_{k/2} & k \text{ even} \\ 0 & \text{otherwise} \end{cases}, \qquad k \in \mathbb{Z}.$$

The subdivision in (2.1) can be restated as $\mathcal{S}_\alpha c = \alpha * (c \uparrow 2)$.

For a sequence $c \in l(\mathbb{Z})$, we define its *symbol* to be

$$c(z) := \sum_{k\in\mathbb{Z}} c_k z^k, \quad z \in \mathbb{C}.$$

Its even part c_{ev} to be $(c_{ev})_k = c_{2k} = (c \downarrow 2)_k$, $k \in \mathbb{Z}$, and its odd part c_{od} to be $(c_{od})_k = c_{2k+1}$, $k \in \mathbb{Z}$. The symbols of the even and odd parts can be determined by

$$c_{ev}(z^2) = \frac{c(z) + c(-z)}{2} \quad \text{and} \quad c_{od}(z^2) = \frac{c(z) - c(-z)}{2z}.$$

In terms of symbolic computation, it is easily shown that

$$\begin{aligned} &[c \downarrow 2](z) = c_{ev}(z), && [c \uparrow 2](z) = c(z^2), \\ &c(z) = c_{ev}(z^2) + z\, c_{od}(z^2), && [\alpha * c](z) = \alpha(z)c(z). \end{aligned}$$

Moreover, we have $[\mathcal{S}_\alpha c](z) = \alpha(z)c(z^2)$.

In this chapter, we investigate the following multiscale transform based on a subdivision operator $\mathcal{S}_\alpha$:

$$c^{(j-1)} = \mathcal{D}_\gamma c^{(j)} := \gamma * (c^{(j)} \downarrow 2), \quad d^{(j)} := c^{(j)} - \mathcal{S}_\alpha c^{(j-1)}, \tag{2.2}$$

where $\mathcal{D}_\gamma$ is a *decimation operator* associated with a mask γ to be determined. Iterating (2.2) yields a pyramid consisting of the data $\{c^{(0)}; d^{(1)}, \ldots, d^{(j)}\}$, where $c^{(0)}$ is the coarse approximation coefficients and $d^{(\ell)}$ are the detail coefficients at level $\ell = 1, \ldots, j$. The reconstruction (backward transform) from $\{c^{(0)}; d^{(1)}, \ldots, d^{(j)}\}$ is straightforward:

$$c^{(\ell)} = \mathcal{S}_\alpha c^{(\ell-1)} + d^{(\ell)}, \quad \ell = 1, \ldots, j, \tag{2.3}$$

and it has the perfect reconstruction property for any pair (γ, α) of masks. However, the detail coefficients $d^{(\ell)}$, $\ell = 1, \ldots, j$ do not necessarily satisfy (1.3).

Given a subdivision operator $\mathcal{S}_\alpha$ associated with a finitely supported mask $\alpha \in l_0(\mathbb{Z})$, we investigate the decimation operator $\mathcal{D}_\gamma$ associated with a mask γ so that $d^{(\ell)}$ satisfies $d^{(\ell)}_{2k} = 0$, $k \in \mathbb{Z}$, for $\ell = 1, \ldots, j$, as in the case when $\mathcal{S}_\alpha$ is an interpolatory subdivision operator and $\mathcal{D}_\gamma c = c \downarrow 2$. Thus, we are seeking a mask γ such that

$$[(\mathcal{I} - \mathcal{S}_\alpha \mathcal{D}_\gamma)c] \downarrow 2 = 0 \tag{2.4}$$

for any sequence $c \in l_\infty(\mathbb{Z})$, where $\mathcal{I}$ is the identity operator. In such a case, the detail coefficients can be downsampled by a factor of 2 without loss of information, which is a desirable property in data compression.

To solve γ from (2.4), we use Wiener's lemma [16]. First, we introduce Wiener algebra.

Let $\mathbb{T} := \{z \in \mathbb{C} : |z| = 1\}$. The *Wiener algebra* $W(\mathbb{T})$ consists of all complex-valued functions f on $[-\pi, \pi]$ such that f has absolutely convergent Fourier series. That is, $W(\mathbb{T}) := \{f \in C([-\pi, \pi]) : \|f\| := \sum_{n\in\mathbb{Z}} |\widehat{f}(n)| < \infty\}$, where $\widehat{f}(n) := \frac{1}{2\pi}\int_{-\pi}^{\pi} f(x)e^{-inx}dx$ is the nth Fourier coefficient of f. The Wiener algebra $W(\mathbb{T})$ is closed under pointwise multiplication of functions. It is easy to show that

$$\|fg\| \le \|f\| \cdot \|g\| \quad \forall f, g \in W(\mathbb{T}).$$

Thus, the Wiener algebra is a commutative unitary Banach algebra. The Wiener algebra $W(\mathbb{T})$ is isomorphic to the Banach algebra $l_1(\mathbb{Z})$ with the isomorphism given by the Fourier transform: $f \mapsto \{\widehat{f}(n)\}_{n\in\mathbb{Z}}$.

Theorem 2.1 (Wiener's Lemma) *If $f \in W(\mathbb{T})$ and $f(x) \neq 0$ for all $x \in [-\pi, \pi]$, then $1/f \in W(\mathbb{T})$.*

Consequently, for $\alpha \in l_0(\mathbb{Z})$, if $\alpha(z) \neq 0$ for all $z \in \mathbb{T}$, then α has an inverse $\alpha^{-1} =: \gamma \in l_1(\mathbb{Z})$ determined by $\gamma(z) = 1/\alpha(z)$, $z \in \mathbb{T}$.

Any mask α in the Wiener algebra defines a *bi-infinite Toeplitz matrix* of the form $A_\alpha = (\alpha_{j-k})_{j,k\in\mathbb{Z}}$. Then, A_α has an inverse if and only if α has an inverse. In this case, we have that the inverse $(A_\alpha)^{-1}$ of A_α satisfies $(A_\alpha)^{-1} = A_{\alpha^{-1}}$. A Toeplitz matrix A_α is also a linear operator on $l_2(\mathbb{Z})$. One can show that the l_2-operator norm of A_α is given by [13, Section 2.2]

$$\|A_\alpha\|_2 := \sup_{\|c\|_2=1} \|A_\alpha c\|_2 = \sup_{z\in\mathbb{T}} |\alpha(z)|. \tag{2.5}$$

We say that a finitely supported mask $\alpha \in l_0(\mathbb{Z})$ is *s-banded* if $\alpha_k = 0$ for all $|k| > s$. The following theorem gives the decay of the inverse of a banded Hermitian Toeplitz matrix.

Theorem 2.2 (Theorem 2.1 in [27]) *Let A be a bi-infinite Toeplitz matrix acting on $l_2(\mathbb{Z})$ and assume A to be Hermitian, positive definite, and s-banded (i.e. $A_{k,\ell} = 0$ if $|k-\ell| > s$). Set $\kappa = \|A\|_2\,\|A^{-1}\|_2$, $q = (\sqrt{\kappa}-1)/(\sqrt{\kappa}+1)$, and $\lambda = q^{1/s}$. Then,*

$$(A^{-1})_{k,\ell} \le K\lambda^{|k-\ell|}, \quad k,\ell \in \mathbb{Z},$$

where $K = \|A^{-1}\|_2 \max\left\{1, \frac{(1+\sqrt{\kappa})^2}{2\kappa}\right\}$.

A direct consequence of the above results is the decay property of the inverse γ of an invertible mask $\alpha \in l_0(\mathbb{Z})$. For more general results on the decay property of the inverse of a mask, we refer to [16].

Corollary 2.1 *Let $\alpha \in l_0(\mathbb{Z})$ be s-banded (i.e. $\alpha_k = 0$ if $|k| > s$) with $\alpha(z) > 0$ for all $z \in \mathbb{T}$. Set $\kappa = \frac{\sup_{z\in\mathbb{T}}|\alpha(z)|}{\inf_{z\in\mathbb{T}}|\alpha(z)|}$, $q = (\sqrt{\kappa}-1)/(\sqrt{\kappa}+1)$, and $\lambda = q^{1/s}$. Then, $\gamma = \alpha^{-1}$ exists and*

$$|\gamma_\ell| \le K\lambda^{|\ell|}, \quad \ell \in \mathbb{Z},$$

where $K = \frac{1}{\inf_{z\in\mathbb{T}}|\alpha(z)|} \max\left\{1, \frac{(1+\sqrt{\kappa})^2}{2\kappa}\right\}$.

Proof Since $\alpha(z) > 0$ for all $z \in \mathbb{T}$, by Wiener's lemma, $\gamma = \alpha^{-1}$ exists. Now, by Fejér–Riesz lemma [3], there exists a real-valued mask $\beta \in l_0(\mathbb{Z})$ such that $\alpha(z) = \beta(z)\beta(1/z)$ for all $z \in \mathbb{T}$. Then, one can easily show that $A_\alpha = A_\beta(A_\beta)^\top$. Hence, from $\alpha(z) > 0$ for all $z \in \mathbb{T}$, we conclude that A_α is symmetric positive definite. Now, by Theorem 2.2 and the fact that $\|A_\alpha\|_2 = \sup_{z\in\mathbb{T}}|\alpha(z)|$ and $\|A_\alpha^{-1}\|_2 = \|A_{\alpha^{-1}}\|_2 = \inf_{z\in\mathbb{T}}|\alpha(z)|$, we conclude the result. □

3 Even-Reversible Subdivision

In this section, we give a sufficient condition on the mask of a subdivision operator for the existence of the linear multiscale transform defined in (2.2) that satisfies property (2.4). We call such a subdivision operator *even-reversible*. We show that for a large class of even-reversible subdivision operators, the elements of the mask of the decimation operator in (2.4) decay exponentially, and we prove that most of the subdivision operators in the literature are in that class.

3.1 Existence of the Multiscale Transform

We first deduce a sufficient condition on the mask of a subdivision operator for the existence of a linear multiscale transform satisfying (2.4).

Theorem 3.1 *Let $\mathcal{S}_\alpha$ be a subdivision operator with a mask α defined as in* (2.1), *and let $\mathcal{D}_\gamma$ be a decimation operator with a mask γ defined as in* (2.2). *Then,* (2.4) *holds, i.e., $d := (\mathcal{I} - \mathcal{S}_\alpha \mathcal{D}_\gamma)c$ satisfies $(d \downarrow 2) = 0$ for any $c \in l_\infty(\mathbb{Z})$, if and only if γ is the inverse of α_{ev}.*

Proof Note that (2.4) is equivalent to $\frac{c(z)+c(-z)}{2} - \frac{[\mathcal{S}_\alpha \mathcal{D}_\gamma c](z)+[\mathcal{S}_\alpha \mathcal{D}_\gamma c](-z)}{2} = 0$; that is,

$$c_{ev}(z^2) = [\mathcal{S}_\alpha \mathcal{D}_\gamma c]_{ev}(z^2) = \alpha_{ev}(z^2)[\mathcal{D}_\gamma c](z^2) = \alpha_{ev}(z^2)\gamma(z^2)c_{ev}(z^2), \quad z \in \mathbb{T}.$$

Consequently, (2.4) is equivalent to $\alpha_{ev}(z)\gamma(z) = 1$ for all $z \in \mathbb{T}$, i.e., γ is the inverse of the even part α_{ev} of α. □

Now, using Wiener's lemma, we get

Corollary 3.1 *For any mask $\alpha \in l_0(Z)$ satisfying $\alpha_{ev}(z) \neq 0$ for all $z \in \mathbb{T}$,* (2.4) *holds with $\gamma \in l_1(Z)$ determined by $\gamma(z) = \frac{1}{\alpha_{ev}(z)}$, $z \in \mathbb{T}$.*

We call a subdivision operator $\mathcal{S}_\alpha$ with a mask α such that $\alpha_{ev}(z) \neq 0$ for all $z \in \mathbb{T}$ *even-reversible*, and we call the decimation operator $\mathcal{D}_\gamma$ satisfying (2.4) the *even-inverse* of $\mathcal{S}_\alpha$. It is shown in the next subsection that for a large class of finitely supported masks, the condition $\alpha_{ev}(z) \neq 0$ for all $z \in \mathbb{T}$ does hold.

In view of Theorem 3.1, the multiscale transform (2.2) becomes

$$\begin{cases} c^{(\ell-1)}(z) & = \alpha_{ev}^{-1}(z) c_{ev}^{(\ell)}(z) \\ d_{od}^{(\ell)}(z) & = c_{od}^{(\ell)}(z) - \alpha_{od}(z)\alpha_{ev}^{-1}(z)c_{ev}^{(\ell)}(z)\,, \quad \ell = j, \ldots, 1. \\ d_{ev}^{\ell}(z) & \equiv 0 \end{cases} \tag{3.1}$$

We term (3.1) *multiscale transform based on an even-reversible subdivision (MTER)*.

In terms of matrix computation, the MTER can be written as

$$\begin{cases} c^{(\ell-1)} & = A_{\alpha_{ev}^{-1}} c_{ev}^{(\ell)} \\ d_{od}^{(\ell)} & = c_{od}^{(\ell)} - A_{\alpha_{od}} A_{\alpha_{ev}^{-1}} c_{ev}^{(\ell)}\,, \quad \ell = j, \ldots, 1. \\ d_{ev}^{(\ell)} & \equiv 0 \end{cases}$$

In particular, when α is *interpolatory*,

$$\alpha_{ev}(z) \equiv 1 \text{ and } \gamma(z) = \alpha_{ev}^{-1}(z) \equiv 1, \tag{3.2}$$

and the MTER is reduced to an interpolatory pyramid:

$$\begin{cases} c^{(\ell-1)} & = (c^{(\ell)} \downarrow 2) \\ d_{od}^{(\ell)} & = c_{od}^{(\ell)} - \alpha_{od} * (c^{(\ell)} \downarrow 2)\,, \quad \ell = j, \ldots, 1. \\ d_{ev}^{(\ell)} & \equiv 0 \end{cases}$$

3.2 The Even-Inverse of the Subdivision Operator

In case $\alpha_{ev} > 0$ for all $z \in \mathbb{T}$ and α_{ev} is s-banded, then by Corollary 2.1, $\gamma = \alpha_{ev}^{-1}$ exists, and the elements of the mask γ decay exponentially, which is an important feature for the computation of the decimation operation $\mathcal{D}_\gamma c$. More precisely,

$$|\gamma_\ell| \leq K\lambda^{|\ell|} \qquad \forall \ell \in \mathbb{Z},$$

where K and λ are defined as in Corollary 2.1 with α there replaced by α_{ev}.

3.3 Examples

In this subsection, we provide some examples of commonly used subdivision operators with finitely supported masks. We show that for a large class of masks, the corresponding subdivision operators are even-reversible.

First, we give two examples of spline subdivision operators of low order for which we can compute explicitly the even-inverse. We omit the case of the linear spline subdivision operator because it is interpolatory.

Example 1 (Quadratic Spline) Consider the mask α for the (centered) cubic B-spline of order 3: $\alpha(z) = \frac{z^{-1}(1+z)^3}{2^2}$. That is, $\alpha = \{\alpha_{-1}, \alpha_0, \alpha_1, \alpha_2\} = \frac{1}{4}\{1, 3, 3, 1\}_{[-1,2]}$ (i.e. supp$(\alpha) = [-1, 2] \cap \mathbb{Z}$ with $\alpha_{-1} = 1/4$, $\alpha_0 = \alpha_1 = 3/4$, and $\alpha_2 = 1/4$). Then,

$$\alpha_{ev}(z^2) = \frac{1}{2}(\alpha(z) + \alpha(-z)) = \frac{1}{4}(3 + z^2),$$

(continued)

Example 1 (continued)
or as a sequence, $\alpha_{ev} = \frac{1}{4}\{3, 1\}_{[0,1]}$. We have $|\alpha_{ev}(z)| = |\frac{1}{4}(3+z)| \geq \frac{1}{2}$ for all $z \in \mathbb{T}$. Consequently, the inverse γ of α_{ev} exists and is given by

$$\gamma(z) = \frac{1}{\alpha_{ev}(z)} = \frac{4}{3+z} = \frac{4}{3}\left(1 - \frac{1}{3}z + \frac{1}{9}z^2 + \cdots\right) = \frac{4}{3}\sum_{k=0}^{\infty}\left(-\frac{1}{3}\right)^k z^k,$$

which is the symbol of the even-inverse operator $\mathcal{D}_\gamma$. Thus, the quadratic spline subdivision operator is even-reversible. Direct computations show that $\|\gamma\|_1 = 2$, $\|A_\gamma\|_2 = 2$, and $\|\gamma\|_\infty = 4/3$.

Example 2 (Cubic Spline) Consider the mask α for the (centered) B-spline of order 4: $\alpha(z) = \frac{z^{-2}(1+z)^4}{2^3}$. That is, $\alpha = \frac{1}{8}\{1, 4, 6, 4, 1\}_{[-2,2]}$. Then,

$$\alpha_{ev}(z^2) = \frac{1}{2}(\alpha(z) + \alpha(-z)) = \frac{1}{8}(z^{-2} + 6 + z^2),$$

or as a sequence, $\alpha_{ev} = \frac{1}{8}\{1, 6, 1\}_{[-1,1]}$. Note that $\alpha_{ev}(z) = \frac{1}{8}(z^{-1}+6+z) \geq \frac{1}{2}$ for all $z \in \mathbb{T}$. Consequently, the inverse γ of α_{ev} exists. Hence, the cubic spline subdivision operator is even-reversible, and

$$\gamma(z) = \frac{1}{\alpha_{ev}(z)} = \frac{8}{z^{-1}+6+z} = \frac{4}{3} \times \frac{1}{1 + \frac{z^{-1}+z}{6}} = \frac{4}{3}\sum_{n=0}^{\infty}\left(\frac{-1}{6}\right)^n (z^{-1}+z)^n$$

$$= \frac{4}{3}\left(\sum_{n=0}^{\infty} 6^{-2n}(z^{-1}+z)^{2n} - \sum_{n=0}^{\infty} 6^{-2n-1}(z^{-1}+z)^{2n+1}\right).$$

By (2.5), we have $\|A_\gamma\|_2 = 2$. We succeeded to obtain an explicit expression of $\gamma(z)$ of the form

$$\gamma(z) = \sqrt{2} + \sqrt{2}\sum_{k=1}^{\infty}\left(\frac{-1}{3+2\sqrt{2}}\right)^k \times (z^k + z^{-k}). \tag{3.3}$$

With this explicit form of $\gamma(z)$, the decimation operation in the MTER can be implemented. Moreover, from (3.3), we get $\|\gamma\|_1 = 2$, $\|\gamma\|_\infty = \sqrt{2}$, and the exponential decay of the elements of γ. The proof of (3.3) is not straightforward and is given in the Appendix.

From the above examples, one can expect that spline subdivision operators are even-reversible. In fact, it holds more generally. The class of spline subdivision operators can be regarded as a special subclass of a larger class called *pseudo-spline* subdivision operators, which we introduce next.

Let $n, \nu \in \mathbb{N} \cup \{0\}$ satisfy $0 \le \nu \le \lfloor n/2 \rfloor - 1$. Define

$$\alpha^{n,\nu}(z) = \frac{z^{-\lfloor n/2 \rfloor}(1+z)^n}{2^{n-1}} \sum_{j=0}^{\nu} \binom{n/2+j-1}{j} \left(\frac{1}{2} - \frac{z+z^{-1}}{4}\right)^j. \tag{3.4}$$

When $\nu = 0$, the mask $\alpha^{n,0} = \alpha^n$ belongs to the family of masks of spline subdivision operators. When $n = 2k$ and $\nu = k-1$, the mask $\alpha^{2k,k-1}$ is the Deslauries–Dubuc's interpolatory mask [5]. For $n = 2k$ and $0 \le \nu \le k-1$, the masks $\alpha^{2k,\nu}$ are the masks of the primal pseudo-splines (pseudo-splines of type II), while for $n = 2k+1$, the masks $\alpha^{2k+1,\nu}$, $0 \le \nu \le k-1$ are the masks of the dual pseudo-splines. For more about pseudo-splines, see [4, 7, 8, 11] and references therein.

The next theorem states that pseudo-spline subdivision operators are even-reversible, which implies that all spline subdivision operators are even-reversible.

Theorem 3.2 *Let $n, \nu \in \mathbb{N} \cup \{0\}$ satisfying $0 \le \nu \le \lfloor n/2 \rfloor - 1$ and $\alpha^{n,\nu}$ be the pseudo-spline mask defined as in* (3.4)*. Then, the following holds:*

(1) $\alpha_{ev}^{n,\nu}(1) = 1$ *and* $\|A_{\alpha_{ev}^{n,\nu}}\|_2 = \max_{z \in \mathbb{T}} |\alpha_{ev}^{n,\nu}(z)| = 1$.
(2) $\gamma = (\alpha_{ev}^{n,\nu})^{-1}$ *exists with* $\gamma(1) = 1$.
(3) *The l_2-norm of A_γ is given by* $\|A_\gamma\|_2 = \frac{2^{\lfloor \frac{n-1}{2} \rfloor + \nu}}{\sum_{j=0}^{\nu} \binom{n/2+\nu}{j}}$.
(4) *The elements of γ decay exponentially.*

The proof of Theorem 3.2 is postponed to the Appendix. We remark that (i) when $\nu = 0$, we have $\|A_\gamma\|_2 = 2^{\lfloor (n-1)/2 \rfloor}$ and (ii) when $n = 2k$ and $\nu = k-1$, we have $\gamma(z) \equiv 1$.

4 Decay and Stability

We now turn to the study of the decay property of the pyramid sequences and the stability of the transforms.

4.1 Decay

Let $\Delta : l(\mathbb{Z}) \to l(Z)$ denote the *difference operator*: $(\Delta c)_k = c_{k+1} - c_k$ for $c \in l(\mathbb{Z})$. Then, Δ commutes with convolution operators. Indeed, by $[\Delta c](z) =$

$(z^{-1}-1)c(z)$ and $[\gamma * c](z) = \gamma(z)c(z)$, we have

$$[\Delta(\gamma * c)](z) = (z^{-1}-1)(\gamma(z)c(z)) = \gamma(z)[(z^{-1}-1)c(z)] = [\gamma * (\Delta c)](z).$$

Suppose the data sequence $c^{(j)} = f\big|_{2^{-j}\mathbb{Z}}$, where f is a function in $C^1(\mathbb{R})$ with bounded first derivative. Then,

$$[\Delta c^{(j)}]_k = f(2^{-j}(k+1)) - f(2^{-j}k) = f'(\xi)\cdot 2^{-j}$$

for some $\xi \in (2^{-j}k, 2^{-j}(k+1))$. Thus, $\|\Delta c^{(j)}\|_\infty \le K2^{-j}$ with $K = \|f'\|_\infty$.

Applying the scheme in (2.2), we have a pyramid of data consisting of approximation coefficient sequences $c^{(\ell)}$ and detail coefficient sequences $d^{(\ell)}$. The following two results concern the decay property of $\Delta c^{(\ell)}$ and $d^{(\ell)}$.

Theorem 4.1 (Difference of Approximation Coefficients) *Let $c^{(j)}$ be such that $\|\Delta c^{(j)}\|_\infty \le K2^{-j}$. Let $c^{(\ell)} := \gamma * (c^{(\ell+1)} \downarrow 2)$ with $\gamma \in l_1(\mathbb{Z})$ and $0 \le \ell \le j-1$. Then,*

$$\|\Delta c^{(\ell)}\|_\infty \le K\,\|\gamma\|_1^j \cdot (2\|\gamma\|_1)^{-\ell}, \quad 0 \le \ell \le j-1. \tag{4.1}$$

Proof For $c_{ev} = c \downarrow 2$, we have

$$(\Delta c_{ev})_k = c_{2k+2} - c_{2k} = (c_{2k+2} - c_{2k+1}) + (c_{2k+1} - c_{2k}) = (\Delta c)_{2k+1} + (\Delta c)_{2k}, \quad k \in \mathbb{Z},$$

and hence, $\|\Delta c_{ev}\|_\infty \le 2\|\Delta c\|_\infty$. Since Δ commutes with convolution operators, we have

$$\begin{aligned}\|\Delta c^{(j-1)}\|_\infty &= \|\Delta(\gamma * (c^{(j)} \downarrow 2))\|_\infty = \|\gamma * \Delta(c^{(j)} \downarrow 2))\|_\infty \\ &\le \|\gamma\|_1 \cdot \|\Delta(c^{(j)} \downarrow 2)\|_\infty \le \|\gamma\|_1 \cdot 2\|\Delta c^{(j)}\|_\infty.\end{aligned}$$

Iterating the above inequality starting with $\|\Delta c^{(j)}\|_\infty \le K2^{-j}$, we conclude (4.1). □

Under some very mild conditions on the masks α, γ, we can show that the detail coefficients have the same decay property as the differences of the approximation coefficients.

Theorem 4.2 (Detail Coefficients) *Let $c^{(j)} \in l_\infty(\mathbb{Z})$ be a sequence. Define $c^{(\ell-1)}$ and $d^{(\ell)}$ to be*

$$c^{(\ell-1)} := \gamma * (c^{(\ell)} \downarrow 2), \quad d^{(\ell)} := c^{(\ell)} - \mathcal{S}_\alpha c^{(\ell-1)}, \quad 1 \le \ell \le j,$$

where $\alpha, \gamma \in l_1(\mathbb{Z})$ are masks satisfying

$$\sum_k \alpha_{2k} = \sum_k \alpha_{2k+1} = \sum_k \gamma_k = 1 \tag{4.2}$$

and

$$\sum_{k\in\mathbb{Z}} |\alpha_k|\ |k| = K_\alpha < \infty, \quad 2\sum_{k\in\mathbb{Z}} |\gamma_k|\ |k| = K_\gamma < \infty. \tag{4.3}$$

Then,

$$\|d^{(\ell)}\|_\infty \le K_{\alpha,\gamma}\|\Delta c^{(\ell)}\|_\infty, \quad 1 \le \ell \le j \tag{4.4}$$

with $K_{\alpha,\gamma} := K_\gamma\|\alpha\|_1 + K_\alpha\|\gamma\|_1$. *In particular, if* $\|\Delta c^{(j)}\|_\infty \le K \cdot 2^{-j}$ *for some* K *independent of* j, *then*

$$\|d^{(\ell)}\|_\infty \le \left(K \cdot K_{\alpha,\gamma} \cdot \|\gamma\|_1^j\right) \cdot (2\|\gamma\|_1)^{-\ell}, \quad 1 \le \ell \le j. \tag{4.5}$$

Proof Let $\eta = \{\eta_k\}_{k\in\mathbb{Z}} := c^{(\ell-1)} = \gamma * (c^{(\ell)} \downarrow 2)$. Then, $\eta_k = \sum_s \gamma_{k-s} c_{2s}^{(\ell)}$. By (4.2), we have $d_k^{(\ell)} = c_k^{(\ell)} - [S_\alpha \eta]_k = c_k^{(\ell)} - \sum_s \alpha_{k-2s}\eta_s = \sum_s \alpha_{k-2s}\left[\sum_n \gamma_{s-n}(c_k^{(\ell)} - c_{2n}^{(\ell)})\right]$. Consequently,

$$\begin{aligned}
\|d^{(\ell)}\|_\infty &\le \sum_s |\alpha_{k-2s}|\left[\sum_n |\gamma_{s-n}||2n-k|\right]\|\Delta c^{(\ell)}\|_\infty \\
&\le \sum_s |\alpha_{k-2s}|\left[\sum_n |\gamma_{s-n}|(|2n-2s| + |k-2s|)\right]\|\Delta c^{(\ell)}\|_\infty \\
&\le \sum_s |\alpha_{k-2s}|\left[K_\gamma + |k-2s|\|\gamma\|_1\right]\|\Delta c^{(\ell)}\|_\infty \\
&= (K_\gamma\|\alpha\|_1 + K_\alpha\|\gamma\|_1)\|\Delta c^{(\ell)}\|_\infty = K_{\alpha,\gamma}\|\Delta c^{(\ell)}\|_\infty \\
&\le K \cdot K_{\alpha,\gamma} \cdot \|\gamma\|_1^j \cdot (2\|\gamma\|_1)^{-\ell},
\end{aligned}$$

where the last inequality follows from (4.1). □

Theorems 4.1 and 4.2 imply the following corollary.

Corollary 4.1 *Let* $\alpha \in l_0(\mathbb{Z})$ *be a mask of a convergent subdivision scheme satisfying* $\alpha_{ev}(z) > 0$ *for* $z \in \mathbb{T}$, *and let* $c^{(j)}$ *be a sequence satisfying* $\|\Delta c^{(j)}\|_\infty \le K2^{-j}$. *Then, the pyramid generated by the MTER in* (3.1) *satisfies* (4.1) *and* (4.5).

We remark that in the interpolatory case, (3.2) holds. Therefore, inequalities (4.1) and (4.5) depend on the level ℓ and are independent of j. For non-interpolatory subdivision operators and for a fixed j, the decay of the details is faster, but the constant is bigger since $\|\gamma\|_1 > 1$.

4.2 Stability of Reconstruction

In this subsection, we show that the reconstruction is stable.

Theorem 4.3 *Suppose* $\sup_{j\in\mathbb{N}} \|\mathcal{S}_\alpha^j\|_\infty \leq K$ *for some constant* $K > 0$. *Then, the reconstructed data* $c^{(j)}$ *at level* j *from coarse data* $c^{(0)}$ *and details* $d^1, \ldots, d^{(j)}$ *via* (2.3) *are stable; that is,*

$$\|c^{(j)} - \widetilde{c}^{(j)}\|_\infty \leq K \left(\|c^{(0)} - \widetilde{c}^{(0)}\|_\infty + \sum_{\ell=1}^{j} \|d^{(\ell)} - \widetilde{d}^{(\ell)}\|_\infty \right),$$

where $\widetilde{c}^{(j)}$ *is reconstructed from the data* $\widetilde{c}^{(0)}, \widetilde{d}^1, \ldots, \widetilde{d}^{(j)}$ *via* (2.3).

Proof From $c^{(\ell)} = \mathcal{S}_\alpha c^{(\ell-1)} + d^{(\ell)}$, we have

$$\begin{aligned}
\|c^{(j)} - \widetilde{c}^{(j)}\|_\infty &= \|\mathcal{S}_\alpha (c^{(j-1)} - \widetilde{c}^{(j-1)}) + (d^{(j)} - \widetilde{d}^{(j)})\|_\infty \\
&= \|\mathcal{S}_\alpha^2 (c^{(j-2)} - \widetilde{c}^{(j-2)}) + \mathcal{S}_\alpha (d^{(j-1)} - \widetilde{d}^{(j-1)}) + (d^{(j)} - \widetilde{d}^{(j)})\|_\infty \\
&\ \ \vdots \\
&= \left\| \mathcal{S}_\alpha^j (c^{(0)} - \widetilde{c}^{(0)}) + \sum_{\ell=1}^{j} \mathcal{S}_\alpha^{j-\ell} (d^{(\ell)} - \widetilde{d}^{(\ell)}) \right\|_\infty \\
&\leq \|\mathcal{S}_\alpha^j\|_\infty \|c^{(0)} - \widetilde{c}^{(0)}\|_\infty + \sum_{\ell=1}^{j} \|\mathcal{S}_\alpha^{j-\ell}\|_\infty \|d^{(\ell)} - \widetilde{d}^{(\ell)}\|_\infty \\
&\leq K \left(\|c^{(0)} - \widetilde{c}^{(0)}\|_\infty + \sum_{\ell=1}^{j} \|d^{(\ell)} - \widetilde{d}^{(\ell)}\|_\infty \right).
\end{aligned}$$

□

It is well known and easy to see from the uniform boundedness principle that the condition $\sup_{j\in\mathbb{N}} \|\mathcal{S}_\alpha^j\|_\infty \leq K$ holds whenever the subdivision scheme based on $\mathcal{S}_\alpha$ is convergent.

4.3 Stability of Decomposition

We now show stability of the decomposition of the MTER in (3.1) for fixed j.

Theorem 4.4 *Suppose* $c^{(j)}, \widetilde{c}^{(j)} \in l_p(\mathbb{Z})$ *for* $p \in [1, \infty]$. *Let the pyramid data* $\{c^{(0)}; d^{(1)}, \ldots, d^{(j)}\}$ *and* $\{\widetilde{c}^{(0)}; \widetilde{d}^{(1)}, \ldots, \widetilde{d}^{(j)}\}$ *be obtained from* $c^{(j)}$ *and* $\widetilde{c}^{(j)}$, *respectively, by the scheme* (2.2). *Then, we have for* $\ell = 1, \ldots, j$,

$$\begin{aligned} &\|c^{(0)} - \widetilde{c}^{(0)}\|_p \le \|\mathcal{D}_\gamma\|_p^j \cdot \|c^{(j)} - \widetilde{c}^{(j)}\|_p, \\ &\|d^{(\ell)} - \widetilde{d}^{(\ell)}\|_p \le \left(\|\mathcal{I} - \mathcal{S}_\alpha \mathcal{D}_\gamma\|_p \cdot \|\mathcal{D}_\gamma\|_p^j \cdot \|c^{(j)} - \widetilde{c}^{(j)}\|_p\right) \cdot \|\mathcal{D}_\gamma\|_p^{-\ell}. \end{aligned} \tag{4.6}$$

Proof By $\|c^{(\ell-1)} - \widetilde{c}^{(\ell-1)}\|_p = \|\mathcal{D}_\gamma(c^{(\ell)} - \widetilde{c}^{(\ell)})\|_p \le \|\mathcal{D}_\gamma\|_p \|c^{(\ell)} - \widetilde{c}^{(\ell)}\|_p$, we have

$$\|c^{(0)} - \widetilde{c}^{(0)}\|_p \le \|\mathcal{D}_\gamma\|_p^j \|c^{(j)} - \widetilde{c}^{(j)}\|_p.$$

Similarly, by

$$\|d^{(\ell)} - \widetilde{d}^{(\ell)}\|_p = \|(\mathcal{I} - \mathcal{S}_\alpha \mathcal{D}_\gamma)(c^{(\ell)} - \widetilde{c}^{(\ell)})\|_p \le \|\mathcal{I} - \mathcal{S}_\alpha \mathcal{D}_\gamma\|_p \|c^{(\ell)} - \widetilde{c}^{(\ell)})\|_p,$$

we have

$$\|d^{(\ell)} - \widetilde{d}^{(\ell)}\|_p \le \|(\mathcal{I} - \mathcal{S}_\alpha \mathcal{D}_\gamma)\|_p \|\mathcal{D}_\gamma\|_p^{j-\ell} \|c^{(j)} - \widetilde{c}^{(j)}\|_p.$$

We are done. □

Remarks

(i) For $\mathcal{S}_\alpha$ an interpolatory subdivision operator, (3.2) holds. Therefore, in the corresponding MTER, $\mathcal{D}_\gamma c$ is simply $\mathcal{D}_\gamma c = c \downarrow 2$ and $\|\mathcal{D}_\gamma\|_p \equiv 1$. It follows from Theorem 4.4 that the decomposition of MTERs based on interpolatory subdivision operators is stable for all $p \in [1, \infty]$; namely, the constants in (4.6) are independent of j.

(ii) For non-interpolatory subdivision operators corresponding to convergent even-reversible subdivision schemes, the even-inverse $\mathcal{D}_\gamma$ satisfies $\|\mathcal{D}_\gamma\|_p \ge 1$ since $\gamma(1) = 1$. In such a case, the constant $\|\mathcal{D}_\gamma\|_p^j$ depends on the level j of the data.

In the following, we give bounds on $\|\mathcal{D}_\gamma\|_\infty = \|\gamma\|_1$ for the family of primal pseudo-spline subdivision operators.

Theorem 4.5 *Let* $\alpha^{2k,\nu}$ *be the mask as defined in* (3.4) *with* $0 \le \nu \le k-1$ *and* $k \ge 2$, *and let* $\gamma = (\alpha_{ev}^{2k,\nu})^{-1}$. *Then,*

$$\|\mathcal{D}_\gamma\|_\infty = \|\gamma\|_1 \le C(k, \nu),$$

where

$$C(k, \nu) = \kappa \cdot \max\left\{1, \frac{(1 + \sqrt{\kappa})^2}{2\kappa}\right\} \cdot \frac{1 + \lambda}{1 - \lambda} \tag{4.7}$$

is a constant with

$$\kappa = \frac{2^{k+\nu-1}}{\sum_{j=0}^{\nu}\binom{k+\nu}{j}}, \quad \lambda = \left(\frac{\sqrt{\kappa}-1}{\sqrt{\kappa}+1}\right)^{1/s}, \quad \textit{and } s = \lfloor (k+\nu)/2 \rfloor.$$

The proof of Theorem 4.5 is given in the Appendix. We remark that

(i) For the case $\nu = k-1$, $\alpha^{2k,k-1}$ corresponds to the family of Deslauries–Dubuc's interpolatory masks [5]. In such a case, the constant $C(k, k-1)$ is exact in the sense that $C(k, k-1) = 1$.
(ii) For the case $\nu = 0$, $\alpha^{2k,0}$ is the mask of degree $2k-1$ (order $2k$) spline subdivision operator. In such a case, it can be shown that $C(k, 0) = O(k \cdot 2^{\frac{3k}{2}})$ for $k \to \infty$.

Combining the above results, we have the following result regarding the stability of decomposition of the MTER based on primal pseudo-spline masks in $l_p(\mathbb{Z})$ for two important cases $p = 2$ and $p = \infty$.

Corollary 4.2 *Suppose $c^{(j)}, \widetilde{c}^{(j)} \in l_p(\mathbb{Z})$ for $p \in [1, \infty]$. Let the pyramid data $\{c^{(0)}; d^{(1)}, \ldots, d^{(j)}\}$ and $\{\widetilde{c}^{(0)}; \widetilde{d}^{(1)}, \ldots, \widetilde{d}^{(j)}\}$ be obtained from $c^{(j)}$ and $\widetilde{c}^{(j)}$, respectively, by the scheme* (2.2) *with $\alpha = \alpha^{2k,\nu}$ for $0 \le \nu \le k-2$ and $k \ge 2$. Then,*

(1) *for $p = \infty$, we have*

$$\|c^{(0)} - \widetilde{c}^{(0)}\|_\infty \le C(k,\nu)^j \cdot \|c^{(j)} - \widetilde{c}^{(j)}\|_\infty,$$
$$\|d^{(\ell)} - \widetilde{d}^{(\ell)}\|_\infty \le \left(\|\mathcal{I} - \mathcal{S}_{\alpha^{2k,\nu}}\mathcal{D}_\gamma\|_\infty \cdot C(k,\nu)^j \cdot \|c^{(j)} - \widetilde{c}^{(j)}\|_\infty\right) \cdot C(k,\nu)^{-\ell} \tag{4.8}$$

for $\ell = 1, \ldots, j$, where $C(k,\nu)$ is the constant defined in (4.7);
(2) *for $p = 2$,* (4.6) *holds with $\|\mathcal{D}_\gamma\|_2 = \|A_\gamma\|_2 = \frac{2^{k+\nu-1}}{\sum_{j=0}^{\nu}\binom{k+\nu}{j}}$.*

Proof This is a direct consequence of Theorems 4.4 and 4.5. □

5 Final Remarks

In this section, we give some further remarks.

1. Most examples and results in this chapter deal with finitely supported masks for the purpose of simplicity of presentation. We point out that the MTER in (3.1) applies for any mask α provided $\alpha_{ev}(z) \neq 0$ for all $z \in \mathbb{T}$.
2. Using the weighted version of Wiener's lemma [16], one can study classes of masks of infinite support such as masks with polynomial or sub-exponential decay and masks corresponding to rational symbols. The extension of the class of masks might lead to even-inverse operators of smaller norm, and by that to the improvement of the stability of the decomposition and the decay rate of the details as given in Theorems 4.4 and 4.2, respectively.
3. The existence of MTER relies on Wiener's lemma. Since a high-dimensional version of Wiener's lemma holds, our MTER in (3.1) can be generalized to any dimension $d \in \mathbb{N}$.
4. Without involving Wiener's lemma, one can also analyze the decay property of the Laurent polynomials $\gamma(z) = \alpha_{ev}^{-1}(z)$ using the Fourier series techniques. One can show that $|\gamma(k)| := |\frac{1}{2\pi}\int_0^{2\pi} \gamma(e^{-i\xi})e^{-ik\xi}d\xi| \leq Ce^{-r_1|k|}$ for some $0 < r_1 < r$ with r being the radius of convergence of the Laurent polynomials of $\gamma(z)$. Such techniques were employed in [17].
5. One can consider more general decimation operator in (2.2), e.g.

$$c^{(j-1)} = \mathcal{D}_{\gamma_{ev},\gamma_{od}} c^{(j)} := \gamma_{ev} * c_{ev}^{(j)} + \gamma_{od} * c_{od}^{(j)}.$$

When $\gamma_{od} = 0$, we see that $\mathcal{D}_{\gamma_{ev},\gamma_{od}} = \mathcal{D}_{\gamma_{ev}}$, which goes back to our setting. More general results could be obtained with more analysis.

Acknowledgments The authors thank Felipe Cucker for his support which initiated this chapter. The authors also thank anonymous reviewer for pointing out the alternative proof of (3.3). The research of X. Zhuang and the work described in this chapter were partially supported by a grant from the Research Grants Council of the Hong Kong Special Administrative Region, China (Project No. CityU 11302218).

Appendix

Proof of (3.3) Note that

$$\gamma(z) = \frac{8}{z^{-1}+6+z} = \frac{4}{3}\left(\sum_{n=0}^{\infty} 6^{-2n}(z^{-1}+z)^{2n} - \sum_{n=0}^{\infty} 6^{-2n-1}(z^{-1}+z)^{2n+1}\right).$$

For $\sum_{n=0}^{\infty}(\frac{1}{6})^{2n}(z^{-1}+z)^{2n}$, we have

$$\sum_{n=0}^{\infty} 6^{-2n}(z^{-1}+z)^{2n} = \sum_{n=0}^{\infty} 6^{-2n} \sum_{k=0}^{2n} \binom{2n}{k} z^{2(n-k)} = \sum_{n=0}^{\infty} 6^{-2n} \sum_{k=-n}^{n} \binom{2n}{n-k} z^{2k}$$

$$= \sum_{n=0}^{\infty} \binom{2n}{n} 6^{-2n} + \sum_{k=1}^{\infty} \left[\sum_{n=k}^{\infty} \binom{2n}{n-k} 6^{-2n} \right] (z^{2k} + z^{-2k})$$

$$= \sum_{n=0}^{\infty} \binom{2n}{n} 6^{-2n} + \sum_{k=1}^{\infty} \left[\sum_{n=0}^{\infty} \binom{2(n+k)}{n} 6^{-2(n+k)} \right] (z^{2k} + z^{-2k}).$$

Similarly, for the second summation, we have

$$\sum_{n=0}^{\infty} 6^{-2n-1}(z^{-1}+z)^{2n+1} = \sum_{k=0}^{\infty} \left[\sum_{n=0}^{\infty} \binom{2(n+k)+1}{n} 6^{-2(n+k)-1} \right] (z^{2k+1} + z^{-2k-1}).$$

Define for $k = 0, 1, 2, \ldots,$

$$a_k := \sum_{n=0}^{\infty} \binom{2(n+k)}{n} 6^{-2(n+k)} \quad \text{and} \quad b_k := \sum_{n=0}^{\infty} \binom{2(n+k)+1}{n} 6^{-2(n+k)-1}.$$

Then,

$$\gamma(z) = \frac{4}{3} \sum_{n=0}^{\infty} \left(\frac{-1}{6} \right)^n (z^{-1}+z)^n$$

$$= \frac{4}{3} \left[a_0 + \sum_{k=1}^{\infty} a_k (z^{2k} + z^{-2k}) - \sum_{k=0}^{\infty} b_k (z^{2k+1} + z^{-2k-1}) \right].$$

Using the formula $\binom{n+1}{m} = \binom{n}{m} + \binom{n}{m-1}$, it is easy to check that

$$a_k = \frac{1}{6}(b_{k-1} + b_k) \quad \text{and} \quad b_k = \frac{1}{6}(a_k + a_{k+1}).$$

Hence,

$$a_{k+1} = 6b_k - a_k \quad \text{and} \quad b_k = 6a_k - b_{k-1}. \tag{A.1}$$

Next, we proceed to prove the following identities using the recurrence relations (A.1):

$$a_k = \frac{3\sqrt{2}}{4}(3-2\sqrt{2})^{2k} \quad \text{and} \quad b_k = \frac{3\sqrt{2}}{4}(3-2\sqrt{2})^{2k+1}, \quad k = 0, 1, 2, \ldots. \tag{A.2}$$

First, we compute directly a_0 and b_0. Using the identities

$$(1-x)^{-t} = \sum_{n=0}^{\infty} \binom{t-1+n}{n} x^n \tag{A.3}$$

and $\binom{2n}{n} = 4^n \binom{n-1/2}{n}$, we have

$$a_0 = \sum_{n=0}^{\infty} \binom{2n}{n} 6^{-2n} = \sum_{n=0}^{\infty} \binom{1/2-1+n}{n} 9^{-n} = (1-1/9)^{-1/2} = \frac{3\sqrt{2}}{4}.$$

Similarly,

$$b_0 = \sum_{n=0}^{\infty} \binom{2n+1}{n} 6^{-2n-1} = 3\left[\sum_{n=0}^{\infty} \binom{1/2-1+n}{n} 9^{-n} - 1\right] = \frac{3\sqrt{2}}{4}(3-2\sqrt{2}).$$

Now, recursively using the relation of a_k and b_k in (A.1), we get

$$\begin{aligned} a_{k+1} = 6b_k - a_k &= \frac{3\sqrt{2}}{4}(6(3-2\sqrt{2})^{2k+1} - (3-2\sqrt{2})^{2k}) \\ &= \frac{3\sqrt{2}}{4}(3-2\sqrt{2})^{2k}(6(3-2\sqrt{2})-1) = \frac{3\sqrt{2}}{4}(3-2\sqrt{2})^{2k}(3-2\sqrt{2})^2 \\ &= \frac{3\sqrt{2}}{4}(3-2\sqrt{2})^{2(k+1)}, \end{aligned}$$

and

$$\begin{aligned} b_k = 6a_k - b_{k-1} &= \frac{3\sqrt{2}}{4}(6(3-2\sqrt{2})^{2k} - (3-2\sqrt{2})^{2k-1}) \\ &= \frac{3\sqrt{2}}{4}(3-2\sqrt{2})^{2k-1}(6(3-2\sqrt{2})-1) = \frac{3\sqrt{2}}{4}(3-2\sqrt{2})^{2k-1}(3-2\sqrt{2})^2 \\ &= \frac{3\sqrt{2}}{4}(3-2\sqrt{2})^{2k+1}. \end{aligned}$$

Therefore, (A.2) holds. In summary, we conclude that

$$\begin{aligned} \gamma(z) &= \frac{4}{3}\left[a_0 + \sum_{k=1}^{\infty} a_k(z^{2k} + z^{-2k}) - \sum_{k=0}^{\infty} b_k(z^{2k+1} + z^{-2k-1})\right] \\ &= \sqrt{2}\left[1 + \sum_{k=1}^{\infty}(3-2\sqrt{2})^{2k}(z^{2k} + z^{-2k}) - \sum_{k=0}^{\infty}(3-2\sqrt{2})^{2k+1}(z^{2k+1} + z^{-2k-1})\right], \end{aligned}$$

which proves (3.3).

Alternative, one can use standard complex analysis technique to compute the Laurent series of the function $\frac{8}{z^{-1}+6+z}$. Note that $z^2+6z+1=0$ has two real roots $-3\pm 2\sqrt{2}$ with one outside $\mathbb{T}$ and the other inside $\mathbb{T}$. Hence, one can write

$$\frac{8}{z^{-1}+6+z}=\frac{8z}{z^2+6z+1}=\frac{\sqrt{2}z}{z+3-2\sqrt{2}}-\frac{\sqrt{2}z}{z+3+2\sqrt{2}}.$$

Since $|-3+2\sqrt{2}|<1$, we have

$$\frac{\sqrt{2}z}{z+3-2\sqrt{2}}=\frac{\sqrt{2}}{1-(-3+2\sqrt{2})z^{-1}}=\sqrt{2}\sum_{k=0}^{\infty}(-3+2\sqrt{2})^k z^{-k} \tag{A.4}$$

holds for $|z|>|-3+2\sqrt{2}|$. Similarly, since $|-3-2\sqrt{2}|>1$, we have

$$-\frac{\sqrt{2}z}{z+3+2\sqrt{2}}=-\frac{\sqrt{2}z(3-2\sqrt{2})}{1-(-3+2\sqrt{2})z}=\sqrt{2}\sum_{k=1}^{\infty}(-3+2\sqrt{2})^k z^{k} \tag{A.5}$$

holds for $|z|<3+2\sqrt{2}$. Putting (A.4) and (A.5) together, we get (12). □

Proof of Theorem 3.2 Let $z=e^{-i\theta}$ and $x=\sin^2(\theta/2)$. One can show that

$$\alpha^{n,\nu}(e^{-i\theta})=2e^{i(\lfloor n/2\rfloor-n/2)\theta}\cos^n(\theta/2)Q_{n,\nu}(\sin^2(\theta/2)),$$

with

$$Q_{n,\nu}(x):=\sum_{j=0}^{\nu}\binom{n/2-1+j}{j}x^j=\frac{1}{(1-x)^{n/2}}+O(x^{\nu+1}),$$

where in the last equality, we use (A.3). Note that $Q_{n,\nu}(x)\geq 1$ for all $x\geq 0$. By that $\alpha_{ev}^{n,\nu}(z^2)=\frac{1}{2}(\alpha^{n,\nu}(z)+\alpha^{n,\nu}(-z))$, we have

$$\alpha_{ev}^{n,\nu}(e^{-2i\theta})=e^{i(\lfloor n/2\rfloor-n/2)\theta}\Big[(1-x)^{n/2}Q_{n,\nu}(x)+i^{2\lfloor n/2\rfloor-n}(-1)^n x^{n/2}Q_{n,\nu}(1-x)\Big].$$

It is easy to see that $\alpha_{ev}^{n,\nu}(1)=1$.

For $n=2k$, we have

$$\alpha_{ev}^{2k,\nu}(e^{-2i\theta})=(1-x)^kQ_{2k,\nu}(x)+x^kQ_{2k,\nu}(1-x)=:R(x)+R(1-x), \tag{A.6}$$

where $R(x):=(1-x)^kQ_{2k,\nu}(x)=(1-x)^k\sum_{j=0}^{\nu}\binom{k-1+j}{j}x^j$. Define $g(x):=R(x)+R(1-x)$. By using $(j+1)\binom{k+j}{j+1}-j\binom{k-1+j}{j}=k\binom{k-1+j}{j}$, one can show that

$$R'(x) = -(k+\nu)\binom{k-1+\nu}{\nu}(1-x)^{k-1}x^{\nu}.$$

Thus,

$$g'(x) = R'(x) - R'(1-x) = (k+\nu)\binom{k-1+\nu}{\nu}x^{\nu}(1-x)^{\nu}\left[x^{k-1-\nu} - (1-x)^{k-1-\nu}\right].$$

It is easily seen that $g'(x) \leq 0$ for $x \in [0, 1/2]$ and $g'(x) \geq 0$ for $x \in [1/2, 1]$. Consequently,

$$\begin{aligned} \min_{z\in\mathbb{T}} |\alpha_{ev}^{2k,\nu}(z)| &= \min_{z\in\mathbb{T}} \alpha_{ev}^{2k,\nu}(z) = \min_{x\in[0,1]} g(x) = g(1/2) = 2R(1/2) \\ &= 2^{1-k}\sum_{j=0}^{\nu}\binom{k-1+j}{j}2^{-j} > 0, \end{aligned}$$

where the last equation can be shown to be equivalent to

$$\min_{z\in\mathbb{T}} |\alpha_{ev}^{2k,\nu}(z)| = 2^{1-k-\nu}\sum_{j=0}^{\nu}\binom{k+\nu}{j} > 0. \tag{A.7}$$

Moreover, by (2.5), we have

$$\|A_{\alpha_{ev}^{2k,\nu}}\|_2 = \max_{z\in\mathbb{T}} |\alpha_{ev}^{2k,\nu}(z)| = \max_{x\in[0,1]} g(x) = g(0) = g(1) = 1.$$

For $n = 2k+1$, we have

$$\alpha_{ev}^{2k+1,\nu}(e^{-2i\theta}) = e^{-i\theta/2}\left[(1-x)^{k+1/2}Q_{2k+1,\nu}(x) + i\cdot x^{k+1/2}Q_{2k+1,\nu}(1-x)\right],$$

from which, we have

$$\begin{aligned} |\alpha_{ev}^{2k+1,\nu}(e^{-2i\theta})|^2 &= (1-x)^{2k+1}(Q_{2k+1,\nu}(x))^2 + x^{2k+1}(Q_{2k+1,\nu}(1-x))^2 \\ &=: R(x)^2 + R(1-x)^2, \end{aligned}$$

where

$$R(x) := (1-x)^{k+1/2}Q_{2k+1,\nu}(x) = (1-x)^{k+1/2}\sum_{j=0}^{\nu}\binom{k-1/2+j}{j}x^j.$$

Define $g(x) = R(x)^2 + R(1-x)^2$. Then, by using $(j+1)\binom{k+1/2+j}{j+1} - j\binom{k-1/2+j}{j} = (k+1/2)\binom{k-1/2+j}{j}$, we can show similarly that

$$R'(x) = -(k+1/2+\nu)\binom{k-1/2+\nu}{\nu}(1-x)^{k-1/2}x^{\nu}.$$

Consequently,

$$\begin{aligned} g'(x) &= 2\big[R(x)R'(x) - R(1-x)R'(1-x)\big] \\ &= -2cx^{\nu}(1-x)^{\nu}\left[\sum_{j=0}^{\nu}\binom{k-1/2+j}{j}(1-x)^j x^j\Big[(1-x)^{2k-\nu-j} - x^{2k-\nu-j}\Big]\right], \end{aligned}$$

where $c = (k+1/2+\nu)\binom{k-1/2+\nu}{\nu}$. Now, it is easy to see that each term

$$t_j(x) := \binom{k-1/2+j}{j}(1-x)^j x^j\big[(1-x)^{2k-\nu-j} - x^{2k-\nu-j}\big]$$

in the above summation for $j = 0, \dots, \nu$ satisfies

$$t_j(x) \geq 0 \text{ for } x \in [0, 1/2] \text{ and } t_j(x) \leq 0 \text{ for } x \in [1/2, 1].$$

Hence, $g'(x) \leq 0$ for $x \in [0, 1/2]$ and $g'(x) \geq 0$ for $x \in [1/2, 1]$. Consequently,

$$\begin{aligned} \min_{x\in[0,1]} g(x) = g(1/2) = 2R(1/2)^2 &= 2\left(2^{-k-1/2}\sum_{j=0}^{\nu}\binom{k-1/2+j}{j}2^{-j}\right)^2 \\ &= 2^{-2k-2\nu}\left(\sum_{j=0}^{\nu}\binom{k+1/2+\nu}{j}\right)^2. \end{aligned}$$

Therefore,

$$\min_{z\in\mathbb{T}} |\alpha_{ev}^{2k+1,\nu}(z)| = \min_{x\in[0,1]} \sqrt{g(x)} = 2^{-k-\nu}\sum_{j=0}^{\nu}\binom{k+1/2+\nu}{j} > 0,$$

and by (2.5), we have

$$\|A_{\alpha_{ev}^{2k+1,\nu}}\|_2 = \max_{z\in\mathbb{T}} |\alpha_{ev}^{2k+1,\nu}(z)| = g(0) = g(1) = 1.$$

Combining the above results for n even and odd, we see that $|\alpha_{ev}^{n,\nu}(z)| > 0$ for all $z \in \mathbb{T}$ (in particular, $\alpha_{ev}^{2k,\nu}(z) > 0$ for all $z \in \mathbb{T}$). Hence, by Wiener's lemma, its inverse $\gamma = (\alpha_{ev}^{n,\nu})^{-1}$ exists and $\gamma(1) = \frac{1}{\alpha_{ev}^{n,\nu}(1)} = 1$. Moreover, by (2.5), we have

$$\|A_{\alpha_{ev}^{n,\nu}}\|_2 = \max_{z\in\mathbb{T}} |\alpha_{ev}^{n,\nu}(z)| = 1$$

and

$$\|A_\gamma\|_2 = \max_{z\in\mathbb{T}} |\gamma(z)| = \frac{1}{\min_{z\in\mathbb{T}} |\alpha_{ev}^{n,\nu}(z)|} = \frac{2^{\lfloor\frac{n-1}{2}\rfloor+\nu}}{\sum_{j=0}^{\nu}\binom{n/2+\nu}{j}}.$$

The exponential decay of the elements of γ in the case that $n = 2k$ follows directly from Corollary 2.1 since $\alpha_{ev}^{2k,\nu}(z) > 0$ for all $z \in \mathbb{T}$. In case $n = 2k+1$, the exponential decay of the elements of γ follows from the weighted version of Wiener's lemma [16]. □

Proof of Theorem 4.5 By (A.6) and (A.7), we have $\alpha_{ev}^{2k,\nu}(z) > 0$ for all $z \in \mathbb{T}$. Applying Corollary 2.1 and Theorem 3.2, we have

$$|\gamma_n| \le K\lambda^{|n|}, \quad n \in \mathbb{Z}, \tag{A.8}$$

where $K = \kappa \cdot \max\left\{1, \frac{(1+\sqrt{\kappa})^2}{2\kappa}\right\}$ with $\kappa = \frac{\sup_{z\in\mathbb{T}}|\alpha_{ev}^{2k,\nu}(z)|}{\inf_{z\in\mathbb{T}}|\alpha_{ev}^{2k,\nu}(z)|} = \|A_\gamma\|_2 = \frac{2^{k+\nu-1}}{\sum_{j=0}^{\nu}\binom{k+\nu}{j}}$, and $\lambda = q^{1/s}$ with $q = (\sqrt{\kappa}-1)/(\sqrt{\kappa}+1)$, $s = \lfloor(k+\nu)/2\rfloor$. Thus, by (A.8), we have

$$\|\gamma\|_1 = |\gamma_0| + 2\sum_{n=1}^{\infty}|\gamma_n| \le K\Big(1 + 2\sum_{n=1}^{\infty}\lambda^n\Big) = K\frac{1+\lambda}{1-\lambda} = C(k,\nu).$$

□

References

1. Amat, S., Donat, R., Liandrat, J., Trillo, J. C.: A fully adaptive PPH multiresolution scheme for image processing, *Math. Comput. Model.* **46** (1–2), 2–11 (2007).
2. Chui, C. K., De Villiers, J., Zhuang, X.: Multirate systems with shortest spline-wavelet filters, *Appl. Comput. Harmon. Anal.* **41** (1), 266–296 (2016).
3. Daubechies, I.: Ten Lectures on Wavelets. CBMS-NSF Regional Conference Series in Applied Mathematics, **61**, SIAM, Philadelphia, PA (1992).
4. Daubechies, I., Han, B., Ron, A., Shen, Z.: Framelets: MRA-based constructions of wavelet frames, Appl. Comput. Harmon. Anal. **14**, 1–46 (2003).
5. Deslauriers, G., Dubuc, S.: Symmetric iterative interpolation processes, Constr. Approx. **5**, 49–68 (1989).

6. Donoho, D. L., Yu, T. P.-Y.: Nonlinear Pyramid transforms based on median-interpolation, SIAM J. Math. Anal. **31** (5), 1030–1061 (2000).
7. Dong, B., Shen, Z.: Pseudo-splines, wavelets and framelets, Appl. Comput. Harmon. Anal. **22** (1), 78–104 (2007).
8. Dong, B., Dyn, N., Hormann, K.: Properties of dual pseudo-splines, Appl. Comput. Harmon. Anal. **29**, 104–110 (2010).
9. Dyn, N.: Subdivision Schemes in Computer-Aided Geometric Design, Advances in Numerical Analysis II, Wavelets, Subdivision Algorithms and Radial Basis Functions, W. Light (ed.), Clarendon Press, Oxford, 36–104 (1992).
10. Dyn, N., Goldman, R.: Convergence smoothness of nonlinear Lane-Riesenfeld algorithms in the functional functional setting, Found. Comput. Math. **11**, 79–94 (2011).
11. Dyn, N., Hormann, K., Sabin, M. A., Shen, Z.: Polynomial reproduction by symmetric subdivision schemes, J. Approx. Theory **155** (1), 28–42 (2008).
12. Dyn, N., Oswald, P.: Univariate subdivision and multiscale transforms: The nonlinear case, in Multiscale, Nonlinear, and Adaptive Approximation, R.A. DeVore, A. Kunoth (eds.), pp. 203–247, Springer, Berlin (2009).
13. Gohberg, I., Goldberg, S., Kaashoek, M. A.: Basic Classes of Linear Operators, Birkhäuser (2003).
14. Grohs P., Wallner, J.: Interpolatory wavelets for manifold-valued data, Appl. Comput. Harmon. Anal. **27**, 325–333 (2009).
15. Grohs P., Wallner, J.: Definability stability of multiscale decompositions for manifold-valued data, data, J. Franklin Inst. **349**, 1648–1664 (2012).
16. Gröchenig, K.: Wiener's lemma: theme and variations. An introduction to spectral in-variance and its applications, in Four Short Courses on Harmonic Analysis: Wavelets, Frames, Time-Frequency Methods, and Applications to Signal and Image Analysis, B. Forster, P. Massopust (eds.), Applied and Numerical Harmonic Analysis. Birkhäauser Boston, ch. 5 (2009).
17. Han, B.: Refinable functions and cascade algorithms in weighted spaces with Hölder continuous masks, SIAM J. Math. Anal. **40** (1), 70–102 (2008).
18. Han, B., Jiang, Q. T., Shen, Z. W., Zhuang, X.: Symmetric canonical quincunx tight framelets with high vanishing moments and smoothness, Math. Comput. **87** (309), 347–379 (2018).
19. Han, B., Kwon, S. G., Zhuang, X.: Generalized interpolating refinable function vectors, J. Comput. Appl. Math. **227**, 254–270 (2009).
20. Han, B., Zhuang, X.: Algorithms for matrix extension orthogonal wavelet filter banks over algebraic algebraic number fields, Math. Comput. **82** (281), 459–490 (2013).
21. Harten, A.: Multiresolution representation of data: A general framework, SIAM J. Numer. Anal. **33** (3), 1205–1256 (1996).
22. Hassan M. F., Dodgson, N. A.: Reverse subdivision, in Advances in Multiresolution for Geometric Modelling, N. A. Dodgson, M. S. Floater, M. A. Sabin (eds.), 271–283 (2005).
23. Lanquetin, S., Neveu, M.: Reverse Catmull-Clark Subdivision, Conference proceedings WSCG (2006).
24. Rahman, I. U., Drori, I., Stodden, V. C., Donoho, D. L., Schröder, P.: Multiscale representations for manifold-valued data, Multiscale Model. Simul., **4** (4), 1201–1232 (2005).
25. Sadeghi, J., Samavati, F. F.: Smooth reverse subdivision, Computers and Graphics, **33** (3), 217–225 (2009).
26. Samavati, F. F., Bartels, R. H.: Multiresolution curve and surface representation: reversing subdivision rules by least-squares data fitting, Computer Graphics Forum, **18** (2), 97–119 (1999).
27. Strohmer, T.: Four short stories about Toeplitz matrix calculations, Lin. Alg. Appl. **343–344**, 321–344 (2002).
28. Zhuang, X.: Quincunx fundamental refinable functions in arbitrary dimensions, Axioms **6** (3), 20 (2017).

Part V
Compressed Sensing and Optimization

We conclude with a part on compressed sensing and optimization. Compressed sensing is a novel approach in sampling theory for acquisition of high dimensional signals with low dimensional structure. Since its inception in early 2000s, it attracted interest from a multitude of disciplines ranging from pure and applied mathematics to computer science, engineering, and applied sciences. In this final part, we present some recent developments on mathematical as well as algorithmic aspects of compressed sensing and optimization.

The first chapter in this part, by A. Abtahi and F. Marvasti, focuses on the use of compressed sensing and related techniques, together with sparse recovery methods based on convex optimization, for reducing the sampling rate, thus reducing cost, in MIMO radar systems.

In the next chapter, J. Cahill and D. G. Mixon introduce the "robust width property", which they use to provide a characterization of the (compressive) sensing operators that allow uniformly stable and robust recovery via convex optimization.

In the next chapter, S. B. Damelin, D. L. Ragozin, and M. Werman study best uniform affine approximants of real-valued convex or concave functions together with some interesting applications in computer graphics.

In the next chapter of this part, C. Hedge, F. Keinert, and E. S. Weber propose a modified Kaczmarz algorithm for solving systems of linear equations that are distributed over multiple nodes of a network.

In the last chapter of this part, W. Czaja, W. Li, Y. Li, and M. Pekala study pooling operators in convolutional neural networks (CNNs). Inspired by, on the one hand the Hardy–Littlewood maximal function from harmonic analysis, and on the other hand max pooling and average pooling operators used in CNNs, they introduce a new pooling operation called maxfun and study its properties both theoretically and empirically.

Sparsity-Based MIMO Radars

Azra Abtahi and Farokh Marvasti

Abstract With advances in technology, Multiple-Input Multiple-Output (MIMO) radars have attracted a lot of attention in different modern military and civilian applications. As there are multiple receivers in a MIMO radar system, the cost can significantly be reduced if we can reduce the sampling rate and send fewer samples to the common processing center. Sometimes, the problem is not even the cost. It is the technology issues of high sampling rates such as the necessity of high-rate Analog-to-Digital (A/D) converters. The reduction in sampling rate can be achieved using Compressive Sensing (CS) or, in a much simpler form, Random Sampling (RS). In CS, we take a number of linear combinations of sparse signal samples which is smaller than what is necessary according to Shannon–Nyquist sampling theory. The sparse signal can be recovered from these linear combinations exploiting sparse recovery methods. By using sparse recovery methods, not only the sampling rate can be reduced but also the performance of the radars in detection and estimation procedures can be improved. In this chapter, we discuss the use of CS, RS, and sparse recovery methods in a MIMO radar system, the main challenges we face in this concept, and the solutions to these challenges proposed up to now.

She has been supported by Iran National Science Foundation under grant #97006816.

A. Abtahi (✉) · F. Marvasti
Advanced Communications Research Institute (ACRI), Electrical Engineering Department, Sharif University of Technology, Tehran, Iran
e-mail: azra_abtahi@ee.sharif.edu; marvasti@ee.sharif.ir

M. Hirn et al. (eds.), *Excursions in Harmonic Analysis, Volume 6*, Applied and Numerical Harmonic Analysis, https://doi.org/10.1007/978-3-030-69637-5_17

1 Introduction

A radar[1] system is an electromagnetic system that radiates radio waves into space and receives the echo signals reflected from an object to determine some parameters such as its range, angle, and velocity. A Multiple-Input Multiple-Output (MIMO) radar is a radar system that consists of multiple transmitters, multiple receivers, and a common processing center called fusion center. Nowadays, due to the advancement of technology, these radars which are superior in detection and target parameter estimation attract a lot of attention [44].

There are two different kinds of MIMO radar systems: colocated MIMO radars [43, 44] and distributed MIMO radars [35]. In a colocated one, the distances between the antennas are small compared to the distances between the antennas and the targets. Hence, in this type, all the transmitter–receiver pairs view a target from the same angle. On the other hand, in a distributed MIMO radar, the antennas are distributed in a large area. Thus, if one transmitter–receiver pair is unable to view a target, it can be compensated by the other pairs. In both of these groups, the receivers send the received samples to the fusion center. Hence, the costs can significantly be reduced by decreasing the sampling rate at each receiver. This reduction can be accomplished by Compressive Sensing (CS) or, in a much simpler form, Random Sampling (RS).

CS is a sampling method for sparse signals. A sparse signal is a signal, in which most of its entries are zero in some domain. If only K entries of a sparse signal are non-zero, we call it a K-sparse signal [16]. A special case of sparsity is block sparsity. Consider $\mathbf{s}$ as a vector with L blocks of length d:

$$\mathbf{s} = [\underbrace{s_1, \cdots, s_d}_{\mathbf{s}[1]}, \cdots, \underbrace{s_{L\times(d-1)+1}, \cdots, s_{L\times d}}_{\mathbf{s}[L]}]^T.$$

The vector $\mathbf{s}$ is called block K-sparse when only K blocks of $\mathbf{s}$ have non-zero norms.

In CS, we take a number of linear combinations of a sparse signal entries, which is smaller than what is necessary according to Shannon–Nyquist sampling theory for signal recovery, instead of all its entries. These linear combinations of signal entries are called measurements. Then, exploiting sparse recovery methods, the signal can be recovered from the measurements. Due to the sparsity of targets in the position–velocity space, the received signal at the fusion center of a MIMO radar system is sparse (see Sect. 2), and CS can be useful in this concept. We can also use Random Sampling (RS) in these systems. RS is another sparsity-based sampling method which randomly selects some entries of a sparse signal and can be considered as a special case of CS.

Using CS or RS, we can eliminate the necessity of high-rate Analog-to-Digital (A/D) converters and send fewer samples to the fusion center. Then, at the fusion

[1] RAdio Detection And Ranging.

center, the received signal can be recovered using a sparse recovery method. Exploiting sparse recovery methods, not only the sampling rate can be reduced but also the performance of the radars in the detection and the estimation procedures can be improved. References [18, 66], and [67] were among the first papers that evaluate the exploiting of CS and sparse recovery methods in colocated MIMO radars. They have shown the superiority of sparse recovery methods against matched filters and some other target parameter estimation schemes in these radars. Many other references have tried to improve the performance of sparsity-based MIMO radars that are discussed in this chapter [1–3, 22, 25, 27, 28, 32, 36, 41, 42, 51, 54, 57, 59, 65–69].

This chapter is organized as follows: in Sect. 2, the received signal of a MIMO radar system is modeled. Then, the use of CS and RS in these radars is discussed in Sect. 3. Section 4 is allocated to the sparse recovery methods, target detection, and target parameter estimation in sparsity-based MIMO radars. In Sect. 5, we discuss the main challenges we face in sparsity-based MIMO radars and the solutions to these challenges proposed up to now. Finally, we make some concluding remarks in Sect. 6.

Notations

The imaginary unit is denoted by j. The transpose, the conjugate transpose, the pseudoinverse, and the convolution operations are denoted by $(.)^T$, $(.)^H$, $(.)^\dagger$, and $*$, respectively. The spectral norm of matrix B, which is the largest eigenvalue of the matrix $B^H B$, is denoted by $\rho(B)$. Let us define $0^0 = 0$. Then, l_p norm of vector $\mathbf{x} = [x_1, x_2, \ldots, x_U]^T$ is denoted by $\|\mathbf{x}\|_p$ and defined as $\|\mathbf{x}\|_p = (\sum_{i=1}^{U} x_i{}^p)^{\frac{1}{p}}$. Furthermore, we define function $diag\{.\}$ for input matrices $\mathbf{A}^p_{m_p \times n_p}$, where $p \in \{1, 2, \ldots, P\}$ as follows:

$$diag\{\mathbf{A}^1_{m_1\times n_1}, \mathbf{A}^2_{m_2\times n_2}, \ldots, \mathbf{A}^P_{m_N\times n_P}\} = \begin{bmatrix} \mathbf{A}^1_{m_1\times n_1} & \mathbf{0}_{m_1\times n_1} & \cdots & \mathbf{0}_{m_1\times n_1} \\ \mathbf{0}_{m_2\times n_2} & \mathbf{A}^2_{m_2\times n_2} & \cdots & \mathbf{0}_{m_2\times n_2} \\ \vdots & \vdots & \ddots & \vdots \\ \mathbf{0}_{m_P\times n_P} & \mathbf{0}_{m_P\times n_P} & \cdots & \mathbf{A}^P_{m_P\times n_P} \end{bmatrix},$$

2 Received Signal Model

2.1 Received Signal Model in a Distributed MIMO Radar

We consider a distributed MIMO radar with M_t transmitters and N_r receivers. The transmitters and the receivers are respectively located at $\mathbf{t}_i = [t_{x_i}, t_{y_i}]$ for $i = 1, \ldots, M_t$ and $\mathbf{r}_l = [r_{x_l}, r_{y_l}]$ for $l = 1, \ldots, N_r$ on a Cartesian coordinate system. The ith transmitter transmits the baseband signal $x_i(t)$ with the duration of T_P and the carrier frequency of f_c. The Pulse Repetition Interval (PRI) and the period of sampling at the receivers are denoted by T and T_s, respectively. Let us assume that there are K targets at the search area located at $\mathbf{p}_k = [p_x^k, p_y^k]$ for $k = 1, 2, \ldots, K$ and moving with the velocities of $\mathbf{v}_k = [\widetilde{v}_x^k, \widetilde{v}_y^k]$ for $k = 1, 2, \ldots, K$. For simplicity, we have considered that all the targets and antennas are located on a two-dimensional plane. However, the modeling can easily be extended to the three-dimensional case.

After down-converting the received band-pass signal from the radio frequency and under a narrow band assumption on the waveforms, the received baseband signal at the lth receiver from the ith transmitter can be expressed as

$$z_{il}(t) = \sum_{k=1}^{K} \beta_k^{il} x_i(t - \tau_k^{il}) e^{j2\pi(f_k^{il} t - f_c \tau_k^{il})} + \overline{n}_{il}(t), \tag{2.1}$$

where $\overline{n}_{il}(t)$ is the received noise corresponding to the ith transmitter and lth receiver, and β_k^{il}, f_k^{il}, and τ_k^{il} are the attenuation coefficient, the Doppler shift, and the delay corresponding to the kth target between the ith transmitter and the lth receiver, respectively. We have

$$f_k^{il} = \frac{f_c}{c}(\mathbf{v}_k.\mathbf{u}_{r_l}^k - \mathbf{v}_k.\mathbf{u}_{t_i}^k), \tag{2.2}$$

$$\tau_k^{il} = \frac{1}{c}(\|\mathbf{p}_k - \mathbf{t}_i\|_2 + \|\mathbf{p}_k - \mathbf{r}_l\|_2), \tag{2.3}$$

where $\mathbf{u}_{t_i}^k$ and $\mathbf{u}_{r_l}^k$ are, respectively, the unit vector from the ith transmitter to the kth target and the unit vector from the kth target to the lth receiver. The speed of light is denoted by c.

Let us consider the estimation space as the position–velocity in direction x and y and discretize it into L points as (x_h, y_h, v_x^h, v_y^h), $h = 1, \ldots, L$. At each receiver, the received signal passes through a bank of M_t filters with impulse responses of $h_i(t)$ for $i = 1, \ldots, M_t$ to separate the signals related to different transmitters. Hence, considering N_p pulses for the estimation process, the samples of the received signal at the mth pulse of the estimation process at the fusion center can be represented as

$$\mathbf{z}^m(n) = \widetilde{\boldsymbol{\psi}}^m(n)\mathbf{s} + \mathbf{e}^m(n), \tag{2.4}$$

where the basis matrix for the sampling time of nT_s at the mth pulse, $\widetilde{\boldsymbol{\psi}}^m(n)$, is defined as

$$\widetilde{\boldsymbol{\psi}}^m(n) = [\boldsymbol{\psi}_k^m(n)|_{(p_x^k, p_y^k, \widetilde{v}_x^k, \widetilde{v}_y^k)=(x_1, y_1, v_x^1, v_y^1)}, \boldsymbol{\psi}_k^m(n)|_{(p_x^k, p_y^k, \widetilde{v}_x^k, \widetilde{v}_y^k)=(x_2, y_2, v_x^2, v_y^2)}, \ldots, \boldsymbol{\psi}_k^m(n)|_{(p_x^k, p_y^k, \widetilde{v}_x^k, \widetilde{v}_y^k)=(x_L, y_L, v_x^L, v_y^L)}], \tag{2.5}$$

$$\boldsymbol{\psi}_k^m(n) = diag\left\{\boldsymbol{\psi}_{1,1}^m(n), \boldsymbol{\psi}_{2,1}^m(n), \ldots, \boldsymbol{\psi}_{N_r,1}^m(n)\right\} \tag{2.6}$$

$$\psi^m_{l,k}(n) = diag\left\{e^{j2\pi(f_k{}^{1l}t-f_c(\tau_k^{1l}+(m-1)T))}x_1(t-\tau_k^{1l}-(m-1)T)*h_1(t)\,|_{t=(m-1)T+T_p+nT_s},\right.$$
$$e^{j2\pi(f_k{}^{2l}t-f_c(\tau_k^{2l}+(m-1)T))}x_2(t-\tau_k^{2l}-(m-1)T)*h_2(t)\,|_{t=(m-1)T+T_p+nT_s},\ldots,$$
$$\left.e^{j2\pi(f_k^{M_tl}t-f_c(\tau_k^{M_tl}+(m-1)T))}x_{M_t}(t-\tau_k^{M_tl}-(m-1)T)*h_{M_t}(t)\,|_{t=(m-1)T+T_p+nT_s}\right\}. \tag{2.7}$$

The noise vector at the mentioned sampling time, $\mathbf{e}^m(n)$, is

$$\mathbf{e}^m(n)=[(\mathbf{e}_1^m(n))^T,\ldots,(\mathbf{e}_{N_r}^m(n))^T]^T, \tag{2.8}$$

$$\mathbf{e}_l{}^m(n)=[\overline{n}_{1l}(t)*h_1(t)\,|_{t=(m-1)T+T_p+nT_s},\ldots,\overline{n}_{M_tl}(t)*h_{M_t}(t)\,|_{t=(m-1)T+T_p+nT_s}], \tag{2.9}$$

and vector $\mathbf{s}$ is also defined as

$$\mathbf{s}=[(\overline{\boldsymbol{\beta}}_1)^T,(\overline{\boldsymbol{\beta}}_2)^T,\ldots,(\overline{\boldsymbol{\beta}}_L)^T]^T, \tag{2.10}$$

$$\overline{\boldsymbol{\beta}}_h=\begin{cases}[\beta_k^{11},\ldots,\beta_k^{M_t1},\ldots,\beta_k^{1N_r},\ldots,\beta_k^{M_tN_r}]^T, & \text{if the } k\text{th target is at } (x_h,y_h,v_x^h,v_y^h)\\ \mathbf{0}_{(M_tN_r)\times 1}, & \text{otherwise.}\end{cases} \tag{2.11}$$

We have assumed that β_k^{il} does not vary within the estimation process duration. If N_s denotes the number of samples in each pulse related to a transmitter–receiver pair, stacking $\{\{z^m(n)\}_{n=1}^{N_s}\}_{m=1}^{N_p}$ into each other, we have

$$\begin{aligned}\mathbf{z}_{(N_pN_sM_tN_r)\times 1}&=[(\mathbf{z}^1(0))^T,\ldots,(\mathbf{z}^1(N_s-1))^T,\ldots,(\mathbf{z}^{N_p}(0))^T,\ldots,(\mathbf{z}^{N_p}(N_s-1))^T]^T\\&=\boldsymbol{\psi}_{(N_pN_sM_tN_r)\times(LM_tN_r)}\mathbf{s}_{(LM_tN_r)\times 1}+\mathbf{e}_{(N_pN_sM_tN_r)\times 1},\end{aligned} \tag{2.12}$$

where

$$\boldsymbol{\psi}=[(\widetilde{\boldsymbol{\psi}}^1(0))^T,\ldots,(\widetilde{\boldsymbol{\psi}}^1(N_s-1))^T,\ldots,(\widetilde{\boldsymbol{\psi}}^{N_p}(0))^T,\ldots,(\widetilde{\boldsymbol{\psi}}^{N_p}(N_s-1))^T]^T, \tag{2.13}$$

$$\mathbf{e}=[(e^1(0))^T,\ldots,(e^1(N_s-1))^T,\ldots,(e^{N_p}(0))^T,\ldots,(e^{N_p}(N_s-1))^T]^T. \tag{2.14}$$

As it is clear, if there is a target at (x_h,y_h,v_x^h,v_y^h), the energy of the hth block of $\mathbf{s}$ is non-zero. Hence, $\mathbf{s}$ is a block K-sparse signal with block length of $d=M_tN_r$.

2.2 *Received Signal Model in a Colocated MIMO Radar*

We can also model the received signal of a colocated MIMO radar considering the procedure in Sect. 2.1. The main difference is that all the transmitter–receiver pairs in a colocated MIMO radar view a target from the same angle. Hence, their corresponding target attenuation coefficients are the same and we can write $\beta_k^{il}=\beta_k$

for $i = 1, \ldots, M_t$ and $l = 1, \ldots, N_r$. The received signal at the fusion center can be shown as

$$\mathbf{z}_{(N_p N_s M_t N_r)\times 1} = \boldsymbol{\psi}_{(N_p N_s M_t N_r)\times L} \mathbf{s}_{L\times 1} + \mathbf{e}, \tag{2.15}$$

where

$$\mathbf{s} = [\overline{\boldsymbol{\beta}}_1, \overline{\boldsymbol{\beta}}_2, \ldots, \overline{\boldsymbol{\beta}}_L]^T, \tag{2.16}$$

$$\overline{\boldsymbol{\beta}}_h = \begin{cases} \boldsymbol{\beta}_k & \text{if the } k\text{th target is at } (x_h, y_h, v_x^h, v_y^h), \\ 0 & \text{otherwise.} \end{cases} \tag{2.17}$$

As it is clear, $\mathbf{s}$ is a K-sparse vector. Furthermore, $\boldsymbol{\psi}$ is made according to (2.5) where

$$\boldsymbol{\psi}_k^m(n) = \sum_{i=1}^{M_t} \sum_{l=1}^{N_r} e^{j2\pi(f_k{}^{il} t - f_c(\tau_k^{il} + (m-1)T))} x_i(t - \tau_k^{il} - (m-1)T)$$
$$* \, h_i(t)\, |_{t=(m-1)T+T_p+nT_s} \, . \tag{2.18}$$

As the distances between the antennas are much less than the distances between the antennas and the targets, (2.18) can be simplified.

3 Compressive Sensing for MIMO Radar Systems

Since the received signal in a MIMO radar system is sparse, we can use CS at each receiver to decrease the sampling rate. Hence, at each receiver, we take a number of linear combinations of the received signal entries which is smaller than what is necessary when the sparsity of the signal is ignored and send fewer samples to the fusion center [12, 30, 61]. In other words, at the lth receiver, the received signal from the ith transmitter is multiplied by a measurement matrix, $\boldsymbol{\varphi}_{i,l}$, that is a $M \times N_s N_p$ matrix ($M \leq N_s N_p$), and the measurements are sent to the fusion center. Hence, at the fusion center, the received signal is also multiplied by $\boldsymbol{\varphi}_{MM_tN_r \times NM_tN_r}$, and we have

$$\mathbf{y} = \boldsymbol{\theta}\mathbf{s} + \boldsymbol{\varphi}\mathbf{e}, \tag{3.1}$$

where $\boldsymbol{\theta} = \boldsymbol{\varphi}\boldsymbol{\psi}$,

$$\boldsymbol{\varphi} = \left[\boldsymbol{\varphi}[1]\ \boldsymbol{\varphi}[2] \ldots \boldsymbol{\varphi}[N]\right], \tag{3.2}$$

$$\varphi[h] = \begin{bmatrix} diag\{\phi_{1,1}^{1,h}, \phi_{2,1}^{1,h}, \ldots, \phi_{M_t,1}^{1,h}, \ldots, \phi_{1,N_r}^{1,h}, \phi_{2,N_r}^{1,h}, \ldots, \phi_{M_t,N_r}^{1,h}\} \\ diag\{\phi_{1,1}^{2,h}, \phi_{2,1}^{2,h}, \ldots, \phi_{M_t,1}^{2,h}, \ldots, \phi_{2,N_r}^{2,h}, \phi_{2,N_r}^{2,h}, \ldots, \phi_{M_t,N_r}^{2,h}\} \\ \vdots \\ diag\{\phi_{1,1}^{M,h}, \phi_{2,1}^{M,h}, \ldots, \phi_{M_t,1}^{M,h}, \ldots, \phi_{1,N_r}^{M,h}, \phi_{2,N_r}^{M,h}, \ldots, \phi_{M_t,N_r}^{M,h}\} \end{bmatrix}, \quad (3.3)$$

and $\phi_{i,l}^{k_1,k_2}$ is the entry of $\varphi_{i,l}$ related to its k_1th row and k_2th column. Matrix θ is called sensing matrix.

3.1 Random Sampling: A Simple Form of CS

RS [8, 9, 49] is another sparsity-based sampling scheme which can be considered as a special case of compressed sensing. In this sampling scheme, instead of taking a weight sum of several entries, a random selection of its entries is considered as the measurements. RS is much simpler than the other CS-based sampling schemes. Furthermore, the simulation results show that it can perform almost better than CS using a random Gaussian matrix [5].

4 Sparse Recovery Methods for Target Detection and Target Parameter Estimation in Sparsity-Based MIMO Radars

At the fusion center, first, the received signal should be recovered. For the recovery of the compressively sampled signal in noiseless environments, the following problem gives us the unique answer under some assumptions on matrix $\boldsymbol{\theta}$ [48]:

$$\min_{\mathbf{s}} \parallel \mathbf{s} \parallel_0 \quad s.t. \quad \mathbf{y} = \boldsymbol{\theta}.\mathbf{s} \quad (4.1)$$

Since problem (4.1) is an NP-hard problem, to recover the sparse vector in noisy or noiseless environments and for a suitably-chosen sensing matrix, some other sparse recovery methods are proposed. One of these methods which is the basic method for a category of sparse recovery methods called l_1 methods [13, 17, 19, 47, 48, 60] is Basis Pursuit (BP) [11]. BP solves (4.1), where l_0 is replaced by l_1. Among other sparse recovery methods, we can mention greedy algorithms [21, 46, 48] which are iterative, and in each iteration, try to find the position of non-zero entries and then estimate their values.

The proposed sparse recovery methods can achieve accurate solutions under some conditions for the sensing matrix. The coherence of the sensing matrix is one of the parameters that can be used for the evaluation of the sensing matrix qualification. This parameter is defined as the maximum absolute value of the correlation between the sensing matrix columns. For ideal sparse recovery, the coherence of the sensing matrix should be small. It has been shown that a zero-mean random Gaussian matrix or a Bernoulli matrix is a good choice for the measurement matrix with high probability [16].

As we discussed before, the received signal in a distributed MIMO radar is block sparse in some domain. For the recovery of a block sparse signal, we can achieve higher efficiency using block sparse recovery methods [23, 24, 26, 34, 58] instead of the traditional ones. This superiority is achieved by using the knowledge of signal structure in block sparse recovery methods. A critical parameter for block sparse recovery methods is block coherence (μ_B) which should be small for having an accurate recovery [24]. This parameter is defined as

$$\mu_B = \max_{l \neq r} \frac{1}{\sqrt{\|\boldsymbol{\theta}[l]\| \|\boldsymbol{\theta}[r]\|}} \rho(\boldsymbol{\theta}[l]\boldsymbol{\theta}[r]), \tag{4.2}$$

where $\boldsymbol{\theta}[r]$ is the rth block of the sensing matrix which consists of its $(r-1)d+1$th column to rdth column.

As the non-zero entries (or blocks for distributed MIMO radars) of $\mathbf{s}$ are corresponding to the target parameters, while $\mathbf{s}$ is recovered, target detection and target parameter estimation can be done by thresholding the entries (or blocks) of $\mathbf{s}$.

5 Main Challenges in Sparsity-Based MIMO Radars

5.1 Measurement Matrix Design

For a proper sparse recovery, the sensing matrix should be suitable. Hence, designing the measurement matrix is one of the challenges in sparsity-based MIMO radars. References [66] and [67] have tried to improve the sparsity-based colocated MIMO radar by designing the measurement matrix instead of using a zero-mean random Gaussian matrix. In [66], a modification of a random measurement matrix is designed to maximize the average signal to jammer ratio, and the results of CS method using this measurement matrix are compared with some other estimation methods such as Capon [44, 62, 63], APES[2] [44, 62, 63], GLRT[3] [44, 63], and MUSIC[4] [44**?**]. Figure 1 shows the estimated target attenuation coefficient versus the estimated Direction Of Arrival (DOA)[5] for the CS method, Capon, APES, and GLRT in a colocated MIMO radar with the parameters mentioned in Table 1. The mentioned CS method exploits CS for reducing the sampling rate and uses a sparse recovery method, called the Dantzig selector [15], for the target parameter

[2]Amplitude and Phase EStimation.

[3]Generalized Likelihood Ratio Test.

[4]MUltiple SIgnal Classification.

[5]The direction of a target from which a propagating wave arrives.

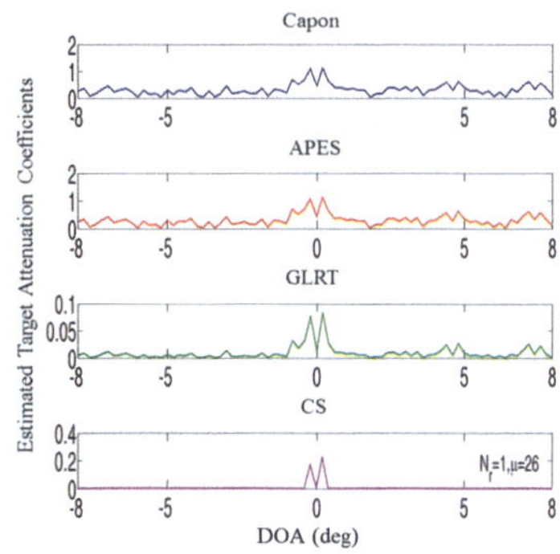

Fig. 1 Estimated target attenuation coefficient versus estimated DOA for two targets in $-0.2°$ and $0.2°$ and a jammer in $7°$ in the SNR of 0 dB [66]

Table 1 MIMO radar parameters in Fig. 1

Parameter	Value
The number of transmitters and receivers	$M_t = 30$ and $N_r = 1$
The locations of the antennas	Uniformly distributed in a disc of radius 10m
The carrier frequency	$f_c = 5GHz$
Pulse Repetition Interval (PRI)	$T = 250\mu s$
The period of sampling	$T_s = \frac{1}{20}\mu s$
The number of samples in Capon, APES, and GLRT methods	$N_p N_s = 512$
The estimation space	$\{-8°, -7.8°, \ldots, 7.8°, 8°\}$

estimation. As the MUSIC method needs the receiver number to be greater than the target number, it cannot be used in this case. In this figure, there are two targets in $-0.2°$ and $0.2°$ and a jammer in $7°$. The target attenuation coefficients are 1. The SNR, the measurement number for the CS method, and the sample number of the other methods are 0 dB, 30, and 512, respectively. As you can see, the CS method even with fewer samples outperforms the other methods.

In [67], two measurement matrix design methods are proposed. The first one tries to minimize the coherence of the sensing matrix plus the inverse of the signal to interference ratio by designing a proper measurement matrix. In the second one, for reducing the complexity, a particular form for measurement matrix is considered, and its parameters are determined to enhance the signal-to-interference ratio. The second method can only be used for a particular form of the transmitted waveforms. However, for these waveforms, it outperforms the first one. Let us denote the jammer-to-signal ratio by β. Figure 2 shows the Receiver Operating Characteristic (ROC) curves of the Matched Filter Method (MFM) [41] and the CS methods using the first designed measurement matrix ($\Phi_{\#1}$), the second designed measurement matrix ($\Phi_{\#2}$), and Gaussian random measurement matrix (G) where the transmitted waveforms are Hadamard waveforms and the percentage of measurements is 25% for CS methods. The ROC curve illustrates the detection probability (P_d) versus the false alarm probability (P_{fa}). Pd and Pfa are defined as the division of the number

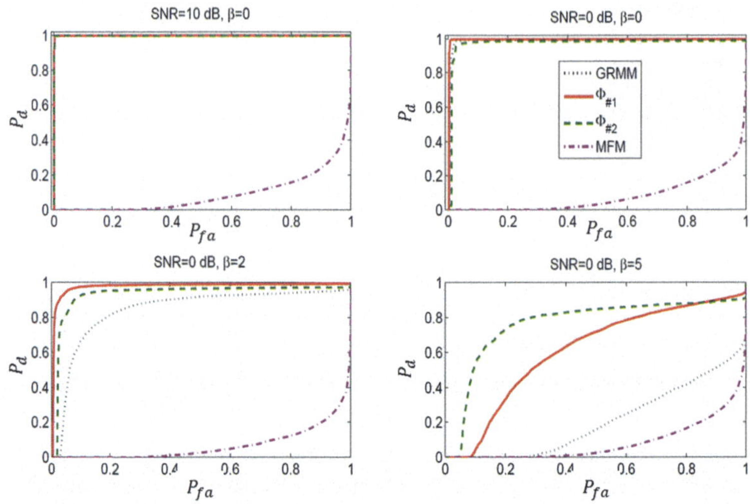

Fig. 2 The ROC curves of Matched Filter Method (MFM) and the CS methods using the first designed measurement matrix ($\Phi_{\#1}$), the second designed measurement matrix ($\Phi_{\#2}$), and the Gaussian random measurement matrix (G) where the transmitted waveforms are Hadamard waveforms and the percentage of measurements is 25% for CS methods [67]

Table 2 MIMO radar parameters in Fig. 2

Parameter	Value
The number of transmitters and receivers	$M_t = N_r = 10$
The locations of the antennas	Uniformly distributed in a disc of radius 10m
The carrier frequency	$f_c = 5GHz$
The number of samples in MFM	$N_p N_s = 100$
The estimation space	Angel $\in \{0°, 0.2°, \ldots, 1°\}$
	Range $\in \{1000, 1015, \ldots, 1090\}$

of correct estimations of all parameters of all targets to the total run number and the devision of the number of false estimations to the total run number, respectively. Note that if only one parameter of a target is falsely estimated, we call it a false estimation. The parameters of the assumed colocated MIMO radar are mentioned in Table 2. There are three targets that are uniformly distributed at the search area in 100 runs. As you can see, the CS methods outperform the MFM and $\Phi_{\#2}$ is better than $\Phi_{\#1}$ for this kind of waveforms.

In [1], a measurement matrix design is proposed for a distributed MIMO radar. In this paper, it is assumed that the sum of the signals of different receivers is received at the fusion center (there is no possibility of separating the signals of different receivers), and the measurement matrix is designed to minimize an upper bound of the sum of the block coherences of the sensing matrix blocks.

5.2 *Optimal Waveform Design for the Transmitters*

Designing the waveforms for the transmitters is an important challenge in a sparsity-based MIMO radar as it affects the sensing matrix. Optimal energy allocation to the transmitters is a simple form of this design. References [68] and [28] design step frequency waveforms and a generalization of linear frequency-modulated (LFM) waveforms for sparsity-based colocated MIMO radars, respectively. References [69] and [7] try to reduce the coherence of the sensing matrix by both transmit waveform design and optimal energy allocation to the transmitters in these systems. For the waveform design, [69] solves a non-convex problem by applying iterative descent methods, which could achieve unfeasible suboptimal solutions. In [7], the problem is solved by showing that the coherence measure depends only on the covariance matrix of the waveforms and reformulating the waveform design as convex optimization problems.

The performance of a distributed MIMO radar is also improved by proper energy allocation schemes. In [27], an adaptive energy allocation scheme is proposed to improve the performance of this system. This energy allocation method tries to maximize the minimum target returns after the first recovery by solving the following problem:

$$\max_{\{E_i\}_{i=1}^{M_t}} \min_{k} \sum_{i=1}^{M_t} \sum_{l=1}^{N_r} E_i \|\widehat{\beta}_k^{il}\|^2, \tag{5.1}$$

where E_i is the energy allocated to the ith transmitter, and we have $\sum_{i=1}^{M_t} E_i = M_t$. Furthermore, $\widehat{\beta}_k^{il}$ is a non-zero element of the recovered signal related to the ith transmitter and lth receiver. The optimization problem (5.1) can easily be solved by CVX[6] [29].

As it was mentioned in Sect. 4, a critical parameter for block sparse recovery methods is block coherence. Reference [69] tries to allocate energy to the transmitters of a distributed MIMO radar in a way that the coherence of the sensing matrix becomes as small as possible. By decreasing the coherence, the block coherence may also decrease. In [1], the authors propose a superior energy allocation method for these systems. In this paper, it is assumed that there is not the possibility of separating the signals of different receivers at the fusion center. By allocating proper energy to the transmitters, [1] tries to minimize an upper bound of the sum of the block coherences of the sensing matrix blocks. As it was mentioned in the previous subsection, [1] also proposes a proper measurement matrix using the mentioned upper bound as the cost function. It is also shown that the accuracy of target parameter estimation can be significantly increased combining the proposed energy allocation scheme and the proposed measurement matrix design.

[6] A software for solving convex optimization problems.

Table 3 MIMO radar parameters in [1]

Parameter	Value
The number of transmitters and receivers	$M_t = N_r = 2$
The locations of the transmitters	$\mathbf{t}_1 = [100, 0]$ and $\mathbf{t}_2 = [200, 0]$
The locations of the receivers	$\mathbf{r}_1 = [0, 200]$ and $\mathbf{r}_2 = [0, 100]$
The carrier frequency	$f_c = 1GHz$
Pulse Repetition Interval (PRI)	$T = 200ms$
The number of pulses in the estimation process	$N_p = 4$
The period of sampling	$T_s = 0.2ms$
Sample number related to each pulse	$N_s = 10$

Table 4 Target parameters in [1]

Target	Parameters
1	$\mathbf{p}_1 = [100, 260], \mathbf{v}_1 = [120, 100]$
2	$\mathbf{p}_2 = [80, 280], \mathbf{v}_2 = [110, 120]$

Let us consider a distributed MIMO radar with parameters mentioned in Table 3 and 2 targets with the parameters depicted in Table 4. The search area is divided into $L = 144$ points with the following positions and velocities:

$$\begin{aligned} x_h &\in \{80, 90, 100\} && \text{for } h = 1, \cdots, 144 \\ y_h &\in \{260, 270, 280\} && \text{for } h = 1, \cdots, 144 \\ {v_x}^h &\in \{100, 110, 120, 130\} && \text{for } h = 1, \cdots, 144 \\ {v_y}^h &\in \{100, 110, 120, 130\} && \text{for } h = 1, \cdots, 144. \end{aligned}$$

For the simulations, Block Matching Pursuit (BMP) and Block Orthogonal Matching Pursuit (BOMP) [24], which are two block greedy sparse recovery methods, are used. BMP and BOMP are the block forms of Matching Pursuit (MP) and Orthogonal Matching Pursuit (OMP) [50], which are fast and can perform well when the SNR is not low. Let us use BOMP-E and BMP-E, respectively, for BOMP and BMP when the transmitted energy is allocated to the transmitters according to [1]; BOMP and BMP for the mentioned methods when the uniform energy allocation is used; and BOMP-CE and BMP-CE, respectively, for BOMP and BMP in the case of exploiting energy allocation of [69]. We define ENR as the ratio of total transmitted energy to the noise energy.

Figure 3 shows the success rate of BOMP, BMP, BOMP-E, BMP-E, BOMP-CE, and BMP-CE versus different percentages of measurements in the ENR of 10 dB. The success rate versus ENR for the mentioned methods in the percentage of measurements of 60% is also shown in Fig. 4.

The notations BOMP-M and BMP-M are respectively used for BOMP and BMP when the measurement matrix is designed exploiting the proposed method of [1] and with the assumption of uniform energy allocation. Furthermore, the notations BOMP-EM and BMP-EM are exploit when the designed measurement matrix is used, and energy allocation is also done according to the proposed method of this

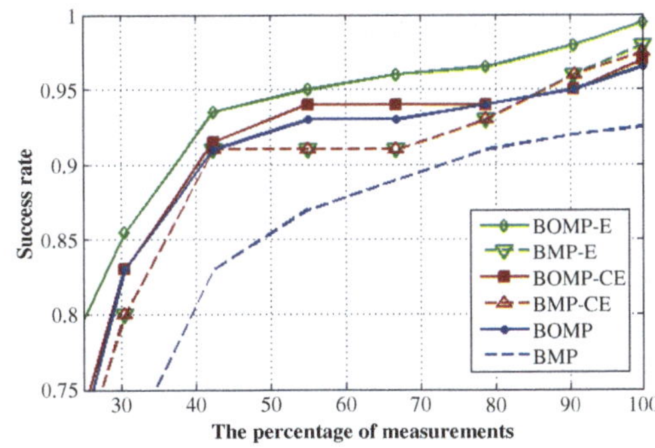

Fig. 3 Success rate versus the percentage of measurements in the ENR of 10 dB [1]

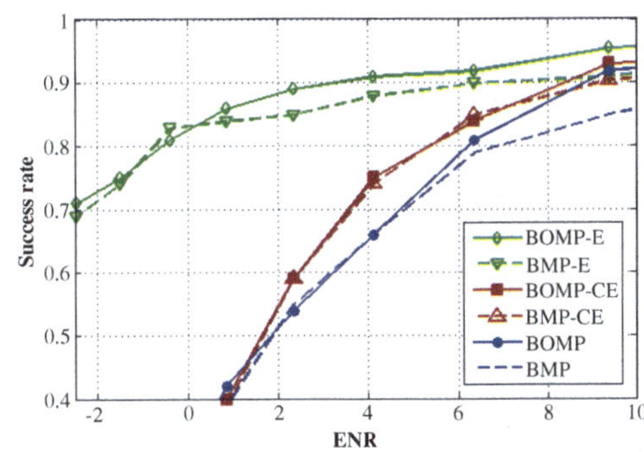

Fig. 4 Success rate versus ENR in the percentage of measurements of 60% [1]

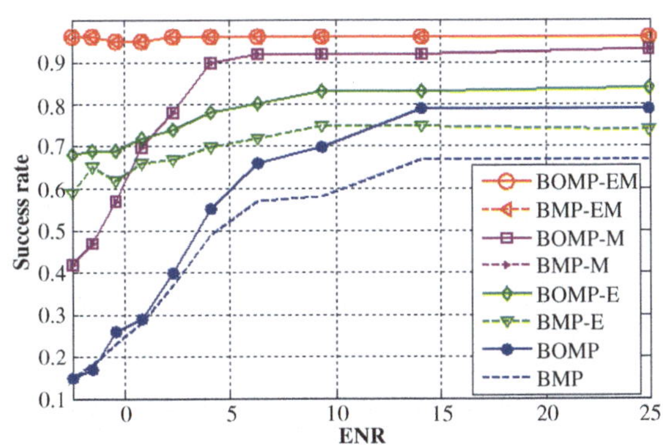

Fig. 5 Success rate versus ENR in the percentage of measurements of 50% [1]

reference. The success rates of BOMP, BMP, BOMP-E, BMP-E, BOMP-M, BMP-M, BOMP-EM, and BMP-EM for different ENR values are shown in Fig. 5 where the percentage of measurements is 50%. The assumptions for the simulations can be found in [1]. As seen, BOMP-EM and BMP-EM are significantly better than the other methods.

5.3 Suitable Sparse Recovery Methods

At the fusion center of a sparsity-based MIMO radar, the received signal should be recovered using a sparse recovery method. Then, targets are detected and their parameters are estimated. As the sparsity number of the received signal, which is related to the number of targets, is unknown beforehand, we shall use a sparse recovery method that does not require this knowledge for accurate recovery. Besides, the complexity is of paramount importance. Two popular sparse recovery methods in this concept are OMP and Least Absolute Shrinkage and Selection Operator (LASSO) [60], in which their block forms are BOMP and group LASSO [70], respectively.

In [45], a fast OMP algorithm for MIMO radar systems is proposed, and it is shown that its performance can approximately be the same as the one of the traditional OMP, while its complexity is less. The OMP algorithm itself has a low complexity but does not have a good performance in low SNRs when an inaccurate sparsity number is used.

LASSO is a l_1 sparse recovery method that does not need the knowledge of sparsity number and has an acceptable complexity and performance. In this method, the Basis Pursuit (BP) problem, which is a complicated problem, is solved by exploiting Lagrangian relaxation method. As it was mentioned, the group LASSO method is the block form of LASSO. LASSO and group LASSO are the recommended methods in many references related to the sparsity-based MIMO radars such as [27, 51, 54], and [69]. Reference [41] has also proposed a solution for group LASSO problem based on the Alternating Direction Method of Multipliers (ADMMs). By using this proposed method, the computational complexity of group LASSOs significantly decreases and the target parameter estimation is improved.

In [5], a block sparse recovery method called Block Iterative Method with Adaptive Thresholding for Sparse Recovery (BIMATSR) is proposed, and it is shown that under some sufficient conditions, the proposed method converges to a stable solution. The Block Iterative Method with Adaptive Thresholding for Compressive Sensing (BIMATCS), which is proposed in [2], can be a special case of BIMATSR under some conditions. BIMATCS is the block form of IMATCS [8, 10]. Both BIMATSR and BIMATCS are somehow similar to the Block Iterative Hard Thresholding (BIHT) algorithm [26]. The BIHT algorithm is the block form of Iterative Hard Thresholding (IHT) [14] and needs the prior knowledge of block sparsity number for having a good performance. However, BIMATSR and BIMATCS do not need this knowledge and exploit a decreasing thresholding scheme to induce the block sparsity. The simulation results show the superiority of BIMATSR and BIMATCS over their counterparts (BOMP,BIHT, and group LASSO) for distributed MIMO radars in different scenarios [2, 5].

It should be mentioned that under some conditions, matrix completion techniques can also be applied to sparsity-based MIMO radar for recovering the received signal [40, 55].

5.3.1 Sparse Recovery Methods for Strong Clutter Environments

The unwanted echoes returned from rain, animal/insects, atmospheric turbulence, ground, sea, etc. are called clutter. In environments with strong clutter, usual sparse recovery method may not work properly. Hence, finding a suitable sparse recovery method for these environments is a challenge.

In [3], a block adaptive CS method called L-BGT with tolerable data acquisition time is proposed for the target parameter estimation in a distributed MIMO radar. Furthermore, the essential changes in a distributed MIMO radar for exploiting this method are presented. In adaptive CS methods, each row of the measurement matrix is determined according to the previous measurements. As adaptive CS methods outperform the non-adaptive ones in the presence of noise [33, 37, 38], they can be useful in strong clutter environments. These methods mostly suffer from the large data acquisition time.

Reference [64] is also related to the imaging in strong clutter environments and tries to remove the effect of clutter before the sparse recovery.

We can also model clutter as a complex random Gaussian vector [31]. Thus, it seems that Bayesian compressive sensing methods [38] can be suitable for the strong clutter environments. Reference [52] has also considered strong clutter for a colocated MIMO radar and used the TR principle [39] for joint estimation of the directions of arrival/departure (DOA/DOD) and Doppler information in strong clutter.

In [6], a Maximum Likelihood (ML)-based block sparse recovery scheme called ML-BSR is proposed for target localization in clutter environments. The complexity of ML estimation is very high especially for the case of an unknown target number. Hence, in [6], first, a block sparse recovery is exploited to reduce the size of the estimation space; then, a modified version of ML method is applied on the new estimation space to estimate target locations accurately. Thus, the complexity of the proposed scheme is acceptable, and its accuracy is high.

Let us consider a MIMO radar with the parameters mentioned in Table 5. There are 2 targets at [2000, 4400] and [1600, 4200], and clutter is a zero-mean Gaussian vector with the covariance matrix mentioned in [6]. Figure 6 shows the ROC curves of the ML estimation method, group LASSO, and the ML-BSR scheme based on group LASSO where the sampling rate is 50%. L_r is the size of new estimation space after exploiting the sparse recovery method. If we consider a maximum value for the target number, the complexity order of the ML-BSR is almost $(\frac{L}{L_r})^{K_{\max}}$ times less than the modified ML method with $K_{\max}$.

5.4 *Off-Grid Case for Targets*

In the most references related to the sparsity-based MIMO radars, it is assumed that all targets lay exactly on the estimation space points. However, this assumption is far from the reality, and for targets out of the estimation grid, there is always

Table 5 MIMO radar parameters in Fig. 6

Parameter	Value
The number of transmitters and receivers	$M_t = 2$ and $N_r = 2$
The locations of the transmitters	[100, 0] and [200, 0]
The locations of the receivers	[0, 100] and [0, 200]
The carrier frequency	$f_c = 10GHz$
	$T_p = 0.1s$, $N_p = 1$, and $N_s = 48$
The estimation space	$\{[x_l, y_l] \| x_l \in \{1600, 1800, \cdots, 2600\}$ and $y_l \in \{4000, 4200, \cdots, 5400\}\}$

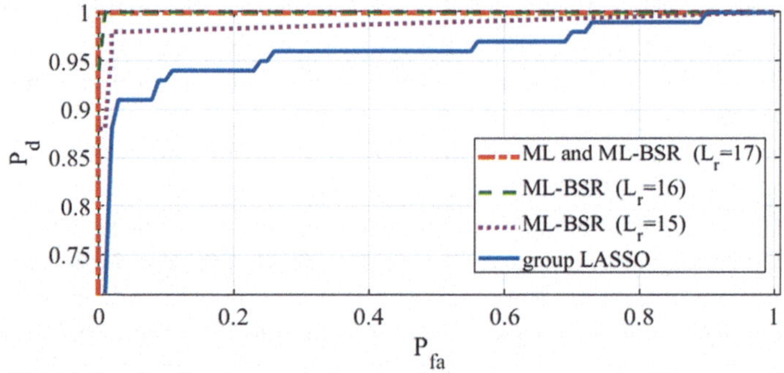

Fig. 6 The ROC curves for the sampling rate of 50%

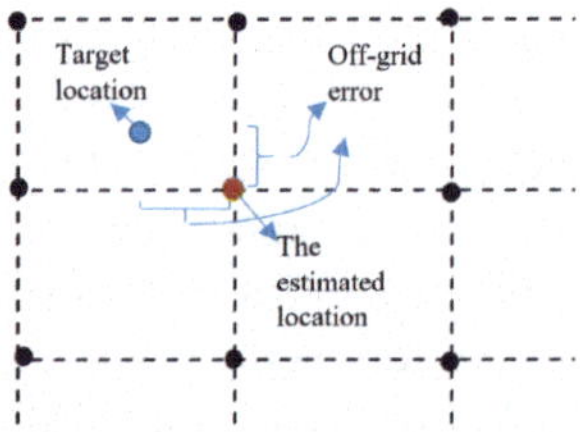

Fig. 7 Off-grid localization error when the nearest grid point to the target is chosen at the fusion center

an error in target parameter estimation even when the nearest grid points to the target parameters are chosen at the fusion center as you can see in Fig. 7 where the estimation space is the location space. We call this error off-grid error. It seems that the easiest way to reduce the off-grid error is to increase the number of the grid points. However, this increases the complexity and the coherence of the sensing matrix [20, 53]. In [25, 36], the authors propose some post-processing methods to refine the output of the on-grid algorithms around the detected target positions; and in [42], it is proposed to alternatively refine the grid after the sparse recovery.

As we do not have perfect information about the basis matrix (or appropriate grid points) at the fusion center, we can consider the following signal instead of $\mathbf{y} = \boldsymbol{\theta}\mathbf{s} + \mathbf{e}$ for reducing the off-grid error:

$$\mathbf{y} = \boldsymbol{\varphi}(\boldsymbol{\psi} + \mathbf{A})\mathbf{s} + \mathbf{e}, \tag{5.2}$$

where the matrix $\mathbf{A}$ is unknown. For solving (5.2), different sparse recovery methods have been proposed considering different structured dictionary mismatch models [4, 22, 32, 56, 57, 59, 65]. In most of these references, linear Taylor expansion is used and $\mathbf{A} = \boldsymbol{\psi}'\boldsymbol{\delta}$, where $\boldsymbol{\delta}$ entries are the deviations of the target parameter (for example, DOA) from the grid points, and $\boldsymbol{\psi}'$ is the derivative of matrix $\boldsymbol{\psi}$ with respect to this parameter.

In [4], instead of Taylor expansion, a linear convex expansion around the neighbor grid points is used. In this reference, any on-grid block sparse recovery algorithm can be used to estimate both the target parameters and the deviation errors from the grid points by using the proposed model and solving a convex optimization problem. Hence, the proposed scheme has a low complexity, while its accuracy is high. Furthermore, it is shown that the proposed scheme is superior to the traditional on-grid approach, the post-processing remedies, and the Taylor expansion approach in detection ability.

6 Conclusion

The received signal of a radar system is a sparse signal as the targets are sparse at the estimation spaces like position–velocity space. Hence, we can use CS for sampling with a sub-Nyquist rate in these systems. By reducing the sampling rate, the costs are also reduced. This reduction for a MIMO radar is more significant due to the existence of multiple receivers that send their received samples to a common processing center. By exploiting a sparse recovery method, the received signal at the fusion center is recovered and target parameters are estimated. Sparse recovery methods can outperform other target parameter estimations even when the sampling rate is reduced. In this chapter, we discussed the possibility of using CS in a MIMO radar system, the main challenges in a sparsity-based MIMO radar for accurate target parameter estimation, and the proposed solutions to these challenges. The discussed challenges are measurement matrix design, optimal waveform design for the transmitters (which includes the optimal energy allocation to the transmitters), suitable sparse recovery methods (especially for strong clutter environments), and the off-grid case for the targets.

References

1. Abtahi, A., Modarres-Hashemi, M., Marvasti, F., and Tabataba, F.: Power allocation and measurement matrix design for block CS-based distributed MIMO radars. Aerosp. Sci. Technol. 53, 128–135
2. Abtahi, A., Azghani, M., and Marvasti, F.: Iterative block-sparse recovery method for distributed MIMO radar. In 2016 Iran Workshop on Communication and Information Theory (IWCIT), 1–4

3. Abtahi, A., Hamidi, S. M., and Marvasti, F.: Block adaptive compressive sensing for distributed MIMO radars in clutter environment. In 2016 17th International Radar Symposium (IRS), 1–5
4. Abtahi, A., Gazor, S., and Marvasti, F.: Off-grid localization in MIMO radars using sparsity. IEEE Signal Process. Lett. 25(2), 313–317
5. Abtahi, A., Azghani, M., and Marvasti, F.: An adaptive iterative thresholding algorithm for distributed MIMO radars. IEEE Trans. Aerosp. Electron. Syst., 55(2), 523–533
6. Abtahi, A., Kamjoo, M.M., Gazor, S., and Marvasti, F.: ML-Based Block Sparse Recovery for distributed MIMO Radars in Clutter Environments. In 7th IEEE Global Conference on Signal and Information Processing (GlobalSIP)
7. Ajorloo, A., Amini, A., and Bastani, M.H.: A Compressive Sensing Based Colocated MIMO Radar Power Allocation and Waveform Design. IEEE Sens. J. 18(22), 9420–9429
8. Azghani, M. and Marvasti, F.: Iterative methods for random sampling and compressed sensing recovery. In Proceedings of 10th International Conference on Eurasip
9. Azghani, M. and Marvasti, F. : Sparse Signal Processing, In New Perspectives on Approximation and Sampling Theory, Springer, 189–213
10. Azghani, M., Kosmas, P., and Marvasti, F.: Microwave medical imaging based on sparsity and an iterative method with adaptive thresholding. IEEE trans. med. imag. 34, 357–365
11. Baraniuk, R.: Compressive sensing. IEEE signal process. mag. 24(4)
12. Baraniuk, R. and Steeghs, P.: Compressive radar imaging. In 2007 IEEE Radar Conference, 128–133
13. Beck, A. and Teboulle, M.: A Fast Iterative Shrinkage-Thresholding Algorithm for Linear Inverse Problems. SIAM J. Imaging Sci. 2, 183–202
14. Blumensath, T. and Davies, M.: Iterative Thresholding for Sparse Approximations. J Fourier Anal Appl. 14, 629–654
15. Candes, E. and Tao, T.: The Dantzig selector: Statistical estimation when p is much larger than n. Ann. Statist., 35, 2313–2351
16. Candes, E. and Wakin, M.: An Introduction To Compressive Sampling. IEEE Signal Process. Mag. 25, 21–30
17. Cevher, V.: Learning with compressible priors. In Advances in Neural Information Processing Systems, 261–269
18. Chen, C. Y. and Vaidyanathan, P. P.: Compressed sensing in MIMO radar. In 2008 42nd Asilomar Conference on Signals, Systems and Computers, 41–44
19. Chen, S., Donoho, D., and Saunders, M.: Atomic Decomposition by Basis Pursuit. SIAM Rev. 43, 129–159
20. Chi, Y., Scharf, L., Pezeshki, A., and Calderbank, A.: Analysis of Orthogonal Matching Pursuit Using the Restricted Isometry Property. IEEE Trans. Signal Process. 59, 2182–2195
21. Davenport, M. and Wakin, M.: Analysis of Orthogonal Matching Pursuit Using the Restricted Isometry Property. IEEE Trans. Inform. Theory. 56, 4395–4401
22. Ekanadham, C., Tranchina, D., and Simoncelli, E. P.: Recovery of sparse translation-invariant signals with continuous basis pursuit. IEEE Trans. Signal Process. 59(10), 4735–4744
23. Eldar, Y. and Mishali, M.: Robust Recovery of Signals From a Structured Union of Subspaces. IEEE Trans. Inform. Theory. 55, 5302–5316
24. Eldar, Y., Kuppinger, P., and Bolcskei, H.: Block-Sparse Signals: Uncertainty Relations and Efficient Recovery. IEEE Trans. Signal Process. 58, 3042–3054
25. Feng, Ch., Valaee, Sh., and Tan, Zh.: Multiple target localization using compressive sensing. In Global Telecommunications Conference, 1–6
26. Garg, R. and Khandekar, R.: Block-sparse Solutions using Kernel Block RIP and its Application to Group Lasso. In AISTATS, 296–304
27. Gogineni, S. and Nehorai, A.: Target Estimation Using Sparse Modeling for Distributed MIMO Radar. IEEE Trans. Signal Process. 59, 5315–5325
28. Gogineni, S. and Nehorai, A.: Frequency-hopping code design for MIMO radar estimation using sparse modeling. IEEE Trans. Signal Process. 60, 3022–3035
29. Grant, M., Boyd, S., and Ye, Y.: CVX: Matlab software for disciplined convex programming

30. Gurbuz, A. C., McClellan, J. H., and Scott, W. R.: Compressive sensing for GPR imaging. In 2007 Conference Record of the Forty-First Asilomar Conference on Signals, Systems and Computers, 2223–2227
31. He, Q., Lehmann, N., Blum, R., and Haimovich, A.: MIMO Radar Moving Target Detection in Homogeneous Clutter. IEEE Trans. Aerosp. Electron. Syst. 46, 1290–1301
32. He, X., Liu, Ch., Liu, B., and Wang, D.: Sparse frequency diverse MIMO radar imaging for off-grid target based on adaptive iterative MAP. Remote Sens. 5, 631–647
33. Haupt, J., Castro, R., and Nowak, R.: Distilled Sensing: Adaptive Sampling for Sparse Detection and Estimation. IEEE Trans. Inform. Theory. 57, 6222–6235
34. Huang, A., Guan, G., Wan, Q., and Mehbodniya, A.: A block orthogonal matching pursuit algorithm based on sensing dictionary. Int. J. Phys. Sci. 6(5), 992–999
35. Ibernon-Fernandez, R., Molina-Garcia-Pardo, J., and Juan-Llacer, L.: Comparison Between Measurements and Simulations of Conventional and Distributed MIMO System. Antennas Wirel. Propag. Lett. 7, 546–549
36. Ibrahim, M., Romer, F. , Alieiev, R., Del Galdo, G., and Thoma, R.: On the estimation of grid offsets in CS-based direction-of-arrival estimation. In International Conference on Speech and Signal Processing (ICASSP), 6776–6780
37. Iwen, M. and Tewfik, A.: Adaptive Strategies for Target Detection and Localization in Noisy Environments. IEEE Trans. Signal Process. 60, 2344–2353
38. Ji, S., Xue, Y., and Carin, L.: Bayesian Compressive Sensing. IEEE Trans. Signal Process. 56, 2346–2356
39. Jin, Y., Moura, J., and O'Donoughue, N.: Time reversal in multiple-input multiple-output radar. IEEE J-STSP. 4, 210–225
40. Kalogerias, D. and Petropulu, A.: Matrix completion in colocated MIMO radar: Recoverability, bounds & theoretical guarantees, IEEE Trans. Signal process. 62, 309–321
41.]Li, B. and Petropulu, A.: Efficient target estimation in distributed MIMO radar via the ADMM. In 2014 48th Annual Conference on Information Sciences and Systems (CISS), 1–5
42. Li, J. and Stoica, P.: A sparse signal reconstruction perspective for source localization with sensor arrays. IEEE Trans. Signal Process. 53, 3010–3022
43. Li, J. and Stoica, P.: MIMO Radar with Colocated Antennas. IEEE Signal Process. Mag. 24, 106–114
44. Li, J. and Stoica, P.: MIMO radar signal processing. J. Wiley & Sons, Hoboken, NJ
45. Liu, Y., Wu, MY, and Wu, S.J.: Fast OMP algorithm for 2D angle estimation in MIMO radar. IET Electron. lett. 46, 444–445
46. Needell, D. and Tropp, J.: CoSaMP: Iterative signal recovery from incomplete and inaccurate samples. Applied and Computational Harmonic Analysis. 26, 301–321
47. Nowak, R. D. and Wright, S. J.: Gradient projection for sparse reconstruction: Application to compressed sensing and other inverse problems. IEEE J-STSP. 1(4), 586–597
48. Marvasti, F., Amini, A., Haddadi, F., Soltanolkotabi, M., Khalaj, B., Aldroubi, A., Sanei, S., and Chambers, J.: A unified approach to sparse signal processing. EURASIP J. Adv. Signal Process. 2012, 44
49. Marvasti, F., Azghani, M., Imani, P., Pakrouh, P., Heydari, S. J., Golmohammadi, A., and Khalili, M. M.: Sparse signal processing using iterative method with adaptive thresholding (IMAT). In 2012 19th International Conference on Telecommunications (ICT), 1–6
50. Pati, Y. C., Rezaiifar, R., and Krishnaprasad, P. S.: Orthogonal matching pursuit: Recursive function approximation with applications to wavelet decomposition. In Proceedings of 27th Asilomar conference on signals, systems and computers, 40–44
51. Petropulu, A. P., Yu, Y., and Huang, J.: On exploring sparsity in widely separated MIMO radar. In 2011 Conference Record of the Forty Fifth Asilomar Conference on Signals, Systems and Computers (ASILOMAR), 1496–1500
52. Sajjadieh, M. and Asif, A.: Joint sparse recovery method for compressed sensing with structured dictionary mismatches. IEEE Signal Process. Lett., 22, 1283–1287
53. Scharf, L., Chong, E., Pezeshki, A., and Luo, J.: Sensitivity considerations in compressed sensing. In 2011 Conference Record of the Forty Fifth Asilomar Conference on Signals, Systems and Computers (ASILOMAR), 744–748

54. Strohmer, T. and Friedlander, B.: Analysis of sparse MIMO radar. Applied and Computational Harmonic Analysis. 37(3), 361–388
55. Sun, Sh., Bajwa, W., and Petropulu, A.: MIMO-MC radar: A MIMO radar approach based on matrix completion. IEEE Trans. Aerosp. Electron. Syst. 51, 1839–1852
56. Tan, Z. and Nehorai, A.: Sparse direction of arrival estimation using co-prime arrays with off-grid targets. IEEE Signal Process. Lett. 21(1), 26–29
57. Tan, Z., Yang, P., and Nehorai, A.: Joint sparse recovery method for compressed sensing with structured dictionary mismatches. IEEE Trans. Signal Process. 62(19), 4997–5008
58. Tang, G. and Nehorai, A.: Computable performance analysis of block-sparsity recovery. In 2011 4th IEEE International Workshop on Computational Advances in Multi-Sensor Adaptive Processing (CAMSAP).
59. Teke, O., Gurbuz, A., and Arikan, O.: Perturbed orthogonal matching pursuits. IEEE Trans. Signal Process., 61(24), 6220–6231
60. Tibshirani, R.: Regression shrinkage and selection via the lasso: a retrospective. J. R. Stat. Soc. Series B Stat. 73, 273–282
61. Tropp, J. A., Wakin, M. B., Duarte, M. F., Baron, D., and Baraniuk, R. G.: Random filters for compressive sampling and reconstruction. In 2006 IEEE International Conference on Acoustics Speech and Signal Processing Proceedings. 3, III-III
62. Xu, L., Li, J., and Stoica, P.: Adaptive techniques for MIMO radar. In 2006 Fourth IEEE Workshop on Sensor Array and Multichannel Processing, 258–262
63. Xu, L., Li, J., and Stoica, P.: Radar imaging via adaptive MIMO techniques. In 2006 14th European Signal Processing Conference, I-V
64. Yang, J., Jin, T., Huang, X., Thompson, J., and Zhou, Z.: Sparse MIMO array forward-looking GPR imaging based on compressed sensing in clutter environment. IEEE Trans. Geosci. Remote Sens. 52(7), 4480–4494
65. Yang, Z., Zhang, C., and Xie, L.: Robustly stable signal recovery in compressed sensing with structured matrix perturbation. IEEE Trans. Signal Process. 60(9), 4658–4671
66. Yu, Y., Petropulu, A., and Poor, H.: MIMO Radar Using Compressive Sampling. IEEE J. Sel. Top. Signal Process. 4, 146–163
67. Yu, Y., Petropulu, A., and Poor, H.: Measurement Matrix Design for Compressive Sensing Based MIMO Radar. IEEE Trans. Signal Process. 59, 5338–5352
68. Yu, Y., Petropulu, A., and Poor, H.: CSSF MIMO RADAR: Compressive-Sensing and Step-Frequency Based MIMO Radar. IEEE Trans. Aerosp. Electron. Syst. 48, 1490–1504
69. Yu, Y., Sun, S., Madan, R., and Petropulu, A.: Power allocation and waveform design for the compressive sensing based MIMO radar. IEEE Trans. Aerosp. Electron. Syst. 50, 898–909
70. Yuan, M. and Lin, Y.: Model selection and estimation in regression with grouped variables. J. R. Stat. Soc. Series B Stat. 68, 49–67

Robust Width: A Characterization of Uniformly Stable and Robust Compressed Sensing

Jameson Cahill and Dustin G. Mixon

Abstract Compressed sensing seeks to invert an underdetermined linear system by exploiting additional knowledge of the true solution. Over the last decade, several instances of compressed sensing have been studied for various applications, and for each instance, reconstruction guarantees are available provided the sensing operator satisfies certain sufficient conditions. In this paper, we completely characterize the sensing operators which allow uniformly stable and robust reconstruction by convex optimization for many of these instances. The characterized sensing operators satisfy a new property we call the *robust width property*, which simultaneously captures notions of *widths* from approximation theory and of *restricted eigenvalues* from statistical regression. We provide a geometric interpretation of this property, we discuss its relationship with the restricted isometry property, and we apply techniques from geometric functional analysis to find random matrices which satisfy the property with high probability.

1 Introduction

Let $x^\natural$ be some unknown member of a finite-dimensional Hilbert space $\mathcal{H}$, and let $\Phi\colon \mathcal{H} \to \mathbb{F}^M$ denote some known linear operator, where $\mathbb{F}$ is either $\mathbb{R}$ or $\mathbb{C}$. In a general form, concerns the task of estimating $x^\natural$ provided

(i) we are told that $x^\natural$ is close in some sense to a particular subset $\mathcal{A} \subseteq \mathcal{H}$, and
(ii) we are given data $y = \Phi x^\natural + e$ for some unknown $e \in \mathbb{F}^M$ with $\|e\|_2 \leq \varepsilon$.

Intuitively, if the subset $\mathcal{A}$ is "small," then (i) offers more information about $x^\natural$, and so we might allow M to be small, accordingly; we are chiefly interested in

J. Cahill (✉)
University of North Carolina Wilmington, Wilmington, NC, USA
e-mail: cahillj@uncw.edu

D. G. Mixon
The Ohio State University, Columbus, OH, USA
e-mail: mixon.23@osu.edu

M. Hirn et al. (eds.), *Excursions in Harmonic Analysis, Volume 6*, Applied and Numerical Harmonic Analysis, https://doi.org/10.1007/978-3-030-69637-5_18

cases where M can be smaller than the dimension of $\mathcal{H}$, as suggested by the name "compressed sensing." A large body of work over the last decade has shown that for several natural choices of $\mathcal{A}$, there is a correspondingly natural choice of norm $\|\cdot\|_\sharp$ over $\mathcal{H}$ such that

$$\Delta_{\sharp,\Phi,\varepsilon}(y) \quad := \quad \arg\min \quad \|x\|_\sharp \quad \text{subject to} \quad \|\Phi x - y\|_2 \leq \varepsilon$$

is an impressively good estimate of $x^\natural$, provided the sensing operator Φ satisfies certain properties. However, the known sufficient conditions on Φ are not known to be necessary.

In this paper, we consider a broad class of triples $(\mathcal{H}, \mathcal{A}, \|\cdot\|_\sharp)$, and for each member of this class, we completely characterize the sensing operators Φ for which

$$\|\Delta_{\sharp,\Phi,\varepsilon}(\Phi x^\natural + e) - x^\natural\|_2 \leq C_0\|x^\natural - a\|_\sharp + C_1\varepsilon \quad \forall x^\natural \in \mathcal{H}, e \in \mathbb{F}^M, \|e\|_2 \leq \varepsilon, a \in \mathcal{A} \tag{1.1}$$

for any given C_0 and C_1. In the left-hand side above, we use $\|\cdot\|_2$ to denote the norm induced by the inner product over $\mathcal{H}$. A comment on terminology: Notice that the above guarantee is *uniform* over all $x^\natural \in \mathcal{H}$. Also, the $\|x^\natural - a\|_\sharp$ term ensures in the sense that we allow $x^\natural$ to deviate from the signal model $\mathcal{A}$, whereas the ε term ensures in the sense that we allow for noise in the sensing process.

The next section describes how our main result fits with the current compressed sensing literature in the traditional sparsity case. Next, Sect. 3 gives the main result: that (1.1) is equivalent to a new property we call the *robust width property (RWP)*. This guarantee holds for a variety of instances of compressed sensing, specifically, whenever the triple $(\mathcal{H}, \mathcal{A}, \|\cdot\|_\sharp)$ forms something we call a *CS space*. We identify several examples of CS spaces in Sect. 4 to help illustrate the extent of the generality. In the special case where $\Phi\Phi^* = I$, the matrix Φ satisfies RWP precisely when a sizable neighborhood of its null space in the Grassmannian is contained in the set of null spaces of matrices which satisfy a natural generalization of the *width property* in [27]; we make this equivalence rigorous in Sect. 5. In Sect. 6, we provide a direct proof that the restricted isometry property (RIP) implies RWP in the traditional sparsity case. Section 7 then applies techniques from geometric functional analysis to show (without appealing to RIP) that certain random matrices satisfy RWP with high probability. In fact, taking inspiration from [42], we produce a sensing matrix that satisfies RWP for the traditional sparsity case, but does not satisfy RIP, thereby proving that RIP is strictly stronger than (2.3). We conclude in Sect. 8 with some remarks.

2 Perspective: The Traditional Case of Sparsity

Since the release of the seminal papers in compressed sensing [10, 11, 18], the community has traditionally focused on the case in which $\mathcal{H} = \mathbb{R}^N$, $\mathcal{A}$ is the set

Σ_K of vectors with at most K nonzero entries (hereafter, K-), and $\|\cdot\|_\sharp$ is taken to be $\|\cdot\|_1$, defined by

$$\|x\|_1 := \sum_{i=1}^{N} |x_i|.$$

In this case, the *(NSP)* characterizes when $\Delta_{1,\Phi,0}(\Phi x^\natural) = x^\natural$ for every K-sparse $x^\natural$ (see Theorem 4.5 in [22], for example). Let x_K denote the *best K-term approximation of* x, gotten by setting all entries of x to zero, save the K of largest magnitude. Then NSP states that

$$\|x\|_1 < 2\|x - x_K\|_1 \qquad \forall x \in \ker(\Phi) \setminus \{0\}. \tag{K-NSP}$$

One may pursue some notion of stability by strengthening NSP. Indeed,

$$\|x\|_1 \leq c\|x - x_K\|_1 \qquad \forall x \in \ker(\Phi) \tag{(K,c)-NSP}$$

is equivalent to

$$\|\Delta_{1,\Phi,0}(\Phi x^\natural) - x^\natural\|_1 \leq \frac{2c}{2-c}\|x^\natural - x_K^\natural\|_1 \qquad \forall x^\natural \in \mathbb{R}^N \tag{2.1}$$

whenever $1 < c < 2$ (see Theorem 4.12 in [22], for example). Observe that $\|x^\natural - x_K^\natural\|_1 \leq \|x^\natural - a\|_1$ for every $a \in \Sigma_K$, and so this is similar to (1.1) in the case where $\varepsilon = 0$. However, we prefer a more isotropic notion of stability, and so we seek error bounds in ℓ_2. Of course, the estimate $\|u\|_2 \leq \|u\|_1$ converts (2.1) into such an error bound. Still, we should expect to do better (by a factor of $\sqrt{K}$), as suggested by Theorem 8.1 in [16], which gives that having

$$\|x\|_2 \leq \frac{c}{\sqrt{2K}}\|x - x_{2K}\|_1 \qquad \forall x \in \ker(\Phi) \tag{$(2K,c)$-NSP$_{2,1}$}$$

for some $c > 0$ is equivalent to the existence of a decoder $\Delta\colon \mathbb{R}^M \to \mathbb{R}^N$ such that

$$\|\Delta(\Phi x^\natural) - x^\natural\|_2 \leq \frac{C}{\sqrt{K}}\|x^\natural - x_K^\natural\|_1 \qquad \forall x^\natural \in \mathbb{R}^N \tag{2.2}$$

for some $C > 0$. (Observe that the C in (2.2) differs from the C_0 in (1.1) by a factor of $\sqrt{K}$; as one might expect, the C_0 in (1.1) depends on $\mathcal{A}$ in general.) Unfortunately, the decoder Δ that [16] constructs for the equivalence is computationally inefficient. Luckily, Kashin and Temlyakov [27] remark that (2.2) holds for $\Delta = \Delta_{1,\Phi,0}$ if and only if Φ satisfies the

$$\|x\|_2 \leq \frac{c}{\sqrt{K}}\|x\|_1 \qquad \forall x \in \ker(\Phi) \qquad ((K,c)\text{-WP})$$

for some $c > 0$. The name here comes from the fact that $c/\sqrt{K}$ gives the maximum radius (i.e., "width") of the portion of the unit ℓ_1 ball that intersects $\ker(\Phi)$. We note that while (K, c)-WP is clearly implied by $(2K, c)$-NSP$_{2,1}$ (and may appear to be a strictly weaker assumption), it is a straightforward exercise to verify that the two are in fact equivalent up to constants. The moral here is that a sensing matrix Φ has a uniformly stable decoder Δ precisely when $\Delta_{1,\Phi,0}$ is one such decoder, thereby establishing that ℓ_1 minimization is particularly natural for the traditional compressed sensing problem.

The robustness of ℓ_1 minimization is far less understood. Perhaps the most popular sufficient condition for robustness is the

$$(1-\delta)\|x\|_2^2 \leq \|\Phi x\|_2^2 \leq (1+\delta)\|x\|_2^2 \qquad \forall x \in \Sigma_{2K}. \qquad ((2K,\delta)\text{-RIP})$$

As the name suggests, an RIP matrix Φ acts as a near-isometry on $2K$-sparse vectors, and if Φ satisfies $(2K, \delta)$-RIP, then

$$\|\Delta_{1,\Phi,\varepsilon}(\Phi x^\natural + e) - x^\natural\|_2 \leq \frac{C_0}{\sqrt{K}}\|x^\natural - x^\natural_K\|_1 + C_1\varepsilon \quad \forall x^\natural \in \mathbb{R}^N, e \in \mathbb{R}^M, \|e\|_2 \leq \varepsilon, \tag{2.3}$$

where, for example, we may take $C_0 = 4.2$ and $C_1 = 8.5$ when $\delta = 0.2$ (Theorem 1.2 in [9]). As such, in the traditional sparsity case, RIP implies the condition (1.1) we wish to characterize. Note that the lower RIP bound $\|\Phi x\|_2^2 \geq (1-\delta)\|x\|_2^2$ is important since otherwise we might fail to distinguish a certain pair of K-sparse vectors given enough noise. On the other hand, the upper RIP bound $\|\Phi x\|_2^2 \leq (1+\delta)\|x\|_2^2$ is not intuitively necessary for (2.3).

In pursuit of a characterizing property, one may be inclined to instead seek an NSP-type sufficient condition for (2.3). To this end, Theorem 4.22 in [22] gives that (2.3) is implied by the following robust version of the null space property:

$$\|x_K\|_2 \leq \frac{c_0}{\sqrt{K}}\|x - x_K\|_1 + c_1\|\Phi x\|_2 \qquad \forall x \in \mathbb{R}^N \qquad ((K,c_0,c_1)\text{-RNSP}_{2,1})$$

provided $0 < c_0 < 1$ and $c_1 > 0$; in (2.3), one may take $C_0 = 2(1+c_0)^2/(1-c_0)$ and $C_1 = (3+c_0)c_1/(1-c_0)$. This property appears to be a fusion of sorts between NSP and the lower RIP bound, which seems promising, but a proof of necessity remains elusive.

As a special case of the main result in this paper, we characterize the sensing matrices Φ which satisfy (2.3) as those which satisfy a new property called the

$$\|x\|_2 \leq \frac{c_0}{\sqrt{K}}\|x\|_1 \qquad \forall x \in \mathbb{R}^N \text{ such that } \|\Phi x\|_2 < c_1\|x\|_2. \qquad ((K,c_0,c_1)\text{-RWP})$$

In particular, C_0 and C_1 scale like c_0 and $1/c_1$, respectively, in both directions of the equivalence (and with reasonable constants). We note that (K, c_0)-WP follows directly from (K, c_0, c_1)-RWP. Also, the contrapositive statement gives that $\|\Phi x\|_2 \geq c_1\|x\|_2$ whenever $\|x\|_2 > (c_0/\sqrt{K})\|x\|_1$, which not only captures the lower RIP bound we find appealing, but also bears some resemblance to the *restricted eigenvalue property* that Bickel, Ritov and Tsybakov [6] use to produce guarantees for Lasso [48] and the Dantzig selector [12]:

$$\|\Phi x\|_2 \geq c_0\|x\|_2 \qquad \forall x \in \mathbb{R}^N \text{ such that } \|x - x_K\|_1 \leq c_1\|x_K\|_1. \tag{(K, c_0, c_1)-REP}$$

Indeed, both RWP and REP impose a lower RIP-type bound on all nearly sparse vectors. The main distinction between RWP and REP is the manner in which "nearly sparse" is technically defined.

3 Main Result

In this section, we present a characterization of stable and robust compressed sensing. This result can be applied to various instances of compressed sensing, and in order to express these instances simultaneously, it is convenient to make the following definition:

Definition 3.1 A $(\mathcal{H}, \mathcal{A}, \|\cdot\|_\sharp)$ *with bound* L consists of a finite-dimensional Hilbert space $\mathcal{H}$, a subset $\mathcal{A} \subseteq \mathcal{H}$, and a norm $\|\cdot\|_\sharp$ on $\mathcal{H}$ with the following properties:

(i) $0 \in \mathcal{A}$.
(ii) For every $a \in \mathcal{A}$ and $z \in \mathcal{H}$, there exists a decomposition $z = z_1 + z_2$ such that

$$\|a + z_1\|_\sharp = \|a\|_\sharp + \|z_1\|_\sharp, \qquad \|z_2\|_\sharp \leq L\|z\|_2.$$

The first property above ensures that $\mathcal{A}$ is not degenerate, whereas the second property is similar to the notion of *decomposability*, introduced by Negahban et al. [39]. Note that one may always take $z_1 = 0$ and $z_2 = z$, leading to a trivial bound $L = \sup\{\|z\|_\sharp/\|z\|_2 : z \in \mathcal{H} \setminus \{0\}\}$. Written differently, this bound satisfies $L^{-1} = \min\{\|z\|_2 : z \in \mathcal{H}, \|z\|_\sharp = 1\}$, which is the smallest width of the unit $\sharp$-ball $B_\sharp$ (this is an example of a *Gelfand width*). As our main result will show, any substantial improvement to this bound will allow for uniformly stable and robust compressed sensing. Such improvement is possible provided one may always decompose $z = z_1 + z_2$ so that z_2 either has small ℓ_2 norm or satisfies $\|z_2\|_\sharp \ll \|z_2\|_2$, which is to say that $z_2/\|z_2\|_\sharp$ lies in proximity to a "pointy" portion of $B_\sharp$. However, we can expect such choices for z_2 to be uncommon, and so the set of all x satisfying

$$\|a + x\|_\sharp = \|a\|_\sharp + \|x\|_\sharp \tag{3.1}$$

should be large, accordingly. For the sake of intuition, denote the *of* $\|\cdot\|_\sharp$ *at* a by

$$\mathcal{D} := \{y : \exists t > 0 \text{ such that } \|a + ty\|_\sharp \leq \|a\|_\sharp\}.$$

In the appendix, we show that if $\|v\|_\sharp = \|a\|_\sharp$ and $v - a$ generates a bounding ray of $\mathcal{D}$, then every nonnegative scalar multiple x of v satisfies (3.1). As such, the set of x satisfying (3.1) is particularly large, for example, when $B_\sharp$ is at $a/\|a\|_\sharp$, that is, the neighborhood of $a/\|a\|_\sharp$ in $B_\sharp$ is identical to the neighborhood of 0 in $\mathcal{D}$ (when translated by $a/\|a\|_\sharp$). Note that B_1 is locally conic at any sparse vector, whereas B_2 is nowhere locally conic.

For the record, we make no claim that ours is the ultimate definition of a CS space; indeed, our main result might be true for a more extensive class of spaces, but we find this definition to be particularly broad. We demonstrate this in the next section with a series of instances, each of which having received considerable attention in the literature.

Next, we formally define our characterizing property:

Definition 3.2 We say a linear operator $\Phi\colon \mathcal{H} \to \mathbb{F}^M$ satisfies the (ρ, α)- *(RWP) over* $B_\sharp$ if

$$\|x\|_2 \leq \rho\|x\|_\sharp$$

for every $x \in \mathcal{H}$ such that $\|\Phi x\|_2 < \alpha\|x\|_2$.

If Φ satisfies (ρ, α)-RWP, then its null space necessarily intersects the unit $\sharp$-ball $B_\sharp$ with maximum radius $\leq \rho$. The RWP gets its name from this geometric feature; "robust" comes from the fact that points in $B_\sharp$ which are sufficiently close to the null space exhibit the same concentration in ℓ_2. We further study this geometric meaning in Sect. 5. In the meantime, we give the main result:

Theorem 3.1 *For any CS space* $(\mathcal{H}, \mathcal{A}, \|\cdot\|_\sharp)$ *with bound* L *and any linear operator* $\Phi\colon \mathcal{H} \to \mathbb{F}^M$*, the following are equivalent up to constants:*

(a) Φ *satisfies the* (ρ, α)*-robust width property over* $B_\sharp$*.*
(b) For every $x^\natural \in \mathcal{H}$*,* $\varepsilon \geq 0$ *and* $e \in \mathbb{F}^M$ *with* $\|e\|_2 \leq \varepsilon$*, any solution* $x^\star$ *to*

$$\arg\min \quad \|x\|_\sharp \quad \text{subject to} \quad \|\Phi x - (\Phi x^\natural + e)\|_2 \leq \varepsilon$$

satisfies $\|x^\star - x^\natural\|_2 \leq C_0\|x^\natural - a\|_\sharp + C_1\varepsilon$ *for every* $a \in \mathcal{A}$*.*

In particular, (a) implies (b) with

$$C_0 = 4\rho, \qquad C_1 = \frac{2}{\alpha}$$

provided $\rho \leq 1/(4L)$*. Also, (b) implies (a) with*

$$\rho = 2C_0, \qquad \alpha = \frac{1}{2C_1}.$$

Notice that C_0 scales with ρ, while C_1 scales with α^{-1}. In the next section, we provide a variety of examples of CS spaces for which $L = O(\sqrt{K})$ for some parameter K, and since C_0 scales with $\rho = O(1/L)$, we might expect to find reconstruction guarantees for these spaces with $C_0 = O(1/\sqrt{K})$; indeed, this has been demonstrated in [9, 19, 36–38, 44]. Also, the fact that C_1 scales with α^{-1} is somewhat intuitive: First, suppose that

$$\|a\|_\sharp \le L\|a\|_2 \qquad \forall a \in \mathcal{A}.$$

(This occurs for every CS space considered in the next section.) If $L < \rho^{-1}$, then the contrapositive of RWP gives that every point in $\mathcal{A}$ avoids the null space of Φ. The extent to which these points avoid the null space is captured by α, and so we might expect more stability when α is larger (as is the case).

Proof of Theorem 3.1

(b)⇒(a) Pick $x^\natural$ such that $\|\Phi x^\natural\|_2 < \alpha\|x^\natural\|_2$, and set $\varepsilon = \alpha\|x^\natural\|_2$ and $e = 0$. Due to the feasibility of $x = 0$, we may take $x^\star = 0$, and so

$$\|x^\natural\|_2 = \|x^\star - x^\natural\|_2 \le C_0\|x^\natural\|_\sharp + C_1\varepsilon = C_0\|x^\natural\|_\sharp + \alpha C_1\|x^\natural\|_2,$$

where the inequality applies (b) with $a = 0$, which is allowed by property (i). Isolating $\|x^\natural\|_2$ then gives

$$\|x^\natural\|_2 \le \frac{C_0}{1-\alpha C_1}\|x^\natural\|_\sharp = \rho\|x^\natural\|_\sharp,$$

where we take $\alpha = (2C_1)^{-1}$ and $\rho = 2C_0$.

(a)⇒(b) Pick $a \in \mathcal{A}$, and decompose $x^\star - x^\natural = z_1 + z_2$ according to property (ii), i.e., so that $\|a + z_1\|_\sharp = \|a\|_\sharp + \|z_1\|_\sharp$ and $\|z_2\|_\sharp \le L\|x^\star - x^\natural\|_2$. Then

$$\begin{aligned}
\|a\|_\sharp + \|x^\natural - a\|_\sharp &\ge \|x^\natural\|_\sharp \\
&\ge \|x^\star\|_\sharp \\
&= \|x^\natural + (x^\star - x^\natural)\|_\sharp \\
&= \|a + (x^\natural - a) + z_1 + z_2\|_\sharp \\
&\ge \|a + z_1\|_\sharp - \|(x^\natural - a) + z_2\|_\sharp \\
&\ge \|a + z_1\|_\sharp - \|x^\natural - a\|_\sharp - \|z_2\|_\sharp \\
&= \|a\|_\sharp + \|z_1\|_\sharp - \|x^\natural - a\|_\sharp - \|z_2\|_\sharp.
\end{aligned}$$

Rearranging then gives $\|z_1\|_\sharp \le 2\|x^\natural - a\|_\sharp + \|z_2\|_\sharp$, which implies

$$\|x^\star - x^\natural\|_\sharp \le \|z_1\|_\sharp + \|z_2\|_\sharp \le 2\|x^\natural - a\|_\sharp + 2\|z_2\|_\sharp. \tag{3.2}$$

Assume $\|x^\star - x^\natural\|_2 > C_1\varepsilon$, since otherwise we are done. With this, we have

$$\|\Phi x^\star - \Phi x^\natural\|_2 \le \|\Phi x^\star - (\Phi x^\natural + e)\|_2 + \|e\|_2 \le 2\varepsilon < 2C_1^{-1}\|x^\star - x^\natural\|_2 = \alpha\|x^\star - x^\natural\|_2,$$

where we take $C_1 = 2\alpha^{-1}$. By (a), we then have

$$\|x^\star - x^\natural\|_2 \le \rho\|x^\star - x^\natural\|_\sharp. \tag{3.3}$$

Next, we appeal to a property of z_2:

$$\|z_2\|_\sharp \le L\|x^\star - x^\natural\|_2 \le \rho L\|x^\star - x^\natural\|_\sharp,$$

where the last step follows from (3.3). Substituting into (3.2) and rearranging then gives

$$\|x^\star - x^\natural\|_\sharp \le \frac{2}{1 - 2\rho L}\|x^\natural - a\|_\sharp \le 4\|x^\natural - a\|_\sharp,$$

provided $\rho \le 1/(4L)$. Finally, we apply (3.3) again to get

$$\|x^\star - x^\natural\|_2 \le \rho\|x^\star - x^\natural\|_\sharp \le 4\rho\|x^\natural - a\|_\sharp = C_0\|x^\natural - a\|_\sharp \le C_0\|x^\natural - a\|_\sharp + C_1\varepsilon,$$

taking $C_0 := 4\rho$. □

Notice that the above proof does not make use of every property of the norm $\|\cdot\|_\sharp$. In particular, it suffices for $\|\cdot\|_\sharp \colon \mathcal{H} \to \mathbb{R}$ to satisfy

(i) $\|x\|_\sharp \ge \|0\|_\sharp$ for every $x \in \mathcal{H}$, and
(ii) $\|x + y\|_\sharp \le \|x\|_\sharp + \|y\|_\sharp$ for every $x, y \in \mathcal{H}$.

For example, one may take $\|\cdot\|_\sharp$ to be , i.e., a function which satisfies every norm property except positive definiteness, meaning $\|x\|_\sharp$ is allowed to be zero when x is nonzero. As another example, in the case where $\mathcal{H} = \mathbb{R}^N$, one may take

$$\|x\|_\sharp = \|x\|_p^p := \sum_{i=1}^N |x_i|^p, \qquad 0 < p < 1.$$

This choice of objective function was first proposed by Chartrand [14]. Since $B_\sharp \subseteq B_1$, then for any α, we might expect Φ to satisfy (ρ, α)-RWP over $B_\sharp$ with a smaller ρ than for B_1. As such, minimizing $\|x\|_\sharp$ instead of $\|x\|_1$ could very well yield more stability or robustness, though potentially at the price of computational efficiency since this alternative minimization is not a convex program.

4 CS Spaces

In this section, we identify a variety of examples of CS spaces to illustrate the generality of our main result from the previous section. For each example, we use the following lemma as a proof technique:

Lemma 1 *Consider a finite-dimensional Hilbert space $\mathcal{H}$, subsets $\mathcal{A}, \mathcal{B} \subseteq \mathcal{H}$, and a norm $\|\cdot\|_\sharp$ on $\mathcal{H}$ satisfying the following:*

(i) $0 \in \mathcal{A}$.
(ii) For every $a \in \mathcal{A}$ and $z \in \mathcal{H}$, there exists a decomposition $z = z_1 + z_2$ with $\langle z_1, z_2 \rangle = 0$ such that

$$\|a + z_1\|_\sharp = \|a\|_\sharp + \|z_1\|_\sharp, \qquad z_2 \in \mathcal{B}.$$

(iii) $\|b\|_\sharp \leq L\|b\|_2$ *for every* $b \in \mathcal{B}$.

Then $(\mathcal{H}, \mathcal{A}, \|\cdot\|_\sharp)$ is a CS space with bound L.

In words, the lemma uses an auxiliary set $\mathcal{B}$ to orthogonally decompose any $z \in \mathcal{H}$. The conclusion follows from the fact that, since $z_2 \in \mathcal{B}$ and z_2 is orthogonal to z_1, we have

$$\|z_2\|_\sharp \leq L\|z_2\|_2 \leq L\sqrt{\|z_1\|_2^2 + \|z_2\|_2^2} = L\|z\|_2.$$

The remainder of this section uses this lemma to verify several CS spaces, as summarized in Table 1.

4.1 Weighted Sparsity

Take $\mathcal{H}$ to be either $\mathbb{R}^N$ or $\mathbb{C}^N$. Let W be a diagonal $N \times N$ matrix of positive weights $\{w_i\}_{i=1}^N$, and take $\|x\|_\sharp = \|Wx\|_1$ for every $x \in \mathcal{H}$. The of a vector x is defined to be $S_W(x) := \sum_{i \in \text{supp}(x)} w_i^2$, and we take $\mathcal{A} = \mathcal{B} = \Sigma_{W,K} := \{x : S_W(x) \leq K\}$.

To verify property (ii), fix any vector $a \in \mathcal{A}$. Then for any vector z, pick z_2 to be the restriction of z to the support of a, i.e.,

Table 1 Examples of CS spaces

Instance	$\mathcal{H}$	$\mathcal{A}$	$\|x\|_\sharp$	L	Location
Weighted sparsity	$\mathbb{R}^N$ or $\mathbb{C}^N$	$\Sigma_{W,K}$	$\|Wx\|_1$	$\sqrt{K}$	Sect. 4.1
Block sparsity	$\bigoplus_{j \in J} \mathcal{H}_j$	$B^{-1}(\Sigma_K)$	$\|B(x)\|_1$	$\sqrt{K}$	Sect. 4.2
Gradient sparsity	$\ker(\nabla)^\perp$	$\nabla^{-1}(\Sigma_K)$	$\|\nabla x\|_1$	$2\Delta\sqrt{K}$	Sect. 4.3
Low-rank matrices	$\mathbb{R}^{m \times n}$ or $\mathbb{C}^{m \times n}$	$\sigma^{-1}(\Sigma_K)$	$\|x\|_*$	$\sqrt{2K}$	Sect. 4.4

$$z_2[i] := \begin{cases} z[i] & \text{if } i \in \operatorname{supp}(a) \\ 0 & \text{otherwise.} \end{cases}$$

Also, pick $z_1 := z - z_2$. Then z_1 and z_2 have disjoint support, implying $\langle z_1, z_2 \rangle = 0$. Next, since Wa and Wz_1 have disjoint support, we also have $\|W(a + z_1)\|_1 = \|Wa\|_1 + \|Wz_1\|_1$. Finally, $S_W(z_2) = S_W(a) \le K$, meaning $z_2 \in \mathcal{B}$. Note that we can take the bound of this CS space to be $L = \sqrt{K}$ since for every $b \in \mathcal{B} = \Sigma_{W,K}$, Cauchy–Schwarz gives

$$\|b\|_\sharp = \sum_{i=1}^{N} w_i |b_i| = \sum_{i \in \operatorname{supp}(b)} w_i |b_i| \le \Big(\sum_{i \in \operatorname{supp}(b)} w_i^2 \Big)^{1/2} \|b\|_2 \le \sqrt{K} \|b\|_2.$$

Weighted sparsity has been applied in a few interesting ways. If one is given additional information about the support of the desired signal $x^\natural$, he might weight the entries in the optimization to his benefit. Suppose one is told of a subset of indices T such that the support of $x^\natural$ is guaranteed to overlap with at least 10% of T (say). For example, if $x^\natural$ is the wavelet transform of a natural image, then the entries of x which correspond to lower frequencies tend to be large, so these indices might be a good choice for T. When minimizing the ℓ_1 norm of x, if we give less weight to the entries of x over T, then the weighted minimizer will exhibit less ℓ_2 error, accordingly [23, 51]. As another example, suppose $x^\natural$ denotes coefficients in an orthonormal basis over some finite-dimensional subspace $F \subseteq L^2([0, 1])$, and suppose $y = \Phi x^\natural + e$ are noisy samples of the corresponding function at M different points in [0, 1]. If one wishes to interpolate these samples with a smooth function that happens to be a sparse combination of basis elements in F, then he can encourage smoothness by weighting the entries appropriately [44]. Uniformly stable and robust compressed sensing with weighted sparsity was recently demonstrated by Rauhut and Ward [44].

4.2 Block Sparsity

Let $\{\mathcal{H}_j\}_{j \in J}$ be a finite collection of Hilbert spaces, and take $\mathcal{H} := \bigoplus_{j \in J} \mathcal{H}_j$. For some applications, it is reasonable to model interesting signals as $x \in \mathcal{H}$ such that, for most indices j, the component x_j of the signal in $\mathcal{H}_j$ is zero; such signals are called . Notationally, we take $B \colon \bigoplus_{j \in J} \mathcal{H}_j \to \ell(J)$ to be defined entrywise by

$$(B(x))[j] = \|x_j\|_2,$$

and then write $B^{-1}(\Sigma_K)$ as the set of K-block sparse vectors. (Here, $\ell(J)$ denotes the set of real-valued functions over J.) In this case, we set $\mathcal{A} = \mathcal{B} = B^{-1}(\Sigma_K)$ and $\|x\|_\sharp = \|B(x)\|_1$. Property (ii) then follows by an argument which is analogous to the weighted sparsity case. To find the bound of this CS space, pick $b \in \mathcal{B}$ and note that $B(b)$ is K-sparse. Then

$$\|b\|_\sharp = \|B(b)\|_1 \leq \sqrt{K}\|B(b)\|_2 = \sqrt{K}\|b\|_2,$$

and so we may take $L = \sqrt{K}$.

Block sparsity has been used to help estimate multi-band signals, measure gene expression levels, and perform various tasks in machine learning (see [21] and references therein). Similar to minimizing $\|B(x)\|_1$, Bakin [2] (and more recently, Yuan and Lin [52]) proposed the *group lasso* to facilitate model selection by partitioning factors into blocks for certain real-world instances of the multifactor analysis-of-variance problem. Eldar and Mishali [19] proved that minimizing $\|B(x)\|_1$ produces uniformly stable and robust estimates of block sparse signals, and the simulations in [20] demonstrate that this minimization outperforms standard ℓ_1 minimization when $x^\natural$ is block sparse.

4.3 Gradient Sparsity

Take $G = (V, E)$ to be a directed graph, and let $\ell(V)$ and $\ell(E)$ denote the vector spaces of real-valued (or complex-valued) functions over V and E, respectively. Then the *gradient* $\nabla\colon \ell(V) \to \ell(E)$ is the linear operator defined by

$$(\nabla x)[(i, j)] = x(j) - x(i) \qquad \forall (i, j) \in E$$

for each $x \in \ell(V)$. Take $\mathcal{H} = \ker(\nabla)^\perp$ and consider the *total variation norm* $\|x\|_\sharp = \|\nabla x\|_1$. We will verify property (ii) for $\mathcal{A} = \nabla^{-1}(\Sigma_K)$ and $\mathcal{B} = \nabla^{-1}(\Sigma_{2K\Delta})$, where Δ denotes the maximum total degree of G.

To this end, pick any a such that ∇a is K-sparse, and let H denote the subgraph $(V, \mathrm{supp}(\nabla a))$. Take C to be the subspace of all $x \in \mathcal{H}$ such that $x[i] = x[j]$ whenever i and j lie in a common (weak) component of H. Then given any $z \in \mathcal{H}$, decompose z using orthogonal projections: $z_1 = P_C z$ and $z_2 = P_{C^\perp} z$. We immediately have $z = z_1 + z_2$ and $\langle z_1, z_2\rangle = 0$. Next, since $z_1 \in C$, we know that $(i, j) \in \mathrm{supp}(\nabla a)$ implies $z_1[i] = z_1[j]$, and so $(\nabla z_1)[(i, j)] = 0$. As such, ∇a and ∇z_1 have disjoint supports, and so

$$\|\nabla(a + z_1)\|_1 = \|\nabla a\|_1 + \|\nabla z_1\|_1.$$

Finally, H has K edges, and so H has at least $N - 2K$ isolated vertices. For each isolated vertex i, we have $z_1[i] = z[i]$, implying $z_2[i] = 0$. As such, z_2 is at most $2K$-sparse. Since $(\nabla z_2)[(i, j)]$ is nonzero only if $z_2[i]$ or $z_2[j]$ is nonzero, then ∇z_2 is only supported on the edges which are incident to support of z_2. The easiest upper bound on this number of edges is $2K\Delta$, and so $z_2 \in \mathcal{B}$.

We claim that this CS space has bound $L = 2\Delta\sqrt{K}$. To see this, first note that $\nabla^*\nabla = D - A$, where D is the diagonal matrix of vertex total degrees, and where A is the adjacency matrix of the underlying undirected graph. Then

$$\|\nabla\|_2^2 = \|\nabla^*\nabla\|_2 = \|D - A\|_2 \leq \|D\|_2 + \|A\|_2 \leq 2\Delta,$$

where the last step follows from Gershgorin's circle theorem. As such, if ∇a is $2K\Delta$-sparse, we have

$$\|\nabla a\|_1 \leq \sqrt{2K\Delta}\|\nabla a\|_2 \leq 2\Delta\sqrt{K}\|a\|_2.$$

In particular, $L = O(\sqrt{K})$ when Δ is bounded.

One important example of gradient sparsity is total variation minimization for compressive imaging. Indeed, if the pixels of an image are viewed as vertices of a grid graph (where each internal pixel has four neighbors: up, down, left, and right), then the total variation of the image x is given by $\|x\|_\sharp$, which is often the objective function of choice (see [32], for example). It might also be beneficial to consider a 3-dimensional image of voxels for applications like magnetic resonance imaging. In either setting, the maximum degree is bounded ($\Delta = 4, 6$), and the uniform stability and robustness of compressed sensing in these settings has been demonstrated by Needell and Ward [37, 38].

4.4 Low-Rank Matrices

Let $\mathcal{H}$ be the Hilbert space of real (or complex) $n \times m$ matrices with inner product $\langle X, Y\rangle = \mathrm{Tr}[XY^*]$, and consider the *nuclear norm* $\|X\|_\sharp = \|X\|_*$, defined to be the sum of the singular values of X. Letting $\mathcal{A} = \sigma^{-1}(\Sigma_K)$ and $\mathcal{B} = \sigma^{-1}(\Sigma_{2K})$ denote the sets of matrices of rank at most K and $2K$, respectively, then $L = \sqrt{2K}$, and property (ii) follows from a clever decomposition originally due to Recht, Fazel, and Parrilo (Lemma 3.4 in [45]).

For any matrix A of rank at most K, consider its singular value decomposition:

$$A = U\begin{bmatrix}\Sigma & 0\\ 0 & 0\end{bmatrix}V^*,$$

where Σ is $K \times K$. Given any matrix Z, we then take $Y := U^*ZV$ and consider the partition

$$Y = \begin{bmatrix}Y_{11} & Y_{12}\\ Y_{21} & Y_{22}\end{bmatrix},$$

where Y_{11} is $K \times K$. We decompose Z as the sum of the following matrices:

$$Z_1 := U\begin{bmatrix}0 & 0\\ 0 & Y_{22}\end{bmatrix}V^*, \qquad Z_2 := U\begin{bmatrix}Y_{11} & Y_{12}\\ Y_{21} & 0\end{bmatrix}V^*.$$

It is straightforward to check that $\langle Z_1, Z_2\rangle = 0$, and also that $AZ_1^* = 0$ and $A^*Z_1 = 0$. The latter conditions imply that A and Z_1 have orthogonal row spaces and orthogonal column spaces, which in turn implies that $\|A + Z_1\|_* = \|A\|_* + \|Z_1\|_*$ (see Lemma 2.3 in [45]). Finally, $[Y_{11}; Y_{21}]$ and $[Y_{12}; 0]$ each have rank at most K, and so $Z_2 \in \mathcal{B}$.

This setting of compressed sensing also has a few applications. For example, suppose you enter some signal into an unknown linear time-invariant system and then make time-domain observations. If the system has low order, then its *Hankel matrix* (whose entries are populated with translates of the system's impulse response) will have low rank. As such, one can hope to estimate the system by minimizing the nuclear norm of the Hankel matrix subject to the observations (along with the linear constraints which define Hankel matrices) [45]. For another application, consider *quantum state tomography*, in which one seeks to determine a nearly pure quantum state (i.e., a low-rank self-adjoint positive semidefinite matrix with complex entries and unit trace) from observations which are essentially inner products with a collection of known matrices. Then one can recover the unknown quantum state by minimizing the nuclear norm subject to the observations (and the linear constraint that the trace must be 1) [26]. We note that the set $\mathbb{H}^{n\times n}$ of self-adjoint $n \times n$ matrices with complex entries is a real vector space of n^2 dimensions, which is slightly different from the setting of this subsection, but still forms a CS space with

$$\mathcal{A} = \sigma^{-1}(\Sigma_K) \cap \mathbb{H}^{n\times n}, \quad \mathcal{B} = \sigma^{-1}(\Sigma_{2K}) \cap \mathbb{H}^{n\times n}, \quad \|\cdot\|_\sharp = \|\cdot\|_*, \quad L = \sqrt{2K}$$

by the same proof. For the original setting of not-necessarily-self-adjoint matrices, uniformly stable and robust compressed sensing was demonstrated by Mohan and Fazel [36].

5 Geometric Meaning of RWP

The previous section characterized stable and robust compressed sensing in terms of a new property called the robust width property. In this section, we shed some light on what this property means geometrically. We start with a definition which we have adapted from [27]:

Definition 5.1 We say a linear operator $\Phi\colon \mathcal{H} \to \mathbb{F}^M$ satisfies the *ρ- over $B_\sharp$* if

$$\|x\|_2 \le \rho\|x\|_\sharp$$

for every x in the null space of Φ.

Notice that the width property is actually a property of the null space of Φ. This is not the case for the robust width property since, for example, multiplying Φ by a scalar will have an effect on α, and yet the null space remains unaltered.

In this section, we essentially mod out such modifications by focusing on a subclass of "normalized" sensing operators. In particular, we only consider Φ's satisfying $\Phi\Phi^* = I$, which is to say that the measurement vectors are orthonormal. For this subclass of operators, we will show that the robust width property is an intuitive property of the null space. For further simplicity, we focus on the case in which $\mathcal{H} = \mathbb{R}^N$.

Definition 5.2 Let $\mathrm{Gr}(N, N-M)$ denote the Grassmannian, that is, the set of all subspaces of $\mathbb{R}^N$ of dimension $N-M$. Given subspaces $X, Y \in \mathrm{Gr}(N, N-M)$, consider the corresponding orthogonal projections P_X and P_Y. Then the d over $\mathrm{Gr}(N, N-M)$ is defined by $d(X, Y) := \|P_X - P_Y\|_2$.

Theorem 5.1 *A linear operator $\Phi\colon \mathbb{R}^N \to \mathbb{R}^M$ with $\Phi\Phi^* = I$ and null space Y satisfies the (ρ, α)-robust width property over $B_\sharp$ if and only if for every subspace X with $d(X, Y) < \alpha$, every linear operator $\Psi\colon \mathbb{R}^N \to \mathbb{R}^M$ with $\Psi\Psi^* = I$ and null space X satisfies the ρ-width property over $B_\sharp$.*

Imagine the entire Grassmannian, and consider the subset WP corresponding to the subspaces satisfying the ρ-width property. Then by Theorem 5.1, a point $Y \in$ WP satisfies the (ρ, α)-robust width property precisely when the entire open ball centered at Y of radius α lies inside WP. One can also interpret this theorem in the context of compressed sensing. First, observe that proving Theorem 3.1 with $\varepsilon = 0$ gives that the width property is equivalent to stable compressed sensing (cf. [27]). As such, if Φ allows for stable and robust compressed sensing with robustness constant C_1, then every Ψ whose null space is within $\alpha = (2C_1)^{-1}$ of the null space of Φ will necessarily enjoy stability. The remainder of this section proves Theorem 5.1 with a series of lemmas:

Lemma 2 *For any subspaces $X, Y \subseteq \mathbb{R}^N$, we have*

$$\min_{\substack{x\in X\\ \|x\|_2=1}} \max_{\substack{y\in Y\\ \|y\|_2=1}} \langle x, y\rangle = \sqrt{1 - \big(d(X, Y)\big)^2}.$$

Proof Pick $x \in X$ and $y \in Y$ with $\|x\|_2 = \|y\|_2 = 1$. Then Cauchy–Schwarz gives

$$\langle x, y\rangle = \langle P_Y x, y\rangle + \langle P_{Y^\perp} x, y\rangle = \langle P_Y x, y\rangle \le \|P_Y x\|_2,$$

with equality precisely when $y = P_Y x / \|P_Y x\|_2$. As such, the left-hand side of the claimed identity can be simplified as

$$\mathrm{LHS} := \min_{\substack{x\in X\\ \|x\|_2=1}} \max_{\substack{y\in Y\\ \|y\|_2=1}} \langle x, y\rangle = \min_{\substack{x\in X\\ \|x\|_2=1}} \|P_Y x\|_2.$$

Next, we appeal to the Pythagorean theorem to get

$$\sqrt{1 - \mathrm{LHS}^2} = \max_{\substack{x\in X\\ \|x\|_2=1}} \|P_{Y^\perp} x\|_2 = \max_{\substack{z\in \mathbb{R}^N\\ \|z\|_2=1}} \|P_{Y^\perp} P_X z\|_2 = \|P_{Y^\perp} P_X\|_2. \tag{5.1}$$

Finally, we appeal to Theorem 2.6.1 in [24], which states that $d(X, Y) = \|P_{Y^\perp} P_X\|_2$. Substituting into (5.1) and rearranging then gives the result. □

Lemma 3 *For any subspaces $X, Y \subseteq \mathbb{R}^N$, we have*

$$X \subseteq \{x : \|P_{Y^\perp} x\|_2 < \alpha \|x\|_2\}$$

if and only if for every $x \in X$, there exists $y \in Y$ of unit ℓ_2 norm such that

$$\langle x, y \rangle > \sqrt{1 - \alpha^2} \|x\|_2. \tag{5.2}$$

Proof

($\Rightarrow$) Given $x \in X$, pick $y := P_Y x / \|P_Y x\|_2$. Then

$$\langle x, y \rangle = \left\langle x, \frac{P_Y x}{\|P_Y x\|_2} \right\rangle = \|P_Y x\|_2 = \sqrt{\|x\|_2^2 - \|P_{Y^\perp} x\|_2^2} > \sqrt{1 - \alpha^2} \|x\|_2,$$

where the inequality follows from the assumed containment.

($\Leftarrow$) Pick $x \in X$. Then there exists $y \in Y$ of unit ℓ_2 norm satisfying (5.2). Recall that for any subspaces $A \subseteq B$, the corresponding orthogonal projections satisfy $P_A P_B = P_A$. Since $Y^\perp$ is contained in the orthogonal complement of y, we then have

$$\|P_{Y^\perp} x\|_2^2 = \|P_{Y^\perp} P_{y^\perp} x\|_2^2 \leq \|P_{y^\perp} x\|_2^2 = \|x\|_2^2 - |\langle x, y \rangle|^2 < \alpha^2 \|x\|_2^2.$$

Since our choice for x was arbitrary, this proves the claim.

□

We now use Lemmas 2 and 3 to prove the following lemma, from which Theorem 5.1 immediately follows:

Lemma 4 *Pick a subspace $Y \in \mathrm{Gr}(N, N - M)$. Then*

$$\bigcup_{\substack{X \in \mathrm{Gr}(N, N-M) \\ d(X, Y) < \alpha}} X = \{x : \|P_{Y^\perp} x\|_2 < \alpha \|x\|_2\}.$$

Proof Let U and E denote the left- and right-hand sides, respectively. Lemma 2 gives that $d(X, Y) < \alpha$ is equivalent to having that for every $x \in X$, there exists $y \in Y$ of unit ℓ_2 norm such that $\langle x, y \rangle > \sqrt{1 - \alpha^2} \|x\|_2$, which in turn is equivalent to $X \subseteq E$ by Lemma 3. The fact that $d(X, Y) < \alpha$ implies $X \subseteq E$ immediately gives $U \subseteq E$. We will show that the converse implies the reverse containment, thereby proving set equality.

To this end, pick any $e \in E$. If $e = 0$, then $e \in Y$ lies in U. Otherwise, we may assume $\|e\|_2 = 1$ without loss of generality since both U and E are closed under scalar multiplication. Also, $U = \mathbb{R}^N = E$ if $\alpha > 1$, and so we may assume $\alpha \leq 1$.

Denote $Z := Y \cap \mathrm{span}\{e\}^\perp$, and take $X := Z + \mathrm{span}\{e\}$. Since $\alpha \leq 1$, we know that $\|P_{Y^\perp} e\|_2 < \alpha\|e\|_2 \leq \|e\|_2$, i.e., $e \notin Y^\perp$, and so $\dim(Z) = \dim(Y) - 1$, which in turn implies $X \in \mathrm{Gr}(N, N-M)$. To see that $X \subseteq E$, pick any $x \in X$. Then

$$P_{Y^\perp} x = P_{Y^\perp} P_e x + P_{Y^\perp} P_Z x = P_{Y^\perp} P_e x = \langle x, e\rangle P_{Y^\perp} e.$$

Taking norms of both sides then gives $\|P_{Y^\perp} x\|_2 = \|P_{Y^\perp} e\|_2 |\langle x, e\rangle| < \alpha\|x\|_2$, as desired. Since $X \subseteq E$, we then have $d(X, Y) < \alpha$ by Lemmas 2 and 3, and so $e \in X \subseteq U$. Finally, since our choice for e was arbitrary, we conclude that $E \subseteq U$. □

6 A Direct Proof That RIP Implies RWP

Recall from Sect. 2 that the restricted isometry property (RIP) implies part (b) of Theorem 3.1 in the traditional sparsity case [9]. As such, one could pass through the equivalence to conclude that RIP implies the robust width property (RWP). For completeness, this section provides a direct proof of this result. (Instead of presenting different versions of RIP for different CS spaces, we focus on the traditional sparsity case here; the proofs for other cases are similar.)

For the direct proof, it is particularly convenient to use a slightly different version of RIP: We say Φ satisfies the (J, δ)- if

$$(1-\delta)\|x\|_2 \leq \|\Phi x\|_2 \leq (1+\delta)\|x\|_2$$

for every J-sparse vector x. We note that this "no squares" version of RIP is equivalent to the original version up to constants, and is not unprecedented (see [45], for example).

Theorem 6.1 *Suppose Φ satisfies the (J, δ)-restricted isometry property with $\delta < 1/3$. Then Φ satisfies the (ρ, α)-robust width property over B_1 with*

$$\rho = \frac{3}{\sqrt{J}}, \qquad \alpha = \frac{1}{3} - \delta.$$

The proof makes use of the following lemma:

Lemma 5 *Suppose Φ satisfies the right-hand inequality of the (J, δ)-restricted isometry property. Then for every x such that $\|x\|_2 > \rho\|x\|_1$, we have*

$$\|x - x_J\|_2 < \frac{1}{\rho\sqrt{J}}\|x\|_2, \qquad \|\Phi(x - x_J)\|_2 < \frac{1+\delta}{\rho\sqrt{J}}\|x\|_2.$$

Proof Let T_0 denote the indices of the J largest entries of x, and for each $j \geq 1$, let T_j denote the indices of the J largest entries of x not covered by T_i for $i < j$. Then

$$\|x_{T_{j+1}}\|_2 \leq \sqrt{J} \max_{i \in T_{j+1}} |x[i]| \leq \sqrt{J} \min_{i \in T_j} |x[i]| \leq \frac{1}{\sqrt{J}} \|x_{T_j}\|_1$$

for every $j \geq 0$. As such,

$$\|x - x_J\|_2 \leq \sum_{j \geq 1} \|x_{T_j}\|_2 \leq \frac{1}{\sqrt{J}} \sum_{j \geq 0} \|x_{T_j}\|_1 = \frac{1}{\sqrt{J}} \|x\|_1 < \frac{1}{\rho\sqrt{J}} \|x\|_2,$$

where the last step uses the hypothesis. Similarly,

$$\|\Phi(x - x_J)\|_2 \leq \sum_{j \geq 1} \|\Phi x_{T_j}\|_2 \leq (1 + \delta) \sum_{j \geq 1} \|x_{T_j}\|_2 < \frac{1 + \delta}{\rho\sqrt{J}} \|x\|_2,$$

as claimed. □

Proof of Theorem 6.1 We will prove a contrapositive of sorts. In particular, we will assume Φ satisfies the right-hand inequality of the restricted isometry property, but violates the robust width property, and we will show that Φ necessarily violates the left-hand inequality of the restricted isometry property.

To this end, pick x such that $\|\Phi x\|_2 < \alpha\|x\|_2$ but $\|x\|_2 > \rho\|x\|_1$. Then

$$\|\Phi x_J\|_2 \leq \|\Phi x\|_2 + \|\Phi(x - x_J)\|_2 < \alpha\|x\|_2 + \|\Phi(x - x_J)\|_2 < \frac{2}{3}(1 - \delta)\|x\|_2, \tag{6.1}$$

where the last step uses Lemma 5 along with our choices for ρ and α. Another application of Lemma 5 gives

$$\|x\|_2 \leq \|x_J\|_2 + \|x - x_J\|_2 < \|x_J\|_2 + \frac{1}{3}\|x\|_2.$$

Isolating $\|x\|_2$ and substituting into (6.1) then gives $\|\Phi x_J\|_2 < (1 - \delta)\|x_J\|_2$, as desired. □

One nice feature of the above proof is that it reveals asymmetry between the upper and lower RIP bounds. Indeed, the upper RIP bound is only used to control $\|\Phi(x - x_J)\|_2$, whereas the lower RIP bound is directly related to α. Notice that the contrapositive of RWP posits that

$$\|\Phi x\|_2 \geq \alpha\|x\|_2 \tag{6.2}$$

whenever $\|x\|_2 > \rho\|x\|_1$, which is the case when x is J-sparse with $J \leq 1/\rho^2$. As such, RWP implies lower RIP with $\delta = 1 - \alpha$. Going the other direction, we could take $\alpha = 1 - \delta$ and conclude from lower RIP that (6.2) holds for every sparse x, but in order to get (6.2) for the remaining "nearly sparse" vectors, namely, those for which $\|x\|_2 > \rho\|x\|_1$, the above proof appeals to an upper RIP bound. The next section presents some alternative methods for demonstrating RWP.

7 RIP-Free Approaches to RWP

Considering the previous section, one can certainly appeal to the restricted isometry property (RIP) in order to establish the robust width property (RWP) for a given sensing operator. The purpose of this section is to provide alternative proof techniques for RWP. Here, it will be convenient to use the contrapositive statement of RWP: A linear operator $\Phi\colon \mathcal{H} \to \mathbb{F}^M$ satisfies the (ρ, α)-robust width property over $B_\sharp$ if and only if

$$\|\Phi x\|_2 \geq \alpha\|x\|_2$$

for every $x \in \mathcal{H}$ such that $\|x\|_2 > \rho\|x\|_\sharp$. By scaling, RWP is further equivalent to having $\|\Phi x\|_2 \geq \alpha$ for every x with $\|x\|_2 = 1$ and $\|x\|_\sharp < \rho^{-1}$. Since $\|\Phi x\|_2$ is a continuous function of x, we deduce the following lemma:

Lemma 6 *A linear operator $\Phi\colon \mathcal{H} \to \mathbb{F}^M$ satisfies the (ρ, α)-robust width property over $B_\sharp$ if and only if $\|\Phi x\|_2 \geq \alpha$ for every $x \in \rho^{-1}B_\sharp \cap \mathbb{S}$, where $\mathbb{S}$ denotes the unit sphere in $\mathcal{H}$.*

In the following subsections, we demonstrate RWP by leveraging Lemma 6 along with ideas from geometric functional analysis. Interestingly, such techniques are known to outperform RIP-based analyses in certain regimes [7, 46]. To help express how this section interacts with the remainder of the paper, we have included a "cheat sheet for the practitioner" that explains how one might apply this paper to future instances of structured sparsity. The remainder of this section takes $\mathcal{H} = \mathbb{R}^N$ and $\mathbb{F} = \mathbb{R}$.

CHEAT SHEET FOR THE PRACTITIONER

Given an instance of structured sparsity for compressed sensing, apply the following process:

1. **Verify that you have a CS space.** Consult Sect. 4 for examples and proofs.
2. **Determine robust width parameters.** Consult Theorem 3.1 to determine the pairs (ρ, α) that make the ℓ_2 error of reconstruction acceptably low.
3. **Estimate a Gaussian width.** For each acceptable ρ, estimate the Gaussian width of $\rho^{-1}B_\sharp \cap \mathbb{S}$. This can be done in two ways:
   (a) *Analytically.* Mimic the proof of Lemma 7 or consult [31, 34, 41].
   (b) *Numerically.* Observe that $\sup_{x\in S}\langle x, g\rangle$ is a convex program for each g. As such, take a random sample of iid $N(0, I)$ vectors $\{g_j\}_{j\in J}$, run the convex program for each $j \in J$, and produce an upper confidence bound on the parameter $w(S) := \mathbb{E}\sup_{x\in S}\langle x, g\rangle$.
4. **Calculate the number of measurements.** Depending on the type of measurement vectors desired, consult Proposition 7.1, Proposition 7.2, or more generally Theorem 6.3 in [50]. The number of measurements will depend on ρ (implicitly through the Gaussian width) and on α. Minimize this number over the acceptable pairs (ρ, α).

7.1 Gordon Scheme

In this subsection, as in Sect. 5, we focus on the case in which Φ satisfies $\Phi\Phi^* = I$. Letting Y denote the null space of Φ, we then have $\|\Phi x\|_2 = \|P_{Y^\perp}x\|_2 = \operatorname{dist}(x, Y)$. As such, by Lemma 6, Φ satisfies RWP if its null space is of distance at least α from $S := \rho^{-1}B_\sharp \cap \mathbb{S}^{N-1}$. In other words, it suffices for the null space to have empty intersection with an α-thickened version of S. As we will see, a random subspace of sufficiently small dimension will do precisely this.

The following theorem is originally due to Gordon (see Theorem 3.3 in [25]); this particular version is taken from [35] (namely, Proposition 2). A few definitions are needed before stating the theorem. Define the of $S \subseteq \mathbb{R}^k$ to be

$$w(S) := \mathbb{E}\sup_{x\in S}\langle x, g\rangle,$$

where g has iid $N(0, 1)$ entries. Also, we take a_k to denote the Gaussian width of $\mathbb{S}^{k-1}$, i.e., $a_k := \mathbb{E}\|g\|_2$. Overall, the Gaussian width is a measure of the size of a given set, and its utility is illustrated in the following theorem:

Theorem 7.1 (Escape Through a Thickened Mesh) *Take a closed subset $S \subseteq \mathbb{S}^{N-1}$. If $w(S) < (1-\varepsilon)a_M - \varepsilon a_N$, then a subspace Y drawn uniformly from* $\mathrm{Gr}(N, N-M)$ *satisfies*

$$\Pr\Big(Y \cap (S + \varepsilon B_2) = \emptyset\Big) \geq 1 - \frac{7}{2}\exp\bigg(-\frac{1}{2}\bigg(\frac{(1-\varepsilon)a_M - \varepsilon a_N - w(S)}{3+\varepsilon+\varepsilon a_N/a_M}\bigg)^2\bigg).$$

It is well known that $(1-1/k)\sqrt{k} < a_k < \sqrt{k}$ for each k. Combined with Lemma 6 and Theorem 7.1, this quickly leads to the following result:

Proposition 7.1 *Fix $\lambda = M/N$, pick $\alpha < 1/(1+1/\sqrt{\lambda})$, and denote $S = \rho^{-1}B_\sharp \cap \mathbb{S}^{N-1}$. Suppose*

$$M \geq C\big(w(S)\big)^2$$

for some $C > 1/(1-(1+1/\sqrt{\lambda})\alpha)^2$. Let G be an $M \times N$ matrix with iid $N(0,1)$ entries. Then $\Phi := (GG^*)^{-1/2}G$ satisfies the (ρ,α)-robust width property over $B_\sharp$ with probability $\geq 1-4e^{-cN}$ for some $c = c(\lambda,\alpha,C)$.

For the sake of a familiar example, we consider the case where $\|\cdot\|_\sharp = \|\cdot\|_1$. The following estimates $w(S)$ in this case (this is essentially accomplished in [33]):

Lemma 7 *There exists an absolute constant c such that*

$$w(\sqrt{J}B_1 \cap \mathbb{S}^{N-1}) \leq c\sqrt{J\log(cN/J)}$$

for every positive integer J.

Proof For any fixed $g \in \mathbb{R}^N$, the $x \in \sqrt{J}B_1 \cap \mathbb{S}^{N-1}$ which maximizes $\langle x, g\rangle$ has the same sign pattern as g and the same order of entry sizes, i.e., $|g[i]| \geq |g[j]|$ if and only if $|x[i]| \geq |x[j]|$. As such, we may assume without loss of generality that g and x have all nonnegative entries in nonincreasing order. Then

$$\langle x, g\rangle = \langle x_J, g_J\rangle + \langle x - x_J, g - g_J\rangle \leq \|x_J\|_2\|g_J\|_2 + \sum_{i>J} x[i]g[i].$$

Note that $\|x_J\|_2 \leq \|x\|_2 = 1$ and $g[i] \leq g[J] \leq \|g_J\|_2/\sqrt{J}$ for every $i > J$, and so

$$\langle x, g\rangle \leq \|g_J\|_2 + \frac{1}{\sqrt{J}}\|g_J\|_2 \sum_{i>J} x[i] \leq 2\|g_J\|_2,$$

where the last step uses the fact that $x \in \sqrt{J}B_1$. Next, we note that

$$\sup_{x\in\Sigma_J\cap B_2} \langle x, g\rangle = \bigg\langle \frac{g_J}{\|g_J\|_2}, g\bigg\rangle = \|g_J\|_2.$$

Letting g have iid $N(0,1)$ entries, we then have

$$w(\sqrt{J}B_1 \cap \mathbb{S}^{N-1}) = \mathbb{E} \sup_{x \in \sqrt{J}B_1 \cap \mathbb{S}^{N-1}} \langle x, g \rangle \leq 2\mathbb{E}\|g_J\|_2 = 2w(\Sigma_J \cap B_2).$$

At this point, we appeal to Lemma 3.3 in [33] (equivalently, Lemma 4.4 in [46]), which gives

$$w(\Sigma_J \cap B_2) \leq c\sqrt{J \log(cN/J)}$$

for some absolute constant c. □

Overall, if $M = \lambda N$ and we take an integer J such that $M \geq 5c^2 J \log(cN/J)$, then Proposition 7.1 and Lemma 7 together imply that $\Phi := (GG^*)^{-1/2}G$ satisfies (ρ, α)-RWP over B_1 with high probability, where

$$\rho := \frac{1}{\sqrt{J}}, \qquad \alpha := \frac{1}{2(1 + 1/\sqrt{\lambda})}.$$

For the sake of comparison, consider $\Psi := (1/\sqrt{\lambda})\Phi$. (This random matrix is known to be a *Johnson–Lindenstrauss projection* [17], and so it satisfies RIP with high probability [5], which in turn implies stable and robust compressed sensing [9].) Then Ψ satisfies $(\rho, \alpha/\sqrt{\lambda})$-RWP with high probability. Taking $\mathcal{A} = \Sigma_K$ with $K \leq J/16$, Theorem 3.1 then gives that ℓ_1 minimization produces an estimate $x^\star$ of $x^\natural$ from noisy measurements $\Phi x^\natural + e$ with $\|e\|_2 \leq \varepsilon$ such that

$$\|x^\star - x^\natural\|_2 \leq \frac{1}{\sqrt{K}}\|x^\natural - x^\natural_K\|_1 + 4(1 + \sqrt{\lambda})\varepsilon.$$

Note that $4(1 + \sqrt{\lambda}) \leq 8$, and so these constants are quite small, even though we have not optimized them.

7.2 Bowling Scheme

In the previous subsection, we were rather restrictive in our choice of sensing operators. By contrast, this subsection will establish similar performance with a much larger class of random matrices. The main tool here is the so-called *bowling scheme*, coined by Tropp [50], which exploits the following lower bound for nonnegative empirical processes, due to Koltchinskii and Mendelson:

Theorem 7.2 (Proposition 5.1 in [50], cf. Theorem 2.1 in [28]) *Take a set $S \subseteq \mathbb{R}^N$. Let φ be a random vector in $\mathbb{R}^N$, and let Φ be an $M \times N$ matrix with rows $\{\varphi_i^\top\}_{i=1}^M$ which are independent copies of $\varphi^\top$. Define*

$$Q_\xi(S;\varphi) := \inf_{x\in S} \Pr\Big(|\langle x,\varphi\rangle| \geq \xi\Big), \qquad W_M(S;\varphi) := \mathbb{E}\sup_{x\in S}\Big\langle x, \frac{1}{\sqrt{M}}\sum_{i=1}^M \varepsilon_i\varphi_i\Big\rangle,$$

where $\{\varepsilon_i\}_{i=1}^M$ *are independent random variables which take values uniformly over* $\{\pm 1\}$ *and are independent from everything else. Then for any* $\xi > 0$ *and* $t > 0$*, we have*

$$\inf_{x\in S} \|\Phi x\|_2 \geq \xi\sqrt{M}\, Q_{2\xi}(S;\varphi) - 2W_M(S;\varphi) - \xi t$$

with probability $\geq 1 - e^{-t^2/2}$.

As an example of how Theorem 7.2 might be applied, consider the case where φ has distribution $N(0,\Sigma)$. We will take $\sigma_{\max}^2$ and $\sigma_{\min}^2$ to denote the largest and smallest eigenvalues of Σ, respectively. First, we seek a lower bound on $Q_{2\xi}(S;\varphi)$. We will exploit the fact that φ has the same distribution as $\Sigma^{1/2}g$, where g has distribution $N(0,I)$, and that

$$\langle x, \Sigma^{1/2}g\rangle = \langle \Sigma^{1/2}x, g\rangle = \|\Sigma^{1/2}x\|_2\Big\langle \frac{\Sigma^{1/2}x}{\|\Sigma^{1/2}x\|_2}, g\Big\rangle.$$

Indeed, taking Z to have distribution $N(0,1)$, then for any $x \in \mathbb{S}^{N-1}$, we have

$$\Pr\Big(|\langle x,\varphi\rangle| \geq \xi\Big) = \Pr\Big(|Z| \geq \xi/\|\Sigma^{1/2}x\|_2\Big) \geq \Pr\Big(|Z| \geq \xi/\sigma_{\min}\Big) \geq \frac{\sigma_{\min}}{\xi}\cdot\frac{1}{\sqrt{2\pi}}e^{-\xi^2/2\sigma_{\min}^2},$$

where the last step assumes $\xi/\sigma_{\min} \geq 1$. Next, we pursue an upper bound on $W_M(S;\varphi)$. For this, we first note that

$$\Pr\Big(|\langle x,\varphi\rangle| \geq \xi\Big) = \Pr\Big(|Z| \geq \xi/\|\Sigma^{1/2}x\|_2\Big) \leq \Pr\Big(|Z| \geq \xi/\sigma_{\max}\Big) \leq e^{-\xi^2/2\sigma_{\max}^2}$$

for any $x \in \mathbb{S}^{N-1}$. Furthermore, φ has the same distribution as $\frac{1}{\sqrt{M}}\sum_{i=1}^M \varepsilon_i\varphi_i$, and so

$$\Pr\Big(\Big|\Big\langle u-v, \frac{1}{\sqrt{M}}\sum_{i=1}^M \varepsilon_i\varphi_i\Big\rangle\Big| \geq \xi\Big) \leq e^{-\xi^2/2\sigma_{\max}^2\|u-v\|_2^2} \qquad \forall u,v \in \mathbb{R}^N.$$

As such, φ satisfies the hypothesis of the generic chaining theorem (Theorem 1.2.6 in [47]), which, when combined with the majorizing measure theorem (Theorem 2.1.1 in [47]), gives

$$W_M(S;\varphi) = \mathbb{E}\sup_{x\in S}\Big\langle x, \frac{1}{\sqrt{M}}\sum_{i=1}^M \varepsilon_i\varphi_i\Big\rangle \leq C\sigma_{\max}\cdot\mathbb{E}\sup_{x\in S}\langle x,g\rangle = C\sigma_{\max}\cdot w(S). \tag{7.1}$$

All together, we have

$$\inf_{x\in S}\|\Phi x\|_2 \geq \xi\sqrt{M}Q_{2\xi}(S;\varphi) - 2W_M(S;\varphi) - \xi t$$
$$\geq \underbrace{\sqrt{M}\cdot(\sigma_{\min}/\sqrt{2\pi})e^{-\xi^2/2\sigma_{\min}^2}}_{a} - \underbrace{2C\sigma_{\max}\cdot w(S)}_{b} - \xi t.$$

At this point, we pick $\xi = \sigma_{\min}$, M such that $a = 2b$, and t such that $\xi t = (a-b)/2$ to get the following result:

Proposition 7.2 *Take $\rho > 0$ and denote $S = \rho^{-1}B_\sharp \cap \mathbb{S}^{N-1}$. Let φ be distributed $N(0, \Sigma)$, and take $\sigma_{\max}^2$ and $\sigma_{\min}^2$ to denote the largest and smallest eigenvalues of Σ, respectively. Set*

$$M = c_0\cdot\frac{\sigma_{\max}^2}{\sigma_{\min}^2}\cdot\big(w(S)\big)^2, \qquad \alpha = c_1\cdot\sigma_{\min}\sqrt{M},$$

and let Φ be an $M\times N$ matrix whose rows are independent copies of $\varphi^\top$. Then Φ satisfies the (ρ,α)-robust width property over $B_\sharp$ with probability $\geq 1 - e^{c_2\sigma_{\min}^2 M}$.

This result is essentially a special case of Theorem 6.3 in [50], which considers a more general notion of subgaussianity, and indeed, by this result, every matrix with iid subgaussian rows satisfies RWP. However, we note that this result is suboptimal in certain regimes, in part thanks to the sledgehammers we applied in the estimate (7.1). To see the suboptimality here, consider the special case where $B_\sharp = B_1$. Then for every $x \in S$, we have

$$\langle x,\varphi\rangle \leq \|x\|_1\|\varphi\|_\infty \leq \rho^{-1}\|\varphi\|_\infty.$$

Also, known results on maxima of Gaussian fields (e.g., equation (2.13) in [31]) imply that

$$\mathbb{E}\|\varphi\|_\infty = \mathbb{E}\|\Sigma^{1/2}g\|_\infty \leq 3\sqrt{v\log N},$$

where v denotes the largest diagonal entry of Σ. Putting things together, we have

$$W_M(S;\varphi) = \mathbb{E}\sup_{x\in S}\left\langle x, \frac{1}{\sqrt{M}}\sum_{i=1}^M \varepsilon_i\varphi_i\right\rangle = \mathbb{E}\sup_{x\in S}\langle x,\varphi\rangle \leq \rho^{-1}\mathbb{E}\|\varphi\|_\infty \leq 3\rho^{-1}\sqrt{v\log N},$$

which leads to the following result:

Proposition 7.3 *Take $\rho = 1/\sqrt{J}$. Let φ be distributed $N(0,\Sigma)$, and take v and $\sigma_{\min}^2$ to denote the largest diagonal entry and smallest eigenvalue of Σ, respectively. Set*

$$M = c_0 \cdot \frac{v}{\sigma_{\min}^2} \cdot J \log N, \qquad \alpha = c_1 \cdot \sigma_{\min}\sqrt{M},$$

and let Φ be an $M \times N$ matrix whose rows are independent copies of $\varphi^\top$. Then Φ satisfies the (ρ, α)-robust width property over B_1 with probability $\geq 1 - e^{c_2\sigma_{\min}^2 M}$.

We note that this result could also have been deduced from Theorem 1 in [42], whose proof is a bit more technical. Overall, this proposition exchanges $\sigma_{\max}^2$ for v (which is necessarily smaller) and $\log(N/J)$ for $\log N$. However, this is far from an even trade, as we illustrate in the following subsection.

7.3 RWP Does Not Imply RIP

In this subsection, we consider a random matrix from Example 2 in [42]. This example will help to compare the performance of Propositions 7.2 and 7.3, as well as provide a construction of an RWP matrix, no scaling of which satisfies RIP.

Pick $\Sigma := \frac{1}{M}(I + \mathbf{1}_N\mathbf{1}_N^\top)$; we selected the $1/M$ scaling here so that $\|\varphi\|_2^2 = \Theta(N/M)$ with high probability, as is typical for RIP matrices. Then

$$\sigma_{\max}^2 = \frac{N+1}{M}, \qquad \sigma_{\min}^2 = \frac{1}{M}, \qquad v = \frac{2}{M}.$$

In this extreme case, Proposition 7.2 uses $M \gg N$ rows to satisfy RWP, whereas Proposition 7.3 uses only $O(J \log N)$ rows, and so the latter performs far better. In either case, Φ is (ρ, α)-RWP with $\rho = 1/\sqrt{J}$ and $\alpha = O(1)$, mimicking the performance of an RIP matrix. However, as we will show, this performance is *logically independent* of the restricted isometry property, that is, the degree to which Φ satisfies RIP is insufficient to conclude that ℓ_1 minimization exactly recovers all sparse signals, let alone with stability or robustness.

To see this, we start by following the logic of Example 2 in [42]. Take any $M \times J$ submatrix Φ_J of Φ, and notice that the rows of Φ_J are iid with distribution $N(0, \Sigma_{JJ})$, where $\Sigma_{JJ} = \frac{1}{M}(I + \mathbf{1}_J\mathbf{1}_J^\top)$. Take $u := \mathbf{1}_J/\sqrt{J}$. Then $\langle u, \varphi_J\rangle$ has distribution $N(0, \lambda_{\max}(\Sigma_{JJ}))$, and so $\|\Phi_J u\|_2^2/\lambda_{\max}(\Sigma_{JJ})$ has chi-squared distribution with M degrees of freedom. As such, Lemma 1 in [30] gives that

$$\Pr\left(\frac{\|\Phi_J u\|_2^2}{\lambda_{\max}(\Sigma_{JJ})} \leq \frac{M}{2}\right) \leq e^{-cM}.$$

Similarly, for any unit vector v which is orthogonal to u, we have that $\|\Phi_J v\|_2^2/\lambda_{\min}(\Sigma_{JJ})$ also has chi-squared distribution with M degrees of freedom, and so the other bound of Lemma 1 in [30] gives

$$\Pr\left(\frac{\|\Phi_J v\|_2^2}{\lambda_{\min}(\Sigma_{JJ})} \geq 2M\right) \leq e^{-cM}.$$

Overall, we have that

$$\frac{\lambda_{\max}(\Phi_J^* \Phi_J)}{\lambda_{\min}(\Phi_J^* \Phi_J)} \geq \frac{\|\Phi_J u\|_2^2}{\|\Phi_J v\|_2^2} > \frac{1}{4} \cdot \frac{\lambda_{\max}(\Sigma_{JJ})}{\lambda_{\min}(\Sigma_{JJ})} = \frac{J+1}{4}$$

with high probability.

Now suppose there is some scaling $\Psi = C\Phi$ such that Ψ satisfies (J, δ)-RIP. Then

$$\frac{1+\delta}{1-\delta} \geq \frac{\lambda_{\max}(\Phi_J^* \Phi_J)}{\lambda_{\min}(\Phi_J^* \Phi_J)} > \frac{J+1}{4},$$

or equivalently, $\delta > (J-3)/(J+5)$. At this point, we appeal to the recently proved Cai–Zhang threshold (Theorem 1 in [8]), which states that, whenever $t \geq 4/3$, (tK, δ)-RIP implies exact recovery of all K-sparse signals by ℓ_1 minimization if and only if $\delta < \sqrt{(t-1)/t}$; here, we are using the traditional "with squares" version of RIP. As such, for Ψ, RIP guarantees the recovery of all K-sparse signals only if

$$\frac{tK-3}{tK+5} < \sqrt{\frac{t-1}{t}}.$$

However, in order for this to hold for any t, a bit of algebraic manipulation reveals that we must have $K \leq 25$, a far cry from the RWP-based guarantee, which allows $K = \Omega(M/\log N)$.

8 Discussion

This paper establishes that in many cases, uniformly stable and robust compressed sensing is equivalent to having the sensing operator satisfy the robust width property (RWP). We focused on the reconstruction algorithm denoted in the introduction by $\Delta_{\sharp,\Phi,\varepsilon}$, but it would be interesting to consider other algorithms. For example, the Lasso [48] and the Dantzig selector [12] are popular alternatives in the statistics community. The restricted isometry property (RIP) is known to provide reconstruction guarantees for a wide variety of algorithms, but does RWP share this ubiquity, or is it optimized solely for $\Delta_{\sharp,\Phi,\varepsilon}$?

We note that recently, ideas from geometric functional analysis have also been very successful in producing non-uniform compressed sensing guarantees [1, 13, 50]. In this regime, one is concerned with a Gaussian width associated with the descent cone at the signal $x^\natural$ instead of a dilated version of the entire $B_\sharp$ ball. In either case, the Gaussian width of interest is the expected value of a random variable $\sup_{x\in S}\langle x,g\rangle$ for some fixed subset S of the unit sphere. Notice that this supremum can instead be taken over the convex hull of S, and so for every instance of g, one may compute $\sup_{x\in S}\langle x,g\rangle$ as a convex program. As such, the desired expected value of this random variable can be efficiently estimated from a random sample. The computational efficiency of this estimation is not terribly surprising in the non-uniform case, since one can alternatively attempt $\sharp$-norm minimization with a fixed $x^\natural$ and empirically estimate the probability of reconstruction. This is a bit more surprising in the uniform case since for any fixed matrix, certifying a uniform compressed sensing guarantee is known to be NP-hard [3, 49]. Of course, there is no contradiction here since (when combined with Proposition 7.1, Proposition 7.2, or more generally Theorem 6.3 in [50]) this randomized algorithm merely certifies a uniform guarantee for most instances of a random matrix distribution. Still, the proposed numerical scheme may be particularly useful in cases where the Gaussian width of $\rho^{-1}B_\sharp\cap\mathbb{S}$ is cumbersome to estimate analytically.

One interesting line of research in compressed sensing has been to find an assortment of random matrices (each structured for a given application, say) that satisfy RIP [4, 29, 40, 43, 46]. In this spirit, the previous section showed how the bowling scheme can be leveraged to demonstrate RWP for matrices with iid subgaussian rows. We note that the bowling scheme (as described in [50] in full detail) is actually capable of analyzing a much broader class of random matrices, though it is limited by the weaknesses of Theorem 7.2. In particular, the bowling scheme requires $Q_\xi(S;\varphi)$ to be bounded away from zero, but this can be small when the distribution of φ is "spiky," e.g., when φ is drawn uniformly from the rows of a discrete Fourier transform. As such, depending on the measurement constraints of a given application, alternatives to the bowling scheme are desired. Along these lines, Koltchinskii and Mendelson provide an alternative estimate of $\inf_{x\in S}\|\Phi x\|_2$ which depends on the VC dimension of a certain family of sets determined by S (see Theorem 2.5 in [28]). For the sake of a target, we pose the following analog to Problem 3.2 in [46]:

Problem 1 What is the smallest $M = M(N,\rho,\alpha,\varepsilon)$ such that drawing M independent rows uniformly from the $N\times N$ discrete Fourier transform matrix produces a random matrix which satisfies the (ρ,α)-robust width property over B_1 with probability $\geq 1-\varepsilon$?

As a benchmark, it is known [15] that taking

$$M\geq\frac{C\log(1/\varepsilon)}{\delta^2}K\log^3 K\log N$$

ensures that the properly scaled version of this random matrix satisfies (K,δ)-RIP with probability $\geq 1-\varepsilon$, and so Theorem 6.1 gives a corresponding upper bound on $M(N,\rho,\alpha,\varepsilon)$; for the record, this uses the "with squares" version of RIP, but the difference in M may be buried in the constant C. Since RWP is strictly weaker than RIP, one might anticipate an improvement from an RIP-free approach.

Acknowledgments The original idea for this work was conceived over mimosas in Pete Casazza's basement; we thank Pete for his hospitality and friendship. This work was supported in part by NSF DMS 1321779, NSF DMS 1829955, and AFOSR FA9550-18-1-0107.

Appendix

Proposition 8.1 *Let $\mathcal{D}$ denote the descent cone of $\|\cdot\|_\sharp$ at some nonzero $a \in \mathcal{H}$. Take v such that $\|v\|_\sharp = \|a\|_\sharp$ and $v - a \in \overline{\mathcal{D}^c}$, where $\overline{\mathcal{D}^c}$ denotes the topological closure of the set complement of $\mathcal{D}$. Then*

$$\|a + x\|_\sharp = \|a\|_\sharp + \|x\|_\sharp$$

for every $x = cv$ with $c \geq 0$.

Proof Notice that

$$\begin{aligned}\overline{\mathcal{D}^c} &= \overline{\bigcap_{t>0}\{y : \|a + ty\|_\sharp > \|a\|_\sharp\}} \\ &\subseteq \bigcap_{t>0}\overline{\{y : \|a + ty\|_\sharp > \|a\|_\sharp\}} = \bigcap_{t>0}\{y : \|a + ty\|_\sharp \geq \|a\|_\sharp\}.\end{aligned}$$

As such, $v - a \in \overline{\mathcal{D}^c}$ implies that

$$\|a + t(v - a)\|_\sharp \geq \|a\|_\sharp \qquad \forall t \geq 0.$$

Also, for every $t \in [0, 1]$, convexity implies

$$\|a + t(v - a)\|_\sharp = \|(1 - t)a + tv\|_\sharp \leq (1 - t)\|a\|_\sharp + t\|v\|_\sharp = \|a\|_\sharp.$$

Combining the last two displays then gives

$$\|(1 - t)a + tv\|_\sharp = \|a\|_\sharp \qquad \forall t \in [0, 1].$$

With this, we get

$$\|a+x\|_\sharp=(1+c)\left\|\frac{1}{1+c}\cdot a+\frac{c}{1+c}\cdot v\right\|_\sharp=(1+c)\|a\|_\sharp=\|a\|_\sharp+c\|v\|_\sharp = \|a\|_\sharp+\|x\|_\sharp,$$

as desired. □

References

1. Amelunxen, D, Lotz, M, McCoy, M. B., Tropp, J. A.: Living on the edge: Phase transitions in convex programs with random data. Available online: arXiv:1303.6672
2. Bakin, S.: Adaptive regression and model selection in data mining problems. Ph.D. Thesis, Australian National University, Canberra.
3. Bandeira, A. S., Dobriban, E., Mixon, D. G., Sawin, W. F.: Certifying the restricted isometry property is hard. IEEE Trans. Inf. Theory **59** 3448–3450 (2013).
4. Bandeira, A. S., Fickus, M., Mixon, D. G., Moreira, J.: Derandomizing restricted isometries via the Legendre symbol. Available online: arXiv:1406.4089
5. Baraniuk, R., Davenport, M., DeVore, R., Wakin, M.: A simple proof of the restricted isometry property for random matrices. Constr. Approx. **28**, 253–263 (2008).
6. Bickel, P. J., Ritov, Y., Tsybakov, A.B.: Simultaneous analysis of Lasso and Dantzig selector. Ann. Stat. **37**, 1705–1732 (2009).
7. Blanchard, J. D., Cartis, C., Tanner, J.: Compressed sensing: How sharp is the restricted isometry property?. SIAM Rev. **53**, 105–125 (2011).
8. Cai, T. T., Zhang, A.: Sparse representation of a polytope and recovery of sparse signals and low-rank matrices. IEEE Trans. Inf. Theory **60**, 122–132 (2014).
9. Candès, E. J.: The restricted isometry property and its implications for compressed sensing. C. R. Acad. Sci. Paris, Ser. I **346** 589–592 (2008).
10. Candès, E. J., Romberg, J., Tao, T.: Robust uncertainty principles: Exact signal reconstruction from highly incomplete frequency information. IEEE Trans. Inf. Theory **52**, 489–509 (2006).
11. Candès, E. J., Tao, T.: Decoding by linear programming. IEEE Trans. Inf. Theory **51**, 4203–4215 (2005).
12. Candès, E. J., Tao, T.: The Dantzig selector: Statistical estimation when p is much smaller than n. Ann. Stat. **35**, 2313–2351 (2007).
13. Chandrasekaran, V., Recht, B., Parrilo, P. A., Willsky, A. S.: The convex geometry of linear inverse problems. Found. Comput. Math. **12** 805–849 (2012).
14. Chartrand, R.: Exact reconstruction of sparse signals via nonconvex minimization. IEEE Signal Proc. Let. **14**, 707–710 (2007).
15. Cheraghchi, M., Guruswami, V., Velingker, A.: Restricted isometry of Fourier matrices and list decodability of random linear codes. SIAM J. Comput. **42**, 1888–1914 (2013).
16. Cohen, A., Dahmen, W., DeVore, R.: Compressed sensing and best k-term approximation. J. Am. Math. Soc. **22**, 211–231 (2009).
17. Dasgupta, S., Gupta, A.: An elementary proof of a theorem of Johnson and Lindenstrauss. Random Struct. Algor. **22**, 60–65 (2003).
18. Donoho, D. L.: Compressed sensing. IEEE Trans. Inf. Theory **52**, 1289–1306 (2006).
19. Eldar, Y. C., Mishali, M.: Robust recovery of signals from a structured union of subspaces. IEEE. Trans. Inf. Theory **55**, 5302–5316 (2009).
20. Eldar, Y. C., Kuppinger, P., Bölcskei, H.: Block-sparse signals: Uncertainty relations and efficient recovery. IEEE Trans. Signal Process. **58**, 3042–3054 (2010).
21. Elhamifar, E., Vidal, R.: Block-sparse recovery via convex optimization. IEEE Trans. Signal Process. **60**, 4094–4107 (2012).
22. Foucart, S., Rauhut, H.: A Mathematical Introduction to Compressive Sensing. Birkäuser (2013).
23. Friedlander, M., Mansour, H., Saab, R., Ö. Yılmaz, Recovering compressively sampled signals using partial support information. IEEE Trans. Inf. Theory **58**, 112–1134 (2012).
24. Golub, G. H., Van Loan, C. F.: Matrix Computations, 3rd ed., Johns Hopkins U. Press (1996).
25. Gordon, Y.: On Milman's inequality and random subspaces which escape through a mesh in $\mathbb{R}^n$. Geometric aspects of functional analysis (1986/87), 84–106. Lecture Notes in Mathematics, 1317. Springer, Berlin (1988).
26. Gross, D., Liu, Y.-K., Flammia, S. T., Becker, S., Eisert, J.: Quantum state tomography via compressed sensing. Phys. Rev. Lett. **105**, 150401 (2010).

27. Kashin, B. S., Temlyakov, V. N.: A remark on compressed sensing. Math. Notes **82**, 748–755 (2007).
28. Koltchinskii, V., Mendelson, S.: Bounding the smallest singular value of a random matrix without concentration. Available online: arXiv:1312.3580
29. Krahmer, F., Mendelson, S., Rauhut, H.: Suprema of chaos processes and the restricted isometry property. Comm. Pure. Appl. Math., **67**, 1877–1904 (2014).
30. Laurent, B., Massart, P.: Adaptive estimation of a quadratic functional by model selection. Ann. Stat. **28**, 1302–1338 (2000).
31. Ledoux, M., Talagrand, M.: Probability in Banach Spaces: Isoperimetry and Processes. Springer-Verlag, New York (1991).
32. Lustig, M., Donoho, D., Pauly, J. M.: Sparse MRI: The application of compressed sensing for rapid MRI imaging. Magn. Reson. Med. **58**, 1182–1195 (2007).
33. Mendelson, S., Pajor, A., Tomczak-Jaegermann, N.: Reconstruction and subgaussian processes. Available online: arXiv:math/0506239
34. Milman, V. D., Schechtman, G.: Asymptotic theory of finite dimensional normed spaces. Lecture Notes in Mathematics, Springer (1986).
35. Mixon, D. G.: Gordon's escape through a mesh theorem, Short, Fat Matrices, Available online: http://dustingmixon.wordpress.com/2014/02/08/gordons-escape-through-a-mesh-theorem/
36. Mohan, K., Fazel, M.: New restricted isometry results for noisy low-rank recovery. ISIT 2010, 1575–1577.
37. Needell, D., Ward, R.: Near-optimal compressed sensing guarantees for total variation minimization. IEEE Trans. Image Process. **22**, 3941–3949 (2013).
38. Needell, D., Ward, R.: Stable image reconstruction using total variation minimization. SIAM J. Imaging Sci. **6**, 1035–1058 (2013).
39. Negahban, S. N., Ravikumar, P., Wainwright, M. J., Yu, B.: A unified framework for high-dimensional analysis of M-estimators with decomposable regularizers. Stat. Sci. **27**, 538–557 (2012).
40. Nelson, J., Price, E., Wootters, M.: New constructions of RIP matrices with fast multiplication and fewer rows. SODA 2014, 1515–1528.
41. Pisier, G.: The volume of convex bodies and Banach space geometry. Cambridge University Press (1989).
42. Raskutti, G., Wainwright, M. J., Yu, B.: Restricted eigenvalue properties for correlated Gaussian designs. J. Mach. Learn. Res. **11**, 2241–2259 (2010).
43. Rauhut, H.: Compressive sensing and structured random matrices. Theoretical foundations and numerical methods for sparse recovery **9**, 1–92 (2010).
44. Rauhut, H., Ward, R.: Interpolation via weighted ℓ_1 minimization. Available online: arXiv:1308.0759
45. Recht, B., Fazel, M., Parrilo, P. A.: Guaranteed minimum-rank solutions of linear matrix equations via nuclear norm minimization. SIAM Rev. **52**, 471–501 (2010).
46. Rudelson, M., Vershynin, R.: On sparse reconstruction from Fourier and Gaussian measurements. Comm. Pure. Appl. Math. **61**, 1025–1045 (2008).
47. Talagrand, M.: The generic chaining. Upper and lower bounds of stochastic processes. Springer Monographs in Mathematics, Springer-Verlag, Berlin (2005).
48. Tibshirani, R.: Regression shrinkage and selection via the lasso. J. Royal. Statist. Soc B. **58**, 267–288 (1996).
49. Tillmann, A. M., Pfetsch, M. E.: The computational complexity of the restricted isometry property, the nullspace property, and related concepts in compressed sensing. IEEE Trans. Inf. Theory **60**, 1248–1259 (2014).
50. Tropp, J. A.: Convex recovery of a structured signal from independent random linear measurements. Available online: arXiv:1405.1102
51. Yu, X., Baek, S.: Sufficient conditions on stable recovery of sparse signals with partial support information. IEEE Signal Proc. Lett. **20**, 539–542 (2013).
52. Yuan, M., Lin, Y.: Model selection and estimation in regression with grouped variables. J. R. Statist. Soc. B **68**, 49–67 (2006).

On Min-Max Affine Approximants of Convex or Concave Real-Valued Functions from $\mathbb{R}^k$, Chebyshev Equioscillation and Graphics

Steven B. Damelin, David L. Ragozin, and Michael Werman

Abstract We study min-max affine approximants of a continuous convex or concave function $f : \Delta \subseteq \mathbb{R}^k \to \mathbb{R}$, where Δ is a convex compact subset of $\mathbb{R}^k$. In the case when Δ is a *simplex*, we prove that there is a *vertical* translate of the supporting hyperplane in $\mathbb{R}^{k+1}$ of the graph of f at the vertices which is the unique best affine approximant to f on Δ. For $k = 1$, this result provides an extension of the Chebyshev equioscillation theorem for linear approximants. Our result has interesting connections to the computer graphics problem of rapid rendering of projective transformations.

1 Introduction

We will work in $\mathbb{R}^k$, $k \geq 1$, where $\mathbf{x} \in \mathbb{R}^k$ is the column vector $[x_1\, x_2\, \cdots\, x_k]^T$ and T denotes transpose. A function $g : \mathbb{R}^k \to \mathbb{R}$ is an *affine* function provided there exists $\boldsymbol{\alpha} \in \mathbb{R}^k$, and $\beta \in \mathbb{R}$ such that $g(\mathbf{x}) = \boldsymbol{\alpha}^T \mathbf{x} + \beta$.

1.1 Min-Max Approximation

In this chapter, we are interested in min-max approximation to a continuous $f : \Delta \subseteq \mathbb{R}^k \to \mathbb{R}$ by affine approximants to f; see, for example, [3, 4, 7, 9, 11].

S. B. Damelin (✉)
Department of Mathematics, University of Michigan, Ann Arbor, MI, USA
e-mail: damelin@umich.edu

D. L. Ragozin
Department of Mathematics, University of Washington, Seattle, WA, USA
e-mail: rag@uw.edu

M. Werman
Department of Computer Science, The Hebrew University, Jerusalem, Israel
e-mail: michael.werman@mail.huji.ac.il

M. Hirn et al. (eds.), *Excursions in Harmonic Analysis, Volume 6*, Applied and Numerical Harmonic Analysis, https://doi.org/10.1007/978-3-030-69637-5_19

By imposing the restriction Δ is a simplex with non-empty interior, we obtain a complete characterization of these approximants and an explicit formula for a unique best approximant. Even in the case of an interval, $k = 1$, our main result, Theorem 2.1, provides an extension of [3, Corollary 7.6.3]. For $k = 1$, our main result also provides an extension of the Chebyshev equioscillation theorem for linear approximants with an explicit unique formula for the best approximant; see Sect. 4. We show interesting connections of Theorem 2.1 to graphics; see Sect. 5.

In order to state our results, we need the following notation. For a continuous function $g : \mathbb{R}^k \to \mathbb{R}$ and simplex Δ, we adopt the usual convention of

$$\|g(\mathbf{x})\|_{\infty(\Delta)} := \sup_{\mathbf{x}\in\Delta} |g(\mathbf{x})| .$$

We will answer the following.

1.2 Problem

Let $f : \Delta \subseteq \mathbb{R}^k \to \mathbb{R}$ be a continuous function where Δ is a compact domain in $\mathbb{R}^k$. Find conditions on Δ and f which allow for the construction of the min-max affine approximation problem (1) to f over Δ. Equivalently, find conditions on Δ and f which allow for the explicit construction of an $\boldsymbol{\alpha} \in \mathbb{R}^k$ and a $\beta \in \mathbb{R}$, which solve

$$\min_{\{\boldsymbol{\alpha},\beta\}} \left\| f(\mathbf{x}) - (\boldsymbol{\alpha}^T\mathbf{x} + \beta) \right\|_{\infty(\Delta)} . \tag{1}$$

2 Main Result: Theorem 2.1

Our **main result** is as follows.

Theorem 2.1 *Let* $\{\mathbf{a}_1, \ldots, \mathbf{a}_{k+1}\}$ *be* $k + 1$ *affinely independent points in* $\mathbb{R}^k$ *so that their convex hull* $\Delta = \mathrm{CH}(\mathbf{a}_1 \ldots \mathbf{a}_{k+1})$ *is a k-simplex, and assume that* $f : \Delta \subseteq \mathbb{R}^k \to \mathbb{R}$ *is a continuous convex or concave function over* Δ*. Then, the min-max affine approximant to* f *over* Δ *is the hyperplane average* $\sigma := \frac{\pi+\rho}{2}$, *where* π *is the affine hyperplane* $\mathrm{AS}((\mathbf{a}_1, f(\mathbf{a}_1)) \ldots (\mathbf{a}_{k+1}, f(\mathbf{a}_{k+1})))$ *in* $\mathbb{R}^{k+1}$ *and* ρ *is the supporting hyperplane to the graph of* f *parallel to* π*.*

Here, AS *denotes the affine span, and any hyperplane* τ *is identified with its graph* $\{(\mathbf{x}, \tau(\mathbf{x})) \in \mathbb{R}^{k+1} : x \in \mathbb{R}^k\}$.

Theorem 2.1 belongs to an interesting class of related optimization problems which can be found, for example, in [1–4, 6–9, 11].

We know of no convex domain other than a k-simplex where we can generate a hyperplane π as in Theorem 2.1 for which we can verify that the graph of f over the domain lies entirely above or entirely below that hyperplane.

We now present two remarks regarding Theorem 2.1.

Remark 2.1 The following stronger version of Theorem 2.1 holds, which has an identical proof as Theorem 2.1. □

Theorem 2.2 *Given a continuous function f on a simplex Δ with the following property: The graph of f lies entirely above or below its secant hyperplane π through the graph points of f over the vertices of Δ. Then, an expression for the min-max affine approximation of f on the simplex Δ has graph given by the hyperplane $\pi + d$, where $2d$ is the non-zero extremum value of $f - \pi$ on Δ. An alternative definition of π is the affine interpolant to f at the $k+1$ vertices of the simplex Δ.*

Remark 2.2 In this remark, we speak to a geometric description of the hyperplane average in Theorem 2.1 and write formulae for the optimal $\boldsymbol{\alpha}$ and β in the notation of Problem 1.2, expression (1).

Let the vector $\boldsymbol{\alpha}'$ and the scalar β' be defined as the solution to the following equation:

$$\begin{bmatrix} \mathbf{a}_1^T & 1 \\ . & 1 \\ . & 1 \\ . & 1 \\ \mathbf{a}_{k+1}^T & 1 \end{bmatrix} \begin{bmatrix} \boldsymbol{\alpha}' \\ \beta' \end{bmatrix} = \begin{bmatrix} f(\mathbf{a}_1) \\ . \\ . \\ . \\ f(\mathbf{a}_{k+1}) \end{bmatrix}.$$

Secondly, define

$$\beta'' = \begin{cases} \min_{\mathbf{x}\in\Delta}(f(\mathbf{x}) - (\boldsymbol{\alpha}'^T\mathbf{x} + \beta')), & \text{if } f \text{ is convex.} \\ \max_{\mathbf{x}\in\Delta}(f(\mathbf{x}) - (\boldsymbol{\alpha}'^T\mathbf{x} + \beta')), & \text{if } f \text{ is concave.} \end{cases}$$

Then, set $\boldsymbol{\alpha} := \boldsymbol{\alpha}'$ and $\beta := \frac{1}{2}(\beta' + \beta'')$. It is easily checked that Problem 1.2, expression (1) is optimized at $(\boldsymbol{\alpha}, \beta)$.

We end this remark by saying that the computation of the optimal β above costs a minimization of a convex function (a linear shift of $\pm f$) over the set Δ. □

3 The Proof of Theorem 2.1

The key ideas in our proof of Theorem 2.1 are the two equivalent definitions of a convex function. $f : \Delta \subseteq \mathbb{R}^k \to \mathbb{R}$ is convex provided for all $\gamma_i \geq 0 : \Sigma_{i=1}^{k+1}\gamma_i = 1$, and any affine independent set of $k+1$ points $\{\mathbf{a}_1', \ldots, \mathbf{a}_{k+1}'\}$, $f(\Sigma_{i=1}^{k+1}\gamma_i\mathbf{a}_i') \leq$

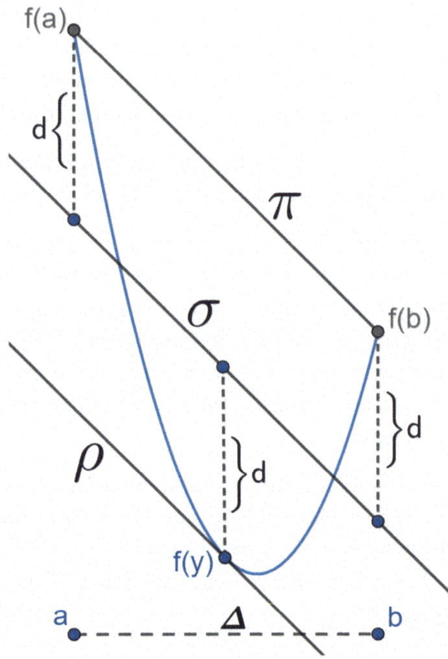

Fig. 1 Illustration of Theorem 2.1. The secant π is the convex hull of $k+1$ values $(\mathbf{x}_i, f(\mathbf{x}_i))$ and ρ the supporting hyperplane. The convexity of f leads to the fact that the plane, σ, with same slope halfway between π and ρ is the best affine approximation

$\Sigma_{i=1}^{k+1}\gamma_i f(\mathbf{a}_i')$, i.e. the graph of f over the convex hull of $k+1$ affinely independent points lies below the convex hull of the $k+1$ image points $\{[\mathbf{a}_i'^T \ f(\mathbf{a}_i')]^T : 1 \leq i \leq k+1\}$, or equivalently, for each $\mathbf{x} \in \Delta$, and each support plane at $[\mathbf{x}^T \ f(\mathbf{x})]^T$ with slope $\boldsymbol{\eta} \in \mathbb{R}^k$, $f(\mathbf{y}) \geq f(\mathbf{x}) + \boldsymbol{\eta}^T(\mathbf{y}-\mathbf{x})$, for all $\mathbf{y} \in \Delta$, i.e. the graph of f lies on or above any supporting hyperplane.

For our proof of Theorem 2.1, we advise the reader to use Fig. 1 for intuition.

Proof First, assume that f is convex, so its graph over Δ lies on or below the hyperplane π. Let $-2d := \inf\{l : \pi + l\mathbf{e}_{k+1} \cap graph(f) \neq \emptyset\}$. Then, any admissible l must satisfy $l \leq 0$, and so this in turn means $d \geq 0$. Here, $\mathbf{e}_{k+1}$ is the $k+1^{st}$ basis vector in $\mathbb{R}^{k+1}$, so each admissible l gives a downward translate of π with non-empty intersection with the graph of f. Since Δ is compact and f is continuous, the inf is actually a min, and so their exists at least one graph point $(y, f(y))$ on $\rho := \pi - 2d$.

Thus, ρ is a supporting hyperplane (tangent plane if f is differentiable at $\mathbf{y}$). Moreover, at the points $\mathbf{a}_i$, an easy computation shows $\pi - d = (2\pi - 2d)/2 = (\pi + \rho)/2 = \rho + d$. These and the construction of σ imply $f(\mathbf{a}_i) - \sigma(\mathbf{a}_i) = d = \sigma(y) - f(y)$. We also observe that $\forall z \in \Delta, |f(z) - \sigma(z)| \leq \sigma(y) - f(y) = d$, as $f(z)$ is between π and ρ whose midplane is σ. Now, with the above in hand, we are able to argue as follows. Assume that μ is a best affine approximating plane. Then, first, $\mu(\mathbf{a}_i) \geq \sigma(\mathbf{a}_i), \quad i \in [1 \ldots k+1]$, otherwise the maximal distance increases. On the other hand, writing $y = \sum_{i=1}^{k+1}\gamma_i \mathbf{a}_i$ for constants $\gamma_i \geq 0 : \Sigma_{i=1}^{k+1}\gamma_i = 1$, we have $\mu(y) = \sum_{i=1}^{k+1} \gamma_i \mu(\mathbf{a}_i) \leq \sigma(y)$ otherwise the maximal distance increases. We deduce that $\mu = \sigma$. This concludes the proof of Theorem 2.1 in case f is convex.

In the case that f is concave, since $\pi = \mathrm{AS}((a_1, f(\mathbf{a}_1)) \ldots (\mathbf{a}_{k+1}, f(\mathbf{a}_{k+1})))$ and the graph of f over Δ lies on or above π, we can repeat the convex argument replacing d by $-d$ and inf by sup and interchanging $\leq$ and $\geq$. Alternatively, note that $-f$ is convex and that one checks that the negative of the solution for $-f$ is just $\frac{\pi+\rho}{2}$. □

Note that even if f is not convex (or concave) and not even smooth, the method produces a min-max approximation for the $k+2$ points comprised of the $k+1$ points $(\mathbf{a}_1, f(\mathbf{a}_1)) \ldots, (\mathbf{a}_{k+1}, f(\mathbf{a}_{k+1}))$ and $(\mathbf{y}, f(\mathbf{y}))$.

4 Theorem 2.1: The Case $k = 1$

For $k = 1$, our main result, Theorem 2.1, provides an extension of the classical Chebyshev equioscillation theorem, see Theorem 4.1 below, for linear approximants with an explicit unique formula for the best approximant.

We now explain this.

4.1 Chebyshev systems and Chebyshev–Markov equioscillation

We will work with the real interval $[p, q]$, where $p < q$ and the space of real-valued continuous functions $f : [p, q] \to \mathbb{R}$. As per convention, we denote this space by $C([p, q])$.

A set of $l+1$, $l \geq 0$ functions $\{u_j(t)\}_{j=0}^{l}$, $u_j \in C([p, q])$ is called a Chebyshev (Haar) system on the interval $[p, q]$ if any linear combination

$$u(t) = \sum_{j=0}^{l} c_j u_j(t), \ t \in [p, q],$$

with not all coefficients c_j zero, has at most l distinct zeros in $[p, q]$.[1]

The $l+1$-dimensional subspace $U_l \subseteq C([p, q)])$ spanned by a Chebyshev system $\{u_j(t)\}_{j=0}^{l}$ on $[p, q]$ defined by

[1]Alfred Haar, 1885–1933, was a Hungarian mathematician. In 1904, he began to study at the University of Göttingen. His doctorate was supervised by David Hilbert.

$$U_l := \left\{ u(t) : u(t) = \sum_{j=0}^{l} c_j u_j(t) \right\}$$

is called a Chebyshev space on $[p, q]$.

The classical Chebyshev equioscillation theorem, see, for example, [5, Chapter 9, Theorem 4.4] and [10], is the following:

Theorem 4.1 *Let $f \in C([p, q])$, and let U_l be a $l + 1$ Chebyshev space on an interval $[p, q]$. Then, $\widehat{u} \in U_l$ is a min-max approximant to f on $[p, q]$, that is, u satisfies*

$$\|f - \widehat{u}\|_{\infty[p,q]} = \inf_{u \in U_l} \|f - u\|_{\infty[p,q]}$$

if and only there exist $l + 2$ points $\{x_1, \ldots, x_{l+2}\}$ with $p \leq x_1 < \ldots < x_{l+2} \leq q$ such that

$$f(x_i) - \widehat{u}(x_i) = w(-1)^i \|f - \widehat{u}\|_{\infty[p,q]}, \; w = \pm 1.$$

Motivated by Theorem 4.1, we have the following.

4.2 Chebyshev Equioscillation

Let $f \in C([p, q)]$. A Chebyshev polynomial $\widehat{h_l}$ (when it exists) of degree at least $l \geq 1$ is the polynomial, which best uniformly approximates f on $[p, q]$, i.e.

$$\widehat{h_l} = \arg\min_{h_l \in \Pi_l} \|f - h_l\|_{\infty[p,q]},$$

where Π_l is the set of polynomials of degree at most l.

The following is often called the Chebyshev equioscillation theorem.

Theorem 4.2 *Let $f \in C([a, b])$. Then, $\widehat{h_l}$ exists if there exist $l + 2$ points $\{x_1, \ldots, x_{l+2}\}$ with $p \leq x_1 < \ldots < x_{l+2} \leq q$ such that*

$$f(x_i) - \widehat{h_l}(x_i) = w(-1)^i \|f - \widehat{h_l}\|_{\infty[p,q]}, \; w = \pm 1.$$

4.3 The Case $l = 1$ of Theorem 4.2

We now show how Theorem 4.2 with $l = 1$ corresponds to Theorem 2.1 for $k = 1$.

We may assume that $p < q$, and for the moment, we do not assume anything about $f : [p, q] \to \mathbb{R}$. Define now a linear function $L : [p, q] \to \mathbb{R}$ from f in terms of parameters d and y to be determined later as follows:

$$L(x) = f(y) + d + m(x - y),\ x \in [p, q]. \tag{2}$$

Now, since $L(y) - f(y) = d$, if we assume $L(p) - f(p) = -d = L(q) - f(q)$, then

$$d = f(p) - L(p) = f(p) - f(y) - d - m(p - y).$$

Thus,

$$d = \frac{f(p) - f(y)}{2} - \frac{m}{2}(p - y).$$

Also,

$$d = f(q) - L(q) = f(q) - f(y) - d - m(q - y).$$

So,

$$\frac{f(p) - f(y)}{2} - \frac{m}{2}(p - y) = \frac{f(q) - f(y)}{2} - \frac{m}{2}(q - y).$$

Thus, $m = \frac{f(q)-f(p)}{p-q}$ in the case when f or $-f$ is a convex and differentiable function, by the mean value theorem.

It is clear that we have proved Theorem 2.1 once we are able to choose r to maximize d if this is possible for the given f. In the case when f is convex, we see that $m = f'(r)$. It is clear that the argument works when f is concave, which implies that $-f$ is convex,

In the case of $k = 1$, we could write

$$\frac{ax + b}{cx + d} == \frac{\left(\left(\frac{a}{c}\right)\left(\mathbf{x} + \frac{d}{c} - \frac{d}{c}\right) + \frac{b}{c}\right)}{\left(\mathbf{x} + \frac{d}{c}\right)}$$

$$= \frac{a}{c} + \frac{\left(\frac{b}{c} - \left(\frac{a}{c}\right)\left(\frac{d}{c}\right)\right)}{\left(\mathbf{x} + \frac{d}{c}\right)},$$

which is either an upward or a downward hyperbola for which the secant between $[p, q]$ lies above (or below) the curve. Then, following the "mean value" argument leads to r with derivative =slope and via our visualization leads to a line with the same slope but halfway between the secant and tangent to r.

5 Connections of Theorem 2.1 to Graphics

One consequence of Theorem 2.1 gives an interesting connection to graphics. We provide our ideas below.

We are given a flat object O and want to render it from a given perspective camera setup. The resulting image is a projective (AKA perspective or homography) transformation of O, $P(O)$. Thus, each pixel (color) in O (an ordered pair (x_1, y_1) in $\mathbb{R}^2$) is transformed to a new pixel location (point) 2 via the action of P

$$P : \begin{bmatrix} x_1 \\ y_1 \end{bmatrix} \mapsto \frac{1}{dx_1 + ey_1 + j} \begin{bmatrix} a_1x_1 + b_1y_1 + c_1 \\ a_2x_1 + b_2y_1 + c_2 \end{bmatrix}. \tag{3}$$

Practically, to render this object that can have 10s of millions of pixels, it is useful to have a good fast approximation of the transformation not entailing division which can cause numerical instabilities. Thus, we seek affine approximants to P. One known method is to simply take the affine approximant to be the linear terms of the Taylor expansion around one point, the tangent approximation.

We aim to provide a better 2d-affine approximant to P than the tangent approximation. From Eq. 3, the components of $P(\mathbf{x})$ are given by

$$f_i(\mathbf{x}) := \frac{a_ix + b_iy + c_i}{dx + ey + j}, \quad i = 1, 2. \tag{4}$$

From our main Theorem 2.1, we know that if we can find a triangle, Δ, containing (a large part of) O such that each $f_i(\mathbf{x})$ is convex or concave, then there exist min-max affine approximations $\boldsymbol{\alpha}_i{}^T\mathbf{x} + \beta_i$ to each f_i. Then, forming the affine transformation A of 2-space given by

$$A(\mathbf{x}) := \begin{bmatrix} \boldsymbol{\alpha}_1{}^T\mathbf{x} + \beta_1 \\ \boldsymbol{\alpha}_2{}^T\mathbf{x} + \beta_2 \end{bmatrix} \tag{5}$$

provides, component-wise, the min-max affine approximants on Δ.

Since a differentiable function is convex or concave on a domain exactly when its Hessian is, respectively, positive or negative semidefinite, the following is a key tool for the application of our main result.

Proposition 5.1 *A symmetric* 2×2 *matrix* $D = \begin{bmatrix} p_{11} & p_{12} \\ p_{12} & p_{22} \end{bmatrix}$ *is* positive or negative *semidefinite according to* $signum(p_{11}) = \pm\text{signum}\,(p_{11}p_{22} - p_{12}^2)$. □

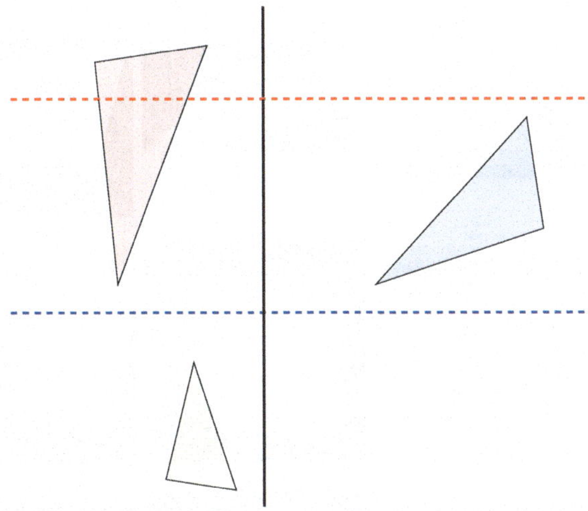

Fig. 2 The black line is where the projective transformation is infinite, $X = -\delta$. The blue line is where the $\frac{\partial}{\partial x^2}$ term of the Hessian of the X transformation is 0, and the red line is where the $\frac{\partial}{\partial x^2}$ term of the Hessian of the Y transformation is 0. The blue and yellow simplices define domains of transformations that are convex (or concave), while the red simplex is not

The Hessian of

$$\frac{\alpha X + \beta Y + \gamma}{X + \delta} = \alpha + \frac{\beta Y - \alpha\delta + \gamma}{X + \delta} \tag{6}$$

is

$$H = \frac{1}{(X+\delta)^3}\begin{bmatrix} 2(\beta Y - \alpha\delta + \gamma) & -\beta(X+\delta) \\ -\beta(X+\delta) & 0 \end{bmatrix}$$

with determinant $\frac{-\beta^2}{(X+\delta)^4}$. The sign of $H[1, 1] = 2(\beta Y - \alpha\delta + \gamma)$ depends on $Y \mathop{\gtrless}\limits \frac{\alpha\delta-\gamma}{\beta}$.

When we combine Proposition 5.1 with the Hessian of Eq. 6, we see that $\frac{\alpha X+\beta Y+\gamma}{X+\delta}$ in each of the connected components of $\mathbb{R}^2 \setminus \{\{X = -\delta\} \cup \{Y = \frac{\alpha\delta-\gamma}{\beta}\}\}$ is either convex or concave. With the 2 coordinate functions of the projective transformation (X and Y), we end with 6 regions in $\mathbb{R}^2$ where Theorem 2.1 holds, Fig. 2.

In order to transform a general projective transformation as in Eq. 5 to the form in Eq. 6, with a simple denominator, we rotate the axes so that X is in the direction of $\begin{bmatrix} d \\ e \end{bmatrix}$ and normalize so that the coefficient of X in the denominator is 1.

In Fig. 3, we show a few examples of the various affine approximations A of projective transformations P using this idea.

The first column is the original image O, the second column is the image transformed by a projective transformation P, modeling a new viewpoint, the third column is the transformation based on the affine approximation A of P using Theorem 2.1 and a user-defined triangle, while the fourth column is computed using

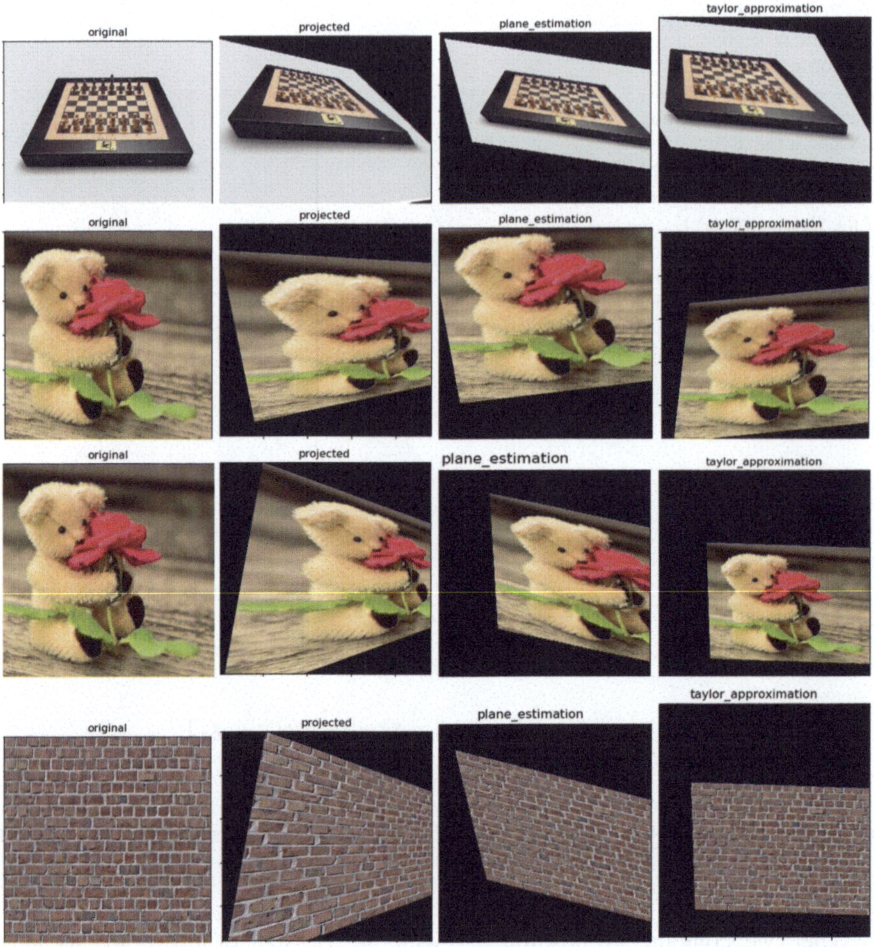

Fig. 3 Original image/projective transformation of the image/min-max affine approximation/Taylor expansion

a Taylor series approximation of P around the image's center. The images in column 3 should visually look closer to those in column 2 than those in column 4, the Taylor approximation.

References

1. Ahlberg, J. H., Nilson, E. N., Walsh, J. F.: Theory of splines and their applications. Acad. Press (1967).
2. Brudnyi, Y. A.: Approximation of functions defined in a convex polyhedron. In: Soviet Math. Doklady, 11. 6, pp. 1587–1590 (2006). Dokl. Akad. Nauk SSSR, **195** , pp. 1007–1009 (1970).
3. Davis, P. J.: Interpolation and approximation. New York (1963).

4. Deutsch, F.: Best approximation in inner product spaces. CMS Books in Mathematics (2001).
5. Krein, M., Nudelman, A.: The Markov Moment problem and extremal problems. AMS translation from the Russian edition of 1973.
6. Korneichuk, N. P.: Extremal problems in approximation theory , Moscow (1976) (In Russian).
7. Laurent, P. J.: Approximation et optimization. Hermann (1972).
8. Miroshichenko, V. L.: Methods of spline functions. Moscow (1980).
9. Nikol'skii, S. M.: Approximation of functions of several variables and imbedding theorems. Springer (1975) (Translated from Russian).
10. Schumaker, L.: Spline Functions: basic theory. Academic Press, NY, (1983).
11. Teml'yakov, V. N.: Best approximations for functions of two variables. In: Soviet Math. Doklady, 16: 4, pp. 1051–1055 (1975). Dokl. Akad. Nauk SSSR, **223**, pp. 1079–1082 (1975).

A Kaczmarz Algorithm for Solving Tree Based Distributed Systems of Equations

Chinmay Hegde, Fritz Keinert, and Eric S. Weber

Abstract The Kaczmarz algorithm is an iterative method for solving systems of linear equations. We introduce a modified Kaczmarz algorithm for solving systems of linear equations in a distributed environment, i.e., the equations within the system are distributed over multiple nodes within a network. The modification we introduce is designed for a network with a tree structure that allows for passage of solution estimates between the nodes in the network. We prove that the modified algorithm converges under no additional assumptions on the equations. We demonstrate that the algorithm converges to the solution, or the solution of minimal norm, when the system is consistent. We also demonstrate that in the case of an inconsistent system of equations, the modified relaxed Kaczmarz algorithm converges to a weighted least-squares solution as the relaxation parameter approaches 0.

1 Introduction

The Kaczmarz method [16] is an iterative algorithm for solving a system of linear equations $A\mathbf{x} = \mathbf{b}$, where A is an $m \times k$ matrix. Written out, the equations are $\mathbf{a}_i^*\mathbf{x} = b_i$ for $i = 1, \ldots, m$, where $\mathbf{a}_i^*$ is the ith row of the matrix A. Given a solution guess $\mathbf{x}^{(n-1)}$ and an equation number i, we calculate $r_i = b_i - \mathbf{a}_i^*\mathbf{x}^{(n-1)}$ (the residual for equation i), and define

$$\mathbf{x}^{(n)} = \mathbf{x}^{(n-1)} + \frac{r_i}{\|\mathbf{a}_i\|^2}\mathbf{a}_i.$$

C. Hegde
Electrical and Computer Engineering, Iowa State University, Ames, IA, USA

Electrical and Computer Engineering, New York University, New York, NY, USA

F. Keinert · E. S. Weber (✉)
Department of Mathematics, Iowa State University, Ames, IA, USA
e-mail: keinert@iastate.edu; esweber@iastate.edu

M. Hirn et al. (eds.), *Excursions in Harmonic Analysis, Volume 6*, Applied and Numerical Harmonic Analysis, https://doi.org/10.1007/978-3-030-69637-5_20

This makes the residual of $\mathbf{x}^{(n)}$ in equation i equal to 0. Here and elsewhere, $\|\cdot\|$ is the usual Euclidean (ℓ^2) norm. We iterate repeatedly through all equations (i.e. we consider $\lim_{n\to\infty}\mathbf{x}^{(n)}$ where $n \equiv i \mod m$, so the equations are repeated cyclically). Kaczmarz proved that if the system of equations has a unique solution , then $\mathbf{x}^{(n)}$ converges to that solution. Later, it was proved in [33] that if the system is consistent (but the solution is not unique), then the sequence converges to the solution of minimal norm . Likewise, it was proved in [7, 21] that if inconsistent, a relaxed version of the algorithm can provide approximations to a weighted least-squares solution .

Obtaining the nth estimate requires knowledge only of the i-th equation ($n \equiv i \mod m$ as above) and the $n-1$-st estimate. We suppose that the equations are indexed by the nodes of a tree, representing a network in which the equations are distributed over many nodes. In our distributed Kaczmarz algorithm, solution estimates can only be communicated when there exists an edge between the nodes. The estimates for the solution will disperse through the tree, which results in several different estimates of the solution. When these estimates then reach the leaves of the tree, they are pooled together into a single estimate. Using this single estimate as a seed, the process is repeated, with the goal that the sequence of single estimates will converge to the true solution. We illustrate the dispersion and pooling processes in Fig. 1.

1.1 Notation

For linear transformations T, we denote by $\mathcal{N}(T)$ and $\mathcal{R}(T)$ the kernel (nullspace) and range, respectively. We use $\rho(T)$ to denote the spectral radius .

When $\mathbf{a}^* = \mathbf{a}_v^*$ corresponds to a row of the matrix A indexed by a node v, we will denote the linear projection onto the subspace $\mathbf{a}_v^*\mathbf{z} = 0$ by

$$P_v(\mathbf{z}) = \left(I - \frac{\mathbf{a}_v\mathbf{a}_v^*}{\mathbf{a}_v^*\mathbf{a}_v}\right)(\mathbf{z}) \tag{1}$$

and the affine projection onto the linear manifold $\mathbf{a}_v^*\mathbf{z} = b_v$ by

$$Q_v(\mathbf{z}) = P_v(\mathbf{z}) + \mathbf{h}_v \tag{2}$$

where $\mathbf{h}_v$ is the vector that satisfies $\mathbf{a}_v^*\mathbf{h}_v = b_v$ and is in $\mathcal{N}(P_v)$.

A tree is a connected graph with no cycles. We denote arbitrary nodes (vertices) of a tree by v, u. Our tree will be rooted; the root of the tree is denoted by r. Following the notation from MATLAB, when v is on the path from r to u, we will say that v is a predecessor of u and write $u \prec v$. Conversely, u is a successor of v. By immediate successor of v we mean a successor u such that there is an edge between v and u (this is referred to as a *child* in graph theory parlance [35]). Similarly, v is an immediate predecessor (i.e. *parent*). We denote the set of all immediate successors of node v by $C(v)$. A node without a successor is called a leaf; leaves of the tree are denoted by ℓ. We will denote the set of all leaves by $\mathcal{L}$. Often we will have need to enumerate the leaves as $\ell_1, \ldots, \ell_t$, hence t denotes the number of leaves.

A weight w is a nonnegative function on the edges of the tree; we denote this by $w(u, v)$, where u and v are nodes that have an edge between them. We assume $w(u, v) = w(v, u)$, though we will typically write $w(u, v)$ when $u \prec v$. When $u \prec v$, but u is not an immediate successor, we write

$$w(u, v) := \prod_{j=1}^{J-1} w(u_{j+1}, u_j),$$

where $v = u_1, \ldots, u_J = u$ is a path from v to u.

When the system of equations $A\mathbf{x} = \mathbf{b}$ has a unique solution, we will denote this by $\mathbf{x}^S$. When the system is consistent but the solution is not unique, we denote the solution of minimal norm by $\mathbf{x}^M$, which is given by

$$\mathbf{x}^M = \text{argmin}\,\{\|\mathbf{x}\| : A\mathbf{x} = \mathbf{b}\}. \tag{3}$$

1.2 The Distributed Kaczmarz Algorithm

The iteration begins with an estimate (say $\mathbf{x}^{(n)}$) at the root of the tree. When node u receives from its immediate predecessor v an input estimate $\mathbf{x}_v^{(n)}$, it generates a new estimate via the Kaczmarz update:

$$\mathbf{x}_u^{(n)} = \mathbf{x}_v^{(n)} + \frac{r_u(\mathbf{x}_v^{(n)})}{\|\mathbf{a}_u\|^2}\mathbf{a}_u, \tag{4}$$

where the residual is given by

$$r_u(\mathbf{x}_v^{(n)}) := b_u - \mathbf{a}_u^* \mathbf{x}_v^{(n)}.$$

The root node updates the estimate that begins the iteration using its equation: $\mathbf{x}_r^{(n)} = \mathbf{x}^{(n)} + \frac{r_r(\mathbf{x}^{(n)})}{\|\mathbf{a}_r\|^2}\mathbf{a}_r$. Node u then passes this estimate to all of its immediate successors, and the process is repeated recursively. We refer to this as the *dispersion stage* . Once this process has finished, each leaf ℓ of the tree now possesses an estimate: $\mathbf{x}_\ell^{(n)}$.

The next stage, which we refer to as the *pooling stage* , proceeds as follows. For each leaf, set $\mathbf{y}_\ell^{(n)} = \mathbf{x}_\ell^{(n)}$. Each node v calculates an updated estimate as

$$\mathbf{y}_v^{(n)} = \sum_{u \in C(v)} w(u, v)\mathbf{y}_u^{(n)}, \tag{5}$$

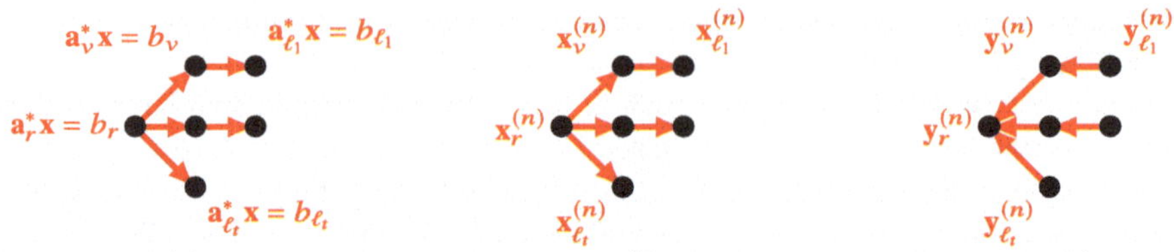

Fig. 1 Illustration of updates in the distributed Kaczmarz algorithm with measurements indexed by nodes of the tree. (**a**) Equations distributed across nodes. (**b**) Updates disperse through nodes. (**c**) Updates pool and pass to next iteration

subject to the constraints that $w(u, v) > 0$ when $u \in C(v)$ and $\sum_{u \in C(v)} w(u, v) = 1$. This process continues until reaching the root of the tree, resulting in the estimate $\mathbf{y}_r^{(n)}$.

We set $\mathbf{x}^{(n+1)} = \mathbf{y}_r^{(n)}$, and repeat the iteration. The updates in the dispersion stage (Eq. 4) and pooling stage (Eq. 5) are illustrated in Fig. 1.

We note that the tree topology is fixed *a priori*, and remains fixed over all iterations.

1.3 Related Work

The Kaczmarz method was originally introduced in [16]. It became popular with the introduction of Computer Tomography, under the name of ART (Algebraic Reconstruction Technique). ART added non-negativity and other constraints to the standard algorithm [8]. Other variations on the Kaczmarz method allowed for relaxation parameters [33], re-ordering equations to speed up convergence [11], or considering block versions of the Kaczmarz method with relaxation matrices Ω_i [7]. Relatively recently, choosing the next equation randomly has been shown to dramatically improve the rate of convergence of the algorithm [28, 32, 40]. Moreover, this randomized version of the Kaczmarz algorithm has been shown to be comparable to the gradient descent method [26]. Recent advances in accelerating the Kaczmarz method include subsampling techniques, meaning subsampling the rows of the matrix [27], or sketching the full matrix with a preconditioner [9].

Our version of the Kaczmarz method differs from these versions in the following crucial sense. Each node has access to only its equation. Therefore, the next equation cannot be chosen randomly or otherwise, since the ordering of the equations is determined *a priori* by the network topology and thus is different from all randomized versions. Similarly, the block versions and sketched versions require access to several (but not necessarily all) of the rows simultaneously; this is also prohibited in our distributed context. Our version here is most similar to the Cimmino method [6], which was extended in [3], as well as the greedy method given in [23]. Both of these methods involve averaging estimates, in addition to applying the Kaczmarz update, as we do here. The proofs in [23] require the system to be consistent, which we do not, and the method (and proofs) in [3] require access to *columns* of the matrix as well as rows. Because of these differences, we make no direct comparisons.

The situation we consider in the present paper can be considered a distributed estimation problem . Such problems have a long history in applied mathematics, control theory, and machine learning. At a high level, similar to our approach, they all involve averaging local copies of the unknown parameter vector interleaved with update steps [2, 15, 25, 29–31, 34, 36, 38, 39]. Recently, a number of protocols for gossip methods, including a variation of the Kaczmarz method, was analyzed in [20]. The protocols analyzed in that paper require the system to be consistent for convergence guarantees.

Following [38], a consensus problem takes the following form. Consider the problem of minimizing:

$$F(\mathbf{x}) = \sum_{v=1}^{m} f_v(\mathbf{x}),$$

where f_v is a function that is known (and private) to node v in the graph. Then, one can solve this minimization problem using decentralized gradient descent, where each node updates its estimate of $\mathbf{x}$ (say $\mathbf{x}_v$) by combining the average of its neighbors with the negative gradient of its local function f_v:

$$\mathbf{x}_v^{(n+1)} = \frac{1}{\deg v} \sum_u m(v,u)\mathbf{x}_u^{(n)} - \omega \nabla f_v(\mathbf{x}_v^{(n)}),$$

where $M = (m(v,u)) \in \{0,1\}^{m \times m}$ represents the adjacency matrix of the graph. Specializing $f_v(\mathbf{x}) = c_v(b_v - \mathbf{a}_v^* \mathbf{x})^2$ yields our least-squares estimation problem that we establish in Theorem 2.4 (where c_v is a fixed weight for each node).

However, our version of the Kaczmarz method differs from previous work in a few aspects: (i) we assume an *a priori* fixed tree topology (which is more restrictive than typical gossip algorithms); (ii) there is no master node as in parallel algorithms, and no shared memory architecture; (iii) as we will emphasize in Theorem 2.4, we make no strong convexity assumptions (which is typically needed for distributed optimization algorithms, but see [22, 24] for a relaxation of this requirement); and (iv) we make no assumptions on the matrix A, in particular we do not assume that it is nonnegative.

On the other end of the spectrum are algorithms that distribute a computational task over many processors arranged in a fixed network. These algorithms are usually considered in the context of parallel processing, where the nodes of the graph represent CPUs in a highly parallelized computer. See [1] for an overview.

The algorithm we are considering does not really fit either of those categories. It requires more structure than the gossip algorithms, but each node depends on results from other nodes, more than the usual distributed algorithms.

This was pointed out in [1]. For iteratively solving a system of linear equations, a Successive Over-Relaxation (SOR) variant of the Jacobi method is easy to parallelize; standard SOR, which is a variation on Gauss-Seidel, is not. The authors also consider what they call the *Reynolds method*, which is similar to a Kaczmarz method with all equations being updated simultaneously. Again, this method is easy to parallelize. A sequential version called RGS (Reynolds Gauss-Seidel) can only be parallelized in certain settings, such as the numerical solution of PDEs.

A distributed version of the Kaczmarz algorithm was introduced in [17]. The main ideas presented there are very similar to ours: updated estimates are obtained from prior estimates using the Kaczmarz update with the equations that are available at the node, and distributed estimates are averaged together at a single node (which the authors refer to as a fusion center, for us it is the root of the tree). In [17], the convergence analysis is limited to the case of consistent systems of equations, and inconsistent systems are handled by Tikhonov regularization [12, 14] rather than by varying the relaxation parameter. Another distributed version was proposed in [19], which has a shared memory architecture.

Finally, the Kaczmarz algorithm has been proposed for online processing of data in [5, 13]. In these papers, the processing is online, so neither distributed nor parallel.

2 Analysis of the Kaczmarz Algorithm for Tree Based Distributed Systems of Equations

In this section, we will demonstrate that the Kaczmarz algorithm for tree based equations as defined in Eqs. (4) and (5) converges. We consider three cases separately: (i) the system is consistent and the solution is unique; (ii) the system is consistent but there are many solutions; and (iii) the system is inconsistent. In Sect. 2.1, we prove that for case (i) the algorithm converges to the solution, and in Sect.2.2, we prove that for case (ii) the algorithm converges to the solution of minimal norm. Also in Sect.2.2, we introduce the relaxed version of the update in Eq. (4). We prove that for every relaxation parameter $\omega \in (0, 2)$, the algorithm converges to the solution of minimal norm. Then in Sect.2.3, we prove that for case (iii) the algorithm converges to a generalized solution $\mathbf{x}(\omega)$ which depends on ω, and $\mathbf{x}(\omega)$ converges to a weighted least-squares solution as $\omega \to 0$.

2.1 Systems with Unique Solutions

For our analysis, we need to trace the estimates through the tree. Suppose that the tree has t leaves; for each leaf ℓ, let $p_\ell - 1$ denote the length of the path between the root r and the leaf ℓ. We will denote the nodes on the path from r to ℓ by $r = (\ell, 1), (\ell, 2), \ldots, (\ell, p_\ell) = \ell$. During the dispersion stage, we have for $p =$

$2, \ldots, p_\ell$:

$$\mathbf{x}_{\ell,p}^{(n)} = \mathbf{x}_{\ell,p-1}^{(n)} + \left(\frac{r_{\ell,p}(\mathbf{x}_{\ell,p-1}^{(n)})}{\|\mathbf{a}_{\ell,p}\|^2} \right) \mathbf{a}_{\ell,p}.$$

Then at the beginning of the pooling stage, we have the estimates $\mathbf{y}_\ell^{(n)} := \mathbf{x}_\ell^{(n)}$ (we denote $\mathbf{x}_\ell^{(n)} := \mathbf{x}_{\ell,p_\ell}^{(n)}$ and $\mathbf{y}_\ell^{(n)} := \mathbf{y}_{\ell,p_\ell}^{(n)}$). These estimates then pool back at the root as follows (the proof is a straightforward induction argument):

Lemma 1 *The estimate at the root at the end of the pooling stage is given by*

$$\mathbf{y}_r^{(n)} = \sum_{\ell \in \mathcal{L}} w(\ell, r) \mathbf{y}_\ell^{(n)}.$$

Note that also by induction, we have that

$$\sum_{\ell \in \mathcal{L}} w(\ell, r) = 1. \tag{6}$$

Theorem 2.1 *Suppose that the equation $A\mathbf{x} = \mathbf{b}$ has a unique solution, denoted by $\mathbf{x}^S$. There exists a constant $\alpha < 1$, such that*

$$\|\mathbf{x}^S - \mathbf{x}^{(n+1)}\| \leq \alpha \|\mathbf{x}^S - \mathbf{x}^{(n)}\|.$$

Consequently,

$$\lim_{n \to \infty} \mathbf{x}^{(n)} = \mathbf{x}^S,$$

and the convergence is linear in order.

Proof Along any path from the root r to the leaf ℓ, the dispersion stage is identical to the classical Kaczmarz algorithm, and so we can write (see [18])

$$\mathbf{x}^S - \mathbf{x}_\ell^{(n)} = P_{\ell,p_\ell}(\mathbf{x}^S - \mathbf{x}_{\ell,p_\ell-1}^{(n)}) = P_{\ell,p_\ell} \ldots P_{\ell,2} P_{\ell,1}(\mathbf{x}^S - \mathbf{x}^{(n)}), \tag{7}$$

from which it follows immediately that

$$\|\mathbf{x}^S - \mathbf{x}_\ell^{(n)}\| \leq \|\mathbf{x}^S - \mathbf{x}^{(n)}\|. \tag{8}$$

We claim that unless $\mathbf{x}^S = \mathbf{x}^{(n)}$, we must have a strict inequality for at least one leaf, say ℓ_0. Indeed, suppose to the contrary that for every leaf ℓ, we had equality in Eq. (8), then by Eq. (7), we must have for every node $v = (\ell, k)$ in the path from the root r to the leaf ℓ:

$$P_v(\mathbf{x}^S - \mathbf{x}^{(n)}) = \mathbf{x}^S - \mathbf{x}^{(n)}.$$

Therefore, we obtain

$$\mathbf{a}_v^*(\mathbf{x}^S - \mathbf{x}^{(n)}) = 0 \text{ for all nodes } v.$$

By our assumption that the equation has a unique solution, we obtain that $\mathbf{x}^S - \mathbf{x}^{(n)} = 0$.

By Eqs. (6) and (8) and our previous claim, we have

$$\|\mathbf{x}^S - \mathbf{x}^{(n+1)}\| < \sum_{\ell \in \mathcal{L}} w(\ell, r)\|\mathbf{x}^S - \mathbf{x}^{(n)}\| = \|\mathbf{x}^S - \mathbf{x}^{(n)}\|.$$

By continuity and compactness, there is a uniform constant α less than 1 that satisfies the claim. This completes the proof. □

As we shall see in the sequel, we can interpret the above proof in the following way: define the mapping

$$\mathcal{P} := \sum_{\ell \in \mathcal{L}} w(\ell, r) P_{\ell, p_\ell} \dots P_{\ell,2} P_{\ell,1},$$

then the mapping $\mathbf{z} \mapsto \mathbf{x}^S - \mathcal{P}(\mathbf{x}^S - \mathbf{z})$ is a contraction with unique fixed point $\mathbf{x}^S$. Moreover, the iteration of the algorithm can be expressed as

$$\mathbf{x}^{(n+1)} = \mathbf{x}^S - \mathcal{P}(\mathbf{x}^S - \mathbf{x}^{(n)}).$$

2.2 *Consistent Systems*

We shall show in this section that the distributed Kaczmarz algorithm as defined in Eqs. (4) and (5) will converge to the solution with minimal norm in the case that there exists more than one solution. We first introduce the relaxed version of the algorithm (required to deal with inconsistent systems); we will show that for any appropriate relaxation parameter, the relaxed algorithm will converge to the solution of minimal norm.

The relaxed distributed Kaczmarz algorithm for tree based equations is as follows. Choose a relaxation parameter $\omega > 0$ (generally, we will require $\omega \in (0, 2)$, though see Sect. 3 for further discussion). At each node w during the dispersion stage of iteration n, the Kaczmarz update becomes

$$\mathbf{x}_w^{(n)} = \mathbf{x}_v^{(n)} + \omega \frac{r_w(\mathbf{x}_v^{(n)})}{\|\mathbf{a}_w\|^2} \mathbf{a}_w. \tag{9}$$

We suppress the dependence of $\mathbf{x}_v^{(n)}$ on ω, but we will consider the limit

$$\mathbf{x}(\omega) := \lim_{n\to\infty} \mathbf{x}^{(n)}$$

which (in general) depends on ω. We will prove in Theorem 2.2 that when the system of equations is consistent, then this limit exists and is in fact independent of ω.

As in Eqs. (1) and (2), we use P_v and Q_v to denote the linear and affine projections, respectively. We will need the fact that Q_v is Lipschitz with constant 1:

$$\|Q_v \mathbf{z}_1 - Q_v \mathbf{z}_2\| \leq \|\mathbf{z}_1 - \mathbf{z}_2\|.$$

The relaxed Kaczmarz update in Eq. (9) can be expressed as

$$\mathbf{x}_u^{(n)} = [(1-\omega)I + \omega Q_u]\mathbf{x}_v^{(n)} =: Q_u^{\omega}\mathbf{x}_v^{(n)}.$$

Thus, the estimate $\mathbf{x}_\ell^{(n)}$ of the solution at leaf ℓ, given the solution estimate $\mathbf{x}^{(n)}$ as input at the root r, is

$$\mathbf{x}_\ell^{(n)} = Q_{\ell,p_\ell}^{\omega} \cdots Q_{\ell,2}^{\omega} Q_{\ell,1}^{\omega} \mathbf{x}^{(n)} =: \mathcal{Q}_\ell^{\omega}\mathbf{x}^{(n)}. \tag{10}$$

We can now write the full update, with both dispersion and pooling stages, of the relaxed Kaczmarz algorithm as

$$\mathbf{x}^{(n+1)} = \sum_{\ell\in\mathcal{L}} w(\ell,r)\mathcal{Q}_\ell^{\omega}\mathbf{x}^{(n)} =: \mathcal{Q}^{\omega}\mathbf{x}^{(n)}. \tag{11}$$

We note that, as above, each Q_v^{ω} is a Lipschitz map with constant 1 whenever $0 < \omega < 1$, but in fact, since $Q_v\mathbf{z}_1 - Q_v\mathbf{z}_2 = P_v\mathbf{z}_1 - P_v\mathbf{z}_2$, we have that Q_v^{ω} is Lipschitz with constant 1 whenever $0 < \omega < 2$. Moreover, as $\sum_{\ell\in\mathcal{L}} w(\ell,r) = 1$, we obtain:

Lemma 2 *For* $0 < \omega < 2$, $\mathcal{Q}_\ell^{\omega}$ *and* $\mathcal{Q}^{\omega}$ *are Lipschitz with constant* 1.

We note that the mappings $Q_{(\cdot)}^{(\cdot)}, \mathcal{Q}_{(\cdot)}^{(\cdot)}$ are affine transformations; we also have use for the analogous linear transformations. Similar to Eqs. (10) and (11), we write

$$\begin{aligned}
P_v^{\omega} &:= (1-\omega)I + \omega P_v;\\
\mathcal{P}_\ell^{\omega} &:= P_{\ell,p_\ell}^{\omega} \cdots P_{\ell,2}^{\omega} P_{\ell,1}^{\omega};\\
\mathcal{P}^{\omega} &:= \sum_{\ell\in\mathcal{L}} w(\ell,r)\mathcal{P}_\ell^{\omega}.
\end{aligned}$$

Theorem 2.2 *If the system of equations given by* $A\mathbf{x} = \mathbf{b}$ *is consistent, then for any* $0 < \omega < 2$, *the sequence of estimates* $\mathbf{x}^{(n)}$ *as given in Eq. (11) converges* *to the solution* $\mathbf{x}^M$ *of minimal norm as given by (3), provided the initial estimate* $\mathbf{x}^{(0)} \in \mathcal{R}(A^*)$.

We shall prove Theorem 2.2 using a sequence of lemmas. We follow the argument as presented in Natterer [21], adapting the lemmas as necessary. For completeness, we will state (without proof) the lemmas that we will use unaltered from [21]. (See also Yosida [37].)

Lemma 3 ([21, Lemma V.3.1]) *Let* T *be a linear map on a Hilbert space* H *with* $\|T\| \leq 1$. *Then,*

$$H = \mathcal{N}(I - T) \oplus \overline{\mathcal{R}(I - T)}.$$

Lemma 4 ([21, Lemma V.3.2]) *Suppose* $\{\mathbf{z}_k\}$ *is a sequence in* $\mathbb{C}^d$ *such that for any leaf* $\ell \in \mathcal{L}$,

$$\|\mathbf{z}_k\| \leq 1 \text{ and } \lim_{k\to\infty} \|\mathcal{P}_\ell^\omega \mathbf{z}_k\| = 1.$$

Then for $0 < \omega < 2$, *we have*

$$\lim_{k\to\infty} (I - \mathcal{P}_\ell^\omega)\mathbf{z}_k = 0.$$

Lemma 5 *Suppose* $\{\mathbf{z}_k\}$ *is a sequence in* $\mathbb{C}^d$ *such that*

$$\|\mathbf{z}_k\| \leq 1 \text{ and } \lim_{k\to\infty} \|\mathcal{P}^\omega \mathbf{z}_k\| = 1,$$

then for $0 < \omega < 2$, *we have*

$$\lim_{k\to\infty} (I - \mathcal{P}^\omega)\mathbf{z}_k = 0.$$

Proof Note that

$$\left(I - \mathcal{P}^\omega\right)\mathbf{z}_k = \sum_{\ell\in\mathcal{L}} w(\ell, r)\left(I - \mathcal{P}_\ell^\omega\right)\mathbf{z}_k,$$

so it is sufficient to show that the hypotheses of Lemma 4 are satisfied. Since $\|\mathcal{P}_\ell^\omega \mathbf{z}_k\| \leq 1$ and $\sum_\ell w(\ell, r) = 1$, we have

$$1 = \lim_{k\to\infty} \|\mathcal{P}^\omega \mathbf{z}_k\| \leq \lim_{k\to\infty} \sum_{\ell\in\mathcal{L}} w(\ell, r)\|\mathcal{P}_\ell^\omega \mathbf{z}_k\| \leq 1.$$

Thus, we must have $\lim \|\mathcal{P}_\ell^\omega \mathbf{z}_k\| = 1$ for every $\ell \in \mathcal{L}$. □

Lemma 6 *For* $0 < \omega < 2$, *we have*

$$\mathcal{N}(I - \mathcal{P}^{\omega}) = \bigcap_{v\text{node}} \mathcal{N}(I - P_v). \tag{12}$$

Proof Suppose $P_v\mathbf{z} = \mathbf{z}$ for every node v. Then

$$\mathcal{P}^{\omega}\mathbf{z} = \sum_{\ell\in\mathcal{L}} w(\ell, r) P^{\omega}_{\ell,p_\ell} \dots P^{\omega}_{\ell,1}\mathbf{z} = \sum_{\ell\in\mathcal{L}} w(\ell, r)\mathbf{z} = \mathbf{z}$$

thus the left containment follows.

Conversely, suppose that $\mathcal{P}^{\omega}\mathbf{z} = \mathbf{z}$. Again, we obtain

$$\|\mathbf{z}\| = \|\mathcal{P}^{\omega}\mathbf{z}\| \leq \sum_{\ell\in\mathcal{L}} w(\ell, r)\| P^{\omega}_{\ell,p_\ell} \cdots P^{\omega}_{\ell,1}\mathbf{z}\| \leq \|\mathbf{z}\|$$

which implies that

$$P^{\omega}_{\ell,p_\ell} \cdots P^{\omega}_{\ell,1}\mathbf{z} = \mathbf{z}$$

for every leaf ℓ. Hence, for every ℓ, and every $j = 1, \dots, p_\ell$, $P^{\omega}_{\ell,j}\mathbf{z} = \mathbf{z}$. □

Lemma 7 ([21, Lemma V.3.5]) *For* $0 < \omega < 2$, $(\mathcal{P}^{\omega})^k$ *converges strongly, as* $k \to \infty$, *to the orthogonal projection onto*

$$\bigcap_{v\text{node}} \mathcal{N}(I - P_v) = \mathcal{N}(A).$$

The proof is identical to that in [21], using Lemmas 3, 5, and 6.

Proof of Theorem 2.2 Let $\mathbf{y}$ be any solution to the system of equations. We claim that for any $\mathbf{z}$,

$$Q^{\omega}\mathbf{z} = \mathcal{P}^{\omega}(\mathbf{z} - \mathbf{y}) + \mathbf{y}. \tag{13}$$

Indeed, for any nodes v and w, and consequently for any leaf ℓ, we have

$$\begin{aligned}
& Q^{\omega}_v\mathbf{z} = \mathbf{y} + P^{\omega}_v(\mathbf{z} - \mathbf{y}) \\
\Rightarrow\quad & Q^{\omega}_w Q^{\omega}_v\mathbf{z} = \mathbf{y} + P^{\omega}_w P^{\omega}_v(\mathbf{z} - \mathbf{y}) \\
\Rightarrow\quad & Q^{\omega}_\ell\mathbf{z} = \mathbf{y} + \mathcal{P}^{\omega}_\ell(\mathbf{z} - \mathbf{y}) \\
\Rightarrow\quad & \sum_{\ell\in\mathcal{L}} w(\ell, r) Q^{\omega}_\ell\mathbf{z} = \sum_{\ell\in\mathcal{L}} w(\ell, r)\left(\mathbf{y} + \mathcal{P}^{\omega}_\ell(\mathbf{z} - \mathbf{y})\right),
\end{aligned}$$

which demonstrates Eq. (13).

Therefore, by Lemma 7, we have that for any $\mathbf{z}$,

$$\left(Q^{\omega}\right)^k \mathbf{z} \to \mathbf{y} + Pr(\mathbf{z} - \mathbf{y}),$$

as $k \to \infty$, where Pr is the projection onto $\mathcal{N}(A)$. If $\mathbf{z} \in \mathcal{R}(A^*)$, we have that $\mathbf{y} + Pr(\mathbf{z} - \mathbf{y})$ is the unique solution to the system of equations that is in $\mathcal{R}(A^*)$, and hence is the solution of minimal norm. □

We can see that for $\mathbf{z} \in \mathcal{R}(A^*)$, the convergence rate of $(Q^{\omega})^k \mathbf{z} \to \mathbf{x}^M$ is linear, but we will formalize this in the next subsection (Corollary 2.1).

2.3 Inconsistent Equations

We now consider the case of inconsistent systems of equations. For this purpose, we must consider the relaxed version of the algorithm, as in the previous subsection. Again, we assume $0 < \omega < 2$ and consider the limit

$$\lim_{n \to \infty} \mathbf{x}^{(n)} = \mathbf{x}(\omega).$$

We will prove in Theorem 2.3 and Corollary 2.1 that the limit exists, but unlike in the case of consistent systems, the limit will depend on ω. Moreover, we will prove in Theorem 2.4 that the limit

$$\lim_{\omega \to 0} \mathbf{x}(\omega) = \mathbf{x}^{LS}$$

exists, and $\mathbf{x}^{LS}$ is a generalized solution which minimizes a weighted least-squares norm. We follow the presentation of the analogous results for the classical Kaczmarz algorithm as presented in [21]. Indeed, we will proceed by analyzing the distributed Kaczmarz algorithm using the ideas from Successive Over-Relaxation (SOR). We need to follow the updates as they disperse through the tree, and also how the updates are pooled back at the root, and so we define the following quantities.

We begin with reindexing the equations, which are currently indexed by the nodes as $a_v^* \mathbf{x} = b_v$. As before, for each leaf ℓ, we consider the path from the root r to the leaf ℓ, and index the corresponding equations as

$$\mathbf{a}_{\ell,1}^* \mathbf{x} = b_{\ell,1}, \ \ldots, \mathbf{a}_{\ell,p_\ell}^* \mathbf{x} = b_{\ell,p_\ell}.$$

For each leaf ℓ, we can define

$$\mathcal{A}_\ell = \begin{pmatrix} \mathbf{a}_{\ell,1}^* \\ \mathbf{a}_{\ell,2}^* \\ \vdots \\ \mathbf{a}_{\ell,p_\ell}^* \end{pmatrix}, \quad \mathbf{b}_\ell = \begin{pmatrix} b_{\ell,1} \\ b_{\ell,2} \\ \vdots \\ b_{\ell,p_\ell} \end{pmatrix}, \quad \mathcal{D}_\ell = \begin{pmatrix} \mathbf{a}_{\ell,1}^* \mathbf{a}_{\ell,1} & 0 & \ldots & 0 \\ 0 & \mathbf{a}_{\ell,2}^* \mathbf{a}_{\ell,2} & \ldots & 0 \\ \vdots & \vdots & \ddots & \vdots \\ 0 & 0 & \ldots & \mathbf{a}_{\ell,p_\ell}^* \mathbf{a}_{\ell,p_\ell} \end{pmatrix}$$

and

$$\mathcal{L}_\ell = \begin{pmatrix} 0 & 0 & \dots & 0 & 0 \\ \mathbf{a}_{\ell,2}^*\mathbf{a}_{\ell,1} & 0 & \dots & 0 & 0 \\ \vdots & \vdots & \ddots & \vdots & \vdots \\ \mathbf{a}_{\ell,p_\ell}^*\mathbf{a}_{\ell,1} & \mathbf{a}_{\ell,p_\ell}^*\mathbf{a}_{\ell,2} & \dots & \mathbf{a}_{\ell,p_\ell}^*\mathbf{a}_{\ell,p_\ell-1} & 0. \end{pmatrix}$$

Then from input $\mathbf{x}^{(n)}$ at the root of the tree, the approximation at leaf ℓ after the dispersion stage in iteration n is given by

$$\mathbf{x}_\ell^{(n)} = Q_\ell^\omega \mathbf{x}^{(n)} = \mathbf{x}^{(n)} + \sum_{j=1}^{p_\ell} u_j \mathbf{a}_{\ell,j} = \mathbf{x}^{(n)} + \mathcal{A}_\ell^* \mathbf{u},$$

where

$$\mathbf{u} := \left(u_1 \dots u_{p_\ell}\right)^T = \omega \left(\mathcal{D}_\ell + \omega\mathcal{L}_\ell\right)^{-1} \left(\mathbf{b}_\ell - \mathcal{A}_\ell \mathbf{x}^{(n)}\right).$$

Therefore, we can write

$$\begin{aligned} \mathbf{x}_\ell^{(n)} &= \mathbf{x}^{(n)} + \omega\mathcal{A}_\ell^* \left(\mathcal{D}_\ell + \omega\mathcal{L}_\ell\right)^{-1} \left(\mathbf{b}_\ell - \mathcal{A}_\ell \mathbf{x}^{(n)}\right) \\ &= \left(I - \omega\mathcal{A}_\ell^* \left(\mathcal{D}_\ell + \omega\mathcal{L}_\ell\right)^{-1} \mathcal{A}_\ell\right) \mathbf{x}^{(n)} + \omega\mathcal{A}_\ell^* \left(\mathcal{D}_\ell + \omega\mathcal{L}_\ell\right)^{-1} \mathbf{b}_\ell. \end{aligned}$$

Combining these approximations back at the root yields:

$$\begin{aligned} \mathbf{x}^{(n+1)} &= \sum_{\ell\in\mathcal{L}} w(\ell, r)\mathbf{x}_\ell^{(n)} \\ &= \sum_{\ell\in\mathcal{L}} w(\ell, r) \left(I - \omega\mathcal{A}_\ell^* \left(\mathcal{D}_\ell + \omega\mathcal{L}_\ell\right)^{-1} \mathcal{A}_\ell\right) \mathbf{x}^{(n)} + \omega \sum_{\ell\in\mathcal{L}} w(\ell, r)\mathcal{A}_\ell^* \left(\mathcal{D}_\ell + \omega\mathcal{L}_\ell\right)^{-1} \mathbf{b}_\ell \\ &= \left(I - \omega \sum_{\ell\in\mathcal{L}} w(\ell, r)\mathcal{A}_\ell^* \left(\mathcal{D}_\ell + \omega\mathcal{L}_\ell\right)^{-1} \mathcal{A}_\ell\right) \mathbf{x}^{(n)} + \omega \sum_{\ell\in\mathcal{L}} w(\ell, r)\mathcal{A}_\ell^* \left(\mathcal{D}_\ell + \omega\mathcal{L}_\ell\right)^{-1} \mathbf{b}_\ell. \end{aligned} \tag{14}$$

We write

$$\mathbf{x}^{(n+1)} = \sum_{\ell\in\mathcal{L}} w(\ell, r)\mathcal{B}_\ell^\omega \mathbf{x}^{(n)} + \sum_{\ell\in\mathcal{L}} w(\ell, r)\mathbf{b}_\ell^\omega,$$

where

$$\mathcal{B}_\ell^\omega := I - \omega\mathcal{A}_\ell^* \left(\mathcal{D}_\ell + \omega\mathcal{L}_\ell\right)^{-1} \mathcal{A}_\ell; \qquad \mathbf{b}_\ell^\omega := \omega\mathcal{A}_\ell^* \left(\mathcal{D}_\ell + \omega\mathcal{L}_\ell\right)^{-1} \mathbf{b}_\ell. \tag{15}$$

Written in this form, for each leaf ℓ, the input at the root undergoes the linearly ordered Kaczmarz algorithm. So, if the input at the root is $\mathbf{x}^{(n)}$, then the estimate at leaf ℓ is

$$\mathbf{x}_\ell^{(n)} = Q_\ell^\omega \mathbf{x}^{(n)} = \mathcal{B}_\ell^\omega \mathbf{x}^{(n)} + \mathbf{b}_\ell^\omega.$$

As we shall see, for each leaf ℓ and $\omega \in (0, 2)$, $\mathcal{B}_\ell^\omega$ has operator norm bounded by 1, and the eigenvalues are either 1 or strictly less than 1 in magnitude. We state these formally in Lemma 8.

We enumerate the leaves of the tree as $\ell_1, \ldots, \ell_t$, and write

$$\mathcal{A} = \begin{pmatrix} \mathcal{A}_{\ell_1} \\ \vdots \\ \mathcal{A}_{\ell_t} \end{pmatrix} \qquad \mathbf{b} = \begin{pmatrix} \mathbf{b}_{\ell_1} \\ \vdots \\ \mathbf{b}_{\ell_t} \end{pmatrix}$$

The system of equations $A\mathbf{x} = \mathbf{b}$ becomes

$$\mathcal{A}\mathbf{x} = \mathbf{b}, \tag{16}$$

where many of the equations are now repeated in Eq. (16). However, we have $\mathcal{N}(\mathcal{A}) = \mathcal{N}(A)$ and $\mathcal{R}(\mathcal{A}^*) = \mathcal{R}(A^*)$.

We also write

$$\mathcal{D} := \begin{pmatrix} \mathcal{D}_{\ell_1} & 0 & \ldots & 0 \\ 0 & \mathcal{D}_{\ell_2} & \ldots & 0 \\ \vdots & \vdots & \ddots & \vdots \\ 0 & 0 & \ldots & \mathcal{D}_{\ell_t} \end{pmatrix} \qquad \mathcal{L} = \begin{pmatrix} \mathcal{L}_{\ell_1} & 0 & \ldots & 0 \\ 0 & \mathcal{L}_{\ell_2} & \ldots & 0 \\ \vdots & \vdots & \ddots & \vdots \\ 0 & 0 & \ldots & \mathcal{L}_{\ell_t} \end{pmatrix}$$

so

$$(\mathcal{D} + \omega\mathcal{L})^{-1} = \begin{pmatrix} \left(\mathcal{D}_{\ell_1} + \omega\mathcal{L}_{\ell_1}\right)^{-1} & 0 & \ldots & 0 \\ 0 & \left(\mathcal{D}_{\ell_2} + \omega\mathcal{L}_{\ell_2}\right)^{-1} & \ldots & 0 \\ \vdots & \vdots & \ddots & \vdots \\ 0 & 0 & \ldots & \left(\mathcal{D}_{\ell_t} + \omega\mathcal{L}_{\ell_t}\right)^{-1} \end{pmatrix}.$$

We also define

$$\mathcal{W} = \begin{pmatrix} w(\ell_1, r)I_{p_{\ell_1}} & 0 & \dots & 0 \\ 0 & w(\ell_2, r)I_{p_{\ell_2}} & \dots & 0 \\ \vdots & \vdots & \ddots & \vdots \\ 0 & 0 & \dots & w(\ell_t, r)I_{p_{\ell_t}} \end{pmatrix}.$$

Note that since $\mathcal{D}+\omega\mathcal{L}$ and $\mathcal{W}$ are block matrices with blocks of the same size, and in $\mathcal{W}$ the blocks are scalar multiples of the identity, we have that the two matrices commute

$$(\mathcal{D}+\omega\mathcal{L})^{-1}\mathcal{W} = \mathcal{W}(\mathcal{D}+\omega\mathcal{L})^{-1} = \mathcal{W}^{1/2}(\mathcal{D}+\omega\mathcal{L})^{-1}\mathcal{W}^{1/2}.$$

We can therefore write Eq. (14) as

$$\begin{aligned} \mathbf{x}^{(n+1)} &= \left(I - \omega\mathcal{A}^*(\mathcal{D}+\omega\mathcal{L})^{-1}\mathcal{W}\mathcal{A}\right)\mathbf{x}^{(n)} + \omega\mathcal{A}^*(\mathcal{D}+\omega\mathcal{L})^{-1}\mathcal{W}\mathbf{b}. \\ &:= \mathcal{B}^{\omega}\mathbf{x}^{(n)} + \mathbf{b}^{\omega}. \end{aligned}$$

Note that $\mathcal{R}(\mathcal{A}^*)$ is an invariant subspace for $\mathcal{B}^{\omega}$, and that $\mathbf{b}^{\omega} \in \mathcal{R}(\mathcal{A}^*)$. We let $\widehat{\mathcal{B}}^{\omega}$ denote the restriction of $\mathcal{B}^{\omega}$ to the subspace $\mathcal{R}(\mathcal{A}^*)$. As we shall see, provided the input $\mathbf{x}^0 \in \mathcal{R}(\mathcal{A}^*)$, the sequence $\mathbf{x}^k$ converges. In fact, we will show that the transformation $\widehat{\mathcal{B}}^{\omega}$ is a contraction, and since $\mathbf{b}^{\omega} \in \mathcal{R}(\mathcal{A}^*)$, then the mapping

$$\mathbf{z} \mapsto \widehat{\mathcal{B}}^{\omega}\mathbf{z} + \mathbf{b}^{\omega}$$

has a unique fixed point within $\mathcal{R}(\mathcal{A}^*)$. We shall do so via a series of lemmas.

Lemma 8 *For each leaf ℓ and for $\omega \in (0, 2)$, $\mathcal{B}^{\omega}_{\ell}$ is Lipschitz continuous with constant at most* 1 *(i.e. it has operator norm at most* 1*). Consequently, $\widehat{\mathcal{B}}^{\omega}$ is also Lipschitz continuous with constant at most* 1.

Moreover, for each leaf ℓ and $\omega \in (0, 2)$, if λ is an eigenvalue of $\mathcal{B}^{\omega}_{\ell}$ with $|\lambda| = 1$, then $\lambda = 1$. Consequently, any eigenvalue $\lambda \neq 1$ has the property $|\lambda| < 1$.

Proof For input $\mathbf{z}_i$, we have that

$$\mathcal{Q}^{\omega}_{\ell}\mathbf{z}_i = \mathcal{B}^{\omega}_{\ell}\mathbf{z}_i + \mathbf{b}^{\omega}_{\ell},$$

hence

$$\|\mathcal{B}^{\omega}_{\ell}\mathbf{z}_1 - \mathcal{B}^{\omega}_{\ell}\mathbf{z}_2\| = \|\mathcal{Q}^{\omega}_{\ell}\mathbf{z}_1 - \mathcal{Q}^{\omega}_{\ell}\mathbf{z}_2\| \le \|\mathbf{z}_1 - \mathbf{z}_2\|$$

by Lemma 2. Since $\mathcal{B}^{\omega}$ is a convex combination of the $\mathcal{B}^{\omega}_{\ell}$, it also has Lipschitz constant at most 1. The last conclusion follows from [21, Lemma V.3.9]. □

Theorem 2.3 *The spectral radius of* $\widehat{\mathcal{B}}^{\omega}$ *is strictly less than* 1.

Proof For each leaf ℓ, Lemma 8 implies that

$$\|\mathcal{B}_{\ell}^{\omega}\| \leq 1, \qquad |\langle \mathcal{B}_{\ell}^{\omega}\mathbf{v}, \mathbf{v}\rangle| \leq \|\mathbf{v}\|^2.$$

Let λ be an eigenvalue for $\widehat{\mathcal{B}}^{\omega}$. We must have $\lambda \neq 1$; if it were not so, then there exists a nonzero $\mathbf{z} \in \mathcal{R}(\mathcal{A}^*)$ with $\widehat{\mathcal{B}}^{\omega}\mathbf{z} = \mathbf{z}$. However, by Lemma 6 we must have $\mathbf{z} \in \mathcal{N}(A) = \mathcal{N}(S)$ which is a contradiction. Let $\mathbf{v}$ be a unit norm eigenvector for λ. We have

$$|\lambda| = |\langle \widehat{\mathcal{B}}^{\omega}\mathbf{v}, \mathbf{v}\rangle| \leq \sum_{\ell \in \mathcal{L}} w(\ell, r)|\langle \mathcal{B}_{\ell}^{\omega}\mathbf{v}, \mathbf{v}\rangle| \leq 1.$$

Now suppose that $|\lambda| = 1$, then we similarly obtain

$$\lambda = \sum_{\ell \in \mathcal{L}} \langle \mathcal{B}_{\ell}^{\omega}\mathbf{v}, \mathbf{v}\rangle$$

from which we deduce that the argument of the complex number $\langle \mathcal{B}_{\ell}^{\omega}\mathbf{v}, \mathbf{v}\rangle$ is independent of the leaf ℓ. Therefore, we must have for every leaf ℓ

$$\langle \mathcal{B}_{\ell}^{\omega}\mathbf{v}, \mathbf{v}\rangle = \lambda. \tag{17}$$

However, we know by the Cauchy–Schwarz inequality that equality in Eq. (17) can only occur when $(\mathbf{v}, \lambda)$ is an eigenvector/eigenvalue pair for $\mathcal{B}_{\ell}^{\omega}$. However, Lemma 8 implies that none of the leaves ℓ has the property that λ is an eigenvalue, so we have arrived at a contradiction. □

Corollary 2.1 *For* $\omega \in (0, 2)$ *and for any initial input* $\mathbf{x}^{(0)} \in \mathcal{R}(\mathcal{A}^*)$, *we have that the sequence given by*

$$\mathbf{x}^{(n+1)} = \widehat{\mathcal{B}}^{\omega}\mathbf{x}^{(n)} + \mathbf{b}^{\omega} \tag{18}$$

converges to a unique point in $\mathcal{R}(\mathcal{A}^*)$, *independent of* $\mathbf{x}^{(0)}$, *and the convergence rate is linear.*

The following can be found in [21, Theorem IV.1.1]:

Lemma 9 *For each* $\omega \in (0, 2)$, *let*

$$\mathbf{x}(\omega) = \lim_{n\to\infty} \mathbf{x}^{(n)},$$

where $\mathbf{x}^{(n)}$ *are as in Eq. (18). Then,* $\mathbf{x}(\omega)$ *is the unique vector that satisfies the conditions*

$$\mathcal{A}^* (\mathcal{D} + \omega\mathcal{L})^{-1} \mathcal{W} (\mathbf{b} - \mathcal{A}\mathbf{z}) = \mathbf{0}; \qquad \mathbf{z} \in \mathcal{R}(\mathcal{A}^*). \tag{19}$$

Theorem 2.4 *For each* $\omega \in (0, 2)$, *let*

$$\mathbf{x}(\omega) = \lim_{n\to\infty} \mathbf{x}^n$$

as in Eq. (18). Then,

$$\lim_{\omega\to 0} \mathbf{x}(\omega) = \mathbf{x}^{LS},$$

where $\mathbf{x}^{LS}$ *minimizes the functional*

$$\mathbf{z} \mapsto \langle \mathcal{D}^{-1}\mathcal{W}(\mathbf{b} - \mathcal{A}\mathbf{z}), (\mathbf{b} - \mathcal{A}\mathbf{z})\rangle. \tag{20}$$

Proof Let $\mathbf{x}^{LS}$ be the unique vector that satisfies the conditions

$$\mathcal{A}^*\mathcal{D}^{-1}\mathcal{W}\left(\mathbf{b} - \mathcal{A}\mathbf{x}^{LS}\right) = \mathbf{0}; \quad \mathbf{x}^{LS} \in \mathcal{R}(\mathcal{A}^*). \tag{21}$$

We have that $\mathbf{x}(\omega)$, as the unique solution of Eq. (19) and $\mathbf{x}^{LS}$, as the unique solution of Eq. (21), satisfy

$$\mathbf{x}(\omega) = \mathbf{x}^{LS} + O(\omega).$$

Indeed, this follows from the fact that $(\mathcal{D} + \omega\mathcal{L})^{-1} \to \mathcal{D}^{-1}$ as $\omega \to 0$, together with the fact that $\mathbf{x}(\omega), \mathbf{x}^{LS} \in \mathcal{R}(\mathcal{A}^*)$. □

We can re-write Eq. (20) in the following way:

$$\mathbf{z} \mapsto \langle D^{-1}V(\mathbf{b} - A\mathbf{z}), (\mathbf{b} - A\mathbf{z})\rangle,$$

where D is the diagonal matrix with entries given by $\|\mathbf{a}_v\|^2$, and V is the diagonal matrix whose entry for node v is given by

$$V_{vv} = \sum_{\substack{\ell\in\mathcal{L}\\ \ell\prec v}} w(\ell, r).$$

2.4 Distributed Solutions

For each node v in the tree, the sequence of approximations $\mathbf{x}_v^{(n)}$ and $\mathbf{y}_v^{(n)}$ will have a limit, i.e. the following limits exist:

$$\lim_{n\to\infty} \mathbf{x}_v^{(n)} = \mathbf{x}_v; \qquad \lim_{n\to\infty} \mathbf{y}_v^{(n)} = \mathbf{y}_v. \tag{22}$$

In the relaxed case, these limits may depend on the relaxation parameter ω; if so we will denote this dependence by $\mathbf{x}_v(\omega)$ and $\mathbf{y}_v(\omega)$.

Corollary 2.2 *If the system of equations $A\mathbf{x} = \mathbf{b}$ is consistent, then for every node v and every $\omega \in (0, 2)$, the limits $\mathbf{x}_v$ and $\mathbf{y}_v$ as in Eq. (22) equal $\mathbf{x}^M$, the solution of minimal norm.*

Proof We have by Theorem 2.2 that $\mathbf{x}(\omega) = \mathbf{x}^M$ for every $\omega \in (0, 2)$. For a node v, let the path from the root r to v be denoted by $r = (v, 1), \ldots, (v, p_v) = v$, where $p_v - 1$ is the length of the path. Then, we have that

$$\lim_{n\to\infty} \mathbf{x}_v^{(n)} = \lim_{n\to\infty} Q^{\omega}_{v,p_v} \cdots Q^{\omega}_{v,1}\mathbf{x}^{(n)} = Q^{\omega}_{v,p_v} \cdots Q^{\omega}_{v,1}\mathbf{x}(\omega) = \mathbf{x}^M.$$

This holds as a consequence of the fact that any solution to the system of equations is invariant under $Q^{\omega}_{(\cdot)}$.

Since we have that $\mathbf{y}_v^{(n)}$ is a convex combination of the vectors $\mathbf{x}_\ell^{(n)}$, which all converge to $\mathbf{x}(\omega)$, we have that $\mathbf{y}_v = \mathbf{x}^M$ also. □

Corollary 2.3 *If the system of equations $A\mathbf{x} = \mathbf{b}$ is inconsistent, then for every node v and every $\omega \in (0, 2)$, the limits $\mathbf{x}_v$ and $\mathbf{y}_v$ as in Eq. (22) exist and depend on ω. Moreover, we have*

$$\lim_{\omega\to 0} \mathbf{x}_v(\omega) = \mathbf{x}^{LS} \qquad \lim_{\omega\to 0} \mathbf{y}_v(\omega) = \mathbf{x}^{LS},$$

where $\mathbf{x}^{LS}$ is the vector as in Theorem 2.4.

Proof We apply the SOR analysis of $\mathbf{x}_v^{(n)} = Q^{\omega}_{(v,p_v)} \cdots Q^{\omega}_{(v,1)}\mathbf{x}^{(n)}$ with input $\mathbf{x}^{(n)}$ to obtain

$$\mathbf{x}_v^{(n)} = \mathcal{B}_v^{\omega}\mathbf{x}^{(n)} + \mathbf{b}_v^{\omega},$$

where $\mathcal{B}_v^{\omega}$ and $\mathbf{b}_v^{\omega}$ are analogous to those in Eq. (15). Taking limits on n, we obtain

$$\mathbf{x}_v(\omega) = \mathcal{B}_v^{\omega}\mathbf{x}(\omega) + \mathbf{b}_v^{\omega}.$$

Since, as $\omega \to 0$, we have that $\mathcal{B}_v^{\omega} \to I$, $\mathbf{b}_v^{\omega} \to 0$, and $\mathbf{x}(\omega) \to \mathbf{x}^{LS}$, we obtain

$$\lim_{\omega\to 0} \mathbf{x}_v(\omega) = \mathbf{x}^{LS}.$$

As previously, $\mathbf{y}_v(\omega)$ is a convex combination of $\mathbf{x}_\ell(\omega)$, so $\mathbf{y}_v(\omega) \to \mathbf{x}^{LS}$ as $\omega \to 0$ also. □

2.5 Error Analysis

We consider the question of how errors propagate through the iterations of the dispersion and pooling stages. We model errors as additive; the sources of errors could be machine errors, transmission errors, errors from compression to reduce communication complexity, etc. Additive errors then take on the form

$$\mathbf{x}_{v,e}^{(n)} = \mathbf{x}_v^{(n)} + \varepsilon_v^{(n)}; \qquad \mathbf{y}_{v,e}^{(n)} = \mathbf{y}_v^{(n)} + \delta_v^{(n)} \tag{23}$$

Here, $\mathbf{x}_{v,e}^{(n)}$ and $\mathbf{y}_{v,e}^{(n)}$ are the error-riddled estimates which are passed to the successor (or predecessor) nodes in the dispersion (or pooling) stage, respectively, with additive errors $\varepsilon_v^{(n)}$ and $\delta_v^{(n)}$. Measurement errors, meaning errors in $\mathbf{b}$, are considered in [4, 10].

We trace the errors during the dispersion stage as follows: for node v on a path between the root r and leaf ℓ, and the path parameterized by $r = (\ell, 1), \ldots, (\ell, p_\ell) = \ell$, suppose that $v = (\ell, k)$. Then, the error introduced at node v (with errors introduced at no other node) results in the estimate

$$\begin{aligned} \mathbf{x}_{\ell,e}^{(n)} &= Q_{\ell,p_\ell}^{\omega} \cdots Q_{\ell,k+1}^{\omega}(\mathbf{x}_v^{(n)} + \varepsilon_v^{(n)}) \\ &= Q_{\ell,p_\ell}^{\omega} \cdots Q_{\ell,k+1}^{\omega}(\mathbf{x}_v^{(n)}) + \widetilde{\varepsilon}_{v,\ell}^{(n)} \\ &= Q_\ell^{\omega}(\mathbf{x}^{(n)}) + \widetilde{\varepsilon}_{v,\ell}^{(n)}. \end{aligned} \tag{24}$$

Equation (24) follows for some $\widetilde{e}_v^{(n)}$ since the $Q_{(\cdot)}^{\omega}$ are affine transformations. We have that

$$\|\widetilde{\varepsilon}_v^{(n)}\| = \|Q_{\ell,p_\ell}^{\omega} \cdots Q_{\ell,k+1}^{\omega}(\mathbf{x}_v^{(n)} + \varepsilon_v^{(n)}) - Q_{\ell,p_\ell}^{\omega} \cdots Q_{\ell,k+1}^{\omega}(\mathbf{x}_v^{(n)})\| \le \|\varepsilon_v^{(n)}\|$$

since the $Q_{(\cdot)}^{\omega}$ have Lipschitz constant 1. The additive errors $\delta_v^{(n)}$ simply sum in the pooling stage, and thus we calculate the total errors from iteration n to iteration $n+1$.

Lemma 10 *Suppose we have additive errors as in Eq. (23) introduced in iteration n. Suppose no errors were introduced in previous iterations. Then the estimate after iteration n is*

$$\mathbf{x}_e^{(n+1)} = \mathbf{x}^{(n+1)} + \sum_v \sum_{\substack{\ell \in \mathcal{L} \\ \ell \prec v}} w(\ell, r)\widetilde{e}_v^{(n)} + \sum_v w(v, r)\delta_{v,e}^{(n)}.$$

The magnitude of the error is bounded by

$$\|\mathbf{x}_e^{(n+1)} - \mathbf{x}^{(n+1)}\| \le K\, max\, \{\|\varepsilon_v^{(n)}\|, \|\delta_v^{(n)}\|\},$$

where K is 2 times the depth of the tree.

We write

$$E^{(n)} = \sum_{v} \sum_{\substack{\ell \in \mathcal{L} \\ \ell \prec v}} w(\ell, r)\widetilde{e}_v^{(n)} + \sum_{v} w(v, r)\delta_{v,e}^{(n)}.$$

Theorem 2.5 *If the additive errors in Eq. (23) are uniformly bounded by M, and the system of equations* $A\mathbf{x} = \mathbf{b}$ *has a unique solution, then the sequence of approximations* $\{\mathbf{x}_e^{(n)}\}$ *has the property that*

$$\limsup_{n\to\infty} \|\mathbf{x}(\omega) - \mathbf{x}_e^{(n)}\| \leq \frac{2KM}{1-\rho(\mathcal{B}^{\omega})}, \tag{25}$$

where K is the depth of the tree.

Proof We have

$$\mathbf{x}_e^{(n)} = \mathbf{x}^{(n)} + \sum_{k=1}^{n} \left(\mathcal{B}^{\omega}\right)^{n-k} E^{(k)}.$$

As noted previously, $\|E^{(k)}\| \leq 2KM$, and if $A\mathbf{x} = \mathbf{b}$ has a unique solution, then $\rho(\mathcal{B}^{\omega}) < 1$ (see proof of Theorem 2.1).

Thus, for any matrix norm $\|\cdot\|$ with $\rho(\mathcal{B}^{\omega}) < \|\mathcal{B}^{\omega}\|$

$$\|\mathbf{x}^{(n)} - \mathbf{x}_e^{(n)}\| \leq \sum_{k=0}^{n-1} 2KM\|\mathcal{B}^{\omega}\|^k$$

from which Eq. (25) follows. □

If the system of equations does not have a unique solution, then the mapping $\mathcal{B}^{\omega}$ has 1 as an eigenvalue, and so the parts of the errors that lie in that eigenspace accumulate. Hence, no stability result is possible in this case.

3 Implementation and Examples

For the standard Kaczmarz algorithm, it is well known that the method converges if and only if the relaxation parameter ω is in the interval $(0, 2)$. For our distributed Kaczmarz, the situation is not nearly as clear. The proofs of Theorems 2.2 and 2.4 require that $\omega \in (0, 2)$, but in numerical experiments, convergence occurred for $\omega \in (0, \Omega)$ for some $\Omega \geq 2$. The largest Ω observed was around 3.8. The precise upper limit depends on the equations themselves. In this section, we perform a preliminary analysis of the computation of Ω and the optimal ω_{opt} for a very simple setup, and give numerical results for several examples.

3.1 Examples

Example 1 We consider the matrix

$$A = \begin{pmatrix} -\sin\alpha & \cos\alpha \\ 0 & 1 \end{pmatrix}.$$

In geometric terms, the Kaczmarz method for this example corresponds to projection onto the x-axis and onto a line forming an angle α with the x-axis.

For standard Kaczmarz, the iteration matrix is

$$B_\omega = I - \omega A^*(D+\omega L)^{-1}A = \begin{pmatrix} 1-\omega\sin^2\alpha & \omega\sin\alpha\cos\alpha \\ \omega(1-\omega)\sin\alpha\cos\alpha & (1-\omega)(1-\omega\cos^2\alpha) \end{pmatrix}.$$

The eigenvalues are

$$\lambda = \left[\frac{\omega^2\cos^2\alpha}{2} + (1-\omega)\right] \pm \omega\cos\alpha\sqrt{(\omega-2)^2 - \omega^2\sin^2\alpha}.$$

For small ω, the eigenvalues are real and decreasing as a function of ω. They become complex at

$$\omega_{opt} = \frac{2}{1+\sin\alpha},$$

which is between 1 and 2. After that point, both eigenvalues have magnitude $\omega - 1$, and the spectral radius increases in a straight line. The dependence of ρ on ω is illustrated below in the left half of Fig. 3. Here $\alpha = \pi/3$, $\omega_{opt} \approx 1.0718$, $\rho_{opt} \approx 0.0718$.

As pointed out in [21], there is a strong connection between the classical Kaczmarz method and Successive Over-Relaxation (SOR). In SOR the relationship between ω and ρ shows the same type of behavior.

The example with two equations is too small to implement as distributed Kaczmarz, but we consider something similar. We project the same $\mathbf{x}^{(n)}$ onto each line, and average the result to get $\mathbf{x}^{(n+1)}$. We will refer to this as the *averaged Kaczmarz method*.

The iteration matrix is

$$B_\omega = \begin{pmatrix} 1 - \frac{\omega}{2}\sin^2\alpha & \frac{\omega}{2}\sin\alpha\cos\alpha \\ \frac{\omega}{2}\sin\alpha\cos\alpha & \frac{\omega}{2}\sin^2\alpha - \omega + 1. \end{pmatrix}$$

(continued)

Example 1 (continued)
The eigenvalues here are always real and vary linearly with ω, namely

$$\lambda_{1,2} = 1 + \frac{\omega}{2} (\pm \cos\alpha - 1) .$$

They both have the value 1 at $\omega = 0$, and are both decreasing with increasing ω. The first one reaches (-1) at

$$\Omega = \frac{4}{1 + \cos\alpha} .$$

Thus, the upper limit Ω is somewhere between 2 and 4, depending on α. In numerical experiments with the distributed Kaczmarz method for larger matrices, we have observed Ω near 4, but never above 4. We conjecture that Ω can never be larger than 4.

The minimum spectral radius occurs at $\omega_{opt} = 2$, independent of α, with $\rho_{opt} = \cos\alpha$. The dependence of ρ on ω is illustrated below in the left half of Fig. 3. In this example, the graph for the averaged Kaczmarz method consists of two line segments, with $\omega_{opt} = 2$, $\rho_{opt} = 0.5$.

Figure 2 illustrates the optimal ω for $\alpha = \pi/2$. The optimal ω for standard Kaczmarz is $\omega = 1$, with $\rho = 0$. Convergence occurs in a single step. For the averaged method, the optimal ω is 2, where again convergence occurs in a single step. The averaged method would still converge for a range of $\omega > 2$.

Numerical experiments with larger sets of equations indicate that the optimal ω for classical Kaczmarz is usually larger than 1, but of course cannot exceed 2. The optimal ω for distributed Kaczmarz is usually larger than 2, sometimes even approaching 4.

Example 2 We used a random matrix of size 8×8, with entries generated using a standard normal distribution. For the distributed Kaczmarz method, we used the 8-node graph as shown on the right in Fig. 4.

For the standard Kaczmarz method, the optimal relaxation parameter was $\omega_{opt} \approx 1.7354$, with spectral radius $\rho_{opt} \approx 0.93147$. For the distributed Kaczmarz method, the results were $\omega_{opt} \approx 3.7888$, with spectral radius $\rho_{opt} \approx 0.99087$. This is illustrated in on the right in Fig. 3.

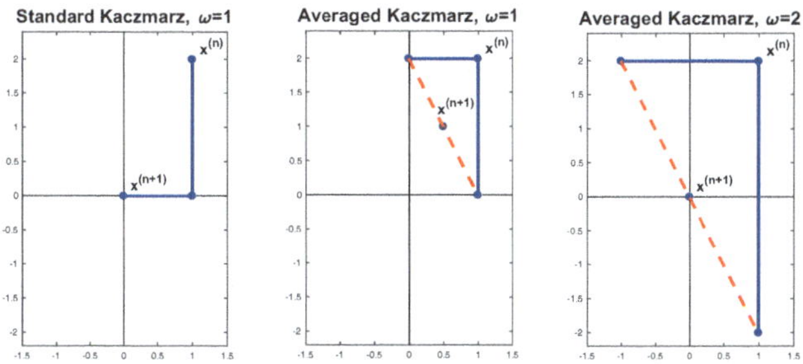

Fig. 2 Example 1 with $\alpha = \pi/2$. The pictures show one step of standard Kaczmarz with $\omega = 1$, and one step of averaged Kaczmarz for $\omega = 1$ and $\omega = 2$. This illustrates the need for a larger ω in the averaged Kaczmarz method

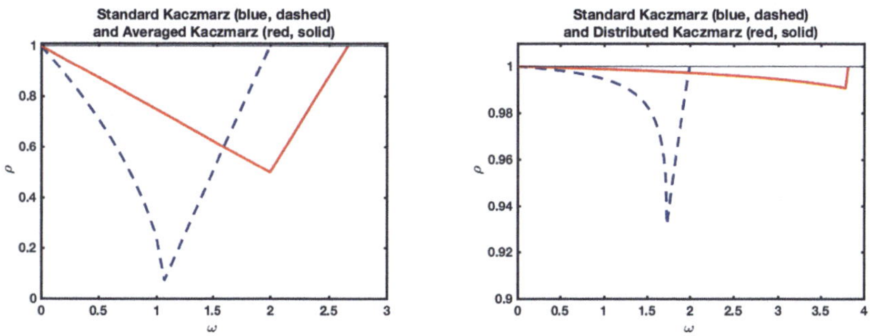

Fig. 3 Dependence of the spectral radius ρ of the iteration matrix on the relaxation parameter ω. The left graph shows Example 1 with $\alpha = \pi/3$. The right graph shows Example 2

3.2 *Implementation*

The implementation of the distributed Kaczmarz algorithm is based on the Matlab Graph Theory toolbox. This toolbox provides support for standard graphs and directed graphs (digraphs), weighted or unweighted. We are using a weighted digraph. The graph is defined by specifying the edges, which automatically also defines the nodes. Specifying nodes is only necessary if there are additional isolated nodes. Both nodes and edges can have additional properties attached to them. We take advantage of that by storing the equations and right-hand sides, as well as the current approximate solution, in the nodes. The weights are stored in the edges. We are currently only considering tree-structured graphs. One node is the root. Each node other than the root has one incoming edge, coming from the predecessor, and zero or more outgoing edges leading to the successors. A node without a successor is called a *leaf*.

The basic Kaczmarz step has the form `x_new = update_node (node,omega,x)`. The graph itself is a global data structure, accessible to all subroutines; it would be very inefficient to pass it as an argument every time.

Fig. 4 The two graphs used in numerical experiments with the distributed Kaczmarz method. For these trees, all of the weights were uniform: $w(u, v) = (C(u))^{-1}$

The `update_node` routine does the following:

- Uses the equation(s) in the node to update `x`
- Executes the `update_node` routine for each successor node
- Combines the results into a new `x`, using the weights stored in the outgoing edges
- Return `x_new`

This routine needs to be called only once per iteration, for the root. It will traverse the entire tree recursively.

3.3 Numerical Experiments

We illustrate the methods with some simple numerical experiments. All experiments were run with three different nonsingular matrices each, of sizes 3×3 and 8×8. All matrices were randomly generated once, and then stored. The right-hand size vectors are also random, and scaled so that the true solution has L^2-norm 1. The test matrices are

- An almost orthogonal matrix, generated from a random orthogonal matrix by truncating to one decimal of accuracy
- A random matrix, based on a standard normal distribution
- A random matrix, based on a uniform distribution in $[-1, 1]$

In each case, we used the optimal ω, based on minimizing the spectral radius of the iteration matrix numerically. The distributed Kaczmarz method used the graphs shown in Fig. 4. Results are shown in Tables 1 and 2. In all cases, we start with $\mathbf{x}_0 = \mathbf{0}$, so the initial L^2-error is $e_0 = 1$. e_{10} refers to the error after 10 iteration steps. For an orthogonal matrix, the standard Kaczmarz method converges in a single step. It is not surprising that it performs extremely well for the almost orthogonal matrices.

In all cases, the distributed Kaczmarz has larger spectral radius (and hence slower convergence). This is to be expected, since the distributed Kaczmarz averages several estimates into one, so bad estimates will increase the error of the average estimate.

Acknowledgments This research was supported by the National Science Foundation and the National Geospatial-Intelligence Agency under awards DMS-1830254 and CCF-1750920.

Table 1 Numerical results for a 3 × 3 system of equations

	Standard Kaczmarz			Distributed Kaczmarz		
	ω_{opt}	ρ_{opt}	e_{10}	ω_{opt}	ρ_{opt}	e_{10}
Orthogonal	1.00030	0.00294	0	1.33833	0.33753	$1.5974 \cdot 10^{-5}$
Normal	1.07213	0.20188	$1.2793 \cdot 10^{-6}$	1.82299	0.29611	$7.2461 \cdot 10^{-6}$
Uniform	1.18634	0.37073	$9.0922 \cdot 10^{-4}$	1.92714	0.82562	$1.49608 \cdot 10^{-1}$

Table 2 Numerical results for an 8 × 8 system of equations

	Standard Kaczmarz			Distributed Kaczmarz		
	ω_{opt}	ρ_{opt}	e_{10}	ω_{opt}	ρ_{opt}	e_{10}
Orthogonal	1.01585	0.04931	$1.53 \cdot 10^{-13}$	1.76733	0.73919	$2.6757 \cdot 10^{-2}$
Normal	1.73543	0.93147	$8.5663 \cdot 10^{-1}$	3.78883	0.99087	$9.0960 \cdot 10^{-1}$
Uniform	1.88188	0.92070	$7.1463 \cdot 10^{-1}$	3.73491	0.99890	$7.7508 \cdot 10^{-1}$

References

1. Bertsekas, D.P. and Tsitsiklis, J.N.: *Parallel and distributed computation: Numerical methods*, Athena Scientific, Nashua, NH, 1997, originally published in 1989 by Prentice-Hall; available for free download at http://hdl.handle.net/1721.1/3719.
2. Boyd, S., Parikh, N., Chu, E., Peleato, B., Eckstein, J., et al.: *Distributed optimization and statistical learning via the alternating direction method of multipliers*, Foundations and Trends® in Machine Learning **3** (2011), no. 1, 1–122.
3. Censor, Y., Gordon, D., and Gordon, R.: *Component averaging: an efficient iterative parallel algorithm for large and sparse unstructured problems*, Parallel Comput. **27** (2001), no. 6, 777–808.
4. Chen, X.: *The Kaczmarz algorithm, row action methods, and statistical learning algorithms*, Frames and harmonic analysis, Contemp. Math., vol. 706, Amer. Math. Soc., Providence, RI, 2018, pp. 115–127.
5. Chi, Y. and Lu, Y.M.: *Kaczmarz method for solving quadratic equations*, IEEE Signal Processing Letters **23** (2016), no. 9, 1183–1187.
6. Cimmino, G.: *Calcolo approssimato per soluzioni dei sistemi di equazioni lineari*, La Ricerca Scientifica XVI, Series II, Anno IX 1 (1938), 326–333.
7. Eggermont, P.P.B., Herman, G.T., and Lent, A.: *Iterative algorithms for large partitioned linear systems, with applications to image reconstruction*, Linear Alg. Appl. **40** (1981), 37–67.
8. Gordon, R., Bender, R., and Herman, G.: *Algebraic reconstruction techniques (ART) for threedimensional electron microscopy and x-ray photography*, Journal of Theoretical Biology **29** (1970), no. 3, 471–481.
9. Gower, R.M. and Richtárik, P.: *Randomized iterative methods for linear systems*, SIAM J. Matrix Anal. Appl. **36** (2015), no. 4, 1660–1690.
10. Haddock, J. and Needell, D.: *Randomized projections for corrupted linear systems*, AIP Conference Proceedings, vol. 1978, AIP Publishing, 2018, p. 470071.
11. Hamaker, C. and Solmon, D.C.: *The angles between the null spaces of X rays*, Journal of Mathematical Analysis and Applications **62** (1978), no. 1, 1–23.
12. Hansen, P.C.: *Discrete inverse problems*, Fundamentals of Algorithms, vol. 7, Society for Industrial and Applied Mathematics (SIAM), Philadelphia, PA, 2010, Insight and algorithms.
13. Herman, G.T., Lent, A., and Hurwitz, H.: *A storage-efficient algorithm for finding the regularized solution of a large, inconsistent system of equations*, J. Inst. Math. Appl. **25** (1980), no. 4, 361–366.

14. Herman, G.T., Hurwitz, H., Lent, A., and Lung, H.P.: *On the Bayesian approach to image reconstruction*, Inform. and Control **42** (1979), no. 1, 60–71.
15. Johansson, B., Rabi, M., and Johansson, M.: *A randomized incremental subgradient method for distributed optimization in networked systems*, SIAM Journal on Optimization **20** (2009), no. 3, 1157–1170.
16. Kaczmarz, S.: *Angenäherte Auflösung von Systemen linearer Gleichungen*, Bulletin International de l'Académie Polonaise des Sciences et des Lettres. Classe des Sciences Mathématiques et Naturelles. Série A, Sciences Mathématiques (1937), 355–357.
17. Kamath, G., Ramanan, P., and Song, W.-Z.: *Distributed randomized Kaczmarz and applications to seismic imaging in sensor network*, 2015 International Conference on Distributed Computing in Sensor Systems, 06 2015, pp. 169–178.
18. Kwapień, S. and Mycielski, J.: emphOn the Kaczmarz algorithm of approximation in infinite-dimensional spaces, Studia Math. **148** (2001), no. 1, 75–86.
19. Liu, J., Wright, S.J., and Sridhar, S.: *An asynchronous parallel randomized Kaczmarz algorithm*, arXiv preprint arXiv:1401.4780 (2014).
20. Loizou, N. and Richtárik, P.: *Revisiting randomized gossip algorithms: General framework, convergence rates and novel block and accelerated protocols*, `arXiv:1905.08645`, 2019.
21. Natterer, F.: *The mathematics of computerized tomography*, Teubner, Stuttgart, 1986.
22. Necoara, I., Nesterov, Y., and Glineur, F.: *Linear convergence of first order methods for non-strongly convex optimization*, Math. Program. **175** (2019), no. 1–2, Ser. A, 69–107.
23. Necoara, I.: *Faster randomized block Kaczmarz algorithms*, `arXiv:1902.09946`, 2019.
24. Necoara, I., Nesterov, Y., and Glineur, F.: *Random block coordinate descent methods for linearly constrained optimization over networks*, J. Optim. Theory Appl. **173** (2017), no. 1, 227–254.
25. Nedic, A. and Ozdaglar, AS.: *Distributed subgradient methods for multi-agent optimization*, IEEE Transactions on Automatic Control **54** (2009), no. 1, 48.
26. Needell, D., Srebro, N., and Ward, R.: *Stochastic gradient descent, weighted sampling, and the randomized Kaczmarz algorithm*, Math. Program. **155** (2016), no. 1–2, Ser. A, 549–573.
27. Needell, D. and Tropp, J.A.: *Paved with good intentions: analysis of a randomized block Kaczmarz method*, Linear Algebra Appl. **441** (2014), 199–221.
28. Needell, D., Zhao, R., and Zouzias,A.: *Randomized block Kaczmarz method with projection for solving least squares*, Linear Algebra Appl. **484** (2015), 322–343.
29. Sayed, A.H.: *Adaptation, learning, and optimization over networks*, Foundations and Trends® in Machine Learning **7** (2014), no. 4–5, 311–801.
30. Scaman, K., Bach, F., Bubeck, S., Massoulié, L., and Lee, Y.T.: *Optimal algorithms for non-smooth distributed optimization in networks*, Advances in Neural Information Processing Systems, 2018, pp. 2740–2749.
31. Shah, D.: *Gossip algorithms*, Foundations and Trends® in Networking **3** (2008), no. 1, 1–125.
32. Strohmer, T. and Vershynin, R.: *A randomized Kaczmarz algorithm with exponential convergence*, Journal of Fourier Analysis and Applications **15** (2009), no. 2, 262–278.
33. Tanabe, K.: *Projection method for solving a singular system of linear equations and its application*, Numer. Math. **17** (1971), 203–214.
34. Tsitsiklis, J., Bertsekas, D., and Athans, M.: *Distributed asynchronous deterministic and stochastic gradient optimization algorithms*, IEEE Transactions on Automatic Control **31** (1986), no. 9, 803–812.
35. West, D.B.: *Introduction to graph theory*, Prentice Hall, Inc., Upper Saddle River, NJ, 1996.
36. Xiao, L., Boyd, S., and Kim, S.J.: *Distributed average consensus with least-mean-square deviation*, Journal of Parallel and Distributed Computing **67** (2007), no. 1, 33–46.
37. Yosida, K.: *Functional analysis*, Second edition. Die Grundlehren der mathematischen Wissenschaften, Band 123, Springer-Verlag New York Inc., New York, 1968.
38. Yuan, K., Ling, Q., and Yin, W.: *On the convergence of decentralized gradient descent*, SIAM Journal on Optimization **26** (2016), no. 3, 1835–1854.

39. Zhang, X., Liu, J., Zhu, Z., and Bentley, E.S.: *Compressed distributed gradient descent: Communication-efficient consensus over networks*, 2018, `arxiv.org/pdf/1812.04048`.
40. Zouzias, A. and Freris, N.M.: *Randomized extended Kaczmarz for solving least squares*, SIAM J. Matrix Anal. Appl. **34** (2013), no. 2, 773–793.

Maximal Function Pooling with Applications

Wojciech Czaja, Weilin Li, Yiran Li, and Mike Pekala

This paper is dedicated to our friend, Professor John Benedetto, on the occasion of his 80th birthday.

Abstract Inspired by the Hardy–Littlewood maximal function, we propose a novel pooling strategy which is called maxfun pooling. It is presented both as a viable alternative to some of the most popular pooling functions, such as max pooling and average pooling, and as a way of interpolating between these two algorithms. We demonstrate the features of maxfun pooling with two applications: first in the context of convolutional sparse coding, and then for image classification.

1 Introduction

In the last decade, the rapid developments in machine learning and artificial intelligence have captured the imagination of many scientists, across a full spectrum of disciplines. Mathematics has not been immune to this phenomenon. In fact, quite the opposite has happened and many mathematicians have been at the forefront of this fundamental research effort. Their contributions ranged from statistics, to optimization, to approximation theory, and last but not least, to harmonic analysis and related representation theory. It is this last aspect that we want to focus on in this paper, as it has led to many intriguing developments associated with the general theory of deep learning, and more specifically, with convolutional neural networks, see [17].

W. Czaja (✉) · Y. Li · M. Pekala
Norbert Wiener Center, Department of Mathematics, University of Maryland College Park, College Park, MD, USA
e-mail: wojtek@math.umd.edu; mpekala@umd.edu

W. Li
Courant Institute of Mathematical Sciences, New York University, New York, NY, USA
e-mail: weilinli@cims.nyu.edu

M. Hirn et al. (eds.), *Excursions in Harmonic Analysis, Volume 6*, Applied and Numerical Harmonic Analysis, https://doi.org/10.1007/978-3-030-69637-5_21

Convolutional neural nets (CNNs) are a popular type of architecture in deep learning, which has shown an outstanding performance in various applications, e.g., in image and video recognition, in image classification [3, 4], or in natural speech processing [6]. CNNs can be effectively modeled by multiscale contractions with wavelet-like operations, applied interchangeably with pointwise nonlinearities [17]. This results in a wealth of network parameters, which can negatively impact the numerical performance of the network. Thus, a form of dimensionality reduction or data compression is needed in order to efficiently process the information through the artificial neural network. For these purposes many examples of CNNs use *pooling* as a type of layer in their networks. Pooling is a dimension reduction technique that divides the input data into subregions and returns only one value as the representative of each subregion. Many examples of such compression strategies have been proposed to-date. Max pooling and average pooling are the two most widely used traditional pooling strategies and they have demonstrated good performance in application tasks [10]. In addition to controlling the computational cost associated with using the network, pooling also helps to reduce overfitting of the training data [12], which is a common problem in many applications.

In addition to the classical examples of maximal and average pooling, many more pooling methods have been proposed and were implemented in neural net architectures. Among those constructions, a significant role has been played by ideas from harmonic analysis, due to their role in providing effective models for data compression and dimension reduction. Spectral pooling was proposed in [21] to perform the reduction step by truncating the representation in the frequency domain, rather than in the original coordinates. This approach claims to preserve more information per parameter than other pooling strategies and to increase flexibility for the size of pooling output. Hartley pooling was introduced in [27] to address the loss of information that happens in the dimensionality reduction process. Inspired by the Fourier spectral pooling, the Hartley transform was proposed as the base, thus avoiding the use of complex arithmetic for frequency representations, while increasing the computational effectiveness and network's discriminability. Transformation invariant pooling based on the discrete Fourier transform was introduced in [22] to achieve translation invariance and shape preservation thanks to the properties of the Fourier transform. Wavelet pooling [25] is another alternative to the traditional pooling procedures. This method decomposes features into a two level decomposition and discards the first-level subbands to reduce the dimension, thus addressing the overfitting, while reducing features in a structurally conscious manner. Multiple wavelet pooling [9] builds upon the wavelet pooling idea, while introducing more sophisticated wavelet transforms such as Coiflets and Daubechies wavelets, into the process. An even more general approach was proposed in [2], where ℓ_p pooling was defined based on the concept of a representation in terms of general frames for $\mathbb{R}^d$, to provide invariance to the system.

In this paper we follow in the footsteps of the aforementioned constructions and propose a novel method for reducing the dimension in CNNs, which is based on a fundamental concept from harmonic analysis. Inspired by the *Hardy–Littlewood maximal function*, [13], cf., [5, 18] for its modern treatment, we introduce a novel pooling strategy, called *maxfun pooling*, which can be viewed as a natural alternative to both max pooling and average pooling. In particular, max pooling takes the maximum value in each pooling region as the scalar output, and average pooling takes the average of all entries in each pooling region as the scalar output. As such, maxfun pooling can be interpreted as a novel and creative way of interpolating between max and average pooling algorithms.

In what follows, we introduce a discrete and computationally feasible analogue of the Hardy–Littlewood maximal function. The resulting operator depends on two integer parameters b and s, corresponding to the size of the pooling region and the stride. We limit the support of this operator to be finite, we discretize it and define the maximal function pooling operation, denoted throughout this paper by *maxfun pooling*. The maxfun pooling computes averages of subregions of different sizes in each pooling region, and it selects the largest average among all. To demonstrate the features of maxfun pooling, we present two different applications. First, we study its properties in the realm of convolutional sparse coding. It has been shown that feedforward convolutional neural networks can be viewed as convolutional sparse coding [19]. Moreover, under the point of view presented by the convolutional sparse coding, stable recovery of the signal contaminated with noise can be achieved, given simple sparsity conditions [19]. Equivalently, it implies that feedforward neural networks maintain stability under noisy conditions. The case of pooling function analyzed via convolutional sparse coding is studied in [15], where the two common pooling functions, max pooling and average pooling are analyzed. We follow the framework presented in [15] and we show that stability of the neural network under the presence of noise is also preserved with maxfun pooling. We close this paper with a different application, presenting illustrative numerical experiments utilizing maxfun pooling for image classification.

2 Preliminaries

In this section, we elaborate upon the role of pooling in neural networks, and discuss two traditional strategies, max and average pooling. We focus on image data as our main application domain, but we mention that the subsequent definitions can be readily generalized.

We view images as functions on a finite lattice $X\colon [M]\times[N] \to \mathbb{R}$, where $[K] := \{0, 1, \dots, K-1\}$, or $X \in \mathbb{R}^{M\times N}$ in short. Its $(i, j)-th$ coordinate is denoted $X_{i,j}$. In practice, it is convenient to fold images into vectors. Slightly abusing notation, for $X \in \mathbb{R}^{M\times N}$, we also let $X \in \mathbb{R}^{MN}$ denote its corresponding vectorization,

$$X = (X_{1,1}, X_{2,1}, \dots, X_{M,1}, X_{1,2}, \dots, X_{M,2}, \dots, X_{M,N})^T. \tag{2.1}$$

Throughout this chapter, we will not make distinctions between vectors and images. As a consequence, we shall assume without any loss of generality that pooling layer's input is always nonnegative.

2.1 Max and Average Pooling

Both the maximum and average pooling operators depend on a collection of sets $Q \subseteq [M] \times [N]$, which we refer to *pooling regions*. There is a standard choice of Q, which we describe below. Fix an odd integer s and set $m = \lfloor M/s \rfloor$ and $n = \lfloor N/s \rfloor$. For each pair of integers (k, ℓ), we define the square $Q_{k,\ell}$ of size $s \times s$ by

$$Q_{k,\ell} := \{(i, j) \colon (k-1)s \leq i < ks,\ (\ell-1)s \leq j < \ell s\}. \tag{2.2}$$

It is common to refer to s as the *stride*. The stride determines the size of each pooling region $Q_{k,\ell}$ and the total number of squares in Q. The use of a stride implicitly reduces the input data dimension since an image of size $M \times N$ is then reduced to one of size $m \times n$.

For this collection Q, the *average pooling* operator $P_{\text{avg}} \colon \mathbb{R}^{M\times N} \to \mathbb{R}^{m\times n}$ is given by

$$(P_{\text{avg}}X)_{k,\ell} := \frac{1}{s^2} \sum_{(i,j)\in Q_{k,\ell}} X_{i,j}.$$

For the same collection, the *max pooling* operator $P_{\max} \colon \mathbb{R}^{M\times N} \to \mathbb{R}^{m\times n}$ is defined as

$$(P_{\max}X)_{k,\ell} := \max_{(i,j)\in Q_{k,\ell}} X_{i,j}.$$

In other words, P_{avg} simply averages the input image on each $Q_{k,\ell}$, while $P_{\max}$ is the supremum of the values of the image restricted to $Q_{k,\ell}$.

2.2 Maximal Function

For a locally integrable function $f \in L^1_{\text{loc}}(\mathbb{R}^d)$, we can define its *Hardy–Littlewood maximal function* $M(f)$ as

$$\forall\, x \in \mathbb{R}^d, \quad M(f)(x) = \sup_{B(x)} \frac{1}{m^d(B(x))} \int_{B(x)} |f|\, dm^d, \tag{2.3}$$

where the supremum is taken over all open balls $B(x)$ centered at x and the integral is taken with respect to the Lebesgue measure m^d. Because of this last property, sometimes we talk about the *centered* maximal function, as opposed to the non-centered analogue. The function M was introduced by Godfrey H. Hardy and John E. Littlewood in 1930 [13]. Maximal functions had a profound influence on the development of classical harmonic analysis in the 20th century, see, e.g., [11] and [23]. Among other things, they play a fundamental role in our understanding of the differentiability properties of functions, in the evaluation of singular integrals, and in applications of harmonic analysis to partial differential equations. One of the key tools in this theory is the following Hardy–Littlewood lemma.

Theorem 2.1 (Hardy–Littlewood Lemma)
Let $(\mathbb{R}^d, \mathcal{M}(\mathbb{R}^d), m^d)$ be the Lebesgue measure space. Then, for any $f \in L^1(\mathbb{R}^d)$,

$$\forall\, \alpha > 0, \quad m^d\left(\{x \in \mathbb{R}^d : M(f)(x) > \alpha\}\right) \leq \frac{3^d}{\alpha} \int_{\mathbb{R}^d} |f|\, dm^d.$$

In view of Theorem 2.1, the maximal function is sometimes interpreted as encoding the worst possible behavior of the signal f. This point of view is further exploited through the *Calderón–Zygmund decomposition theorem*.

Theorem 2.2 (Calderón–Zygmund Decomposition) *Let $(\mathbb{R}^d, \mathcal{M}(\mathbb{R}^d), m^d)$ be the Lebesgue measure space. Then, for any $f \in L^1(\mathbb{R}^d)$ and any $\alpha > 0$, there exists $F, \Omega \subseteq \mathbb{R}^d$ such that $\mathbb{R}^d = F \cup \Omega$, $F \cap \Omega = \emptyset$, and*

1. *$f(x) \leq \alpha$, m^d-a.e. in F*
2. *Ω is the union of cubes Q_k, $k = 1, \ldots,$ whose interiors are pairwise disjoint, edges are parallel to the coordinate axes, and for each $k = 1, \ldots,$ we have*

$$\alpha < \frac{1}{m^d(Q_k)} \int_{Q_k} |f|\, dm^d \leq 2^d \alpha.$$

The Calderón-Zygmund decomposition allows us to split an arbitrary integrable function into its "good" (i.e., small) and "bad" (i.e., large) parts, which is a standard technique in addressing problems in the theory of Calderón–Zygmund operators. In this approach, the maximal function helps us control the undesirable behavior of the signal by constraining the regions (represented by balls or cubes) on which this happens. It is this aspect of the maximal functions that we also aim to exploit in our construction of the *maxfun pooling*, which will be defined in the next section.

3 Maxfun Pooling

For maxfun pooling, we first fix an odd stride s and parameter b such that $2b+1 \leq s$. Consider the same collection Q with $Q_{k,\ell}$ defined in (2.2), and let $q_{k,\ell}$ denote the center of $Q_{k,\ell}$, which is well defined since s is assumed odd. For each $r \leq b$, we define the sub-squares,

$$B_{k,\ell,r} \subseteq Q_{k,\ell},$$

where $B_{k,\ell,r}$ denotes the square of side length $2r + 1$ whose center is also $q_{k,\ell}$.

The *maxfun pooling* operator $M_{b,s} : \mathbb{R}^{M\times N} \to \mathbb{R}^{m\times n}$ is given by

$$(M_{b,s}X)_{k,\ell} := \max_{1\leq r\leq b} \left(\frac{1}{(2r+1)^2} \sum_{(i,j)\in B_{k,\ell,r}} X_{i,j} \right). \tag{3.1}$$

The quantity $(M_{b,s}X)_{k,\ell}$ is the average of X on $B_{k,\ell,r}$, and the optimal choice of r depends on the restriction of X to $Q_{k,\ell}$.

Let us briefly justify the definition of maxfun pooling. Similar to the maximal function given in (2.3) where the supremum is taken over all balls B centered at a fixed x, each $B_{k,\ell,r}$ is centered at $q_{k,\ell}$. Our convention regarding the nonnegative inputs of the pooling layer aligns with the use of the absolute value in the definition of the maximal function. Finally, the computational complexity of evaluating maxfun can be further reduced by choosing the b parameter to be strictly smaller than $(s-1)/2$.

For each region $Q_{k,\ell}$ and any image X, clearly we have the inequalities

$$(P_{\text{avg}}X)_{k,\ell} \leq (M_{b,s}X)_{k,\ell} \leq (P_{\max}X)_{k,\ell}.$$

From this point of view, the output of maxfun on each pooling region is quantitatively between that of average and max pooling. It is not difficult to construct examples of X and Y for which $P_{\text{avg}}X = M_{b,s}X$ and $P_{\max}Y = M_{b,s}Y$.

Another main conceptual difference between these three pooling operators is the amount of information of X that determines their outputs. Average pooling depends on the values of X on the entire region $Q_{k,\ell}$, whereas max pooling selects the largest value of X on $Q_{k,\ell}$, which is not necessarily representative of X. From this perspective, average pooling is more robust to perturbations. Maxfun pooling is adaptive in the sense that the optimal region $B_{k,\ell,r}$ depends on X, which is a different procedure for selecting a representative value.

4 Convolutional Sparse Coding

The sparse coding problem is an important problem in signal processing, where one aims at finding a low dimensional representation using few atoms for high dimensional data [20]. Following standard conventions, we treat an image X as a vector $X \in \mathbb{R}^N$ using the vectorization operator defined in (2.1). Given a vector $X \in \mathbb{R}^N$, and a dictionary $D \in \mathbb{R}^{N\times M}$, the sparse coding problem attempts to find the sparsest vector $\Gamma \in \mathbb{R}^M$ such that $X = D\Gamma$. In other words, for a fixed dictionary $D \in \mathbb{R}^{N\times M}$, the sparse coding problem attempts to solve

$$\min_{\Gamma} \|\Gamma\|_0 \quad s.t. \quad D\Gamma = X, \tag{4.1}$$

where $\|\Gamma\|_0$ is the l_0 pseudo norm, which provides the number of nonzero elements in vector Γ. Each column in D represents one element in the dictionary, and finding the dictionary to represent data X, so that minimal number of dictionary elements are used, solves the sparse coding problem.

Restriction on the sparsity of Γ with respect to the mutual coherence of the dictionary D can guarantee uniqueness of the solution to (4.1). The mutual coherence of a matrix D, see [7], is defined as

$$\mu(D) = \max_{i \neq j} \frac{|di^T d_j|}{\|d_i\|_2 \cdot \|d_j\|_2},$$

where d_i's are the columns of matrix D. However, finding the solution remains NP hard. Relaxation of the model to allow noise and form error bounds leads to the following formulation:

$$\min_{\Gamma} \|\Gamma\|_0 \quad s.t. \quad \|D\Gamma - X\| < \varepsilon.$$

When high dimensional signals are present, an alternative method called the convolutional sparse coding model (CSC) was proposed. One attempts to represent the whole signal $X \in \mathbb{R}^N$ as a multiplication of a global convolutional dictionary $D \in \mathbb{R}^{N \times Nm_1}$ and a sparse vector $\Gamma \in \mathbb{R}^{Nm_1}$. D is constructed by shifting a local matrix of size $n_0 \times m_1$ in all possible positions, as shown in Fig. 1.

We define the jth stripe γ_j of the sparse vector Γ as a group of $2n_0 - 1$ adjacent sparse vectors of length m_1, starting at the jth vector of length m_1. See Fig. 1 for an illustration.

The stripe γ_j gives the representation of a patch of X, x_j of length n_0 by $x_j = \Omega_j \gamma_j$. $\Omega_j \in \mathbb{R}^{n_0 \times (2n_0-1)m_1}$ is a submatrix of D, called a stripe dictionary consisting of n_0 consecutive rows of D and the columns of zeros removed. The $l_{0,\infty}$ norm of the global sparse vector Γ_1 is defined by the maximum number of nonzeros in any stripe of length $(2n_0 - 1)m_1$ extracted from it, i.e.,

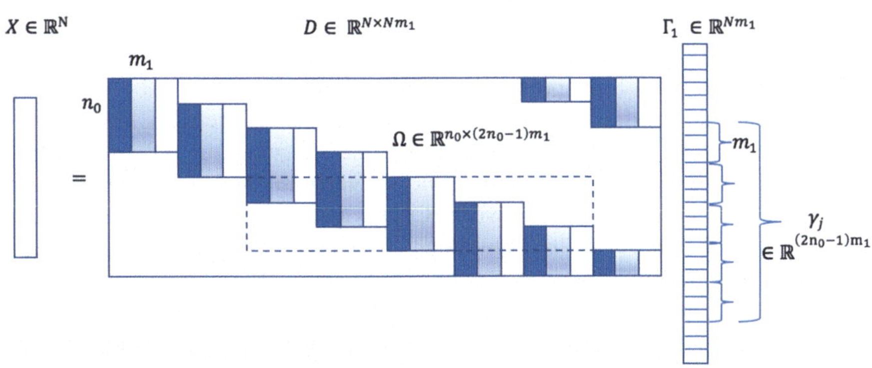

Fig. 1 Convolutional sparse coding, level 1 [19]

$$\|\Gamma\|_{0,\infty} = \max_{i\in\{1,\dots,N\}} \|\gamma_i\|_0.$$

Here $\|\cdot\|_0$ is the zero norm that gives the number of nonzero elements of a vector.

A multi-layer convolutional sparse coding model is defined so that the output sparse vector Γ from the previous layer serves as the input vector in the next layer, and we aim at finding a new representation Γ_2 for a new set of dictionary D_2. Formally, the problem of finding solutions to multi-layer convolutional sparse coding problem is defined as the deep coding problem DCP_λ in [19]:

$$\begin{array}{lll} Find\ \ \{\Gamma_i\}_{i=1}^{L} & s.t. & \\ X = D_1\Gamma_1, & \|\Gamma_1\|_{0,\infty} & \le \lambda_1 \\ \Gamma_1 = D_2\Gamma_2, & \|\Gamma_2\|_{0,\infty} & \le \lambda_2 \\ \vdots & & \\ \Gamma_{L-1} = D_L\Gamma_L, & \|\Gamma_L\|_{0,\infty} & \le \lambda_L \end{array}$$

where λ_i are bounds on sparsity of the output vector Γ_i at each level, and L is the number of layers. Note that we want to find representations of the input vectors at each layer that are sparse in terms of its stripe sparsity, defined by $\|\cdot\|_{0,\infty}$.

In practice, the input signal X can be contaminated with noise, and we have $Y = X + E$ as the input signal instead of X, where E represents noise. In this case, we relax the constraint and allow the representation to vary within some error bounds of the input signal. The deep coding problem when noise is present (DCP_λ^ε) [19] is defined as

$$\begin{array}{lll} Find\ \ \{\Gamma_i, P_i\}_{i=1}^{L} & s.t. & \\ \|Y - D_1\Gamma_1\|_2 \le \varepsilon_1, & [-3pts]\|\Gamma_1\|_{0,\infty} & \le \lambda_1 \\ \|\Gamma_1 - D_2\Gamma_2\|_2 \le \varepsilon_2, & \|\Gamma_2\|_{0,\infty} & \le \lambda_2 \\ \vdots & & \\ \|\Gamma_{L-1} - D_L\Gamma_L\|_2 \le \varepsilon_L, & \|\Gamma_L\|_{0,\infty} & \le \lambda_L. \end{array}$$

Here ε_i is the error bound that are allowed in the ith layer.

Uniqueness of the solution to the DCP_λ model, and the stability of the solution to the DCP_λ^ε problem have been shown in [19]. The connection between deep convolutional sparse coding problem and feed forward neural network has also been demonstrated in [19], where it is proven that one can view the output vector Γ_i from each layer of the DCP_λ problem as the output from one layer of feed forward simplified convolutional neural network (CNN), and thus the deep convolutional sparse coding problem can be viewed as a signal reconstruction problem for simplified CNN models.

Pooling is a common operation included in CNNs that serves as feature extraction method to reduce redundancy of representation of signal and save computational resources. It has been shown that adding max pooling and average pooling in the feed forward path preserves the stability of the neural network [15]. We demonstrate that the maxfun pooling preserves the stability of a convolutional neural network when added in between layers of convolutions.

Given a input signal X, the deep convolutional sparse coding problem with pooling ($DCPP$) is defined by [15]

$$\begin{array}{llll} Find \quad \{\Gamma_i, P_i\}_{i=1}^{L} & s.t. & & \\ X = D_1\Gamma_1, & \|\Gamma_1\|_{0,\infty} & \leq \lambda_1 & P_1 = Pool_{b_1,s_1}(\Gamma_1), \\ P_1 = D_2\Gamma_2, & \|\Gamma_2\|_{0,\infty} & \leq \lambda_2 & P_2 = Pool_{b_2,s_2}(\Gamma_2), \\ \vdots & & & \vdots \\ P_{L-1} = D_L\Gamma_L, & \|\Gamma_L\|_{0,\infty} & \leq \lambda_L, & P_L = Pool_{b_L,s_L}(\Gamma_L), \end{array} \tag{4.2}$$

where $Pool_{i b,s}$ denotes the pooling operation at the step i. We take $Pool_{b_i,s}$ to be the maxfun pooling, i.e., $Pool_{b_i,s} = M_{b,s}$, as defined in (3.1).

Problem (4.2) intends to find a stable sparse representation Γ_1 of X with dictionary elements in D_1, given restriction on the stripe sparsity of Γ_1. Then pooling operation is performed on Γ_1 to get P_1. In second layer, we attempt to find the sparse representation Γ_2 of P_1 with dictionary elements in D_2. The stripe sparsity of Γ_2 is restricted to be no greater than λ_2. We repeat the process L times.

If our input signal X is contaminated by noise E, we are still interested in finding a sparse representation that is stable. Define the deep convolutional sparse coding problem with pooling when noise is present ($DCPP^{\varepsilon}$) by [15]

$$\begin{array}{llll} Find \quad \{\Gamma_i, P_i\}_{i=1}^{L} & s.t. & & \\ \|Y - D_1\Gamma_1\| \leq \varepsilon_1, & \|\Gamma_1\|_{0,\infty} & \leq \lambda_1 & P_1 = Pool_{b_1,s_1}(\Gamma_1), \\ \|P_1 - D_2\Gamma_2\| \leq \varepsilon_2, & \|\Gamma_2\|_{0,\infty} & \leq \lambda_2 & P_2 = Pool_{b_2,s_2}(\Gamma_2), \\ \vdots & & & \vdots \\ \|P_{L-1} - D_L\Gamma_L\| \leq \varepsilon_L, & \|\Gamma_L\|_{0,\infty} & \leq \lambda_L, & P_L = Pool_{b_L,s_L}(\Gamma_L) \end{array} \tag{4.3}$$

It has been shown in [15] that when max pooling and average pooling are used, the stability of solution to the $DCPP^{\varepsilon}$ problem is preserved. We show that when we use maxfun pooling, the stability result also holds. We prove the following theorem:

Theorem 4.1 *Suppose a vector X satisfies the $DCPP$ model in (4.2), but is contaminated with noise E, where $\|E\|_2 \leq \varepsilon$, resulting in $Y = X + E$. Suppose $\{\Gamma_i^*, P_i^*\}_{i=1}^{L}$ solves the problem in (4.2) and $\{\widehat{\Gamma}_i, \widehat{P}_i\}_{i=1}^{L}$ solves the problem in (4.3). If*

$$\|\Gamma_i^*\|_{0,\infty} \leq \lambda_i < \frac{1}{2}\left(1 + \frac{1}{\mu(D_i)}\right), \quad \forall 0 \leq i \leq L,$$

$$\varepsilon_0 = \varepsilon, \quad \varepsilon_i^2 = \frac{4\varepsilon_{i-1}^2}{1 - (2\|\Gamma_i^*\|_{0,\infty} - 1)\mu(D_1)} \quad \forall i \geq 1, \tag{4.4}$$

then for all $\quad 1 \leq i \leq L,$

$$\|P_i^* - \widehat{P}_i\|_2^2 \leq \|\Gamma_i^* - \widehat{\Gamma}_i\|_2^2 \leq \varepsilon_i^2.$$

Here maxfun pooling is used as the pooling operation and we assume that the minimum pooling region size $b \geq 1$.

In order to prove Theorem 4.1, we first prove the following Lemma for maxfun pooling.

Lemma 1 *Let X and $\widehat{X}$ be two functions in $\mathbb{R}^{N\times N}$, and let $P = M_{b,s}(X)$, $\widehat{P} = M_{b,s}(\widehat{X})$ be the outcome of maxfun pooling of X and $\widehat{X}$, respectively, and assume that $s \geq b$. Then $\|P - \widehat{P}\|_2 \leq \|X - \widehat{X}\|_2$, where $\|\cdot\|_2$ denotes the Frobenius norm.*

Proof Let $B_{k,\ell,r}$ be the set of indices that represents points in the centered sub-square of side length $2r+1$ in the (k,ℓ)-th pooling region. Let

$$\gamma_{k,\ell,r} = \frac{1}{(2r+1)^2}\sum_{(i,j)\in B_{k,\ell,r}} X_{i,j} \quad \text{and} \quad \widehat{\gamma}_{k,\ell,r} = \frac{1}{(2r+1)^2}\sum_{(i,j)\in B_{k,\ell,r}} \widehat{X}_{i,j}.$$

We take $r^*_{k,\ell} = \operatorname{argmax}_{1\leq r\leq b}\gamma_{k,\ell,r}$ to indicate the r parameter of the sub-square with the maximum of $\gamma_{k,\ell,r}$ over all r for each (k,ℓ)-th region; and we take $\widehat{r}^*_{k,\ell} = \operatorname{argmax}_{1\leq r\leq b}\widehat{\gamma}_{k,\ell,r}$ to indicate the r parameter of the sub-square with maximum of $\widehat{\gamma}_{k,\ell,r}$ over all r for each (k,ℓ)-th region. We also take the r^*_{min} to be minimum of all $r^*_{k,\ell}$, and $\widehat{r}^*_{min}$ to be minimum of all $\widehat{r}^*_{k,\ell}$. Furthermore, we let K_1 be the set of indices of (k,ℓ) so that $\gamma_{k,\ell,r^*_{k,\ell}} \geq \widehat{\gamma}_{k,\ell,\widehat{r}^*_{k,\ell}}$. Let K_2 be the set of indices of (k,ℓ) so that $\gamma_{k,\ell,r^*_{k,\ell}} < \widehat{\gamma}_{k,\ell,\widehat{r}^*_{k,\ell}}$. Then we have

$$\|P-\widehat{P}\|_2^2$$

$$= \sum_{(k,l)}\left(\max_{1\leq r\leq b}\left(\frac{1}{(2r+1)^2}\sum_{(i,j)\in B_{k,\ell,r}} X_{i,j}\right) - \max_{1\leq r\leq b}\left(\frac{1}{(2r+1)^2}\sum_{(i,j)\in B_{k,\ell,r}}\widehat{X}_{i,j}\right)\right)^2 \tag{4.5}$$

$$= \sum_{(k,\ell)\in K_1}(\gamma_{k,\ell,r^*_{k,\ell}} - \widehat{\gamma}_{k,\ell,\widehat{r}^*_{k,\ell}})^2 + \sum_{(k,\ell)\in K_2}(\widehat{\gamma}_{k,\ell,\widehat{r}^*_{k,\ell}} - \gamma_{k,\ell,r^*_{k,\ell}})^2 \tag{4.6}$$

$$\leq \sum_{(k,\ell)\in K_1}(\gamma_{k,\ell,r^*_{k,\ell}} - \widehat{\gamma}_{k,\ell,r^*_{k,\ell}})^2 + \sum_{(k,\ell)\in K_2}(\widehat{\gamma}_{k,\ell,\widehat{r}^*_{k,\ell}} - \gamma_{k,l,\widehat{r}^*_{k,\ell}})^2 \tag{4.7}$$

$$= \sum_{(k,\ell)\in K_1}\left(\frac{1}{(2r^*_{k,\ell}+1)^2}\sum_{(i,j)\in B_{k,l,r^*_{k,\ell}}} X_{i,j} - \frac{1}{(2r^*_{k,\ell}+1)^2}\sum_{(i,j)\in B_{k,\ell,r^*_{k,\ell}}}\widehat{X}_{i,j}\right)^2$$

$$+\sum_{(k,\ell)\in K_2}\left(\frac{1}{(2\widehat{r}^*_{k,\ell}+1)^2}\sum_{(i,j)\in B_{k,\ell,\widehat{r}^*_{k,\ell}}}\widehat{X}_{i,j}-\frac{1}{(2\widehat{r}^*_{k,\ell}+1)^2}\sum_{(i,j)\in B_{k,\ell,\widehat{r}^*_{k,\ell}}}X_{i,j}\right)^2 \tag{4.8}$$

$$\leq \sum_{(k,\ell)\in K_1}\frac{1}{(2r^*_{k,\ell}+1)^2}\sum_{(i,j)\in B_{k,\ell,r^*_{k,\ell}}}(X_{i,j}-\widehat{X}_{i,j})^2$$

$$+\sum_{(k,\ell)\in K_2}\frac{1}{(2\widehat{r}^*_{k,\ell}+1)^2}\sum_{(i,j)\in B_{k,\ell,\widehat{r}^*_{k,\ell}}}(\widehat{X}_{i,j}-X_{i,j})^2 \tag{4.9}$$

$$\leq \frac{1}{(2r^*_{min}+1)^2}\sum_{(k,\ell)\in K_1,(i,j)\in B_{k,\ell,r^*_{k,\ell}}}(X_{i,j}-\widehat{X}_{i,j})^2$$

$$+\frac{1}{(2\widehat{r}^*_{min}+1)^2}\sum_{(k,\ell)\in K_2,(i,j)\in B_{k,\ell,\widehat{r}^*_{k,\ell}}}(\widehat{X}_{i,j}-X_{i,j})^2$$

$$\leq \sum_{(k,\ell)\in K_1,(i,j)\in B_{k,\ell,b}}(X_{i,j}-\widehat{X}_{i,j})^2+\sum_{(k,\ell)\in K_2,(i,j)\in B_{k,\ell,b}}(\widehat{X}_{i,j}-X_{i,j})^2 \tag{4.10}$$

$$\leq \|X-\widehat{X}\|_2^2.$$

The inequality (4.7) comes from the fact that $\widehat{\gamma}_{k,l,\widehat{r}^*_{k,\ell}}$ is the maximum over all r and thus $\widehat{\gamma}_{k,\ell,\widehat{r}^*_{k,\ell}} \geq \widehat{\gamma}_{k,\ell,r^*_{k,\ell}}$, and similarly $\gamma_{k,\ell,r^*_{k,\ell}} \geq \gamma_{k,\ell,\widehat{r}^*_{k,\ell}}$. In (4.8), B_{k,ℓ,r^*_k} and $B_{k,\ell,\widehat{r}^*_k}$ are the corresponding set of indices for which $\gamma_{k,\ell,r^*_{k,\ell}}$ and $\widehat{\gamma}_{k,\ell,\widehat{r}^*_{k,\ell}}$ are maximums across all r's, respectively. The inequality (4.9) holds based on the inequality $(\sum_{i=1}^n a_i)^2 \leq n\sum_i a_i^2$. Inequality (4.10) follows from the fact that $r \leq b$ and stride size $s \geq 2b+1$. $B_{k,\ell,b}$ represents the indices of the sub-square of length $2b+1$ at the initial position in the (k,l)th pooling region. □

We now prove Theorem 4.1.

Proof By Theorem 3 in [20], we know that for a signal $Y = X + E$, if

$$1. \|\Gamma_1^*\|_{0,\infty} < \frac{1}{2}(1+\frac{1}{\mu(D_1)}) \text{ and } \|E\|_2 = \|Y - D_1\Gamma_1^*\|_2 \leq \varepsilon_0,$$

$$2. \|\widehat{\Gamma}_1\|_{0,\infty} < \frac{1}{2}(1+\frac{1}{\mu(D_1)}) \text{ and } \|Y - D_1\widehat{\Gamma}_1\|_2 \leq \varepsilon_0,$$

then

$$\|\Delta\|_1^2 = \|\Gamma_1^* - \widehat{\Gamma}_1\|_2^2 \leq \frac{4\varepsilon_0^2}{1-(2\|\Gamma_1\|_{0,\infty}-1)\mu(D_1)} = \varepsilon_1^2.$$

Since $\|\Gamma_1^*\|_{0,\infty} \leq \lambda_1$ and $\|\widehat{\Gamma}_1^*\|_{0,\infty} \leq \lambda_1$ by assumption in problem (4.2) and (4.3), and λ is bounded by assumption (4.4) in Theorem 4.1, the first parts of 1 and 2 hold. Since Γ_1^* is the solution to the $DCPP$ problem, it must be true that $\|Y - D_1\Gamma_1^*\| \leq \varepsilon_0$. $\|Y - D_1\widehat{\Gamma}_1\| \leq \varepsilon_0$ by assumption in problem (4.3). Therefore, we have $\|\Delta\|_2^2 \leq \varepsilon_1^2$. And hence by Lemma 1, we have

$$\|P_1^* - \widehat{P}_1\|_2^2 \leq \|\Delta_1\|_2^2 \leq \varepsilon_1^2.$$

At second level, the same argument holds so that $\|\Gamma_2^*\| < \lambda_2 < \frac{1}{2}(1 + \frac{1}{\mu(D_1)})$ and $\|\widehat{\Gamma}_2\| < \lambda_2 < \frac{1}{2}(1 + \frac{1}{\mu(D_1)})$. $\|P_1^* - D_2\Gamma_2^*\|_2 < \varepsilon_2$ by assumption in problem (4.2). $\|\widehat{P}_1 - D_2\widehat{\Gamma}_2\|_2 < \varepsilon_2$ by assumption in problem (4.3). Therefore, by theorem 3 in [20] we have

$$\|\Gamma_2^* - \widehat{\Gamma}_2\|_2^2 \leq \varepsilon_2^2,$$

and by Lemma 1, we get

$$\|P_2^* - \widehat{P}_2\|_2^2 \leq \|\Gamma_2^* - \widehat{\Gamma}_2\|_2^2 \leq \varepsilon_2^2.$$

Following this argument for $1 \leq i \leq L$, we complete the proof and showed that

$$\|P_i^* - \widehat{P}_i\|_2^2 \leq \|\Gamma_i^* - \widehat{\Gamma}_i\|_2^2 \leq \varepsilon_i^2, \quad \forall \quad 1 \leq i \leq L.$$

□

5 Classification Experiments

As described in Sect. 1, pooling as a technique for dimension reduction arises in a variety of signal processing contexts. Accordingly, the sparse coding problem described in Sect. 4 is not the only possible application for maxfun pooling. In this section we consider maxfun for supervised classification problems.

Prior work has examined properties of pooling operators in the context of supervised classification. For example, [1] performed theoretical analyses and conducted empirical experiments comparing maximum and average pooling. Their findings indicated that, depending upon the data and features at hand, either maximum or average pooling may be preferable. The authors also identified potential benefits of pooling methodologies that "interpolate" between maximum and average pooling. As demonstrated in Sect. 3, maxfun pooling also has the property that it resides between maximum and average pooling in a precise sense. Thus it is of interest to explore how maxfun relates not only to average and maximum pooling, but also to other intermediate pooling techniques.

5.1 Approach

Our goal is to highlight differences between different pooling strategies; therefore we construct a classification problem where pooling plays a significant role in developing the feature representation. Instead of the gradual dimension reduction one might observe throughout a standard feedforward deep neural network (where 2x2 or 3x3 pooling regions might be used at any given layer) here we consider an approach akin to very shallow networks followed by a single, aggressive pooling step. The preliminary experiments presented in this section are in the same spirit as those of [1], albeit with a different implementation.

Since natural images often manifest meaningful correlations in the spatial dimension we choose to focus upon image classification problems. We generate raw features by extracting the outputs from the first layer of a standard CNN (prior to pooling) that has been pre-trained on natural images. We then apply various pooling techniques along the spatial dimensions of these feature maps. Finally, the vectorized feature maps are used in conjunction with a traditional multiclass support vector machine (SVM) [14] to compare the utility of the different pooling methods.

For our experiments we use a subset of the Caltech-101 data set [8]. To mitigate the impact of class imbalance, we limit our study to classes that have between 80 and 130 instances. The result is an 18-class classification problem with mild class imbalance. To standardize the spatial dimensions all images are first zero padded to make them square; e.g. a 100x120 pixel image is padded to produce a 120x120 pixel image. Images are kept centered when padding (in the previous example 10 rows would be added to the top and 10 to the bottom of the image). Finally, we resize all images to 128x128 pixels. This provides a collection of natural images with the same dimensions where some attempt has been made to preserve the aspect ratio.

For raw features we use the outputs from the first layer of the Inception-v3 network [24]. This deviates from [1], which used SIFT to construct feature maps (note their study predated a number of key developments in convolutional neural networks). We then spatially decompose the resulting feature tensor into multiple (possibly overlapping) pooling regions. This decomposition is applied independently along each channel, and thus preserves the cardinality of the channel dimension. After identifying the pooling regions, we apply various pooling operators. These pooled tensors are then vectorized and constitute the feature maps used to solve the multiclass classification problem.

5.1.1 Baselines

In addition to maximum and average pooling, we also compare maxfun to other pooling strategies that sit in between these two extremes. In particular, we consider the "stochastic pooling" method of [26] and the "mixed pooling" method of [16]. For stochastic pooling, the authors recommend a particular weighted average motivated by their original stochastic formulation:

$$v_{k,\ell} = c_{k,\ell} \sum_{(i,j)\in Q_{k,\ell}} X_{i,j}^2, \qquad c_{k,\ell} = \frac{1}{\sum_{(i,j)\in Q_{k,\ell}} X_{i,j}}, \tag{5.1}$$

where $v_{k,\ell}$ denotes the pooled value and X is a (nonnegative) two-dimensional feature map, following the notation of Sect. 2. Note X could represent a raw image or some higher-level representation thereof. The non-negativity assumption is consistent with image pixel intensities or with features extracted from the output of a RELU layer within a CNN. See [26] for full details on the stochastic origins of (5.1).

Mixed pooling is defined as a convex combination of maximum and average pooling; that is

$$v_{k,\ell} = \alpha\,(P_{\max}X)_{k,\ell} + (1-\alpha)\,(P_{\text{avg}}X)_{k,\ell}, \tag{5.2}$$

where $\alpha \in (0, 1)$.

The maxfun and the mixed pooling strategies both introduce hyperparameters to our experiments; in the case of the maximal function we elected to add a minimum radius $r_{min} > 1$ as a lower bound on r in (3.1) while for the mixed pooling strategy we must select the scalar α in (5.2). In both cases we use a k-fold cross-validation procedure with $k = 3$ to select these hyperparameters. Our training and testing data sets are of size 975 and 649 with the partition chosen uniformly at random.

5.2 Preliminary Results and Discussion

Results for our pooling experiments are summarized in Table 1. Across all pooling methods we consider two scenarios: one where the pooling regions partition the spatial domain and another where the pooling windows overlap. For this study we implemented both centered and non-centered versions of the maxfun pooling. The centered variant is a realization of (3.1) while the non-centered variant allows the squares to be centered at points other than the center of pooling region.

Table 1 Empirical results for our Caltech-101 classification experiments. Pooling regions have dimensions $b \times b$ and the pooling stride is s. Thus, when $s = b$ (column 1), the pooling regions partition the spatial domain. When $s < b$ (column 2) the pooling regions overlap

	SVM accuracy	
Pooling strategy	($b = 21, s = 21$)	($b = 21, s = 11$)
Average	0.5763	0.6102
Maximum	0.5932	0.5932
Mixed	0.6287	0.6240
Stochastic	0.6502	0.6641
Maxfun	0.6225	0.6102
Centered maxfun	0.6626	0.6579

Table 1 suggests that the centered maxfun pooling strategy provides relatively good results, on par with those of the stochastic pooling method. Note that these results are not state of the art for this problem. The simplified network topology and aggressive pooling we utilize here are sub-optimal from a pure performance standpoint. Recall our goal was to highlight relative differences among pooling strategies; at this stage of this work we are yet not focused on absolute performance.

It is interesting to observe the more flexible non-centered maxfun does not appear as effective as the centered variant. We might speculate that edge effects are causing some issues with the non-centered variant. Alternatively, it is possible that the location of the pooling window may be a more significant driver in feature importance than the spatial extent of the pooling subregion.

Obviously the experiments we describe here are highly preliminary. One direction for future work is to expand the scope of these studies and to further inquire into differences between the centered vs non-centered maxfun variants. Furthermore, there could be value in a more detailed comparison, both theoretically and empirically, of maxfun with other pooling variants that "interpolate" between average and maximum pooling. Ultimately, there is an open question as to whether maxfun can benefit modern CNN architectures if it were to be included as a pooling layer. This would entail defining a suitable notion of a gradient for discrete maxfun pooling and conducting appropriate numerical studies. These studies would also need to address the added computational expense of maxfun over simpler methods, such as stochastic pooling (5.1).

6 Conclusions

In this paper we introduced a novel pooling method, maxfun pooling, inspired by the Hardy–Littlewood maximal function. We proved that this new pooling strategy maintains the stability of a convolutional neural network and we demonstrated its experimental performance by comparing it with state-of-the-art pooling algorithms in classification tasks.

Many functions and transformations originating from harmonic analysis have demonstrated the ability to extract useful features from the input data sets for classification or segmentation purposes. Maxfun pooling is another example of such a strategy that effectively extracts features from outputs of layers of neural networks and produces faithful representation of input data.

References

1. Boureau, Y.-L., Ponce, J., and LeCun, Y. (2010) A theoretical analysis of feature pooling in visual recognition. *Proceedings of the 27th International Conference on Machine Learning (ICML-10)*, pp. 111–118.
2. Bruna, J., Szlam, A., and LeCun, Y. (2014) Signal recovery from pooling representations. *Proceedings of the 31st International Conference on Machine Learning*, **32**, 307–315.

3. Cireşan, D., Meier, U., and Schmidhuber, J. (2012) Multi-column deep neural networks for image classification. *Computer Vision and Pattern Recognition (CVPR), 2012 IEEE conference on*, pp. 3642–3649, IEEE.
4. Cireşan, D. C., Meier, U., Masci, J., Maria Gambardella, L., and Schmidhuber, J. (2011) Flexible, high performance convolutional neural networks for image classification. *IJCAI Proceedings-International Joint Conference on Artificial Intelligence*, vol. 22, p. 1237, Barcelona, Spain.
5. Coifman, R. and Fefferman, C. (1974) Weighted norm inequalities for maximal functions and singular integrals. *Studia Mathematica*, **51**, 241–250.
6. Collobert, R. and Weston, J. (2008) A unified architecture for natural language processing: deep neural networks with multitask learning. *Proceedings of the 25th International Conference on Machine Learning (ICML 2008)*, pp. 160–167.
7. Donoho, D. L., Elad, M., and Temlyakov, V. N. (2006) Stable recovery of sparse overcomplete representations in the presence of noise. *IEEE Transactions on Information Theory*, **52**, 6–18.
8. Fei-Fei, L., Fergus, R., and Perona, P. (2007) Learning generative visual models from few training examples: An incremental Bayesian approach tested on 101 object categories. *Computer vision and Image understanding*, **106**, 59–70.
9. Ferra, A., Aguilar, E., and Radeva, P. (2018) Multiple wavelet pooling for CNNs. *Proceedings of the European Conference on Computer Vision (ECCV 2018)*, pp. 671–685.
10. Goodfellow, I., Bengio, Y., Courville, A., and Bengio, Y. (2016) *Deep Learning*, vol. 1. MIT press Cambridge.
11. Grafakos, L. (2004) *Classical and Modern Fourier Analysis*. Prentice Hall, Upper Saddle River, NJ.
12. Graham, B. (2014) Fractional max-pooling. *arXiv preprint arXiv:1412.6071*.
13. Hardy, G. H. and Littlewood, J. E. (1930) A maximal theorem with function-theoretic applications. *Acta Math.*, **54**, 81–116.
14. Hearst, M. A., Dumais, S. T., Osuna, E., Platt, J., and Scholkopf, B. (1998) Support vector machines. *IEEE Intelligent Systems and Their Applications*, **13**, 18–28.
15. Kabkab, M. (2017) The case for spatial pooling in deep convolutional sparse coding. *Fifty-First Asilomar Conference on Signals, Systems and Computers*.
16. Lee, C.-Y., Gallagher, P. W., and Tu, Z. (2016) Generalizing pooling functions in convolutional neural networks: Mixed, gated, and tree. *Artificial Intelligence and Statistics*, pp. 464–472.
17. Mallat, S. (2016) Understanding deep convolutional networks. *Philos. Trans. R. Soc. London A, Math. Phys. Sci*, **374**.
18. Muckenhoupt, B. (1972) Weighted norm inequalities for the hardy maximal function. *Transactions of the American Mathematical Society*, **165**, 207–226.
19. Papyan, V., Romano, Y., and Elad, M. (2017) Convolutional neural networks analyzed via convolutional sparse coding. *Journal of Machine Learning Research*, **18**, 27.
20. Papyan, V., Sulam, J., and Elad, M. (2016) Working locally thinking globally-part ii: Stability and algorithms for convolutional sparse coding. *arXiv preprint arXiv:1607.02009*.
21. Rippel, O., Snoek, J., and Adams, R. (2015) Spectral representations for convolutional neural networks. *Advances in Neural Information Processing Systems 28 (NIPS 2015)*, pp. 2449–2457.
22. Ryu, J., Yang, M.-H., and Lim, J. (2018) DFT-based transformation invariant pooling layer for visual classification. *Proceedings of the European Conference on Computer Vision (ECCV 2018)*, pp. 84–99.
23. Stein, E. M. (1970) *Singular Integrals and Differentiability Properties of Functions*. Princeton University Press, Princeton, NJ.
24. Szegedy, C., Vanhoucke, V., Ioffe, S., Shlens, J., and Wojna, Z. (2016) Rethinking the inception architecture for computer vision. *Proceedings of the IEEE Conference on Computer Vision and Pattern Recognition*, pp. 2818–2826.
25. Williams, T. and Li, R. (2018) Wavelet pooling for convolutional neural networks. *Proceedings of the International Conference on Learning Representations (ICLR 2018)*.

26. Zeiler, M. D. and Fergus, R. (2013) Stochastic pooling for regularization of deep convolutional neural networks. *1st International Conference on Learning Representations, ICLR 2013*.
27. Zhang, H. and Ma, J. (2018) Hartley spectral pooling for deep learning. *arXiv:1810.04028*.

Index

A
absolutely
  continuous function, 40
adjacency
  spectral, 103
adjacency matrix, 98
admissible observation operator, 273
affine designs, 137
analysis operator, 278
angle set, 166
angular momentum, 59, 62
antiderivative, 38
approximately controllable (ACO), 272
approximately observable (AOB), 274
approximation coefficient sequences, 307
approximation to the identity, 151
$\asymp$, 283
average pooling, 414, 416

B
Banach–Zaretsky Theorem, 45
bases
  mutually unbiased, 132
BEC, 60
Bessel, 278
bi-infinite sequence, 298
bi-infinite Toeplitz matrix, 301
block designs, 134
  symmetric, 135
block sparse, 352
blowup rate, 60, 63
Born–Jordan
  BJ distribution, 224
  BJ distribution of order n, 226
bounded
  variation, 30

C
$\mathbb{C}_+$, 293
Calderón-Zygmund theorem, 417
canonical dual, 164
Cantor–Lebesgue function, 29
Carleson's condition in $\mathbb{D}$, 292
Cayley graph, 100
coherence, 121
compatible dilation, 186, 188, 191
composite dilation, 185
compressed sensing, 343
continuity
  uniform, 39
continuous frame, 278
  tight, 278
continuous linear systems, 271
controllability maps $\Theta(k)$, 272
controllability maps $\Theta(t)$, 271
control operator, 271
control space, 271
convolution, 300
convolutional neural nets, 414
coordinate projections, 132
cospectral, 166
(Cross-)Wigner Distribution, 221
crystallographic group, 185, 187, 188
CS space, 347

M. Hirn et al. (eds.), *Excursions in Harmonic Analysis, Volume 6*, Applied and Numerical Harmonic Analysis, https://doi.org/10.1007/978-3-030-69637-5

cube
  Boolean, 100
cubic B-spline, 304

D
$\mathbb{D}$, 286
decimation operator, 301
descent cone, 348
detail coefficient sequences, 307
Devil's staircase, 29
difference operator, 306
direct integral, 186
discrete linear systems, 271
dispersion stage, 387
downsampling operator, 300
DS, 213
dynamical sampling, 212
dynamical sampling problem, 277
dynamic operator, 271

E
eigenvector
  regular, 176
$\ell^2(\Gamma, \mathcal{U})$, 272
embedded vector, 124
error, 403
estimation problem, 389
even-inverse, 303
even-reversible, 303
exactly controllable (ECO), 272
exactly observable (EOB), 274
exponentially stable, 271

F
fixed point, 399
Fourier Lebesgue wave-front set, 232
Fourier transform
  graph, 98
frame, 211, 278
  dual, 164
  equal norm, 164
  equiangular, 165
  equiangular tight, 164, 165
  k-angle, 166
  tight, 164
  unit norm, 164
frame operator, 164
frames
  equivalent, 168
FTC, 28
function
  singular, 29
  uniformly continuous, 39
Fundamental Theorem of Calculus, 38, 50
fusion frame, 214
  maximal orthoplectic, 131

G
gap metric, 356
Gaussian width, 361
general mild solution, 271
Gerzon's bound, 166
g-frame, 213
glide reflection, 187, 190
gossip method, 389
Gram matrix, 165
ground state, 60, 63, 72, 73

H
Hadamard 3-design, 141
half-plane Carleson's condition, 293
Hardy–Littlewood lemma, 417
Hardy–Littlewood Maximal Theorem, 36
harmonic potential, 59, 60, 63, 73

I
initial state, 271
Intermediate Value Theorem, 49
interpolatory subdivision, 298
iterative system, 212

J
Jordan decomposition, 31

K
Kaczmarz, 385
  method, 388
  update, 392
Kalman's Theorem, 289

L
$L^2([0, \tau), \mathcal{U})$, 271
Laplacian
  graph, 98
lattice, 187, 191, 194, 197
Lebesgue
  Differentiation Theorem, 36
local frame, 213
localized, 212

localized function, 213
locally conic, 348
Luzin's condition, 45

M
maxfun pooling, 415, 417, 422
maximal function, 414, 416
maximal pooling, 414
Maximal Theorem, 36
max pooling, 416
Modulation spaces, 230
monotone function, 31
MTER, 303
multiscale transform, 298
multiwavelet set, 186, 189, 190, 192

N
nonlinear Schrödinger equation, 59, 72
null space property, 345

O
observability Grammian Q_τ, 274
observability map Ψ, 273
observation operator, 271
observation space, 271
optimally spread, 121
orthoplex bound, 122

P
packing constant, 126
phaseless recovery, 213
point group, 187, 188
pooling region, 416
pooling stage, 387
projection
  affine, 386
  linear, 386
projection operators, 213
pseudo-spline, 306

R
reachable states, 272
relaxation parameter, 390
residual, 386
restricted isometry property, 346, 358
Riesz sequence/basis, 282
robustness, 344
robust width property, 346, 348
Rotational NLS, 72

S
semi-continuous frame, 278
semidirect product, 186, 187
seminorm, 350
Short-time Fourier transform (STFT), 229
signal processing, 211
signature matrix, 165
Simple Vitali Lemma, 34
simplex bound, 122
singular
  function, 29
Sobolev space, 60
solution, 386
  of minimal norm, 386
  weighted least-squares, 386
sparse coding, 418
sparse vectors, 345
spectral radius, 386
spectrum limiting, 98
stability, 344
stable reconstruction, 212
state space, 271
Strichartz estimates, 62, 63
stride, 416
strong asymptotically stable, 271
subdivision operator, 299
subdivision scheme, 300
symmorphic, 187, 190
synthesis operator, 164, 278

T
tiling, 189, 190, 192–195, 197
Time-frequency shifts, 222
total
  variation, 30
transformations
  affine, 393

U
unconditional sequence/basis, 282
uniform
  continuity, 39
uniformly continuous one-parameter group, 271
upsampling operator, 300

V
variation of a function, 30
virial identity, 61, 64, 68
Vitali Covering Lemma, 32

W

wallpaper group, 187
wavelet
  crystallographic, 185, 186
  Haar-type, 185
  Parseval, 190
wavelet set, 185, 186
  crystallographic, 188, 190, 191
  simple, 186, 192, 193, 195, 196
  subspace, 190, 191
weight, 387
weighted Sobolev space, 59, 63
weighted sparsity, 351
Welch bound, 163, 165
Welch bound equality set, 164
width property, 345, 355
Wiener algebra, 233, 301
Wiener amalgam spaces, 231
Wiener's Lemma, 301

Applied and Numerical Harmonic Analysis (102 volumes)

1. A. I. Saichev and W. A. Woyczyñski: *Distributions in the Physical and Engineering Sciences* (ISBN: 978-0-8176-3924-2)
2. C. E. D'Attellis and E. M. Fernandez-Berdaguer: *Wavelet Theory and Harmonic Analysis in Applied Sciences* (ISBN: 978-0-8176-3953-2)
3. H. G. Feichtinger and T. Strohmer: *Gabor Analysis and Algorithms* (ISBN: 978-0-8176-3959-4)
4. R. Tolimieri and M. An: *Time-Frequency Representations* (ISBN: 978-0-8176-3918-1)
5. T. M. Peters and J. C. Williams: *The Fourier Transform in Biomedical Engineering* (ISBN: 978-0-8176-3941-9)
6. G. T. Herman: *Geometry of Digital Spaces* (ISBN: 978-0-8176-3897-9)
7. A. Teolis: *Computational Signal Processing with Wavelets* (ISBN: 978-0-8176-3909-9)
8. J. Ramanathan: *Methods of Applied Fourier Analysis* (ISBN: 978-0-8176-3963-1)
9. J. M. Cooper: *Introduction to Partial Differential Equations with MATLAB* (ISBN: 978-0-8176-3967-9)
10. Procházka, N. G. Kingsbury, P. J. Payner, and J. Uhlir: *Signal Analysis and Prediction* (ISBN: 978-0-8176-4042-2)
11. W. Bray and C. Stanojevic: *Analysis of Divergence* (ISBN: 978-1-4612-7467-4)
12. G. T. Herman and A. Kuba: *Discrete Tomography* (ISBN: 978-0-8176-4101-6)
13. K. Gröchenig: *Foundations of Time-Frequency Analysis* (ISBN: 978-0-8176-4022-4)
14. L. Debnath: *Wavelet Transforms and Time-Frequency Signal Analysis* (ISBN: 978-0-8176-4104-7)
15. J. J. Benedetto and P. J. S. G. Ferreira: *Modern Sampling Theory* (ISBN: 978-0-8176-4023-1)

M. Hirn et al. (eds.), *Excursions in Harmonic Analysis, Volume 6*, Applied and Numerical Harmonic Analysis, https://doi.org/10.1007/978-3-030-69637-5

16. D. F. Walnut: *An Introduction to Wavelet Analysis* (ISBN: 978-0-8176-3962-4)
17. A. Abbate, C. DeCusatis, and P. K. Das: *Wavelets and Subbands* (ISBN: 978-0-8176-4136-8)
18. O. Bratteli, P. Jorgensen, and B. Treadway: *Wavelets Through a Looking Glass* (ISBN: 978-0-8176-4280-80
19. H. G. Feichtinger and T. Strohmer: *Advances in Gabor Analysis* (ISBN: 978-0-8176-4239-6)
20. O. Christensen: *An Introduction to Frames and Riesz Bases* (ISBN: 978-0-8176-4295-2)
21. L. Debnath: *Wavelets and Signal Processing* (ISBN: 978-0-8176-4235-8)
22. G. Bi and Y. Zeng: *Transforms and Fast Algorithms for Signal Analysis and Representations* (ISBN: 978-0-8176-4279-2)
23. J. H. Davis: *Methods of Applied Mathematics with a MATLAB Overview* (ISBN: 978-0-8176-4331-7)
24. J. J. Benedetto and A. I. Zayed: *Sampling, Wavelets, and Tomography* (ISBN: 978-0-8176-4304-1)
25. E. Prestini: *The Evolution of Applied Harmonic Analysis* (ISBN: 978-0-8176-4125-2)
26. L. Brandolini, L. Colzani, A. Iosevich, and G. Travaglini: *Fourier Analysis and Convexity* (ISBN: 978-0-8176-3263-2)
27. W. Freeden and V. Michel: *Multiscale Potential Theory* (ISBN: 978-0-8176-4105-4)
28. O. Christensen and K. L. Christensen: *Approximation Theory* (ISBN: 978-0-8176-3600-5)
29. O. Calin and D.-C. Chang: *Geometric Mechanics on Riemannian Manifolds* (ISBN: 978-0-8176-4354-6)
30. J. A. Hogan: *Time?Frequency and Time?Scale Methods* (ISBN: 978-0-8176-4276-1)
31. C. Heil: *Harmonic Analysis and Applications* (ISBN: 978-0-8176-3778-1)
32. K. Borre, D. M. Akos, N. Bertelsen, P. Rinder, and S. H. Jensen: *A Software-Defined GPS and Galileo Receiver* (ISBN: 978-0-8176-4390-4)
33. T. Qian, M. I. Vai, and Y. Xu: *Wavelet Analysis and Applications* (ISBN: 978-3-7643-7777-9)
34. G. T. Herman and A. Kuba: *Advances in Discrete Tomography and Its Applications* (ISBN: 978-0-8176-3614-2)
35. M. C. Fu, R. A. Jarrow, J.-Y. Yen, and R. J. Elliott: *Advances in Mathematical Finance* (ISBN: 978-0-8176-4544-1)
36. O. Christensen: *Frames and Bases* (ISBN: 978-0-8176-4677-6)
37. P. E. T. Jorgensen, J. D. Merrill, and J. A. Packer: *Representations, Wavelets, and Frames* (ISBN: 978-0-8176-4682-0)
38. M. An, A. K. Brodzik, and R. Tolimieri: *Ideal Sequence Design in Time-Frequency Space* (ISBN: 978-0-8176-4737-7)
39. S. G. Krantz: *Explorations in Harmonic Analysis* (ISBN: 978-0-8176-4668-4)

40. B. Luong: *Fourier Analysis on Finite Abelian Groups* (ISBN: 978-0-8176-4915-9)
41. G. S. Chirikjian: *Stochastic Models, Information Theory, and Lie Groups, Volume 1* (ISBN: 978-0-8176-4802-2)
42. C. Cabrelli and J. L. Torrea: *Recent Developments in Real and Harmonic Analysis* (ISBN: 978-0-8176-4531-1)
43. M. V. Wickerhauser: *Mathematics for Multimedia* (ISBN: 978-0-8176-4879-4)
44. B. Forster, P. Massopust, O. Christensen, K. Gröchenig, D. Labate, P. Vandergheynst, G. Weiss, and Y. Wiaux: *Four Short Courses on Harmonic Analysis* (ISBN: 978-0-8176-4890-9)
45. O. Christensen: *Functions, Spaces, and Expansions* (ISBN: 978-0-8176-4979-1)
46. J. Barral and S. Seuret: *Recent Developments in Fractals and Related Fields* (ISBN: 978-0-8176-4887-9)
47. O. Calin, D.-C. Chang, and K. Furutani, and C. Iwasaki: *Heat Kernels for Elliptic and Sub-elliptic Operators* (ISBN: 978-0-8176-4994-4)
48. C. Heil: *A Basis Theory Primer* (ISBN: 978-0-8176-4686-8)
49. J. R. Klauder: *A Modern Approach to Functional Integration* (ISBN: 978-0-8176-4790-2)
50. J. Cohen and A. I. Zayed: *Wavelets and Multiscale Analysis* (ISBN: 978-0-8176-8094-7)
51. D. Joyner and J.-L. Kim: *Selected Unsolved Problems in Coding Theory* (ISBN: 978-0-8176-8255-2)
52. G. S. Chirikjian: *Stochastic Models, Information Theory, and Lie Groups, Volume 2* (ISBN: 978-0-8176-4943-2)
53. J. A. Hogan and J. D. Lakey: *Duration and Bandwidth Limiting* (ISBN: 978-0-8176-8306-1)
54. G. Kutyniok and D. Labate: *Shearlets* (ISBN: 978-0-8176-8315-3)
55. P. G. Casazza and P. Kutyniok: *Finite Frames* (ISBN: 978-0-8176-8372-6)
56. V. Michel: *Lectures on Constructive Approximation* (ISBN: 978-0-8176-8402-0)
57. D. Mitrea, I. Mitrea, M. Mitrea, and S. Monniaux: *Groupoid Metrization Theory* (ISBN: 978-0-8176-8396-2)
58. T. D. Andrews, R. Balan, J. J. Benedetto, W. Czaja, and K. A. Okoudjou: *Excursions in Harmonic Analysis, Volume 1* (ISBN: 978-0-8176-8375-7)
59. T. D. Andrews, R. Balan, J. J. Benedetto, W. Czaja, and K. A. Okoudjou: *Excursions in Harmonic Analysis, Volume 2* (ISBN: 978-0-8176-8378-8)
60. D. V. Cruz-Uribe and A. Fiorenza: *Variable Lebesgue Spaces* (ISBN: 978-3-0348-0547-6)
61. W. Freeden and M. Gutting: *Special Functions of Mathematical (Geo-)Physics* (ISBN: 978-3-0348-0562-9)
62. A. I. Saichev and W. A. Woyczyñski: *Distributions in the Physical and Engineering Sciences, Volume 2: Linear and Nonlinear Dynamics of Continuous Media* (ISBN: 978-0-8176-3942-6)

63. S. Foucart and H. Rauhut: *A Mathematical Introduction to Compressive Sensing* (ISBN: 978-0-8176-4947-0)
64. G. T. Herman and J. Frank: *Computational Methods for Three-Dimensional Microscopy Reconstruction* (ISBN: 978-1-4614-9520-8)
65. A. Paprotny and M. Thess: *Realtime Data Mining: Self-Learning Techniques for Recommendation Engines* (ISBN: 978-3-319-01320-6)
66. A. I. Zayed and G. Schmeisser: *New Perspectives on Approximation and Sampling Theory: Festschrift in Honor of Paul Butzer's 85th Birthday* (ISBN: 978-3-319-08800-6)
67. R. Balan, M. Begue, J. Benedetto, W. Czaja, and K. A. Okoudjou: *Excursions in Harmonic Analysis, Volume 3* (ISBN: 978-3-319-13229-7)
68. H. Boche, R. Calderbank, G. Kutyniok, and J. Vybiral: *Compressed Sensing and its Applications* (ISBN: 978-3-319-16041-2)
69. S. Dahlke, F. De Mari, P. Grohs, and D. Labate: *Harmonic and Applied Analysis: From Groups to Signals* (ISBN: 978-3-319-18862-1)
70. A. Aldroubi: *New Trends in Applied Harmonic Analysis* (ISBN: 978-3-319-27871-1)
71. M. Ruzhansky: *Methods of Fourier Analysis and Approximation Theory* (ISBN: 978-3-319-27465-2)
72. G. Pfander: *Sampling Theory, a Renaissance* (ISBN: 978-3-319-19748-7)
73. R. Balan, M. Begue, J. Benedetto, W. Czaja, and K. A. Okoudjou: *Excursions in Harmonic Analysis, Volume 4* (ISBN: 978-3-319-20187-0)
74. O. Christensen: *An Introduction to Frames and Riesz Bases, Second Edition* (ISBN: 978-3-319-25611-5)
75. E. Prestini: *The Evolution of Applied Harmonic Analysis: Models of the Real World, Second Edition* (ISBN: 978-1-4899-7987-2)
76. J. H. Davis: *Methods of Applied Mathematics with a Software Overview, Second Edition* (ISBN: 978-3-319-43369-1)
77. M. Gilman, E. M. Smith, and S. M. Tsynkov: *Transionospheric Synthetic Aperture Imaging* (ISBN: 978-3-319-52125-1)
78. S. Chanillo, B. Franchi, G. Lu, C. Perez, and E. T. Sawyer: *Harmonic Analysis, Partial Differential Equations and Applications* (ISBN: 978-3-319-52741-3)
79. R. Balan, J. Benedetto, W. Czaja, M. Dellatorre, and K. A. Okoudjou: *Excursions in Harmonic Analysis, Volume 5* (ISBN: 978-3-319-54710-7)
80. I. Pesenson, Q. T. Le Gia, A. Mayeli, H. Mhaskar, and D. X. Zhou: *Frames and Other Bases in Abstract and Function Spaces: Novel Methods in Harmonic Analysis, Volume 1* (ISBN: 978-3-319-55549-2)
81. I. Pesenson, Q. T. Le Gia, A. Mayeli, H. Mhaskar, and D. X. Zhou: *Recent Applications of Harmonic Analysis to Function Spaces, Differential Equations, and Data Science: Novel Methods in Harmonic Analysis, Volume 2* (ISBN: 978-3-319-55555-3)
82. F. Weisz: *Convergence and Summability of Fourier Transforms and Hardy Spaces* (ISBN: 978-3-319-56813-3)

83. C. Heil: *Metrics, Norms, Inner Products, and Operator Theory* (ISBN: 978-3-319-65321-1)
84. S. Waldron: *An Introduction to Finite Tight Frames: Theory and Applications.* (ISBN: 978-0-8176-4814-5)
85. D. Joyner and C. G. Melles: *Adventures in Graph Theory: A Bridge to Advanced Mathematics.* (ISBN: 978-3-319-68381-2)
86. B. Han: *Framelets and Wavelets: Algorithms, Analysis, and Applications* (ISBN: 978-3-319-68529-8)
87. H. Boche, G. Caire, R. Calderbank, M. März, G. Kutyniok, and R. Mathar: *Compressed Sensing and Its Applications* (ISBN: 978-3-319-69801-4)
88. A. I. Saichev and W. A. Woyczyñski: *Distributions in the Physical and Engineering Sciences, Volume 3: Random and Fractal Signals and Fields* (ISBN: 978-3-319-92584-4)
89. G. Plonka, D. Potts, G. Steidl, and M. Tasche: *Numerical Fourier Analysis* (978-3-030-04305-6)
90. K. Bredies and D. Lorenz: *Mathematical Image Processing* (ISBN: 978-3-030-01457-5)
91. H. G. Feichtinger, P. Boggiatto, E. Cordero, M. de Gosson, F. Nicola, A. Oliaro, and A. Tabacco: *Landscapes of Time-Frequency Analysis* (ISBN: 978-3-030-05209-6)
92. E. Liflyand: *Functions of Bounded Variation and Their Fourier Transforms* (978-3-030-04428-2)
93. R. Campos: *The XFT Quadrature in Discrete Fourier Analysis* (978-3-030-13422-8)
94. M. Abell, E. Iacob, A. Stokolos, S. Taylor, S. Tikhonov, J. Zhu: *Topics in Classical and Modern Analysis: In Memory of Yingkang Hu* (978-3-030-12276-8)
95. H. Boche, G. Caire, R. Calderbank, G. Kutyniok, R. Mathar, P. Petersen: *Compressed Sensing and its Applications: Third International MATHEON Conference 2017* (978-3-319-73073-8)
96. A. Aldroubi, C. Cabrelli, S. Jaffard, U. Molter: *New Trends in Applied Harmonic Analysis, Volume II: Harmonic Analysis, Geometric Measure Theory, and Applications* (978-3-030-32352-3)
97. S. Dos Santos, M. Maslouhi, K. Okoudjou: *Recent Advances in Mathematics and Technology: Proceedings of the First International Conference on Technology, Engineering, and Mathematics, Kenitra, Morocco, March 26-27, 2018* (978-3-030-35201-1)
98. Á. Bényi, K. Okoudjou: *Modulation Spaces: With Applications to Pseudodifferential Operators and Nonlinear Schrödinger Equations* (978-1-0716-0330-7)
99. P. Boggiato, M. Cappiello, E. Cordero, S. Coriasco, G. Garello, A. Oliaro, J. Seiler: *Advances in Microlocal and Time-Frequency Analysis* (978-3-030-36137-2)

100. S. Casey, K. Okoudjou, M. Robinson, B. Sadler: *Sampling: Theory and Applications* (978-3-030-36290-4)
101. P. Boggiatto, T. Bruno, E. Cordero, H. G. Feichtinger, F. Nicola, A. Oliaro, A. Tabacco, M. Vallarino: *Landscapes of Time-Frequency Analysis* (978-3-030-56004-1)
102. M. Hirn, S. Li, K. Okoudjou, S. Saliani, O. Yilmaz: *Excursions in Harmonic Analysis, Volume 6* (978-3-030-69636-8)

For an up-to-date list of ANHA titles, please visit http://www.springer.com/series/4968

The manufacturer's authorised representative in the EU is Springer Nature Customer Service Centre GmbH, Europaplatz 3, 69115 Heidelberg, Germany. If you have any concerns regarding our products, please contact ProductSafety@springernature.com

Printed and bound by CPI Group (UK) Ltd, Croydon, CR0 4YY
23/09/2026
02221532-0008